MATERIALS SCIENCE AND TECHNOLOGIES

RHEOLOGY

PRINCIPLES, APPLICATIONS AND ENVIRONMENTAL IMPACTS

MATERIALS SCIENCE AND TECHNOLOGIES

Additional books in this series can be found on Nova's website under the Series tab.

Additional e-books in this series can be found on Nova's website under the e-book tab.

MATERIALS SCIENCE AND TECHNOLOGIES

RHEOLOGY

PRINCIPLES, APPLICATIONS AND ENVIRONMENTAL IMPACTS

EVGENY KARPUSHKIN
EDITOR

nova publishers
New York

Additional color graphics may be available in the e-book version of this book.

Library of Congress Cataloging-in-Publication Data
Rheology (Karpushkin)
Rheology : principles, applications and environmental impacts / editor, Evgeny Karpushkin (Komsomolskii Prospect, Russia). pages cm. -- (Materials science and technologies)
Includes index.
Includes bibliographical references and index.
ISBN 978-1-63482-223-7 (hardcover)
1. Oils and fats--Analysis. 2. Colloids--Analysis. 3. Dairy products--Rheology. 4. Polymers--Rheology. I. Karpushkin, Evgeny. II. Title. TP671.R54 2014 664'.3--dc23 2015004436

Published by Nova Science Publishers, Inc. † New York

CONTENTS

PREFACE

Rheology is a young branch of science; indeed, the very term *rheology* was suggested in 1920. On the other hand, practical concepts of flow have been known and developed since very old times. Even though the science of rheology is sometimes roughly divided into the applied and the theoretical fields, there is no doubt that such division is very vague. Fundamental rheological ideas and equations that may seem extremely complicated at the first glance always readily penetrate into the practice; vice versa, the modern state of development of various industrial and biomedical processes poses sophisticated tasks for theoretical analysis. Hence, theoretical and applied branches of rheology are tightly bound together.

Another special feature of rheology is that its objects are very diverse. Indeed, as the instruments have been developing, more and more materials were discovered to experience flow. Nowadays some scientists believe that there are no materials that cannot flow at all; this means that virtually any material can be studied by methods of rheology. Hence, a professional rheologist should be skilful in many areas of material science and prepared to work with very different substances. However, there are a number of areas where rheology is believed to be of especially outstanding importance: food, cosmetics, building, and oil industries along with biomechanics and geophysics.

This volume covers many of the above-mentioned disciplines. The first chapter tells an amazing story related to interfacial rheology of particle-laden systems. This field is fresh and unusual for many of professional rheologists. However, the authors demonstrate that understanding of the described phenomena opens a way for preparation of advanced dispersed materials and also adds to the knowledge of colloidal stability of various systems.

Dairy products are traditionally very popular objects among the rheologists; due to undoubtful practical importance of milk and related products rheological properties, and because of complexity of the systems showing peculiar features and trends. Three chapters in this volume discuss dairy products rheology. Chapter 2 covers the important practical aspects of rheology of fermented milk products in relation with their utilitarian properties and consumer preferences. Chapters 3 and 4 discuss a similar subject from theoretical point of view; it is demonstrated that study of milk rheological behavior in the course of gelation results in an interesting theoretical concept describing the process. Moreover, the developed model is applicable to many other gelation processes.

Chapter 5 gives another example of a natural product, cress seed gum. Nowadays natural products are very popular among the consumers, and the study reported here presents an excellent example of extensive investigation of properties of a complex natural object in relation to its practical applications. Even though each natural product is somewhat special, this study ends up with important generalizations and also provides a basis for characterization of similar products.

The next chapter is attractive as it again shows the close connection between theoretical concepts of rheology and important flow properties of a widespread product, waxy crude oil. This piece of research is indeed instructive and inspiring, demonstrating that rheology is a field where seemingly pure science can give immediate practical outcome.

Chapter 7 reports on one more brilliant example of how a natural renewable material (polysaccharides) combined with a peculiar medium (ionic liquid) yields a novel material. The ion-gels possess an interesting combination of rheological properties making it a suitable model for further fundamental developments; at the same time, the self-healing nature of such gels makes them promising materials to be used in many fields.

The volume ends with two more contributions devoted to polymer gels. Chapter 8 majorly discusses one of the experimental setup parameters to be considered during rheological testing of soft gels. Surprisingly, it has been scarcely described earlier, however, neglecting it produces highly inaccurate measurement results that cannot be compared between the laboratories. With this consideration kept in mind, a model series of gels varying in the internal structure has been accurately studied, and the relationships between the microscopic as well as macroscopic structural features and the mechanical response of the material has been elucidated. The results of this part of the study are discussed in Chapter 9.

Of course, it was not possible to cover all the topics related to modern rheology within a single volume. On the contrary, our intension was to present just a few striking examples showing that rheology is young but mature field of science; affording both well developed concepts and myriads of novel ideas to be further developed. We hope that you will enjoy the reading.

In: Rheology
Editor: Evgeny Karpushkin

ISBN: 978-1-63482-223-7

Chapter 1

INTERFACIAL RHEOLOGY OF PARTICLE-LADEN INTERFACES AND ITS ROLE IN THE STABILIZATION OF DISPERSED SYSTEMS

Armando Maestro,[1] Eduardo Guzmán[2,3,*] and Eva Santini[3]

[1]Department of Physics, Cavendish Laboratory, University of Cambridge Cambridge, UK
[2]Departamento de Química Física I-Facultad de Químicas, Universidad Complutense de Madrid, Madrid, Spain
[3]Istituto per l'Energetica e le Interfasi-U.O.S. Genova, Consiglio Nazionale delle Ricerche, Genova, Italy

ABSTRACT

Particle-laden interfaces are intriguing systems with implication in many technological and industrial applications. This wide spreading has stimulated an intense research in many fundamental physico-chemical aspects related to the use of particle layers to obtain functional and structural materials from emulsions and foams. However, there are many lacks of knowledge related to the driving forces that govern the stabilization of dispersed systems. In particular, the understanding of the physico-chemical properties governing the stabilization phenomena can be clearly linked to the fabrication of advanced materials. It is well-recognized that among the different aspects involved on the stabilization mechanism of dispersed systems –such as foams and emulsions, the rheological properties of the particle-laden layers play a central role. During the formation and equilibration processes, dispersed systems experience dilation and shear deformations, which define the destabilization profiles through different mechanisms: coalescence, sedimentation, etc. The detailed analysis of the rheological aspects of the particle-laden precursor layers can help to enhance the stability and applicability of emulsions and foams in the design of materials, especially in the field of the highly porous materials. In spite of the critical character of the rheological properties

* e-mail: eduardogs@quim.ucm.es

in the stabilization of dispersed system, its control is not always easy since many physico-chemical parameters such as the wettability of the particles or the structure of the formed layers are key points in the tuning of the rheological response of particle layers. The control of these properties is essential to obtain particle films attached to fluid/fluid interfaces that can applied to stabilized dispersed systems. This chapter analyzes the main aspects of the rheological response of particle-laden interfaces and its correlation to the stabilization of dispersed systems that can be used as precursors of new materials.

1. Introduction: Particle-Laden Interfaces

The interactions of particles with fluid interfaces and the attachment of these particles to the fluid interfaces have been topics of growing interest in recent years due to their recognized implication in many natural and industrial processes [1–3] dealing with scientific and technological aspects of different nature, from material to food science and from biomedicine to renewable energies [1, 4, 5]. This ubiquity of particle-laden layers has driven intense development on the methods and materials dedicated to the interfacial self-assembly of particles at fluid interfaces, either vapor–liquid or liquid-liquid. However, there remain many open challenges to reach a comprehensive understanding of the physico-chemistry of interfacial films composed by particles [6–9]. Among these challenges, those associated with the forces and phenomena involved into the attachment of particles to the fluid interface are probably the most important ones. In general, two different routes can allow the particles to reach the interface [10]: adsorption from suspensions at the interface forming the so called Gibbs monolayers [11, 12] or direct deposition of the particles at the fluid interfaces from a dispersion in a volatile organic solvent to form Langmuir monolayers [13, 14]. The former ones are the most interesting from the point of view of materials fabrication due to the fact that self-assembly of the particles at the fluid interfaces is required where dynamic aspects of the bulk dispersion and the interface must be considered [15], whereas direct deposition at the interface occurs in the case of the Langmuir monolayers, being the phenomena taking place at the interface that controls the average physico-chemical response of these particle-laden layers. This makes the Langmuir monolayer of particles especially interesting to study of fundamental aspects of physico-chemistry of particle-laden interfaces (equilibrium and dynamics properties as well as their structure) [13, 14].

After the seminal work by Pieranski in the early 1980's [16], continuous advancement on the understanding of the physico-chemical bases that govern the formation and response of particle layers has been carried out [1, 6–9]. The formation of particle-laden films requires consideration of the complex interplay between different physico-chemical aspects such as the particles wettability, size, shape, surface charge, and chemical nature [1, 9]. Synergetic performance of the aforementioned aspects determines the attachment of the particles at the fluid interface.

The particle-laden interfacial layers are multiphase regions where three interfaces are involved: one fluid–fluid interface, i.e., air/oil–water, and two solid–fluid interfaces (see Figure 1A) [8, 17, 18]. Formation of particle assemblies at the fluid interfaces can confer adequate stability to the interface, hindering the drops/bubbles coalescence. Under such conditions, it is possible to take advantage of the particular properties of the particle-laden interfaces to stabilize bubbles, thin films, or dispersed systems such as foams and emulsions

[2, 19]. The pioneering works, more than one century ago, by Ramsden [20] and Pickering [21] represent initial steps to the development of a research topic with many theoretical and practical implications in the industrialized society due to their potential application in many fields [2] such as the flotation technology, oil recovery processes, encapsulation, pharmaceutical formulations, and food technology [22–25]. In spite of this recognized importance of the particles in the stabilization of dispersed systems, little has been known about the laws governing the stabilization phenomena. This has raised several physico-chemical questions that are necessary to be answered in order to optimize performance of the particle-laden interfaces in the stabilization of dispersed systems. One of these questions is related to the (ir)reversibility of the particle attachment at the fluid interface that is controlled by the attachment energy, being a balance between the energy of a particle dispersed in the bulk and of that attached at the interface. Considering small particles attached at an arbitrary interface, and neglecting gravity effects, the attachment energy ΔE_p can be simply defined by the classical Young's equation,

$$\Delta E_p = -\pi R^2 \gamma_{f_1 f_2} \left(1 \pm \cos\theta\right)^2 \tag{1}$$

where R is the particle radius, $\gamma_{f_1 f_2}$ is the interfacial tension between the fluid phases, and θ is the contact angle of the particle attached at the fluid interface defined as shown in Figure 1A.

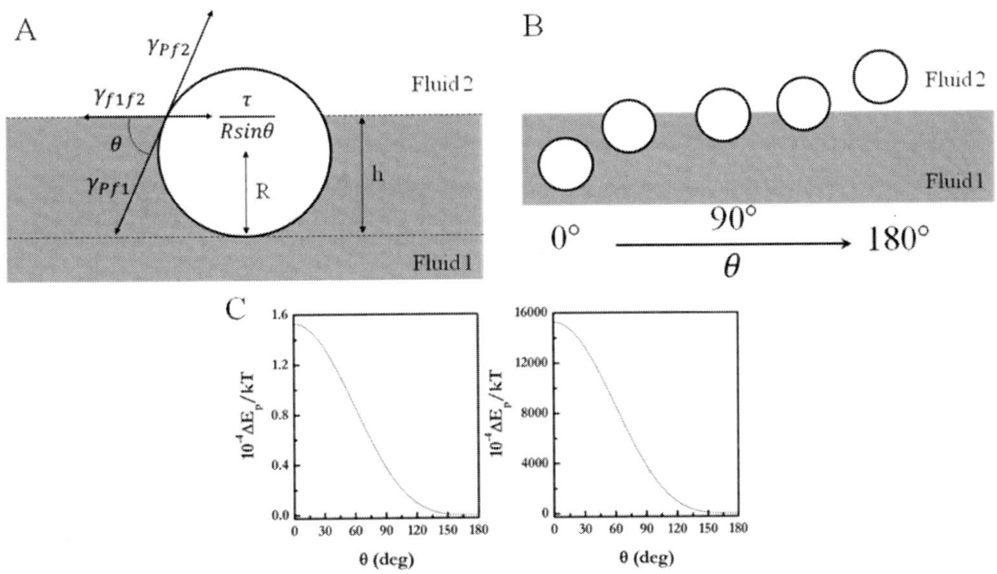

Figure 1. (A) Sketch of the contact angle for a particle at a fluid interface. γ_{Pf_1} is the particle–bottom fluid interfacial tension, γ_{Pf_2} the particle–upper phase interfacial tension, and $\gamma_{f_1 f_2}$ is the fluid–fluid interfacial tension. (B) Sketch of the position of a particle with increasing hydrophobicity at the interface. (C) Energy of the particles attachment to the interface (ΔE_p) as function of θ for particles with R = 10 nm (left) and R = 1 μm (right) at 25°C (γ = 50 mN/m). Reproduced from Ref. 7, Copyright (2014), with permission from Elsevier.

The ± signs account for particles with their centres above (plus) or below (minus) the interfacial plane. Noteworthily, depending on the involved energies, the particles adsorption may occur spontaneously or be hindered due to the existence of strong energetic barriers [9, 26]. In the case of very small particles (several nanometers), additional corrections due to the effect of the contact line on the contact angle should be considered in the energy balance. Generally, the latter effect is accounted by the so called line tension, τ [27–30]. However, for the sake of simplicity the particularities due to the effects of the line tension will not be discussed herein.

The size of the particles also plays an important role in the control of the attachment energy of the particles to the fluid interface due to their important influence on the reversibility and irreversibility of the attachment [7, 9]. This defines different energetic scenarios for micro- and nanoparticles. Figure 1C shows the difference of the energies involved in the cases of micro- and nanoparticles at a water/hexane interface as function of the contact angle. In the case of microparticles, normally $\Delta E_p >> k_B T$ is held, meaning that once microparticles are incorporated to the interface, they remain irreversibly attached to the interface. In the case of nanoparticles, the energies involved are lower, and under certain conditions the adsorption–desorption dynamic of the nanoparticles at the interface can be observed over relatively short timescale [31].

Other aspects to be considered are those related to the mechanical performance of the particle-laden interface [32–34]. Understanding of the rheological response allows one to elucidate the dynamics of exchange and relaxation that leads to the particle till the equilibrium state [6, 35, 36]. It is expected that these phenomena influence the stabilization of dispersed systems since drops and bubbles in dispersed systems are deformed under flow, provoking that the dilational and shear properties of particle-laden interface alter the multiphase flows [9]. Nowadays, there is a strong controversy related to the role of the resistance of particles layers against shear or dilational deformations in the stabilization of dispersed systems arising from the properties of particle-laden interfaces [37]. Several studies have recently attempted to correlate mechanical properties and stability of foams and emulsions [38, 39]. However, a complex picture has been drawn where several questions still remain open [40].

In addition to the above-discussed aspects, the last point to understand in the behaviour of particle-laden interfaces associated with their morphology [8, 9, 13, 14]. The structural features of particle monolayers can limit their application in the fabrication of advanced materials [9]. Particles layers can define different structures at fluid interfaces depending on the interactions present at the interface, from perfect 2D crystals to amorphous patterns with aggregated particles [13, 14]. The importance of the structure of particle-laden layers is also correlated to their influence on the mechanical response (dilational and shear rheology) [41, 42] and the films collapse mechanisms (buckling, jamming, or expulsion of particles) [9, 43]. This requires understanding the interactions involved in particle-laden interfaces (particle–particle, particle–fluids, and fluid–fluid) since they can strongly influence the structural features of the particle monolayers and consequently alter the properties (optical, mechanical, etc.) of the films [44, 45].

The complex interplay of different above-discussed aspects is critical in the application of particle-laden interface. This chapter will thoroughly review the most recent advancements in the field of the rheology of particle-laden interface with special interest to the correlation between this rheological response and the possibility to stabilize dispersed systems using particulate matter.

2. Factors Controlling the Particle Capability in the Stabilization of Disperse Systems

Behaviour of emulsions and foams is governed by mechanisms and physical principles which can be considered comparable for both the disperse systems since pure liquid foams can be thought of as concentrated "emulsions" of air and water. For this reason, many similarities exist among the factors influencing the formation and stability of different disperse systems in the presence of nanoparticles.

Nanoparticles, as mentioned before, are additives that can be used to populate the liquid–fluid interfaces formed during homogenisation. Adsorbing at the oil/air–water interface, nanoparticles lower the amount of energy required to form the interface, since the area of the bare interface is reduced. The free-energy change associated with the transfer of an isolated spherical particle from bulk water (or oil) phase to a planar interface is the highest for a particle which is wet to an intermediate extent by both fluid phases. Such particles are efficiently irreversibly adsorbed at the liquid–fluid interface and, if the drops are covered sufficiently, are responsible for sterically hindered drop coalescence, thus stabilizing the disperse system. In the case of emulsions, the type of drops formed (oil or water) is primarily determined by the particles wettability. It is thought that, similarly to the behaviour of surfactant layers at liquid–liquid interfaces, the particle layer curves so that the larger fraction of the particles surface resides at the external surface of the drop in the liquid which preferentially wets the particles. It is possible, however, to invert the emulsion type through, for example, gradual increase of the dispersed phase volume fraction, causing a so called catastrophic phase inversion.

A variety of factors affect the extent of coalescence during formation of emulsions and foams, including the particle concentration, time of emulsification, and the dispersed phase volume fraction.

Some general properties of the solid-stabilised emulsions/foams concerning the role of the particles can be summarised as follows [46]. The particles have to be partially wettable by the both phases in order to obtain an efficient stabilization. In the case of emulsions, the most stable o/w emulsions are formed for contact angle in the range of 60°–85° and the most stable w/o emulsions for angle ranging between 95° and 120° [47]. This wetting properties of the particles can be affected by the presence of a surfactant which modifies and enhances their ability to reside at the interface [48]. In the case of hydrophilic particles, the presence of surfactant gives them a hydrophobic character, till the complete coverage is achieved. Further increasing of the surfactant concentration renders the particles again hydrophilic, due to the formation of surfactant bilayer at the solid surface. Thus, the stability of the dispersed system increases to a maximum, then decreases as the surfactant concentration is further growing. In this context, it is possible to highlight several key points of the stabilization of disperse systems:

1) The disperse system stability increases with increasing concentration of the particles. This implies that the efficacy of the particles in stabilising the emulsions/foams depends partially on the presence of a sufficiently dense layer of particles around the droplets/bubbles. Moreover, the average size of the droplets decreases as the concentration of solids in the system grows. The decrease of the emulsion droplets size results in an increase in the total droplet surface area, enabling additional particles to be accommodated at the interface. Generally, the relationship between solid concentration and the disperse system stability is not linear; rather, there is an upper limit in the concentration of particles above which the system stability does not increase further.
2) The characteristic particle size is one of the most important factors affecting the efficacy of colloid particles in stabilising emulsions and foams because it controls the ability of the particle to reside at the interface. It has been shown that the stability is increased with the decreasing particles size until a critical particle size is reached. Below this critical value, Brownian effects become significant enough to affect the partitioning of particles at the oil/air–water interface. As a result, the system stability decreases as the size of the particles decreases below the critical size. Besides, particles size is one of the parameter affecting the initial deformation of the droplets/bubbles menisci which can lead to the instability of the system via a coalescence process. It has been observed that for larger particles the deformation of the menisci results from gravitational effects; whereas for particles smaller than 5 μm, gravity effects are negligible and the behaviour cannot be accounted for this way [49].
3) The interactions between particles are important in stabilization of the disperse systems. It has been reported that for stabilisation to be efficient, the solid particles should be in the state of incipient flocculation [50]. This observation indicates that the interaction between particles is an important factor in the stabilization process. The strength of the interfacial film observed at the droplets of colloid-stabilised emulsions, as well as the rheological properties measured for colloid-laden oil-water interfaces is a consequence of the degree of interparticle interactions. The interaction between particles is responsible for the non-Newtonian behaviour observed for concentrated suspensions. This interaction is partly responsible for the mechanical stability of the particles at the oil–water interface.
4) Another important factor that affects the ability of the particles to stabilize the disperse systems is represented by the presence of inorganic ions in the aqueous phase, because the particles tend to coagulate at high electrolyte concentrations. When this happens, the particles form large aggregates and can no longer reside at the oil/air–water interface. These large aggregates of particles have the tendency to settle out of the system.
5) In terms of emulsion stability, the rheology of the interfacial region is particularly important because it governs the rate of drainage of the thin liquid films trapped between coalescing emulsion drops, controlling the coalescence process.

The presence of particles at the fluid interface, moreover, can avoid the coalescence between two droplets. In fact, as discussed above, the interparticle interactions are partly responsible for the disperse system stability, since they have a significant effect on the mechanical stability of the colloidal particles at the fluid interface. The displacement of particles from the interface, which can lead to coalescence, is hindered when particles interact with each other. This means that the reduction of the magnitude of particle interaction can promote the coalescence between droplets. This process can end up with either completely flocculating the particles or completely deflocculating the particles. The latter eliminates the elastic component of the surface viscoelasticity, thereby increasing the rate of the film thinning, and reduces the energy required to displace particles from the drop drop contact region (the magnitude of the steric hindrance). Complete flocculation, on the other hand, renders the particles into larger aggregates thereby reducing their ability to reside at the oil/air–water interface. Moreover, instability can be promoted by decreasing the interfacial tension significantly [50].

The efficacy of colloidal particles in stabilising emulsions and foams depends partly on the formation of a sufficiently dense layer of particles at the interface. Rheological properties of the interfacial layers are also changed as the particles concentration at the interface increases, and complete surface coverage is achieved.

At high particles concentration, the interfaces exhibit viscoelastic behaviour. The viscous properties are predominant rheological properties at the low particles concentrations, while the elastic properties are more prominent at high particle concentrations. The elastic contribution to the viscoelastic behaviour is largely due to the interparticle interactions. These interactions become significant at sufficiently high particles concentration. The viscoelastic nature of the interfaces affects the stability via decreasing the rate of film thinning between coalescing droplets.

A stability mechanism that can be associated with a specific low-particle-concentration interaction is the so called particle bridging. In particular, if the droplets are incompletely coated, the particles adsorbed at one droplet can directly interact with a neighbouring droplet interface (much like in constricted foam systems), causing stable droplet flocs. Several authors (for instance, Stancik and Fuller [51] and Horozov et al. [52]) found that interfaces in close contact could undergo long-range particle bridging, called 'particle zips'. Under certain conditions, when the two networked particle-laden interfaces come into contact, the two interfaces will 'zip' to form one tightly networked monolayer; remarkably, that monolayer would continue to 'zip' nearby particles in the both layers, growing outwards from the initial point of contact, compressing the particles to a finite distance, similarly to crystal formation.

3. Interfacial Rheology Basis

In general, *rheology* deals with the study of deformation and flow of a material in response to the applied stress. Simple solids store energy and provide spring-like elastic response, whereas simple liquids dissipate energy trough viscous flow. For more complex viscoelastic materials, rheological measurements reveal both solid- and fluid-like responses, depending on the time scale at which the sample is probed [53]. Interfacial rheology, like its 3D analogue, defines the relationship between stress and rate of deformation of materials either adsorbed or deposited at fluid interfaces. Boussinesq [54] was the first to describe the interfaces dynamics introducing the concept of surface shear and dilational complex elasticity modules and developing a link between stress and rate of deformation at the interface. Several years later Scriven [55] and Slattery [56] derived mathematical equations required to describe the dynamic conditions at the interface taking into account the interfacial viscosity. As a result, Scriven [55] provided a constitutive equation for a Newtonian surface of a fluid based on the relation between the interfacial stress tensor T_s and the interfacial deformation rate tensor D_s

$$T_s = [(\kappa - \eta)\mathrm{div}_s\, v_s]Q + 2\eta D_s \tag{2}$$

where κ and η represent the interfacial dilational and shear complex viscosities, respectively; v_s accounts for the velocity vector; div_s is the divergence operator; and Q is a projection tensor to provide the tangential component to the interface of each vector [57].

A remarkable difference between the rheology of interfaces and that of bulk fluids is that the latter can be considered incompressible in many practical applications. Yet, supramolecular species attached at fluid surfaces can be compressed and expanded over a large range of areas per molecule [58]. On the contrary to what happens in isotropic bulk systems that are characterized by two deformation modes (compression and shear) because of the uniaxial symmetry, there is an additional surface viscoelastic parameter in interfacial layers, associated with transverse shear [59, 60].

The modes of deformation possible for a film either deposited or adsorbed at a fluid interface can be classified into two main classes [61]: *(i)* out-of-plane (*Transversal*) modes causing curvature motions that are ultimately restored by surface tension-driven Laplace forces; and *(ii)* in-plane modes, including shear and compression contributions, restored by surface tension gradients. In the both cases, due to dissipative effects existing within the film, each of these motions consists of elastic and viscous components.

In general, when a surface-active specie (such as surfactant, polymer, or particle) is adsorbed at a fluid interface, the steady state is reached, being characterized by a relation between the surface pressure Π and the area occupied A at a given temperature ($\Pi - A$ isotherm). However, to address the dynamic behaviour of a material layer adsorbed at a fluid interface, it is necessary to apply an external perturbation. In particular, when the surface is subject to a sinusoidal deformation, the response is linear, and single surface modes are excited. Let us consider a surface deformation mode of wave vector q parallel to x-axis: $u(x,y,z,t)=u_i(t)\, e^{iqx}$; with u_x describing compression deformation modes, u_y describing shear modes, and u_z describing transverse modes (see Figure 2).

The total surface energy is:

$$F = \frac{1}{2}\varepsilon q^2 u_x^2 + \frac{1}{2}G q^2 u_y^2 + \frac{1}{2}\left(\gamma q^2 + \rho g + K q^4\right)u_z^2 \tag{3}$$

where ε and G represent the dilational complex and shear modules. The transversal mode has several contributions: surface tension contribution (γ), gravity contribution (ρg with ρ being density of the bulk fluid), and curvature contribution (K). When q is longer than $q_1 = (\rho g)^{1/2}$, the u_z modes are capillary modes, which are also of interest to describe monolayers rheology.

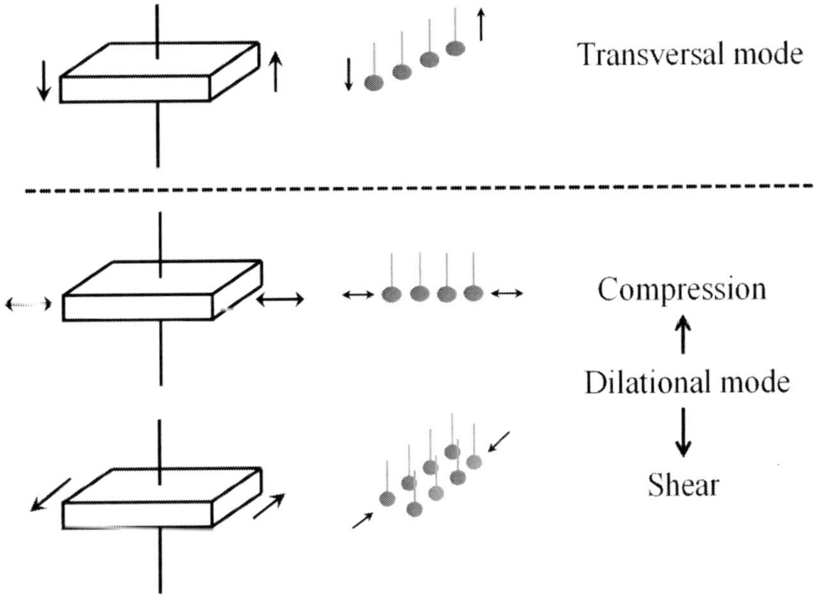

Figure 2. Sketch of different surface relaxation modes: transversal (out-of-plane) and dilational (in-plane), the latter composed of shear and compression. Adapted from Ref. 61, Copyright (1993), with permission from Elsevier.

3.1. Dilational Viscoelasticity

Let us consider an infinitesimal change of the surface area, $\delta A(t)$ produced by an uniaxial stress. The area perturbation induces a change in the surface pressure $\delta \Pi(t)$ which evolves in time until a new equilibrium state is reached. The change in surface pressure depends on the time scale probed in the experiment; according to Ref. 62, it can be expressed as:

$$\delta\Pi(t) = \Pi(t) - \Pi_0 = \frac{\partial \Pi}{\partial A}\delta A = -\varepsilon(t)u(t) \tag{4}$$

where: $\varepsilon(t) = -A_0(\partial \Pi / \partial A)_T$ is the time-dependent dilational modulus that accounts for the elastic energy stored on dilating the film. This equation can be in fact conceived as a general time-dependent response function, where a surface stress $\delta \Pi$ appears as the consequence of an stimulus applied as a compression strain $u(t)=\delta A/A_0$. In this context, the proportionally constant $\varepsilon(t)$ accounts for the material function which usually contains both dilational and shear components: $\varepsilon(t) = \varepsilon_k + G$. This is because uniaxial compression always involves shear. For fluid systems, the shear component vanishes, being nonzero only in the cases of solid or gel-like monolayers. Therefore, for fluid films at equilibrium, the dynamic modulus equals the Gibbs elasticity ε_0:

$$\varepsilon(t) \to \varepsilon_0 = \Gamma\left(\frac{\partial \Pi}{\partial \Gamma}\right)_{eq} \tag{5}$$

where $\Gamma=1/A$ is the surface concentration.

If we consider a small-amplitude oscillatory motion [$\delta A(t) \sim e^{i\omega t}$] of frequency ω, this modulus is a complex quantity:

$$\tilde{\varepsilon}(\omega) = \varepsilon(\omega) + i\omega\kappa(\omega) \tag{6}$$

The viscoelastic dilational modulus contains information about how the surface pressure changes on dilating the film and, hence, about the changes of the adsorption state and the molecular conformation stressed by the external deformation. These changes are, in general, dependent of the time scale probed in the particular rheological experiment. The constitutive viscoelastic parameters $\varepsilon(\omega)$ and $\omega\kappa(\omega)$ obtained as a function of ω contain the time-dependent response of the system subject to a small disturbance, i.e., they allow one to probe the surface dynamics of the adsorbed film.

3.2. Shear Viscoelasticity

The concepts of three-dimensional rheology may be applied to the two-dimensional case, the notable limitations being as follows: (*i*) for a correct analysis, the stress tensor at the interface must be symmetric; (*ii*) due to the drag forces, the interfacial shear properties depend on the geometry of the device, and the eventual coupling of the adjacent phases with the interfacial flow must be considered in the data analysis; (*iii*) to measure interfacial shear properties, each element of the interface must be deformed at constant area and any radial flow in the interface must be avoided [63].

Hydrodynamics equations show that the motion along y evolves independently from the motions along x and z. Shear modes are then completely decoupled from capillary and compression ones: both their amplitude and time evolution are independent [60, 64]. In detail, for deformations in the x–y plane one can define the interfacial shear elasticity G as the proportionality factor between the applied strain (u_{xy}) and the stress response exerted by the monolayer (σ_{xy}). "Solid" films exhibit the elastic (Hookean) behaviour characterized by: $\sigma_{xy} = G\, u_{xy}$. Low-strain amplitudes ($u \leq 1\%$, typically) guarantee measurements of the zero-shear limit ($G = \sigma/u$ at $u \rightarrow 0$). For totally "fluid" films, a viscous (Newtonian) behaviour is observed: $\sigma_{xy} = \eta\, du_{xy} / dt = \eta\, \mathring{u}_{xy}$; where $\mathring{u}_{xy}$ is the strain rate and η is the interfacial shear viscosity.

Attractive interactions between the surface elements (e.g. colloid particles, polymeric chains either adsorbed or spread, surfactant molecules, etc.) yield to an increase of the viscoelastic parameters (G, η) because energy is spent to overcome these interactions and let the surface elements flow. However, most interfacial layers exhibit both elastic and viscous behaviour, and the response of such layer to periodic deformations is viscoelastic. If we consider oscillatory deformation, we can define a complex shear modulus G^* similarly to the bulk rheology case:

$$G^*(\omega) = G'(\omega) + iG''(\omega) \equiv G' + i\omega\eta \tag{7}$$

where the real element G' is the shear storage component and the imaginary part G'' is the loss component. For a small-amplitude oscillatory motion of frequency ω ($u \sim u_0 e^{i\omega t}$), the loss modulus is related to the viscous friction $G''=\omega\eta$ where η is the shear viscosity.

4. General Aspects on the Rheological Behavior of Particles-Laden Interfaces

Understanding of the rheological aspects of particle-laden interfaces plays a central role in the development of systems based on the interactions between particles and fluid interfaces [6, 9]. In recent years, several experimental and theoretical studies on this topic have been carried out to provide insights on the response of particles at fluid interfaces subject to flow or compression/expansion deformation. These aspects are essential in many process of practical interest such as emulsification and foaming, enhanced oil recovery, and phase transfer catalysis [65, 66]. The structural/morphological diversity of particles layers at fluid interfaces provides the basement for remarkable variability of the rheological/mechanical response of these systems to different physico-chemical stimuli [1]. This section will review the most recent advancements in the study of mechanical behaviour of particle-laden layers, paying special attention to the dilational and shear response for both pure particles layers and mixed particles–surfactant layers.

4.1. Shear Rheology

Study of the response of particle-laden interface to the shear deformation has attracted great interest in recent years due to their recognized importance in many phenomena and processes involving this type of systems [1, 6]. This has lead to intense theoretical and experimental development of the knowledge in this field in order to correlate the shear response with morphology of the layers and the potential applications of particle-laden interfaces. Interactions between particles at fluid interface and interfacial structure are the main parameters that govern the response of particles monolayers to shear perturbations [67].

Dependence of the shear response on the coverage fraction of polystyrene latex (3 μm in diameter) particles monolayer at the water/decane interface was analyzed by Cicuta et al. [68] to obtain both the elastic and viscous modules. The authors pointed out the viscous character of the particle layers ($G'' > G'$), independently of the deformation frequency, which was significantly different from the mainly elastic behaviour found for globular proteins, such as β-lactoglobulin. In addition, viscoelasticity of the particle-laden interface increased dramatically from values close to those of the pure water/decane interface at surface coverage going up to around 75–80%, being associated with jamming phenomena. On the other hand, proteins layers showed viscoelastic modulus values tending to an asymptotic value beyond a certain value of the interfacial coverage. Those differences in the shear response were ascribed to the different nature of the material at the interface: whereas nanoparticles could be considered as hard non-deformable disks, proteins behaved as deformable and compressible ones, which determined the formation of different 2D glassy structures. The existence of both long-range and short-range order in the particle-laden interfaces defined important differences in the shear response between the interfacial systems and their bulk analogues [69].

Imperiali et al. [70] dug into the effect of the interfacial coverage fraction on the shear response of particle-laden interfaces. They studied monolayers of graphene oxide particles and found a plastic response for coverage well below that of the close-packed state. On the other side, close-packed monolayers showed a perfectly elastic response. That was related to the different degree of freedom of the particles at the interface and, consequently, to their possible reorganization at the interface. The interplay between the interfacial coverage and the rheological response was also confirmed by Barman and Cristopher [71]. They observed shear thinning at relatively low values of the interfacial coverage, whereas yielding emerged at higher interfacial density. Those differences were ascribed to the structural differences that allowed for different reorganization mechanism to dissipate the viscous stress. Furthermore, the authors found that the response to shear of highly packed particles layers was far from linear.

In order to deeper investigate the relationship between structure and rheological response of particle-laden interfaces, Reynaert el al. [72] compared the interfacial behaviour of 2D colloidal crystal with 2D aggregate suspensions. The results showed that the aggregation of the particles at the interface led to a shear response of the monolayer similar to that found for bulk counterparts [69]. That scenario was in agreement with the Brownian dynamic simulations performed by Wijmans and Dickinson [73]. The rheological response of the 2D aggregate suspensions was characterized by a small linear regime [72], followed by the viscoelastic modulus power law dependence on the surface coverage. Furthermore, the viscoelastic modulus depended strongly on the strength of the inter-particle interaction as was pointed out by Reynaert el al., [72].

The particle morphology plays an important role in the shear response of the interface since it affects to the interfacial morphology. In particular, Madivala et al. [74] pointed out that ellipsoidal polystyrene particles at the water/decane interface formed a structure with individual particles coexisting with linear aggregates of those particles, whereas at the air/water interface the particle were organized in flower-shape aggregates due to the lower electrostatic repulsion. The different particles organization depending on the interface nature determined different response to shear perturbations, being an important evidence of the relation between the particle self-assembling at fluid interfaces and the interfacial mechanical response. It is worth mentioning that the linearity range of the rheological response was short, decreasing with the monolayer packing. When particles at the water/oil interface were sheared, the main elastic response was found, and the viscoelastic modulus increased with the packing density of the monolayer. The increase of the surface concentration led to behaviour similar to that expected for a flocculated network [53]. In the case of monolayers at the air/water interface, the shear elasticity was smaller than that found for the same particles at the water/decane interface.

The results of Madivala et al. [74] showed that the elastic modulus of the monolayers of ellipsoidal particles was very high starting from low surface coverage values. That was in contrast with the fact that elastic modulus increased only after a certain coverage threshold value when spherical particles were considered [68]. Furthermore, ellipsoidal particles showed a storage modulus more than an order of magnitude higher than that reported for spherical particles of the same nature [72]. Those differences were ascribed to the different interfacial organization of the particles.

Basavaraj et al. [75] pointed out that the differences in the mechanism of interfacial reorganization of the interfacial layers led to the different rheological behaviour of spherical and ellipsoidal particles. Whereas for ellipsoidal particles a buckling transition was observed at the highest interfacial coverage values, it was not observed for spherical particles. In general, the phase behaviour for non-spherical particles was more complex than that of spherical ones, defining a richer rheological behaviour of the former type of particles. The particular aggregation pattern of the particles at fluid interfaces was correlated to the existence of an intricate balance between electrostatic and capillary interactions. That balance determined the mechanical response of the particle monolayers [75, 76]. Zahn et al. [76] showed that the shear modulus of colloidal crystals formed by paramagnetic colloidal particles trapped at the air/water interface increased with the strength of the inter-particle interactions.

Furthermore, Brown et al. [77] deepened the knowledge on variation of the shear response due to the particles geometry. To do so, they used four different types of non-spherical particles with different aspect ratio. The results pointed out that the particles showed similar discontinuous shear thickening. An important difference of the shear behaviour of non-spherical particles in comparison to spherical ones was the decrease of the coverage threshold for jamming. In general, the jamming of the particle-laden interface occurred at lower values of the coverage with the increase of the particles asymmetry. Furthermore, the results revealed that the asymmetry of the particles could arrest their dynamics, affecting the rheological response. That was ascribed to the different structural patterns that the particles generated at the interface due to the effect of the shear flows as was pointed by Cui [78] theoretically and confirmed experimentally by Stancick et al. [79] In particular, the work by Cui [78] showed the importance of the coverage and particle charge in determination of the rheological properties of particle-laden interface.

According to the theoretical analysis, the surface roughness of the particles affected strongly the shear response of particle-laden interfaces [80]. The shear viscosity was lowered with increase of the particles roughness in the dilute region, whereas the opposite was true close to the jamming density. That was due to the different importance of the inter-particle friction in the monolayer response. The effect of particle roughness and, consequently, of the inter-particle friction, was demonstrated experimentally by Brown et al. [77] They pointed out that the inter-particle friction was one of the reasons for the shear thickening of particles monolayers.

The particle size also influenced the rheological response against shear of particle-laden interfaces [77]. It is well-known that the particle size affects the phase diagram type of particle monolayers [13]. That could be especially relevant when charged particles were considered; those systems presented rich phase behaviour with a disordered state extended over a large surface density range, followed by a hexagonal solid-like structure. The merging of the different phases was strongly dependent on the particles dimensions [13, 14] Thus, a strong effect of the particle size on the interfacial rheological shear response could be expected, as it was found for the bulk counterparts [81].

Studies of the effect of the particles size on the rheological response of particles-laden interface were extended by the study of behaviour of mixtures of particles differing in size [14]. Keim and Arratia [82] studied jammed monolayers formed by particle mixtures of two different sizes (4.1 μm and 5.6 μm) and pointed out the existence of two different regimes of the rheological response. At low-strain amplitudes, the interfacial structure remained over long time scale, whereas when a threshold strain (yield strain) was overcome, the particles exhibited interfacial rearrangements of plastic origin. Those structural rearrangements of jammed particle-laden interfaces were reversible when cycling shear deformations were imposed. The reversibility on the structural rearrangements occurred in the so-called shear transformation zones [83].

The wettability of particles governed the attachment of particles to fluid interfaces, thus playing a key role in determination of the rheological response of particle-laden interfaces. In particular, Safouane et al. [84] found that at a fixed surface coverage the shear modules of fumed silica particles layers depended strongly on the particle hydrophobicity (in that case, controlled by degree of silanization of the silanol groups on the particle surface), with both *G'* and *G''* increasing with the particle hydrophobicity. The reported values for the shear modules of the particle-laden interface were small (around 10^{-1} mN/m) but in all the cases larger than those reported for dense surfactant layers (around 10^{-3} mN/m). For particles with low hydrophobicity, shear behaviour could be considered mainly elastic, whereas a transition to predominantly viscous behaviour was observed with the increase of the particle hydrophobicity, finding the gel-point ($G' = G''$) for particles with a silanization degree of around 36%. Those particles revealed intermediate hydrophobicity, their contact angle at the air–water interface being around 90º. Zang et al. [85–87] explored the dependence of the shear modulus on the strain amplitude in order to extend the knowledge on the effect of the particle hydrophobicity on the response of the monolayers to shear. The authors found a storage modulus independent of the strain amplitude in the low-strain amplitude region, being in all the case higher than the loss modulus (by almost two orders of magnitude). Melting of the monolayer was observed above a threshold value of the strain amplitude (yield strain amplitude) with *G''* reaching a maximum, whereas *G'* went down when that yield amplitude was overcome. For low deformation frequencies and a fixed strain, $G'' > G'$ was held, whereas a crossover was found with the increase of frequency. That crossover point was reported for frequency values close to those where *G''* reached maximum. Such behavior could be related to a decrease of the structural relaxation time with the increase of the strain-rate amplitude in a similar way than that found for 3D soft solids [88]. The dependence of both *G'* and *G''* on frequency followed the master curves G'/a and G''/a vs. b, a and b being parameters depending exclusively on the shear rate. An important effect observed by the authors was the quasi-linear dependence of the relaxation on the shear rate that was an additional similarity of amorphous 2D particle-laden interfaces with their 3D counterparts. Furthermore, the particle layers revealed self-healing ability when the stress was released, the phenomenon depending on the coverage and wettability properties of the particles. The latter parameter influenced the yield and melting stresses of particle-laden interfaces which reached the highest values for particles of intermediate hydrophobicity (silanization degree of around 36%) [87]. Vandebril et al. [89] found similar dependence of the shear modulus on the strain as reported by Zang et al., [86].

The effect of the coverage of the interface on the response of particle-laden interface to shear has been discussed above from a steady state point of view. However, it is interesting to analyze the same effect observed during the layer formation process [90]. The densification of monolayers of silver nanoparticles (10–50 nm in diameter) at the toluene/water interface increased both the storage and the loss modules till the equilibrium state of the adsorption process was reached. That was associated with the increase of the interfacial concentration in time. Furthermore, the films presented mainly elastic behavior ($G' > G''$) during low-frequency shear deformation. Simultaneously the *G''* dependence on the frequency presented a negative slope with long relaxation times, being an important evidence of the 2D glassy nature of the layer. In addition, the dynamics showed a strong dependence on the strain amplitude. Strain sweep measurements revealed the existence of shear thickening phenomena in the loss modulus behavior at large strain amplitude. On the other side, the storage modulus was almost independent of the strain amplitude before the shear thickening. The shear thickening of *G'* and *G''* followed a power law decay with a 2:1 ratio between the exponents. The above-described scenario was different from that found for gold nanoparticles at the air/water interface [41, 42, 91, 92]. The latter showed a gel-like behavior with a strain-induced softening phenomenon, the mechanical response being independent of the strain till a certain threshold strain value (around 0.1%). Once the strain went beyond the threshold value, a significant drop of *G'* was observed. The viscoelastic modules increased with interfacial coverage, following in that particular case a power law with exponent of 0.65, similar to that found for percolating bulk systems. *G'* was independent of the deformation frequency, whereas *G''* increased at high frequencies.

Behavior of the particles–surfactant mixtures is also important in the stabilization of dispersed systems; therefore, it is of interest to discuss here some features of their response to shear deformations. The shear behavior of the mixtures has been less explored than that of similar systems containing only the particles. Mass et al. [93] studied kinetic of interfacial interaction of lipids and silica nanoparticles of two different hydrophobicities at the water/oil interface by means of shear rheology. The layer formation was found to be divided in two stages. During the first step, within around 1 hour, the elastic modulus increased strongly till formation of a cohesive and mainly elastic layer ($G' > G''$). During the second step (much slower), a slower growing of the modules was observed, the particles and lipid molecules being accumulated at the interface. That process went on till steady state was reached, with rheological response governed by dynamic exchange of material between the bulk phases. The equilibrium values for both *G'* and *G''* were several orders of magnitude higher than those typical of pure surfactant monolayers, in agreement with the scenario found by Safouane et al. [84] for silica nanoparticles monolayers. The linear viscoelasticity region was narrow for those systems. Similar conclusions were made by Degen et al. [94] for monolayers of maghemite nanoparticles (γ-Fe_2O_3) and surfactant molecules of different chemical nature [95].

4.2. Dilational Rheology

The recognized importance of the response of particle-laden interfaces to compression–expansion stimuli has lead to its intense experimental [96] and theoretical [97] development in this field. In spite of this important development, there is a definite lack of knowledge in many aspects of the dilational rheology of particle monolayers, especially concerning the correlation between the experimental results and the theoretical models [6, 8]. Furthermore, the studies devoted to dilational rheology of layers formed exclusively by particles are quite a few, and most of them are focused on two types of particles: fumed silica (dendritic aggregates of primary particles of around 200 nm) [84, 85], and polystyrene latex [98].

The first study focusing on the dilational rheology of particle-laden interface was performed by Miller et al., [97]. They analyzed the mechanical response of particle layers using a theoretical model that allowed explaining the dilational rheology of particles and particle–surfactant layers considering mainly the equilibrium aspects associated with the adsorption isotherm of the system. The model was analogous to that previously developed for the rheology of proteins and protein–surfactant systems [99], introducing the cohesion between the adsorbed particles as an additional contribution to the rheological response. The cohesion was strongly dependent on the wettability properties of the particles. However, the application of the theoretical model to the analysis of experimental data has not been developed yet.

The effect of the hydrophobicity of the particles plays an important role in dilational rheology of particle-laden interface, similarly to the above-discussed for shear rheology. Safouane et al. [84] explored the dilational response of fumed silica monolayers at the air/water interface by means of capillary wave technique in the frequency range of 200–990 Hz. The authors found that the dilational response depended on the hydrophobicity of the particles due to morphological differences between the layers. $E' > E''$ held for all the hydrophobicities and surface coverage degrees analyzed. The elasticity as well as the static compression modulus increased with higher the hydrophobicity of the particles due to their favored incorporation at the interface. Similar behavior was found by Zahn et al.[76] for colloidal crystal formed by paramagnetic particles trapped at the air/water interface; that was associated with increased interaction between the particles. On the other side, the loss modulus did not show any dependence on the wettability or interfacial concentration of the particles. In contrast to the results found for the shear response, the range of linear dilational response of monolayers of silica particles at the air/water interface was broad [85]. Zang et al. [85] studied dilational rheology of silica monolayers using a combination of continuous compression, oscillatory barrier, and stepwise compression experiments. In particular, they evaluated the frequency dependence of mechanical properties of those systems. Negligible frequency effect was found for the dilational response in the low-frequency range (0.016–0.1 Hz). However, dynamic relaxation with characteristic frequency of about 10^{-3} Hz was revealed in the studies over the extended frequency range. The process was related to reorganization of the particles layer, accelerating with the decrease of the particle hydrophobicity [100]. The effect of the particles wettability on the dilational response of silica layers at the air/water interface was also analyzed by Zang et al., [87]. They observed the highest elasticity for particles of intermediate hydrophobicity. Furthermore, the elasticity of layers with similar surface coverage obtained through compression was always higher than those obtained by deposition, the effect being associated with merging of additional relaxation processes due to the non-equilibrium conformations [85, 100].

Kobayashi and Kawaguchi [101] pointed out that latex particles spread at the air/water interface showed viscoelastic behavior almost independent on the strain rate (the frequency being constant). Furthermore, a weak hysteresis was found in the rheological response, evidencing the lost connection between the particles in the monolayers. E' and E'' increased till surface pressure of around 15 mN/m was reached, and once that threshold surface pressure value was exceeded, both E' and E'' decreased steeply with the increase of the surface pressure. The frequency dependence of E' and E'' evidences about a crossover between a fluid-like behavior to a solid-like one ($E' > E''$).

The rheological behavior of polystyrene latex microparticles spread at the air/water showed a strong dependence system position in the phase diagram [102] as a consequence of the structural richness of these systems [13]. In general, $E' > E''$ held over the entire phase diagram. At low surface pressures, *E*' increased with the surface pressure, whereas *E*'' remained close to zero. That was ascribed to the long-range repulsive electrostatic interactions between the particles. When those electrostatic repulsions were screened, both *E*' and *E*'' increased with the surface pressure in the low surface pressure region. In detail, *E*' increased abruptly to quite high values showing a maximum of around 350 mN/m; that allowed claiming the existence of a close-packed monolayer, in agreement with Brewster Angle Microscopy (BAM) images. For systems with strong electrostatic repulsion, the elasticity reached values close to 600 mN/m; that was explained by formation of close-packed monolayers with less of defects. Once the maximum elasticity was reached, further compression led to decrease of both *E*' and *E*''. Such a decrease was caused by collapse of the monolayer and subsequent buckling of the particle-laden interface.

Moreover, particles monolayers could form relatively dense monolayer even when the surface coverage was relatively low [13]. That resulted in a relatively narrow linear region; therefore, the dilational studies should be performed carefully by application of small strains at low frequencies [98].

The monolayers built by the mixtures of surfactant and particles have been considered in some more details recently [8]. We will hereafter discuss these intriguing systems.

The interaction of two oppositely charged components (such as polyelectrolytes and oppositely charged surfactants) in solution leads to formation of adsorption layers where synergetic effects of decrease of the fluid interface surface tension during adsorption of the formed complexes are expected [103, 104]. However, Ravera et al. [17] pointed out that adsorption of mixtures formed by cetyltrimethylammonium bromide (CTAB) and hydrophilic silica nanoparticles did not show any synergism of the surface tension decrease. Nevertheless, important effects due to the particles presence were observed in the dilational rheological response at the low-frequency range (0.005–0.2 Hz) for both air/water and water/hexane interfaces. Those changes were due to the adsorption of complexes of different hydrophobicity that favored their adsorption to the interface. The adsorption of particles resulted in stronger elasticity than was expected for monolayers formed exclusively by surfactant at the same concentration. Furthermore, the viscoelastic response was strongly frequency-dependent. In addition, the effects of the particles adsorption were stronger in the case of the hexane/water interface than those at the air/water interface. The studies over extended frequency range pointed out the existence of two different relaxation processes in the particle-laden interfaces [18]. The first process, observed at lower frequencies, implied diffusive transport of the particles to the interface, the second process being related to the exchange of material between the interface and the bulk. Theoretical analysis of the experimental data taking advantage of the model proposed by Ravera et al. [18] revealed the transition of rheological response from that governed mainly by diffusion to the mixed mechanism of diffusion–reorganization when the particles hydrophobicity was increased. More detailed analysis of the above-described system was performed by Liggieri et al., [15]. The authors investigated the interfacial dilational response of the particle-laden interface over a frequency range from 10^{-3} to 10^{2} Hz. Figure 3 represents a typical example of the frequency dependence of the elastic and viscous modules for a particle-laden interface and the corresponding fittings using the model proposed by Ravera et al., [18].

The studies of silica nanoparticles–CTAB mixed interfacial layers evidenced about the existence of a complex balance of interactions governing the rheological response of such systems. Among those interactions were CTAB depletion, the hydrophobization/

rehydrophilization of the particles due to adsorption of the surfactant molecules at the silica surface due to electrostatic and hydrophobic interactions, and various interactions occurring between the particles in the bulk and at the interface [11, 12]. The scenario discussed above for the system formed by silica nanoparticles and CTAB was also confirmed by Wang et al., [105].

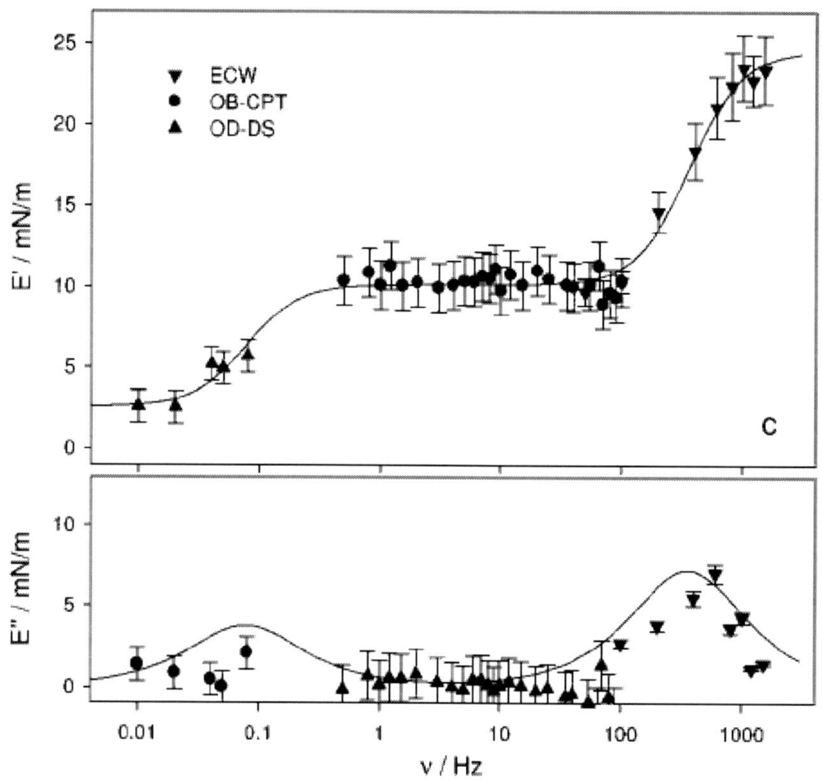

Figure 3. (A) Real part of the viscoelastic modulus and (B) imaginary part of the viscoelastic modulus of the 1 wt% of silica nanoparticles dispersion with 0.5 mmol/L of CTAB. The solids lines correspond to the best fitting with the model proposed by Ravera et al., [18]. Reproduced from Ref. 15 with permission from The Royal Society of Chemistry.

Another important aspect to consider in the mechanical response of particle-laden interfaces is the effect of the interfacial aging [18, 34]. The aging of particle monolayers provokes an increase in the dilational elasticity at both air/water and water/hexane interfaces without any appreciable change of the surface tension, being an evidence of the irreversible attachment of the particles at the interface [34, 106]. Such behavior was ascribed to formation of solid-like layers due to densification process as was evidenced by Yazghur et al. [34] by means of BAM imaging. The high elasticity modulus values found for those particle-laden interfaces were comparable to those found for spread monolayers of latex particles [101, 102]. The high viscoelastic modulus of very dense particle-laden interfaces was further confirmed by Santini et al. [107] by combination of oscillatory barriers experiments and BAM imaging as well as by Whitby et al., [33].

5. Correlations between Rheological Behavior and Stabilization of Foams and Emulsions

Mechanical behaviour of interfaces governs the interfacial stability and deformation of coated drops and bubbles, playing the major role in stabilization of dispersed systems such as thin films, foams, and emulsions, acting against coalescence or Ostwald ripening phenomena and preventing gravitational destabilization via drainage or creaming [108, 109]. In general, attachment of particles at the surface can lead to formation of self-assembled rigid armour that stabilize the interface against deformation, avoiding the drops/bubbles collision and consequently preventing destabilization of the dispersed systems. Stabilization of dispersed systems is specially affected by the jammed layers, occurring when the area available for the particles distribution is close to the close-packing area [40, 110]. The jamming phenomena can arrest interfacial tension-driven coarsening, conferring a special stability to the biphasic air–liquid or liquid–liquid systems. Furthermore, the rheological properties, both shear and dilational ones, allow controlling the stability of emulsions and foams due to their effect on the film drainage. It is expected that the drainage can be hindered by the increased shear viscosity and dilational modulus [111].

Cervantes-Martínez et al. [112] and Stocco et al. [38, 39] attempted to generalize the features of dispersed systems stabilization by nanoparticles on the basis of interfacial measurements (equilibrium surface tension and dilational rheology). Similar problems were considered by Arriaga et al. [113] and Maestro et al., [40]. In particular, Cervantes-Martínez et al. [112] proposed that the destabilization of particles-stabilized foams could be hindered due to formation of layers strong enough to resist compression deformations, thus avoiding the shrinkage of small bubbles and drops. In order to establish the optimal conditions for foam stabilization by means of interfacial measurement, the so-called Gibbs criterion was used. That approach defined foam stability on the basis of a ratio between the elasticity of the formed layers and the surface tension of the particle-laden interface. The Gibbs criterion assumed that a foam was stable when the relationship $E > \gamma/2$ was fulfilled. Furthermore, the authors suggested a potential effect of the surface coverage on the foam stability. Stocco et al. [39] pointed out the validity of the Gibbs criterion for both foams and emulsions. However, they limited the criterion application to spherical drops and bubbles. Furthermore, they demonstrated the role of the particles organization at the interface on stabilization of the foams, suggesting that gel-like particle layers enhanced the stability of dispersed systems.

Stancick et al. [79] and Alvarez et al. [114] confirmed the role of interfacial coverage in stabilization of emulsions by nanoparticles. They pointed out that the role of the particle-laden interface in stabilization of dispersed systems was not limited to the changed mechanical properties, the steric hindrance induced by the nanoparticles being important as well in preventing the coalescence. Similarly, the work by Frith et al.[114] showed the importance of formation of close-packed silica layers to confer rigidity and stabilize the dispersed systems. Sullivan and Kilpatrick [115] showed that stabilization of oil-recovery emulsions was enhanced by the particles forming rigid stagnant films surrounding the drop and preventing the coalescence. The authors investigated the effect of chemically different particles and pointed out that only the particles allowing the accurate interactions at the interface can stabilize emulsions of crude oil. Extending those studied, some recent works have contributed to the controversy about importance of the rheological response in the stability of interfacial layers. Santini et al. [116, 117] pointed out the possibility to stabilize the foams and emulsions with carbon particles, but no evident relationship between the rheological properties of the interfacial layers and the system stabilizing ability was found. Furthermore, Zabiegaj et al. [118, 119] demonstrated fabrication of solid porous materials using carbon-based liquid foams as templates. However, no dependence of the stabilization phenomena on the interfacial viscoelastic properties was elucidated. In contrast, strong correlation between the non-linearity of the interfacial dilational response and the stabilization of the liquid foams used as precursor was pointed out in the case of alumina-based solid foam [118]. The finding was ascribed to irreversible attachment of the material at the interface. The studies by Santini et al. [107, 120] on the silica nanoparticles – palmitic acid system contributed to the controversy related to the relationship between the mechanical properties of the particle-laden interfaces and their ability to stabilize dispersed systems. The authors found a high dilational viscoelastic modulus (>100 mN/m) for silica nanoparticles – palmitic acid mixed monolayers at the air/water interface. However, the system could not be used to stabilize foams [107]. On the other hand, formation of highly stable water-in-oil emulsions was reported for the above-mentioned mixtures [120], but dependence of the ability to stabilize the emulsions on the rheological properties was unclear.

Oppositely to the contradictory role of dilational modulus in the stabilization of dispersed systems, the role of shear modulus seems clearer, being in close relationship with the interfacial structure and coverage. Madivala et al. [121] pointed out that the high interfacial shear modulus of the monolayers formed by ellipsoidal particles and their particular packing characteristics determined formation of interfacial gels. Those gel-like structures surrounding the drops or bubbles contributed to stabilization of the dispersed systems [89] that can be considered an advantage to fabricate ultra-stable emulsions, both water-in-oil and oil-in-water ones [121]. Noteworthily, those emulsions stabilized by ellipsoidal particles were much more stable than those stabilized by spherical ones [121]. Kim et al. [122] confirmed the above-discussed scenario using clays to stabilize oil-in-water emulsions. The clays with hydrophobicity high enough to favour the formation of close-packed layers with high values of the shear modulus showed better performance towards emulsions stabilization than did hydrophilic clays. Moreover, addition of surfactant to hydrophilic clays induced hydrophobization of the latter and favoured the emulsion stabilization. Brugger et al. [123] pointed out that the conditions of increased G' value were optimal for stabilization of emulsions by poly(N-isopropylacrylamide). However, for high values of G'', the emulsion stability was reduced due to formation of softer layers that could be easily broken.

The ability to stabilize foams was also related to large values of yield stress [77]. Under those conditions, it was possible to entrap a large fraction of air in the particles-stabilized system. The high yield stress was also found favourable for stabilization of water-in-oil emulsions by Imperiali et al. [70], due to formation of layers of graphene oxide sheets at the fluid interface. In that case, the drops were non-spherical because of partial coalescence.

Other parameter to consider in the stabilization of dispersed systems is the the particle roughness [124]. The emulsion stability increased with the particle roughness due the increased shear viscosity, thus hindering the drainage processes.

Moreover, understanding of the interfacial rheology of particle-laden interface in relation to the stabilization of dispersed systems is also important to control the bulk rheology of the obtained foams or emulsions [71].

Conclusion

Particle-laden interfaces are intriguing systems due to their numerous industrial applications. Furthermore, the properties of particles monolayers open fascinating challenge to theoretical and experimental studies. Special attention should be paid to the study of mechanical performance of particle-laden interfaces, due to their implications in the stabilization of dispersed systems such as emulsions, foams or thin films. The properties of particle assemblies at fluid interfaces are strongly dependent on multiple aspects involving particles and fluids as well as their interactions. This defines multiple scenarios for the mechanical properties of the particle-laden interface which are closely correlated to structural features of the interfaces. Their deeper investigation allows elaborating of the optimal conditions for stabilization of the dispersed systems. However, there is still poor understanding of the correlation between interfacial rheological properties, the interface morphology, and the stabilization of dispersed systems. In this regard, it is necessary to perform systematic studies of stabilization of the dispersed systems using particle-laden interfaces.

Acknowledgments

This work was supported by MINECO (FIS2012-14008-C02-01), by ESA (FASES and PASTA), and by EU (Marie Curie ITN-COWET), and carried out in the framework of the COST action CM1101. EG is grateful to the MINECO by the Juan de la Cierva contract and AM is grateful to the RSC by the Newton Fellowship. We are grateful to the UIRC of the CAI of Spectroscopy (Universidad Complutense de Madrid) for allow us the use of their facilities. The authors want to thank to R.G. Rubio, F. Ravera, F. Ortega and L. Liggieri for fruitful discussions.

References

[1] Binks, B.P., Horozov, T.S., 2006, Colloidal Particles at Liquid Interfaces. *Cambridge University Press*: Cambridge.

[2] Aveyard, R., Binks, B.P., and Clint, J.H. 2003, *Adv. Colloid Interface Sci.*, 100–102, 503–546.

[3] Binks, B.P. 2002, *Curr. Opin. Colloid Interface Sci.*, 7, 21–41.

[4] Dinsmore, A.D., Hsu, M.F., Nikolaides, M.G., Marquez, M., Bausch, A.R., and Weitz, D.A. 2002, *Science*, 298, 1006–1009.

[5] Velev, O.D. and Gupta, S. 2009, *Adv. Mat.*, 21, 1987–1905.

[6] Mendoza, A.J., Guzmán, E., Martínez-Pedrero, F., Ritacco, H., Rubio, R.G., Ortega, F., Starov, V.M., Miller, R. 2014, *Adv. Colloid Interface Sci.*, 206, 303–319.

[7] Maestro, A., Guzmán, E., Ortega, F., and Rubio, R.G. 2014, *Curr. Opin. Colloid Interface Sci.*, 19, 335–367.

[8] Guzmán, E., Santini, E., Liggieri, L., Ravera, F., Loglio, G., Maestro, A., Rubio, R.G., Krägel, J., Grigoriev, D., and Miller, R. 2013, *In:* Colloid and Interface Chemistry for

Nanotechnology, Kralchevsky, P., Miller, R., and Ravera, F. (Eds.) *CRC Press, Taylor & Francis*: Boca Raton, Florida, pp 77–109.
[9] Garbin, V., Crocker, J.C., and Stebe, K.J. 2012, *J. Colloid Interface Sci.*, 387, 1–11.
[10] Zang, D.Y., Rio, E., Langevin, D., Wei, B., and Binks, B.P., 2010, *Eur. Phys. J. E*, 31, 125–134.
[11] Santini, E., Krägel, J., Ravera, F., Liggieri, L., and Miller, R. 2011, *Colloids Surf. A*, 382, 186–191.
[12] Maestro, A., Guzman, E., Santini, E., Ravera, F., Liggieri, L., Ortega, F., and Rubio, R.G. 2012, *Soft Matter*, 8, 837–843.
[13] Bonales, L.J., Rubio, J.E.F., Ritacco, H., Vega, C., Rubio, R.G., and Ortega, F. 2011, *Langmuir*, 27, 3391–3400.
[14] Bonales, L.J., Martínez-Pedrero, F., Rubio, M.A., Rubio, R.G., and Ortega, F. 2012, *Langmuir*, 28, 16555−16566.
[15] Liggieri, L., Santini, E., Guzman, E., Maestro, A., and Ravera, F. 2011, *Soft Matter* 7, 7699–7709.
[16] Pieranski, P. 1980, *Phys. Rev. Lett.*, 45, 569–572.
[17] Ravera, F., Santini, E., Loglio, G., Ferrari, M., and Liggieri, L. 2006, *J. Phys. Chem. B*, 110, 19543–19551.
[18] Ravera, F., Ferrari, M., Liggieri, L., Loglio, G., Santini, E., and Zanobini, A. 2008, *Colloids Surf. A*, 323, 99–108.
[19] Hunter, T.N., Pugh, R.J., Franks, G.V., and Jameson, G.J. 2008, *Adv. Colloid Interface Sci.*, 137, 57–81.
[20] Ramsden, W. 1903, *Preliminary Account. Proc. R. Soc. Lond.*, 72, 156–164.
[21] Pickering, S.U. 1907, *J. Chem. Soc., Trans.*, 91, 2001–2021.
[22] Zhang, H. and Cooper, A.I. 2005, *Soft Matter*, 1, 107–113.
[23] Studart, A.R., Gonzenbach, U.T., Akartuna, I., Tervoort, E., and Gauckler, L.J. 2007, *J. Mater. Chem.*, 17, 3283–3289.
[24] Dickinson, E. 2010, *Curr. Opin. Colloid Interface. Sci*, 15, 40-49.
[25] Thongprachan, N., Yamamoto, T., Ohmori, T., and Endo, A. 2011, *Adsoprtion*, 17, 205–210.
[26] Garbin, V., Crocker, J.C., and Stebe, K.J. 2012, *Langmuir*, 28(3), 1663–1667.
[27] Ivanov, I.B., Kralchevsky, P.A., and Nikolov, A.D. 1986, *J. Colloid Interface Sci.*, 112, 97–107.
[28] McBride, S.P. and Law, B.M. 2012, *Phys. Rev. Lett.*, 109, 196101.
[29] Aveyard, R. and Clint, J.H. 1996, *J. Chem. Soc., Faraday Trans.*, 92, 85–89.
[30] Bresme, F. and Quirke, N. 1998, *Phys. Rev. Lett.*, 80, 3791–3794.
[31] Levine, S., Bowen, B.D. and Partridge, S.J. 1989, *Colloids Surf.*, 38, 325–343.
[32] Liggieri, L., Santini, E., Guzman, E., Maestro, A., and Ravera, F. 2011, *Soft Matter*, 7, 7699–7709.
[33] Whitby, C.P., Fornasiero, D., Ralston, J., Liggieri, L., and Ravera, F. 2012, *J. Phys. Chem. C*, 116, 3050−3058.
[34] Yazhgur, P.A., Noskov, B.A., Liggieri, L., Lin, S.-Y., Loglio, G., Miller, R. and Ravera, F. 2013, *Soft Matter*, 9, 3305–3314.
[35] Miller, R. and Liggieri, L. 2009, Interfacial Rheology. *Brill*: Leiden-The Netherlands.
[36] Liggieri, L. and Miller, R. 2010, *Curr. Opin. Colloids Interface Sci.*, 15, 256–263.

[37] Martinez, A.C., Rio, E., Delon, G., Saint-Jalmes, A., Langevin, D., and Binks, B.P. 2008, *Soft Matter*, 4, 1531–1535.
[38] Stocco, A., Drenckhan, W., Rio, E., Langevin, D., and Binks, B.P. 2009, *Soft Matter*, 5, 2215–2222.
[39] Stocco, A., Rio, E., Binks, B.P., and Langevin, D. 2011, *Soft Matter*, 7, 1260–1267.
[40] Maestro, A., Rio, E., Drenckhan, W., Langevin, D., and Salonen, A. 2014, *Soft Matter*, doi: 10.1039/C4SM00047A
[41] Orsi, D., Baldi, G., Cicuta, P., and Cristofolini, L. 2012, *Colloids Surf. A*, 413, 71–77.
[42] Orsi, D., Cristofolini, L., Baldi, G., and Madsen, A. 2012, *Phys. Rev. Lett.*, 108(10), 105701.
[43] Guzmán, E., Liggieri, L., Santini, E., Ferrari, M., and Ravera, F. 2011, *J. Phys. Chem. C*, DOI: 10.1021/jp207713x.
[44] Kralchevsky, P. and Nagayama, K. 2001, Particles at Fluid Interfaces and Membranes. *Elsevier Science*: New York, Vol. 10.
[45] Bresme, F., and Oettel, M. 2007, *J. Phys. Condens. Matter*, 19, 413101.
[46] Arditty, S., Whitby, C.P., Binks, B.P., Schmitt, V., and Leal-Calderon, F. 2003, *Eur. Phys. J. E*, 11, 273–281.
[47] Kruglyakov, P. and Nushtaeva, A. 2004 *In:* Emulsions: Structure, Stability and Interactions, Petsev, D. (Ed.), *Elsevier*: Amsterdam, pp 641–676.
[48] Tambe, D. and Sharma, M.M. 1994, *Adv. Colloid Interface Sci.*, 52, 1–63.
[49] Kralchevsky, P.A. and Nagayama, K. 2000, *Adv. Colloid Interface Sci.*, 85, 145–192.
[50] Menon, V.B. and Wasan, D.T., 1988, *Separation Sci. Technol.*, 23, 2131–2142.
[51] Stancik, E.J. and Fuller, G.G. 2004, *Langmuir*, 20(12), 4805–4808.
[52] Horozov, T.S., Aveyard, R., Clint, J.H., and Neumann, B. 2005, *Langmuir*, 21(6), 2330–2341.
[53] Larsson, R.G. 1999, The Structure and Rheology of Complex Fluids. *Oxford University Press*: Oxford, UK.
[54] Boussinesq, M.J. 1913, *Ann. Chim. Phys.*, 29, 349–357.
[55] Scriven, L.E. 1960, *Chem. Eng. Sci.*, 10, 1–13.
[56] Slattery, J.C. 1964, *Chem. Eng. Sci.* 19, 379–385.
[57] Erni, P. 2011, *Soft Matter*, 7, 7586–7600.
[58] Van den Tempel, M. and Lucassen-Reynders, E. 1983, *Adv. Colloid Interface Sci.*, 18, 281–301.
[59] Langevin, D. 2002, *In:* Encyclopedia of Surface and Colloid Science, Hubbard, A. (Ed.), *Marcel Dekker*: New York.
[60] Langevin, D. 1992, Light Scattering by Liquid Surfaces and Complementary Techniques. *Marcel Dekker*: New York.
[61] Kawaguchi, M. 1993, *Prog. Polym. Sci.*, 18, 341–376.
[62] Noskov, B.A. 1996, *Adv. Colloid Interface Sci.*, 69, 63–129.
[63] Krägel, J. and Derkatch, S.R. 2010, *Curr. Opin. Colloids Interface Sci.*, 15, 246–255.
[64] Ferry, J.D. 1980, Viscoelastic Properties of Polymers, *Wiley*: New York.
[65] Crossley, S., Faria, J., Shen, M., and Resasco, D.E. 2010, *Science*, 327, 68–72.
[66] Horozov, T.S., Binks, B.P., Aveyard, R., and Clint, J.H. 2006, *Colloids Surf. A*, 282–283, 377–386.
[67] Gray, J.J. and Bonnecaze, R.T. 1998, *J. Rheol.*, 42, 1121–1151.
[68] Cicuta, P., Stancik, E.J., Fuller, G.G. 2003, *Phys. Rev. Lett.*, 90, 236101.

[69] Trappe, V. and Weitz, D.A. 2000, *Phys. Rev. Lett.*, 85, 449–453.
[70] Imperiali, L., Liao, K.-H., Clasen, C., Fransaer, J., Macosko, C.W., and Vermant, J. 2012, *Langmuir*, 28, 7990–8000.
[71] Barman, S. and Christopher, G.F. 2014, *Langmuir*, DOI: 10.1021/la502329s.
[72] Reynaert, S., Moldenaers, P., and Vermant, J., *Phys. Chem. Chem. Phys.*, 9, 6463–6475.
[73] Wijmans, C.M. and Dickinson, E. 1998, *Langmuir*, 14, 7278–7286.
[74] Madivala, B., Fransaer, J., and Vermant, J. 2009, *Langmuir*, 25(5), 2718–2728.
[75] Basavaraj, M.G., Fuller, G.G., Fransaer, J., and Vermant, J. 2006, *Langmuir*, 22(15), 6605–6612.
[76] Zahn, K., Wille, A., Maret, G., Sengupta, S., and Nielaba, P. 2003, *Phys. Rev. Lett.*, 90(15), 155506.
[77] Brown, E., Zhang, H., Forman, N.A., Maynor, B.W., Betts, D.E., DeSimone, J.M., and Jaeger, H.M. 2011, *Phys. Rev. E*, 84, 031408.
[78] Cui, Z. 2011, *Phys. Rev. E*, 83(3), 031911.
[79] Stancik, E.J., Widenbrant, M.J.O., Laschitsch, A.T., Vermant, J., and Fuller, G.G. 2002, Langmuir, 18, 4372–4375.
[80] Wilson, H.J. and Davis, R.H. 2002, *J. Fluid Mech.*, 452, 425–441.
[81] Keller, D.S. and Keller-Jr, D.V. 1991, *J. Rheol.*, 35, 1583.
[82] Keim, N.C. and Arratia, P.E. 2013, *Soft Matter*, 9, 6222.
[83] Lundberg, M., Krishan, K., Xu, N., and Dennin, M. 2008, *Phys. Rev. E*, 77(4), 041505.
[84] Safouane, M., Langevin, D., and Binks, B.P. 2007, *Langmuir*, 23, 11546–11553.
[85] Zang, D.Y., Rio, E., Langevin, D., Wei, B., and Binks, B.P. 2010, *Eur. Phys. J. E*, 31, 125–134.
[86] Zang, D., Langevin, D., Binks, B.P., and Wei, B. 2010, *Phys. Rev. E*, 81, 011604.
[87] Zang, D.Y., Rio, E., Delon, G., Langevin, D., Wei, B., and Binks, B.P. 2011, *Mol. Phys.*, 109, 1057–1066.
[88] Wyss, H.M., Miyazaki, K., Mattsson, J., Hu, Z., Reichman, D.R., and Weitz, D.A. 2007, *Phys. Rev. Lett.*, 98(23), 238303.
[89] Vandebril, S., Vermant, J., and Moldenaers, P. 2010, *Soft Matter*, 6, 3353–3362.
[90] Krishnaswamy, R., Majumdar, S., Ganapathy, R., Agarwal, V.V., Sood, A.K., Rao, C.N.R. 2007, *Langmuir*, 23, 3084–3087.
[91] Orsi, D., Ruta, B., Chushkin, Y., Pucci, A., Ruggeri, G., Baldi, G., Rimoldi, T., and Cristofolini, L. 2014, *Phys. Rev. E*, 89(4), 042308.
[92] Orsi, D., Vezzani, A., Burioni, R., Pucci, A., Ruggeri, G., and Cristofolini, L. 2014, *Colloids Surf. A*, 441, 912–918.
[93] Maas, M., Ooi, C.C., and Fuller, G.G. 2010, *Langmuir*, 26(23), 17867–17873.
[94] Degen, P., Paulus, M., Maas, M., Kahner, R., Schmacke, S., Struth, B., Tolan, M., and Rehage, H. 2008, *Langmuir*, 24, 12958–12962.
[95] Degen, P., Wieland, D.C.F., Leick, S., Paulus, M., Rehage, H., and Tolan, M. 2011, *Soft Matter*, 7, 7655–7662.
[96] Liggieri, L., Santini, E., Guzmán, E., Maestro, A., and Ravera, F. 2011, *Soft Matter*, 7 6699–7709.
[97] Miller, R., Fainerman, V.B., Kovalchuk, V.I., Grigoriev, D.O., Leser, M.E., and Michel, M. 2006, *Adv. Colloid Interface Sci.*, 128–130, 17–26.

[98] Hilles, H., Monroy, F., Bonales, L.J., Ortega, F., and Rubio, R.G. 2006, *Adv. Colloid Interfaces Sci.*, 122, 67–77.
[99] Miller, R., Fainerman, V.B., Makievski, A.V., Kragel, J., Grigoriev, D.O., Kazakov, V.N., and Sinyachenko, O.V. 2000, *Adv. Colloid Interface Sci.*, 86, 39–82.
[100] Zang, D.Y., Zhang, Y.J., and Hou, Q.W. 2012, *Colloids Surf. A*, 395, 262–266.
[101] Kobayashi, T. and Kawaguchi, M. 2013, *J. Colloid Interface Sci.*, 390, 147–150.
[102] Bykov, A.G., Noskov, B.A., Loglio, G., Lyadinskaya, V.V., and Miller, R. 2014, *Soft Matter*, DOI: 10.1039/c4sm00782d.
[103] Bain, C.D., Claesson, P.M., Langevin, D., Meszaros, R., Nylander, T., Stubenrauch, C., Titmuss, S., and Klitzing, R.V. 2010, *Adv. Colloid Interface Sci.*, 155, 32–49.
[104] Llamas, S., Guzmán, E., Ortega, F., Baghdadli, N., Cazeneuve, C., Rubio, R.G., and Luengo, G.S. 2014, *Adv. Colloid Interface Sci.*, DOI: 10.1016/j.cis.2014.05.007.
[105] Wang, H., Lu, W., and Chen, B. 2008, *Appl. Surf. Sci.*, 254, 3380–3384.
[106] Noskov, B.A., Yazhgur, P.A., Liggieri, L., Lin, S.Y., Loglio, G., Miller, R., and Ravera, F. 2014, *Colloid J.*, 76(2), 127–138.
[107] Santini, E., Guzmán, E., Ravera, F., Ferrari, M., and Liggieri, L. 2012 *Phys. Chem. Chem. Phys.*, 14, 607–615.
[108] Pelipenko, J., Kristl, J., Rosic, R., Baumgartner, S., and Kocbek, P. 2012, *Acta Pharm.*, 62, 123–140.
[109] Salonen, A., Lhermerout, R., Rio, E., Langevin, D., and Saint-Jalmes, A. 2011, *Soft Matter*, DOI: 10.1039/c1sm06537h.
[110] Cheng, H.-L. and Velankar, S.S. 2009, *Langmuir*, 25 4412–4420.
[111] Bhamla, M.S., Giacomin, C.E., Balemans, C., and Fuller, G.G. 2014, *Soft Matter*, DOI: 10.1039/C3SM52934G.
[112] Martinez, A.C., Rio, E., Delon, G., Saint-Jalmes, A., Langevin, D., and Binks, B.P. 2008, *Soft Matter*, 4, 1531–1535.
[113] Arriaga, L.R., Drenckhan, W., Salonen, A., Rodrigues, J.A., Iniguez-Palomares, R., Rio, E., and Langevin, D. 2012, *Soft Matter*, 8, 11085–11097.
[114] Frith, W.J., Pichot, R., Kirkland, M., and Wolf, B. *Ind. Eng. Chem. Res.*, 47, 6434–6444.
[115] Sullivan, A.P. and Kilpatrick, P.K. 2002, *Ind. Eng. Chem. Res.*, 41, 3389–3404.
[116] Santini, E., Ravera, F., Ferrari, M., Alfè, M., Ciajolo, A., and Liggieri, L. 2010, *Colloids Surf. A*, 365, 189–198.
[117] Santini, E., Guzmán, E., Ravera, F., Ciajolo, A., Alfè, M., Liggieri, L., and Ferrari, M. 2012, *Colloids Surf. A*, 413, 216–223.
[118] Zabiegaj, D., Santini, E., Guzmán, E., Ferrari, M., Liggieri, L., Buscaglia, V., Buscaglia, M.T., Battilana, G., Ravera, F., 2013, *Colloids Surf. A*, 438, 132–140.
[119] Zabiegaj, D., Santini, E., Guzmán, E., Ferrari, M., Liggieri, L., and Ravera, F. 2014, *J Nanosci. Nanotech.*, DOI:10.1166/jnn.2014.9730.
[120] Santini, E., Guzmán, E., Ferrari, M., and Liggieri, L. 2014, *Colloids Surf. A*, DOI: 10.1016/j.colsurfa.2014.02.054.
[121] Madivala, B., Vandebril, S., Fransaer, J., and Vermant, J. 2009, *Soft Matter*, 5, 1717–1727.
[122] Kim, J.K., Rühs, P.A., Fischer, P., and Hong, J.S. 2013, *Rheol Acta*, 52, 327–335.
[123] Brugger, B., Rosen, B.A., and Richtering, W. 2008, *Langmuir*, 24(21), 12202–12208.
[124] San-Miguel, A. and Behrens, S. H. 2012, *Langmuir*, 28, 12038–12043.

In: Rheology
Editor: Evgeny Karpushkin
ISBN: 978-1-63482-223-7

Chapter 2

RHEOLOGY AND TEXTURE OF FERMENTED MILK PRODUCTS

Mirela Iličić*[*]*, Spasenija Milanović and Marijana Carić
University of Novi Sad, Faculty of Technology
Novi Sad, Serbia

ABSTRACT

Rheology is a branch of science which defines the relation between forces exerted on a material and the ensuing deformation or flow of the material as a function of time. Food is usually neither liquid nor solid, but has viscoelastic characteristics. Viscoelastic materials reveal a combination of viscous (liquid) and elastic (solid) behavior.

Yoghurt is among the most popular fermented milk products. Yoghurt is a protein network formed by casein micelles entrapping serum and fat globules. The distribution of the other components has an important effect on the rheological behavior of yoghurt. It is a typical example of weak viscoelastic gel. Dynamic rheological measurements of yoghurt are of interest as they are directly related to its viscous and solid properties.

Viscosity measurements can be used to confirm the stability of the fermented milk products. Basically, the viscosity of fermented milk products can be measured by using different types of rheometers: Ostwald, Hoppler, Brookfield and Haake Rheostress ones. Textural properties (firmness, consistency, cohesiveness, and index of viscosity) of fermented milk products are usually analyzed by Texture Analyzer apparatus.

Viscosity, structure, and textural characteristics of the gel are influenced by several factors, including protein concentration, milk fat content, heat treatment of milk, acidity, type of a starter culture, incubation temperature, and addition of flavorings or other functional ingredients.

In the case of low-fat products, behavior of proteins during the gelation process is of particular importance. However, these products somewhat lack the textural and sensory attributes of their full-fat counterparts.

The fortification of the milk base with some ingredients (skimmed milk powder, whey protein concentrates, whey protein isolate, Na-caseinates, inulin as fat replacer, or by incorporating different stabilizers, small-molecule water–soluble surfactants such as

[*] e-mail:panim@uns.ac.rs

monoglycerides or phospholipids) as well as application of transglutaminase (TG) or some other functional components results in increase of elastic and viscous modules and improved stability of fermented milk products. Of all the listed ingredients, TG application is of utmost importance for fermented dairy products manufacture and quality.

The most recent investigations have shown that TG addition in minimal concentrations contributes to the milk coagulation time and to the improvement of physicochemical, rheological, and structural characteristics of the final product.

The milk solids supplementation is a good practice to improve activity of starter culture, especially of probiotic bacteria. It is well documented that probiotic bacteria grow slowly in milk and contribute to sensory and rheological characteristics of the product.

The application of sophisticated techniques: high-pressure processing HPP, high-pressure homogenization HPH, and ultrasonic processing USP allowed elaboration of fermented milk product with rheological properties (flow curve, apparent viscosity, yield stress, and viscoelastic properties) superior to those of untreated samples.

The effect of the addition of different ingredients, type of starter cultures, and application of sophisticated techniques on rheological properties and textural characteristics of fermented milk products will be pointed out in this chapter.

INTRODUCTION

Fermented milk products have been an important component of nutrition and diet. Yoghurt is one of the most popular fermented milk products with ratio of approximately 10% of world dairy production [1]. Economists usually investigate the models of yoghurt consumption in the EU. It is generally believed that a number of factors influence decision-making, including the properties of products, individual consumer preferences, socio-demographic characteristics, and psychological factors as well as previous experience with the product and environmental influences ranging from cultural to economic factors [2]. The authors have suggested that important cultural differences continue to determine yoghurt consumption in the 11 countries of the EU. Also, the level of consumption depends on types and quality of products (chemical composition, rheology, texture, microstructure, and sensory properties).

In addition to yoghurt, there are numerous types of fermented milk products manufactured in different parts of the world. Fermented milk products are classified into the following types:

1. products of lactic fermentation where strains of mesophilic or thermophilic lactic acid bacteria (LAB) are used (acidophilus milk, bioghurt, bifighur, cultured cream, dahi, dadih, filmjölk);
2. products obtained via alcohol-lactic fermentation (kefir, kumys, täjölk);
3. products where molds grow in addition to microorganisms from fermentation of type 1 and 2 is present (villi, yakult) [3]; and
4. products where non-conventional starter is applied (kombucha).

Fermented milk products are normally retailed in one of the three physical states: set (gelled), stirred (medium to high viscosity), or fluid (drinking yoghurt). The texture and rheology of fermented milk products are important quality variables.

The selection of the appropriate starter culture type is of great importance in fermented milk products manufacturing. The employed starter culture may improve the physical attributes of the final product. Ropy and non-ropy starters have been utilized for set and stirred yogurt preparation, respectively. Ropy cultures are capable of producing extracellular polysaccharides during the fermentation process, in contrast to non-ropy cultures [4]. Consequently, researchers have studied the effect of both starter types in addition to incorporation of probiotic bacteria on yoghurt characteristics [5–7]. Milk solids supplementation is a good practice to improve probiotic growth during the fermentation period and flavor bacterial viability in the product. Also, inulin (prebiotic) as a non-digestible ingredient beneficially affects humans, since it selectively stimulates growth and/or activity of limited number bacteria in the colon [8, 9].

Recent studies have been dealing with possibilities of kombucha application as non-conventional starter for milk fermentation [10–12]. Kombucha has been known as a consortium of yeasts (*Pichia, Zygosaccharomyces, Saccharomyces, Schizosaccharomyces, Saccharomycodes, Brettanomyces, Torulaspora* and *Candida*) and acetic acid bacteria (*Acetobacter* and *Gluconobacter*) [13–15]. However, recent research has indicated the significant presence of lactic acid bacteria [14, 16, 17].

It has been claimed by some authors that an improved product could be obtained with the addition of different ingredients (milk proteins MP, whey protein concentrate WPC, whey protein isolates WPI, and transglutaminase enzyme TG) or by application of some sophisticated techniques (high-pressure processing HPP, high-pressure homogenization HPH and ultrasonic processing USP).

Enriched fermented milk products with high nutritive ingredients and application of new techniques contribute not only to better rheological and textural properties and gel stability during storage, but also to appropriate microstructure, sensory characteristics, and health benefits.

Definition and Standard Identity

Codex Alimentarius Comission and its committees together with the International Dairy Federation have established the international standard for fermented milks (Table 1, [18]).

Fermented milks with a minimum protein content of 2.7% are usually classified into yoghurt, alternative cultured yoghurt, kefir, kumys, and fermented special products. Fermented milks include also concentrated fermented milk with minimum 5.6% of protein content (stagisto-strained yoghurt, labneh, ymer, and ylette) and flavored fermented milk as composite milk products which contain a maximum of 50% (w/w) of non-dairy ingredients. The total viable count is proposed to be 10^8 cfu/g at the time of manufacture [3]. Codex Standard for Fermented Milks [18] defines the viable count of yoghurt starter after product has been stored under the conditions specified in the labeling at the end of expired date as a minimum 10^6 cfu/g viable yoghurt starter organisms.

Table 1. Types and composition of fermented milks

	Fermented milk	Yoghurt, Alternate culture yoghurt, and Acidophilus milk	Kefir	Kumys
Milk protein (w/w)	min. 2.7%	min. 2.7%	min. 2.7%	
Milk fat (w/w)	<10%	<15%	<10%	<10%
Titrable acidity expressed as percent lactic acid (w/w)	min. 0.3%	min. 0.6%	min. 0.6%	min. 0.7%
Ethanol				min. 0.5%
Sum of microorganisms constituing the starter clture defined in section 2.1 (cfu/g in total)	min. 10^7	min. 10^7	min. 10^7	min. 10^7
Labeled microorganisms (cfu/g in total)	min. 10^6	min. 10^6		
Yeasts (cfu/g)			min. 10^4	min. 10^4

Source: Food and Agriculture Organization of the United Nations, [18]; reproduced with permission.

Measurement of Rheology and Texture

Rheology is used in food science to define the consistency of different products. The consistency is described by two components: the viscosity ("thickness" or lack of slipperiness) and the elasticity ("stickiness" or structure). Therefore, in practice, rheology implies viscosity measurement, characterization of flow behavior, and determination of the material structure. Basic knowledge of these subjects is essential in process design and product quality evaluation.

Shearing of a substance is the key to knowledge of flow behavior and structure. Shearing between parallel planes is normally used for the basic definition of shear stress and shear rate, corresponding to how much deformation is applied to the material and how fast [19].

Materials which cannot be defined by a single viscosity value at a specified temperature are called non-Newtonian. Fermented milk products of yoghurt types belong to non-Newtonian fluids. The viscosity of these materials should always be stated together with the corresponding temperature and shear rate. The viscosity will change if the shear rate is changed. Generally speaking, high concentration and low temperature induce (or enhance) the non-Newtonian behavior.

Apart from being dependent on shear rate, viscosity of non-Newtonian fluids may also be time-dependent; in this case viscosity is a function not only of the magnitude of the shear rate, but also of duration or frequency of the shear application. Non-Newtonian materials that are rate-independent are defined as shear-thinning, shear-thickening, or plastic. Non-Newtonian materials that are time-dependent can be defined as thixotropic, rheopectic, or anti-thixotropic. Different types of yoghurt are thixotropic fluids. A thixotropic fluid is defined as a shear-thinning system with the viscosity decreases both with increasing shear rate and with time at a constant shear rate. Thixotropic flow behavior is normally studied in a loop test. In this test the material is subject to increasing shear rates followed by the same shear rates in decreasing order. The time-dependent thixotropic flow behavior is seen from the difference between the ascending and descending viscosity and shear stress curves. To recover its structure, the material should rest during a certain period of time, being characteristic of the material. This type of flow behavior is revealed by virtually all gel-forming systems. Beside yoghurt, typical thixotropic profile is characteristic of other food products: ice cream, margarine, and mayonnaise.

For a viscoelastic material, two shear modules are defined: G' and G''. G' is the storage modulus corresponding to the elastic character of material, whereas G" is the loss modulus correlated to the viscous contribution of the material. The elastic modulus (in-phase component) is associated with storage of energy during a cyclic deformation, while the loss modulus (out-of-phase component) is associated with the energy lost as heat. The ratio between the storage and loss modules is called the loss tangent ($G''/G' = \tan\delta$). This value bears information about the overall structure of the tested material. When a material behaves more like a solid (the deformation being essentially elastic and recoverable), the storage modulus G' exceeds the loss modulus G'' and hence $\tan\delta < 1$. On the other hand, when the material is more like a liquid, the viscous character dominates and $\tan\delta > 1$ [4].

Assessment of physical characteristics of set, stirred and fluid (drinking) yoghurt by several techniques and equipment such as penetrometer, texture analyzers, rotational viscometers (Brookfield or Haake), and Bostwick consistometer have been reported by Tamime and Robinson [5] and Tamime [4].

Velez-Ruiz [6] presented an overview of studies of flow properties and thixotropy as well as dynamic characterization of yoghurt.

THE EFFECT OF STARTER CULTURES AND DIFFERENT INGREDIENTS ON RHEOLOGY AND TEXTURE OF FERMENTED MILK PRODUCTS

Yoghurt, the best studied fermented milk product, is a protein network formed by casein micelles entrapping serum and fat globules [20]. The casein matrix of yoghurt is relatively weak; consequently, the distribution of the other components has an important effect on the rheological behavior of yoghurt [6, 21, 22]. Generally, rheological properties of yoghurt affect significantly the product quality. The viscosity and gel structure are influenced by a number of factors, including milk composition, especially contents of fat and proteins.

In the case of low-fat products, behavior of proteins during the gelation process is of particular importance. In recent years, manufacturers have responded to consumers' demand for low-fat and fat-free dairy products. However, these products somewhat lack the textural and sensory attributes of their full-fat counterparts. Attempts have been made to improve low-fat yoghurt gel properties by incorporating different ingredients: exopolysaccharide isolated or producing strains (*Lactococcus lactis* susp. *cremoris, Streptococcus thermophilus, Lacobacillus delbrueckii* subsp. *bulgaricus* or *Bifidobacterim* strains [23–25], soluble dietary fiber from *Pachyrhizus erosus* [26], acetylated starch [27], low-molecular water-soluble surfactants such as monoglycerides or phospholipids [28], skim milk powder, whey protein concentrates, whey protein–pectin [29], sodium caseinate, and inulin as fat replacer as well as by application of enzyme transglutaminase [11, 12, 30–38].

Conventional and Non-Conventional Starters

The selection of the appropriate starter culture type also is of great importance in yoghurt manufacturing. The employed culture may improve the physical attributes of the final product. Ropy and non-ropy starters have been utilized for set and stirred yoghurt preparation, respectively. Ropy cultures are capable of producing extracellular polysaccharides during the fermentation process, in contrast to non-ropy cultures [4]. The effects of the culture type and of incorporation of probiotic bacteria on the rheological properties of stirred yoghurt have been investigated in Refs. [5–7, 39].

Recent researches have been dealing with possibilities of kombucha application as non-conventional starter for milk fermentation [10, 11, 40, 41]. Kombucha is a symbiotic association of yeast (*Pichia, Zygosaccharomyces, Saccharomyces, Schizosaccharomyces, Saccharomycodes,Torulaspora, Candida and Brettanomyces*), acetic acid bacteria (*Acetobacter* and *Gluconobacter*), and lactic acid bacteria (*Lactobacillus* population). Also, it can be cultivated on a dark beer, red vine, white vine, whey, lactose, or some types of tea [10]. Besides nutritive values, kombucha beverage contains components which may have a positive effect on health, physical and psychological state of human organisms [42, 43].

Malbaša et al. [10] investigated the possibility of manufacturing of fermented milk product using different concentrations of evaporated kombucha inoculum. They concluded that fermentation time was similar at 10 to 15% of kombucha inoculum, but the process was two times faster with yoghurt starter. The influence of inoculum concentration on the rate of fermentation was negligible. The measured viscosity was higher for kombucha beverages at lower shear speeds, decreasing with shear rate. Very high sensory scores were achieved for all the beverages after production and after 5 days storage.

Iličić et al. [44] focused on fermentation of lactose from a model system (black tea) and from two types of milk (0.9% w/w and 2.2% w/w of fat) by application of kombucha. Quantities of the applied kombucha starter were of 10% (v/v) and 15% (v/v). Analysis of the fermentation curves showed that inoculum concentration (10% v/v or 15% v/v) did not significantly affect the rate of fermentation. The fermentation time for the samples with 2.2% w/w of fat (M2K10 and M2K15) was slightly shorter than that for low-fat samples (M1K10 and M1K15) in all studied cases. The reduced fat milk turned to be a somewhat better substrate than the low-fat milk. When comparing duration of the processes to achieve pH=4.5 in black tea samples and in milk samples, it was concluded that the process was faster with milk (9–10 h) than with the model system (16 h).

Textural characteristics of the fermented beverages are presented in Figure 1. The following characteristics were measured: firmness, consistency, cohesiveness, and index of viscosity. According to the methods applied, the latter two parameters were expressed as negative numbers (the lower the better). Quite expectedly, products M2K10 and M2K15 that were high in both fat and proteins developed better textural characteristics than the low-fat products M1K10 and M1K15. It seemed that the product labeled as M2K10 possessed the optimal texture, with the best cohesiveness (–8 g) and index of viscosity (–2.7 g s) and with consistency (410 g s) equal that of the M2K15 beverage. Once again, it was proved that concentration of the applied kombucha starter had negligible effect on the texture of fermented products, when compared to the influence of other factors.

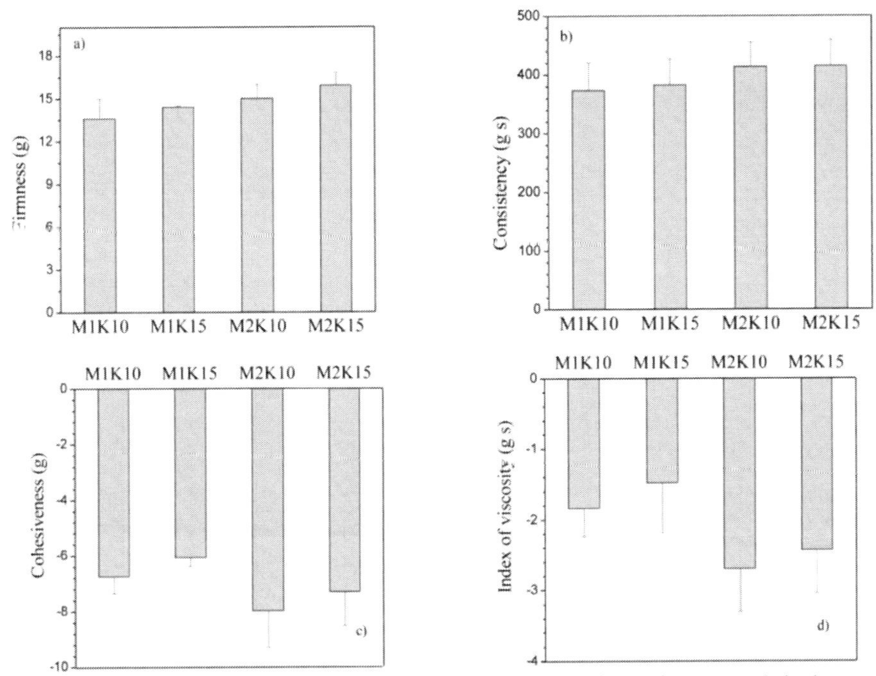

Figure 1. Textural characteristics of fermented beverages: a) firmness, b) consistency, c) cohesiveness and d) index of viscosity [44].

Milanović et al. [11] investigated the possibility of application of kombucha cultivated at two different tea types in combination with probiotics for milk fermentation at different temperatures, and the physico-chemical and textural properties of the so manufactured fermented milk products were assessed. The combination of probiotic starter culture and kombucha inoculums cultivated at *Camellia sinensis* (black tea) and *Thymus serphyllum* (thyme tea) were used for milk fermentation. The results showed that kombucha inoculums cultivated at different tea types could be used to produce fermented milk products in combination with probiotic starter culture.

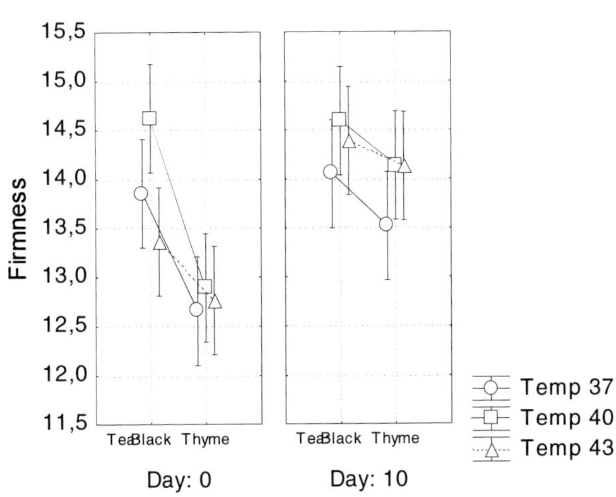

Figure 2. (Continued).

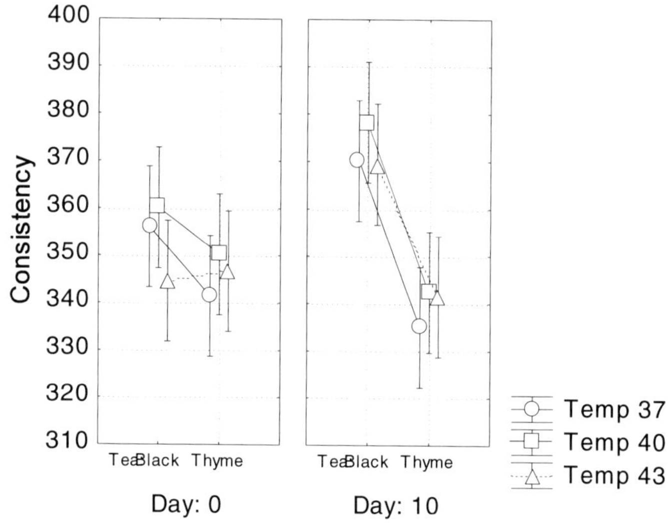

Figure 2. Influence of tea type and duration of storage on the properties of kombucha-fermented products: a) firmness and b) consistency [11].

Different temperatures of milk fermentation had no significant effect on chemical properties of kombucha-fermented dairy products. The results of factor ANOVA test revealed the significant influence of the factors Temperature*Tea type on the L-lactic acid content. The tea type had significant influence on textural characteristics of fermented dairy products, especially after storage. After 10 days of storage, all textural characteristics except firmness were significantly different depending on the tea type used for kombucha inoculum cultivation. The interaction of factors Tea Type*Day of storage showed significant effect on the product textural characteristics (Figure 2).

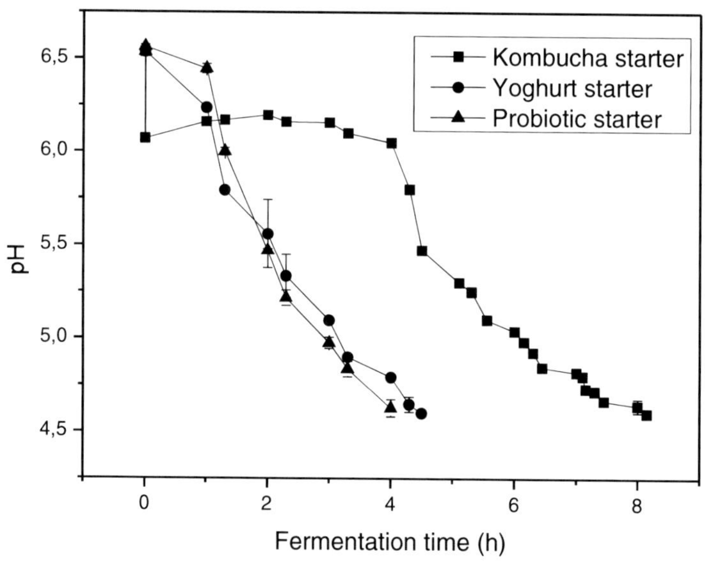

Figure 3. Fermentation process with three different starter cultures [45].

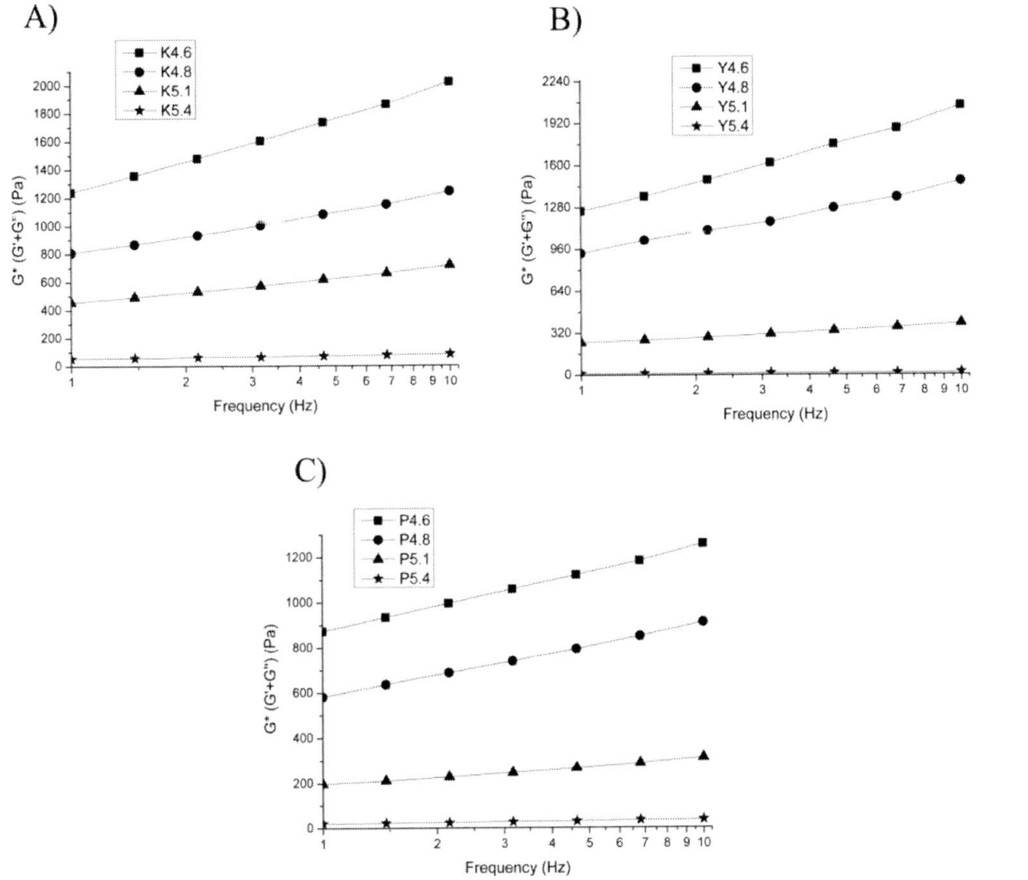

Figure 4. Complex modulus of the samples produced with a) kombucha, b) yoghurt, and c) probiotic as starter [45].

The effect of kombucha starter culture on rheological properties, texture, and microstructure of milk during fermentation as well as on protein profile was investigated by Vukić et al. [45]. The results were compared with the samples obtained with yoghurt starter culture or probiotics. Quality parameters were measured during the fermentation at 42°C at pH of 5.4, 5.1, 4.8, and 4.6. The obtained results of viscosity at pH=5.4 revealed that gelation process was similar for the samples produced with kombucha and probiotics (Figure 3). The samples produced with kombucha had the highest complex modules that caused better rheological characteristics of the final product (Figure 4). Earlier onset of gelation induced significant differences in microstructure at pH=5.4 between the samples (Figure 5).

Hrnjez et al. [46] analyzed protein profile of fermented dairy products obtained with kombucha starter culture at different fermentation temperature. The sample produced at 37ºC showed significantly longer fermentation time than the sample produced at 42ºC. Protein profile analysis revealed more stable α and β casein fractions as compared to other protein fractions during milk fermentation (Table 2).

The results indicated that fermented dairy products produced by kombucha at 37ºC and 42ºC could be less allergenic than milk, and hence suitable for special allergenic nutrition.

Noteworthily, the sample fermented at 37ºC had lower content of α-lactalbumin and β-lactoglobulin than the sample produced at 42ºC.

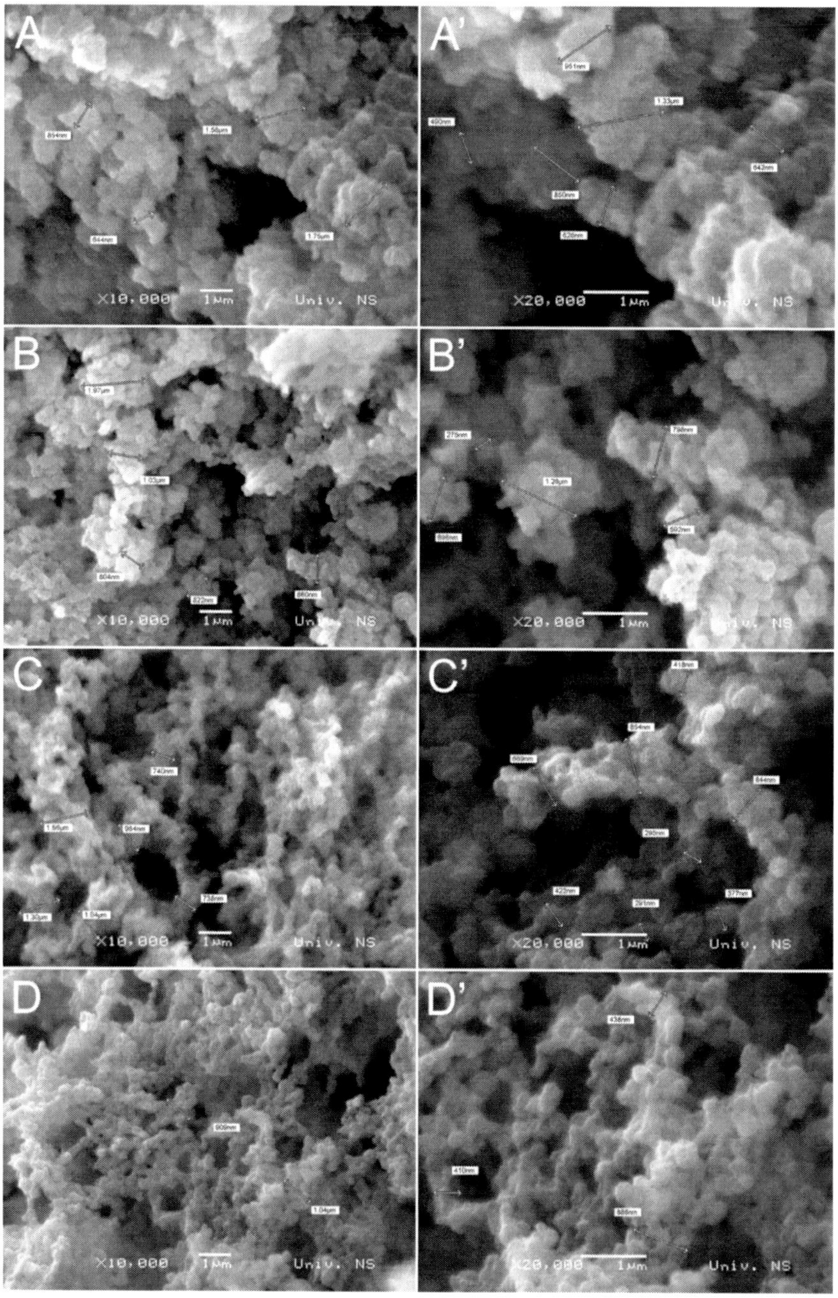

Figure 5. Microstructure of kombucha samples during milk fermentation [45].

Hrnjez et al. [47] analyzed the antioxidant capacity, angiotensin converting enzyme (ACE) inhibiting activity, degree of proteolysis (DP), content of vitamin C as well as sensory properties of kombucha-fermented milk and compared them with the fermented milk products obtained with commercial probiotic (ABT-10) and yoghurt (YF-L812) starter cultures. The kombucha-fermented milk product (K) showed similar trend of changes in pH, DP, and sensory properties as the products obtained by probiotic (P) and yoghurt (Y) starters. Significant ACE inhibiting activity was detected for all the fermented products, increasing during storage.

Table 2. Protein fractions of produced samples [46]

Protein fraction	Sample		
	M (%)	K37 (%)	K42 (%)
α-lactalbumin	4.86±0.04[a]	3.26±0.04[b]	3.88±0.05[c]
lysozyme	0.13±0.02	nd	nd
β-lactoglobulin	6.92±0.04[a]	5.48±0.05[b]	5.81±0.09[c]
β-casein	35.66±0.18[a]	40.57±0.20[b]	38.54±0.01[c]
α-casein	35.17±0.16[a]	38.81±0.14[b]	39.19±0.05[c]
κ-casein	12.04±0.03[a]	8.80±0.02[b]	9.15±0.05[c]
lactoferrin	2.93±0.02[a]	1.42±0.07[b]	1.42±0.05[c]

[nd] – not detected.

[abc] – Means in the same row with marked with different letters are significantly different ($p<0.05$).

Kombucha product had the highest ACE activity (63.43%) at the end of storage as compared with probiotic and yoghurt products. In all the products, higher 2,2'-azino-bis-(3-ethylbenzthiazoline-6-sulphonic acid) (ABTS) radical scavenging activity than that of 1,1-diphenyl-2-picrylhydrazyl (DPPH) was determined, the both activities slightly decreasing during storage. Based on estimated biological activity, the kombucha product can be considered an innovative fermented dairy product suitable for human nutrition.

Different Ingredients Effects

Exopolysacharides (EPS)

Lactic acid bacteria LAB capable of synthesizing extracellular polysaccharides (EPS) have been traditionally used in food processing to improve physical properties and texture of fermented milk products. For yoghurt, the main advantages of LAB application are resulting higher water binding ability and enhanced viscosity. The underlying mechanisms of EPS action on milk gel properties are still not fully understood, but have been frequently associated with their structural characteristics (monosaccharide composition, charge, molar mass, degree of branching, and chain stiffness) and the molecular interaction with milk proteins [24, 48–50]. Some authors observed a positive correlation between viscosity of milk fermented with two strains of *S. thermophilus* and the molar mass [51] of EPS. Charged EPS increased apparent viscosity and the resistance to shear by enhancing the intramolecular repulsion resulting in higher hydrodynamic radius and intrinsic viscosity and via electrostatic interactions between anionic EPS and milk proteins [52].

Mende et al. [24] presented a new approach to investigate the impact of EPS from lactic acid bacteria on the rheological characteristics of acid milk gels by decoupling microbial EPS synthesis from milk acidification. Free and capsular EPS of *Streptococcus thermophilus* ST-

143 were isolated and purified separately from the batch fermentation media to obtain highly pure (>85% EPS) preparations. The authors found that for the stirred yoghurt, apparent viscosity increased significantly with addition of EPS and with storage time at 6°C. Depletion of interactions between uncharged high-molecular EPS and negatively charged casein micelles could be main factor responsible for those effects.

Prasanna et al. [25] systematically examined the physicochemical, rheological, and microstructural properties of low-fat yoghurts produced with EPS-producing bifidobacteria. EPS production reduced syneresis and increased the firmness of the low-fat yoghurt. The yoghurt prepared with EPS-producing *B. longum* subsp. *infantis* (CCUG 52486) showed a well defined porous gel structure. The EPS of the bifidobacteria most likely interacted with the milk proteins, resulting in higher storage modulus (G') of the low-fat yoghurt. No changes in the EPS content of the low-fat yoghurts produced with the two *Bifidobacterium* strains were observed during storage. The study showed that the EPS producing *B. longum* subsp. *infantis* CCUG 52486 was a promising *in situ* EPS-producing probiotic strain which could be used in combination with traditional yoghurt bacteria to produce low-fat set yoghurt with probiotic activity and improved physico-chemical and rheological properties.

Girard and Schaffer-Lequart [23] investigated gelation and resistance of fermented milk with EPS-producing strains *Lactococcus lactis* subsp. *cremoris, Streptococcus thermophilus, Lactobacillus delbrueckii* subsp. *bulgaricus* and the non-EPS-producing strain *L. lactis* subsp. *cremoris.* The milk gelation time depended on the kinetics of acidification of the strains whereas the pH range of gelation depended mostly on the presence of EPS. In the set fermented milk with EPS-producing strains, bacteria were observed in protein-free areas, likely filled with EPS. The phase-separated EPS and caseins contributed to induce the gelation of fermented milk at pH 5.6, and the high resistance to shear of milk fermented with EPS strain *L.lactis* subsp. *cremoris* might be due to the negative charge of exopolysaccharide allowing efficient interaction with caseins. Also, the most important structural features of EPS influencing the texture of fermented milk were its backbone stiffness, molecular weight and size, branching degree, nature of the side groups, the monomeric composition, and the linkages between its monomer [23, 52].

Inulin

Inulin is a non-digestible ingredient which beneficially affects humans, since it selectively stimulates growth and/or activity of bacteria in the colon [53]. Furthermore, inulin is stable at moderate acidity [54], its use being recommended for production of fermented milk derivatives. Technologically, inulin has been applied as a fat substitute in some food products with the advantage of not increasing its caloric value and not giving a residual taste to the product [55, 56]. Besides the selective stimulation of bacteria, Gibson et al. [57] reported that those ingredients were classified as prebiotic because they resisted the processes of digestion, absorption, and adsorption in the gastrointestinal tract. Recent studies have shown that the changes in food with inulin addition favorably affect the acceptance of the product by the consumer [58].

Modifications in the rheological properties of fermented milks can be observed when functional food ingredients are incorporated, for instance, inulin [59, 60].

The objective of many studies was to use the microfiltrate of cow milk with acceptable microbiological characteristics to produce a prebiotic fermented milk product.

The use of microfiltration to replace traditional thermal processes (pasteurization) can reduce the changes in chemical properties of the milk, such as the protein denaturation [61, 62]. Therefore, the production of fermented products from microfiltered milk, fat-free and cold-pasteurized, may result in a product with different properties, mainly regarding the rheological behavior [8].

Debon et al. [8] investigated fermented microfiltered cow milk samples with (prebiotic) and without inulin and evaluated the physico-chemical (day 1 of storage) and rheological (days 1, 7, 14 and 21) properties, at temperatures of 4.0 ± 0.1°C and 6.0 ± 0.1°C. The authors concluded that addition of inulin increased the total solids content of the prebiotic fermented milk as well as its acidity. The Power Law and Mizhari and Berk models were applied successfully to describe the flow of the fermented milks, revealing shear-thinning and non-Newtonian fluid behavior. Similarly to the addition of inulin, the storage time contributed to an increase in the apparent viscosity of the product. The prebiotic fermented milk had greater thixotropy and the hysteresis was less prominent at higher temperature.

Fibers

Dietary fibers (DF) are the edible part of plants or analogous carbohydrates that are resistant to digestion and absorption in the human small intestine with complete or partial fermentation in the large intestine [63]. The DF components are usually grouped into two major classes depending on their simulated intestinal solubility: soluble (pectins, mucilage, loosely bound hemicelluloses, beta-glucans, and a large range of non-digestible oligosaccharides including inulin) and insoluble (cellulose, lignin, and tightly bound celluloses) [64, 65]. The insoluble components of the fiber seem to be related to the intestinal regulation, whereas the soluble components of the fiber are associated with the decrease of cholesterol levels and the adsorption of intestinal glucose [26].

The effect of enrichment of stirred yoghurt (SYYSF) with yam soluble fiber (YSF) extracted from *Pachyrhizus erosus* (1 g per 100 mL) on syneresis, microstructure, and rheological properties was evaluated via comparison with the corresponding properties of a stirred yoghurt (SYC) without added YSF [26].The SYC yoghurt exhibited a more compact casein micelle aggregates network than that of the SYYSF yoghurt which was more open, relaxed and covered with fibrous structures attributed to the YSF components. The rheological analysis showed that the SYYSF yoghurt had lower storage and loss modules in the linear viscoelastic region than the SYC yoghurt, but the flow behavior of the former was characterized by a lower flow index (n), higher consistency index (k), and higher yield stress (τ_0) than that of the SYC yoghurt. Incorporation of the YSF reduced significantly the syneresis and produced a more acceptable mouthfeel of the SYYSF yogurt, indicating the prospects of the process to obtain a commercial product.

Milk Proteins

Manufacturing of probiotic yoghurt involves milk supplementation with dairy ingredients in order to increase protein concentration to 40–50 g of protein per kg [66, 67].

The effect of replacement of skim milk powder (SMP) with whey protein concentrate (WPC) or caseinate on textural and physical properties of yoghurts has been studied by many researchers [68–71].

Marafon et al.[66] optimized the rheological properties of probiotic yoghurts supplemented with SMP, WPC, and sodium caseinate (Na-Cn) by using an experimental simplex-centroid design type for mixture modeling. It included seven batches/trials: three ones were supplemented with each type of the dairy protein used, three ones were corresponding to the binary mixtures, and the latter corresponded to the ternary one in order to increase protein concentration in 1 g per 100 g of final product. A reference experiment was prepared without supplementing the milk base. The processed milk bases were fermented at 42°C until pH 4.5 by using a starter culture blend that consisted of *Streptococcus thermophilus*, *Lactobacillus delbrueckii* subsp. *bulgaricus,* and *Bifidobacterium animalis* subsp. *lactis.* Kinetics of acidification was monitored during the fermentation period along with physico-chemical tests, counting of viable bacteria, and rheological experiments with the yoghurts. The results (kinetic responses, counts of viable bacteria, and rheological parameters) were fitted with three regression models (linear, quadratic, and cubic special) applied to mixtures. The results showed that the addition of milk proteins affected slightly acidification profile and counts of *S. thermophilus* and *B. animalis* subsp. *lactis,* but it was significant for *L. delbrueckii* subsp. *bulgaricus.* Partial replacement of SMP (45 g/100 g) with WPC or Na-Cn simultaneously enhanced the rheological properties of probiotic yoghurts, the kinetics of acidification, and the viable bacteria count.

Reduction of fat and, consequently, reduction in total solids content in low-fat and fat-free yoghurts led to weaker product with poor texture and to whey separation. The problem could be easily and almost routinely solved by milk proteins (MP) supplementation of raw milk prior to heat treatment of milk. Increase in concentration of MP yielded the more compact structure with numerous small pores and a dense network of cross-links, which also increased firmness of the yoghurt [9]. In order to minimize the risk of development of unacceptable aroma in the yoghurt, the level of WPC supplementation should be about 2–3% (w/w) [4].

Furthermore, combination with other ingredients such as honey may improve flavor profile. Honey could be an ideal sweetener for yoghurt due to its sugar concentration, low pH, and a variety of beneficial nutritional properties. Also, it is well known that honey has prebiotic effects, similar to that of fructooligosaccharides and the same rate of production of lactic acid as inulin. In addition, honey improves LAB and bifidobacteria viability and increases the shelf life of plain and probiotic fermented milks [9]. The effect of addition of acacia honey (H: 2% w/w, 4% w/w, and 6% w/w) on the textural and sensory properties of set probiotic yoghurt, after the production as well as during 21 days of storage at 5°C was investigated. The yoghurt was prepared from raw milk (1.5% fat w/w) supplemented with whey protein concentrate (WPC: 1% w/w) and treated at 95ºC during 10 min. Fermentation was performed using probiotic starter culture (70% w/w *Streptococcus thermophilus*, 10% w/w *Lactobacillus bulgaricus*, 10% w/w *Lactobacillus acidophilus*, and 10% w/w *Bifidobacterium* ssp.). All positive effects of the WPC on the syneresis, viscosity and texture of the products remained almost unchanged after addition of honey [9].

A number of researchers have studied the structuring effect of denaturized whey proteins in yoghurt production [9, 34, 72–76]. Addition of whey protein concentrate changed the ratio of casein to whey proteins, which affected the mesh size of the resulting yoghurt gel network. When whey powder was used in yoghurt fortification without TG, better flow properties and enhanced syneresis were observed [77]. Characteristics of yoghurts with various ratios of casein to whey proteins were determined as well; decrease of the ratio increased the highest gel strength of the yoghurt. Also, whey drainage was reduced, the gels became less viscous, and a finer structure with numerous small pores and a dense network of cross-links was obtained [30, 34, 78].

Wang et al. [79] studied consistency, microstructure, and probiotic survivability of goat milk yoghurt using whey proteins (WP), pectin, polymerized whey protein (PWP) and mixture of pectin and polymerized whey protein as a co-thickening agent. Probiotic goat milk yoghurt was developed containing *Lactobacillus acidophilus*, *Lactobacillus casei,* and *Bifidobacterium* spp. and using polymerized whey protein (0.4%) and pectin (0.3%) as gelling agents. There was no significant difference in viscosity over weeks; however, changes in titratable acidity and pH showed significant differences during storage over weeks. The viscosity of the yoghurt samples with different levels of WP, PWP, pectin, and mixture of PWP and pectin in various ratios were investigated (Figure 6). The viscosity results showed that yoghurt fortified with mixture of PWP and pectin had higher viscosity than the samples with a single thickening agent, and the viscosity of yoghurt containing WP was the lowest. Native whey proteins exhibited much lower viscosity than PWP and have not been utilized as a thickening agent due to their low molecular weight and nearly spherical shape. However, PWP had a much larger effective hydrodynamic volume than the native globular proteins. Heating WP solutions under controlled conditions formed soluble WP polymers of high molecular weight, resulting in increased viscosity. Pectin is often added to yoghurt to improve consistency (increase viscosity) and reduce syneresis. Although PWP and pectin increased yoghurt viscosity when used individually, the texture of the goat milk yoghurt was not greatly improved. The textural defects in the yoghurt included weak body, whey separation, and gumminess. The addition of a mixture of PWP and pectin resulted in a desirable texture of yoghurt with increased viscosity and mouth feeling [79]. *L. casei* and *Bifidobacterium* spp. remained viable with populations above 10^6 cfu g^{-1} during storage, but there were no viable counts of *L. acidophilus* at the fourth week. Wang et al. [79] also found significant differences between goats and cow milk yoghurts ($P < 0.001$) and during the storage over weeks ($P < 0.001$) in the case of *Bifidobacterium* spp. For goat milk yoghurt, the difference in properties was significant between the product at weeks 3 and 12 ($P < 0.05$).

However, there was no significant change over all 12 weeks of storage ($P > 0.05$) in the case of cow milk yoghurt. The interaction was also significant ($P < 0.001$) indicating that the rate of *L. casei* number decrease was different for each type. Similarly to *Bifidobacterium* spp., *L. casei* counts remained viable at above 10^6 cfu g^{-1} during the12 week storage. There was no mould or yeast detected in both goat and cow milk yoghurts at any time over the 12 weeks storage indicating long shelf life of the product.

Krzeminski et al., [29] examined the textural impact of non-heated and heated whey protein – pectin mixtures added to the milk formulations. Associative interaction of whey proteins with pectin suppressed whey protein aggregation during yoghurt manufacturing, maintaining the structuring effects of denatured whey protein in yoghurt in terms of increased firmness and flow properties. The structuring effect in yoghurt was reduced when a heated whey protein-pectin mixture was used. The use of a non-heated whey protein-high methoxylated pectin mixture in the skim milk formulations yielded a yoghurt texture resembling the full-fat counterpart. The study demonstrated the ability of whey protein–high methoxyl pectin complexes to act as fat replacers and texturizing elements in reduced-fat yoghurt.

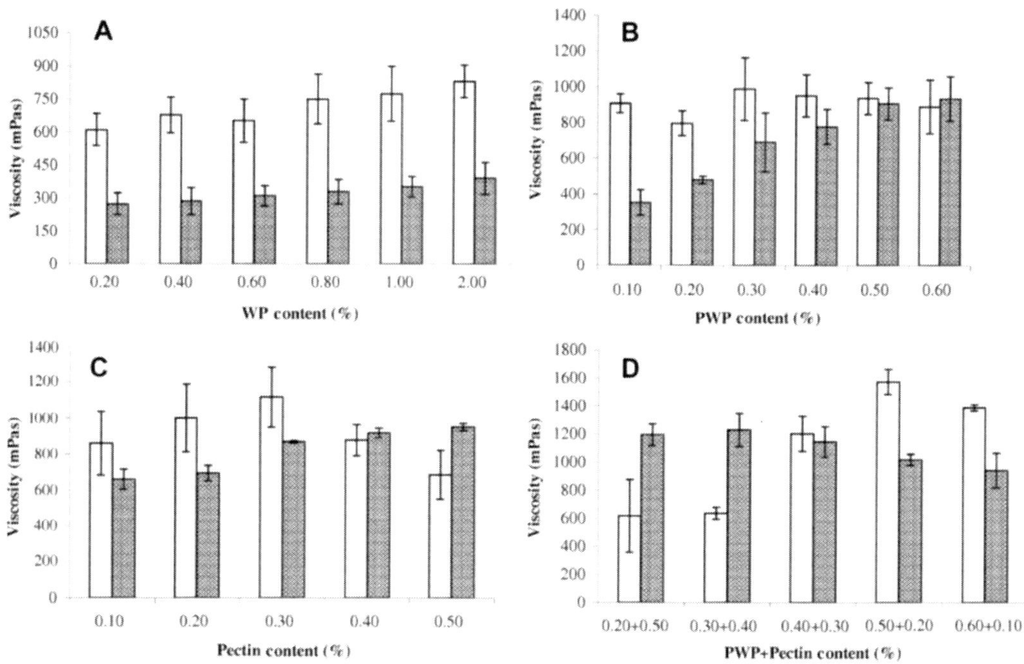

Figure 6. Effects of WP (A), PWP (B), pectin (C), and mixture of PWP and pectin (D) levels on viscosity of yoghurts made with (□) cow milk and (■) goat milk [79]. Error bars represent standard deviation from the mean (n: 3).

Transglutaminase (TG)

The enzyme transglutaminase is a transferase that forms both inter- and intra-molecular isopeptide bonds by cross-linking of amino acid residues of protein-bound glutamine and lysine. Cross-links are formed between the γ-carboxyamide groups of peptide- or protein-bound glutamine residues as acyl donors and several primary amines as acyl acceptors, including the ε-aminogroups of peptide, or protein-bound lysine residues, through an acyl transfer reaction. This reaction leads to the formation of cross-links in the form of ε-(γ-glutamyl) lysine isopeptides, when protein-bound lysine residues act as acyl acceptor, giving rise to novel protein-based gel structures [80–83].

The milk casein is a favorable substrate for cross-linking by transglutaminase (TG), due to the non-globular, easily accessible structure [31, 32, 83]. Enzymatic cross-linking is an important method of the structure modification of food products to control their texture and stability. The kinetics of cross-linking has also been assessed for the individual casein proteins within the caseinate. In solution, the rates were as follows: αs2-casein > β-casein > αs1-casein > κ-casein. In particular, the αs1-casein was cross-linked much slower than expected [84].

In contrast to casein, whey proteins in their native structure are less prone to the cross-linking reaction, mainly due to their globular conformation stabilized by disulphide bonds [34]. The cross-linking of whey proteins can be improved by prior denaturation by heat treatment [31, 83, 85–87].

Transglutaminase is used as commercial microbiological preparation isolated from the *Streptoverticillium* strains purchased from Ajinomoto Co. Inc. (Japan). There are two different ways for the application of TG: either simultaneously or prior to the fermentation.

The application of TG simultaneously with the fermentation implies TG addition together with the starter culture. In this case, the enzyme is gradually inactivated upon acidification. The pre-treatment of the milk with TG prior to fermentation requires additional process time as well as a thermal inactivation step. Besides, the influence of the pH drop during fermentation on the reaction of TG, the protein fortification as well as heat treatment of the yoghurt milk play important role in improving textural and rheological properties. Bönisch et al. [83] found that apparent viscosity and degree of protein polymerization of stirred yoghurt treated with TG were increasing with protein fortification. A novel yoghurt process was investigated in which stirred yoghurt was prepared using milk proteins covalently cross-linked by addition of microbial TG and glutathione as well. The pre-incubation of yoghurt milk with TG and glutathione prior to fermentation did not result in prolonged fermentation time, but protein cross-linking was significantly enhanced by glutathione, and their application in unheated milk and milk products becomes possible. Lorenzen et al. [86], Ozer et al. [88] and Milanović et al. [37] observed increased fermentation times for yoghurt milk treated with TG prior to fermentation. Færgemand et al. [85] attributed that effect to reduced availability of low-molecular peptides for the growth of lactic acid bacteria. Model experiments indicated that the residual activity of TG was drastically reduced with decreasing pH at 40°C, resulting in complete and irreversible inactivation below pH 4.0 [78]. It was found that during yoghurt fermentation 80% of the cross-linking effect was obtained at pH of 6.6–5.7 within the first 150 min. The TG reaction did not negatively affect the fermentation time. No significant increase in degree of protein polymerization (DPP) occurred in cross-linked and non-cross-linked yoghurt at 3.4% or 4.4% protein during shelf life of yoghurt produced with TG addition. The DPP value remained nearly constant (about 28% for the sample with 3.4% protein and 31% at 4.4% of protein) during 6 weeks storage. Indeed, temperature of 4°C during storage was far from the temperature optimum of the enzyme of 50°C [83].

Recent results have shown that UHT treatment of micellar casein suspended in milk improves significantly the cross-linking reaction [78, 83]. Heat treatment of milk prior to cross-linking is necessary because the reactivity of TG is very low in unheated or pasteurized milk, despite the high reactivity of the casein [89, 90].

Attempts have been made to improve low-fat yoghurt gel properties by adding TG in combination with different ingredients (glutathione, milk proteins, or whey protein concentrates) or application of sophisticated techniques (high pressure, high pressure processing, and ultra sound).

Bönisch et al. [30] investigated a novel yoghurt process in which milk proteins were covalently cross-linked by a microbial transglutaminase (TG) preparation containing glutathione (TG+GSH). The authors studied the relationship between the extent of protein cross-linking prior to fermentation and the rheological properties of the resulting yoghurt gel after a stirring process at a constant protein content. Protein cross-linking by TG+GSH was enhanced, resulting in higher apparent viscosity and higher degree of protein polymerization than that produced by TG without GSH. Furthermore, different ratios of casein to whey proteins (CWP ratios) were used to investigate the effect of the both proteins fraction on covalently cross-linked yoghurt gel structures. The results showed a relationship between extent of cross-linking, apparent viscosity, and CWP ratio of the yoghurt gels. During storage for up to 6 weeks at 4°C, no changes in rheological properties and degree of protein polymerization were measurable for stirred yoghurt gels prepared from cross-linked milk proteins.

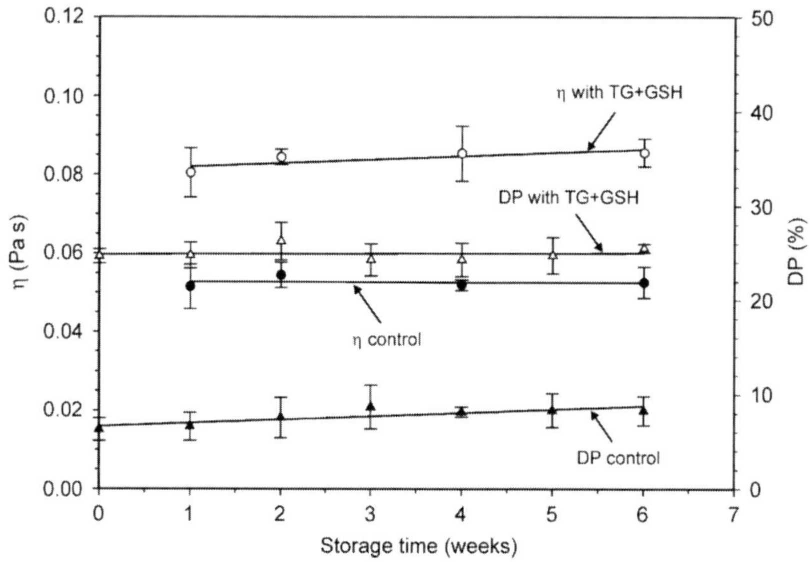

Fig.ure 7. Apparent viscosity η at (○,●) 500 s^{-1} and degree of protein polymerisation (DP, Δ,▲) of yoghurt (4.4% protein) pre-incubated with TG+GSH , (open symbols) or without enzyme (control; closed symbols) as a function of the storage time at 4°C[30].

Figure 7 presents the relation between apparent viscosity (η) and DP of yoghurt samples (protein content of 4.4%) with and without TG as a function of the storage time. Both η and DP for TG+GSH treated and untreated controls remained at a nearly constant level during storage. The addition of TG and TG+GSH resulted in increased η and ΔDP values in comparison with the reference samples without TG.Several authors suggested TG concentrations in the range of 9–14 Ug^{-1} protein to improve the physical, chemical and sensory properties of set style skim milk yoghurt in comparison with untreated samples [86, 88]. According to Farnsworth et al. [91], a four-fold higher TG concentration was necessary to obtain the increased viscosity in stirred yoghurt in comparison with set style yoghurt. In contrast, the application of TG+GSH resulted in significantly higher η levels compared with TG without GSH, even at low enzyme concentrations of 0.6–1.0 Ug^{-1}.

Besides the apparent viscosity, the shear rate and time dependence of the structural breakdown of the yoghurt gel was determined by means of hysteresis loops. Bönisch et al. [30] found that the hysteresis loop area was significantly larger for the TG+GSH-treated yoghurt (19,883 Pa s^{-1}) in comparison with a non-cross-linked reference (13,689 Pa s^{-1}).

In Figure 8, the relation between apparent viscosity η at 500 s^{-1} and ΔDP of yoghurt samples is shown as function of the CWP ratio. It is evident that in reference samples η decreased with increasing CWP ratio. When TG+GSH mixture was added, η increased with increasing CWP ratio. The trend could be attributed to a higher extent of casein cross-linking, which was confirmed by higher ΔDP values; with increasing CWP ratio the ΔDP also increased. The findings clearly showed a relationship between extent of cross-linking, apparent viscosity, and CWP ratio of the yoghurt gels. TG+GSH-treated yoghurt samples with a decreased CWP ratio in the range of 2–3 showed an increased viscosity (40.07 Pa s) and ΔDP (12–16%) in comparison with samples without TG+GSH, but maintained a smooth gel structure.

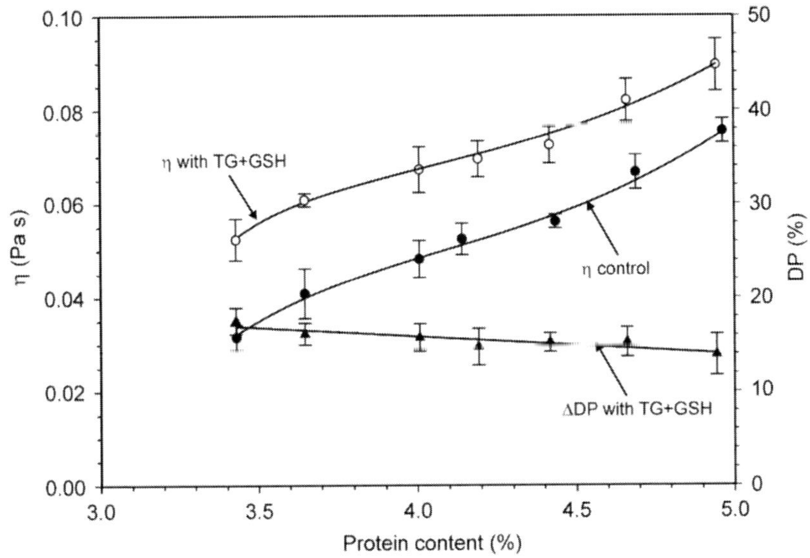

Figure 8. Apparent viscosity η (○,●) at 500 s^{-1} after 1 week at 4°C and TG-induced degree of protein polymerization (ΔDP,▲) of yoghurt (4.4% protein)with different casein/whey protein (CWP) ratio pre-incubated with TG+GSH or without enzyme [30].

Novel method of TG treatment for skim milk yoghurt production was investigated by Guyot and Kulozik [35]. They found that stirred skim milk yoghurts fortified with enzyme-modified skim milk powder (SMP) had higher viscosities and lower serum loss compared to yoghurt samples produced with addition of untreated SMP. Improved properties without undesired structural changes during storage can be realized by addition of TG-pre-treated SMP as protein fortifier [35].

Ercili-Cura et al. [92] examined structural mechanisms leading to improved water retention in acid milk gels induced by TG. Heat-treated skim milk with and without pre-treatment with TG was acidified at 20°C, 30°C, and 40°C at constant glucono-δ-lactone (GDL) level to obtain different acidification rates. TG-treated gels showed decreased tanδ values all through the acidification, which was pronounced around the gelation point. Both acidification temperature and TG treatment had significant effects on the water retention properties of the gels.

Spontaneous syneresis observed at high acidification temperatures (30°C) was prevented upon TG treatment. The authors concluded that water holding capacity (WHC) of TG-treated gels was significantly higher compared to the reference gels over the whole pH range studied. TG-treated milk gels showed a homogeneous network formed of smaller aggregate and pores at the gelation point, and did not show any large-scale re-organization thereafter. Transglutaminase is likely to act as a cross-linker of the protein network at an early stage of gelation and thereby limits network rearrangements that take place in acid milk gels formed at high acidification temperature leading to contraction.

Iličić et al. [12] investigated the effect of transglutaminase addition on rheological properties, textural characteristics, and microstructure of fermented milk products manufactured with different starters (probiotics and kombucha inoculum). Rheological analysis revealed that all the manufactured fermented milk products had higher storage modulus than the loss modulus and exhibited thixotropic and shear-thinning behavior (Figure 9). The addition of TG to milk increased the hysteresis loop area by approximately 10.5%, firmness by 39%, and consistency by 48% in the case of the sample produced with probiotic starter; the product had more firm and stable gel structure than kombucha-fermented milk products (Figure 10). The scanning electron microscopy images revealed that casein matrix of fermented milk products containing TG was continuous except for void spaces occupied by milk serum and starter culture cells (Figure 11).

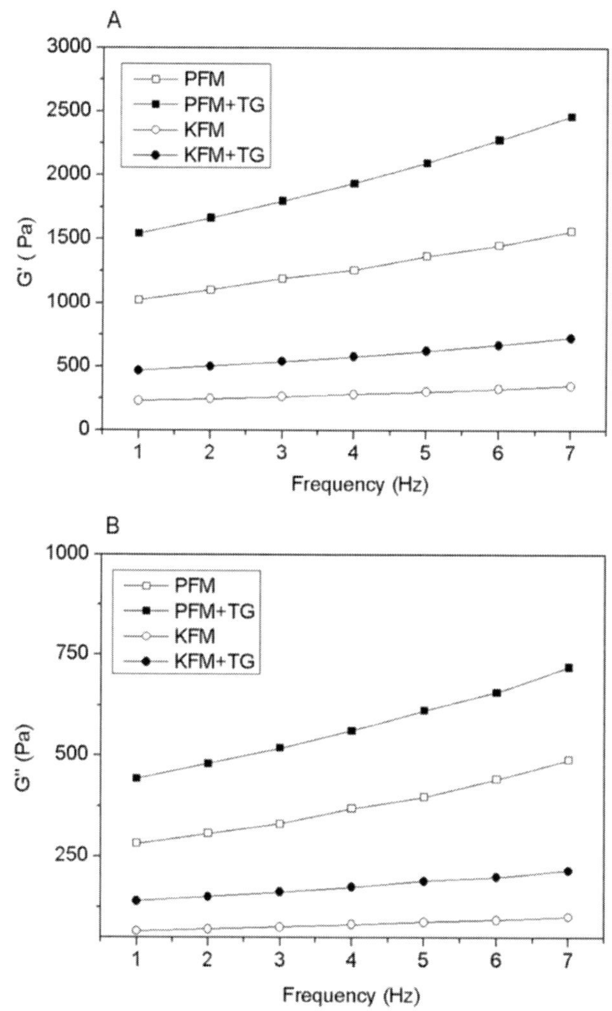

Figure 9. Modulus of fermented milk manufactured with different starters and with and without TG cross-linking prior to inoculation: (A) storage modulus and (B) loss modulus as a function of frequency. KFM, kombucha fermented milk; PFM, probiotic fermented Milk; TG, transglutaminase [12].

The effect of different concentrations of TG inactivated and not inactivated (0.02, 0.06, and 0.12% w/w) on the textural characteristics and flow properties of the stirred probiotic yoghurt prepared from pasteurized skim milk (0.1% w/w fat) was investigated by Iličić et al.

[38]. The physico-chemical characteristics, textural, and rheological properties of the yoghurt samples were determined after production and during 10 days of storage.

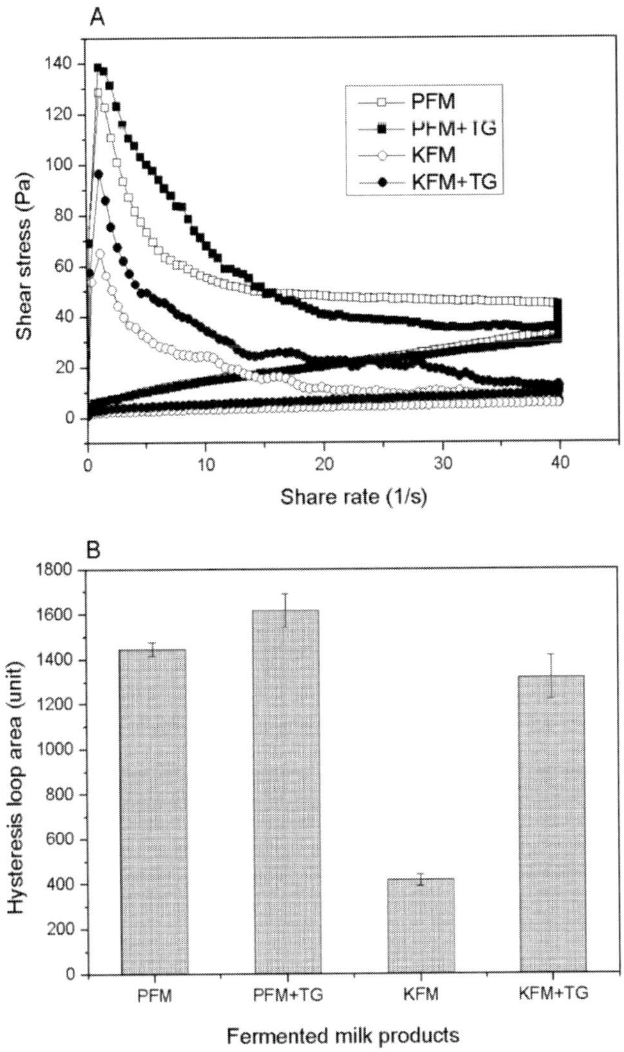

Figure 10. Rheological properties of fermented milk products manufactured with different starters and with and without TG cross-linking prior to inoculation a) flow curves b) hysteresis loop area. KFM, kombucha fermented milk; PFM, probiotic fermented Milk; TG, transglutaminase [12].

The authors found that addition of TG in the production of probiotic yoghurt resulted in significant improvement of textural characteristics (Figure 12). All the samples exhibited thixotropic flow behavior, but the stirred probiotic yoghurt produced with not inactivated TG, at higher enzyme concentrations, showed some additional structure characteristics after production (Figure 13).

From comparison of the presented flow curves it was evident that the structure of samples produced with inactivated TG was more homogenous than that of the samples produced with not inactivated TG. With increased concentration of TG for samples produced with inactivated TG, the shear rate in upward ramp exhibited a steady decline, reaching the value of 200 s^{-1}, and further remained constant (Figure 13). Differences in the flow curves between the samples affected the values of hysteresis loop area for yoghurt produced with different enzyme concentrations during storage.

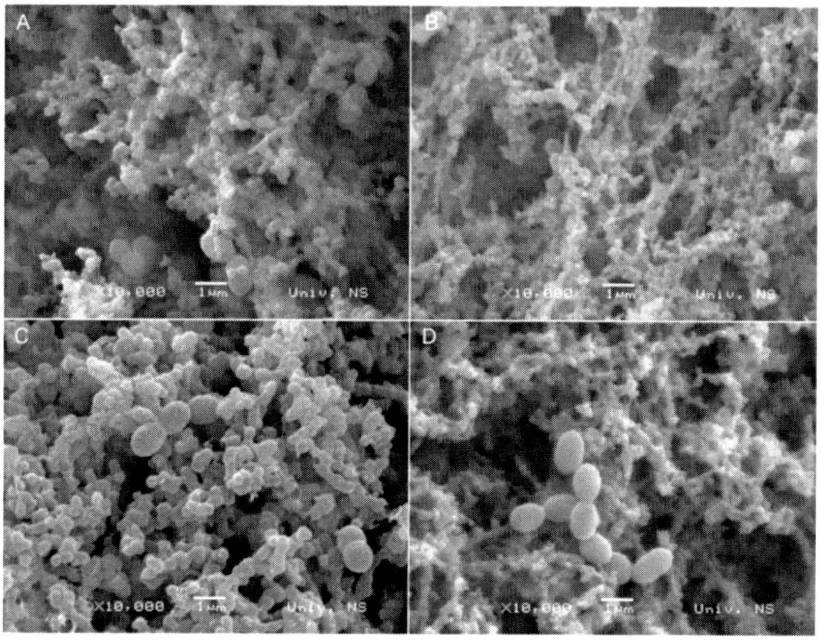

Figure 11. Microstructure of fermented milk products manufactured with different starters and with and without TG cross-linking prior to inoculation (a) PFM,(b) PFM + TG, (c) KFM, and (d) KFM + TG. KFM, kombucha fermented milk, PFM, probiotic fermented milk; TG, transglutaminase [12].

The hysteresis loop area is the indicator of yoghurt structural breakdown and rebuilding during shearing. The addition of inactivated or not inactivated TG at 0.02% (w/w) increased the value of hysteresis loop area (60%) in the samples as compared with the reference yoghurt. During 10 days of storage, a similar trend of change of hysteresis loop area between the samples produced with the lowest concentration of TG (0.02% w/w) was noticed. The yoghurt containing 0.12% (w/w) of not inactivated TG showed the largest hysteresis loop area; after incubation for 2 h at 40°C it was 56% lower compared with samples manufactured with the same concentration of not inactivated TG after 10 days of storage [38]. The authors concluded that the gel structure of the yoghurt samples produced with not inactivated TG was stronger than that of samples produced with inactivated TG and the reference sample.

Milanović et al. [34] studied the effect of addition of transglutaminase (TG non-activated and activated: 0.01% and 0.02% w/w), whey protein concentrate (WPC: 0.3%, 0.4%, and 0.5% w/w), and a combination of 0.01% (w/w) TG and 0.3% (w/w) WPC on the physico-chemical and textural characteristics of low-fat stirred yoghurt. The yoghurt samples were prepared from two batches of pasteurized skim milk (0.1 %w/w fat) and probiotic starter culture. Syneresis and water-holding capacity were determined as relevant physico-chemical characteristics, while the product texture (Table 3) was assessed through firmness, consistency, cohesiveness, and index of viscosity. The addition of TG to milk improved the product quality, while the addition of WPC (Figure 14) improved it even more significantly. However, a combination of TG and WPC decreased the yoghurt textural properties.

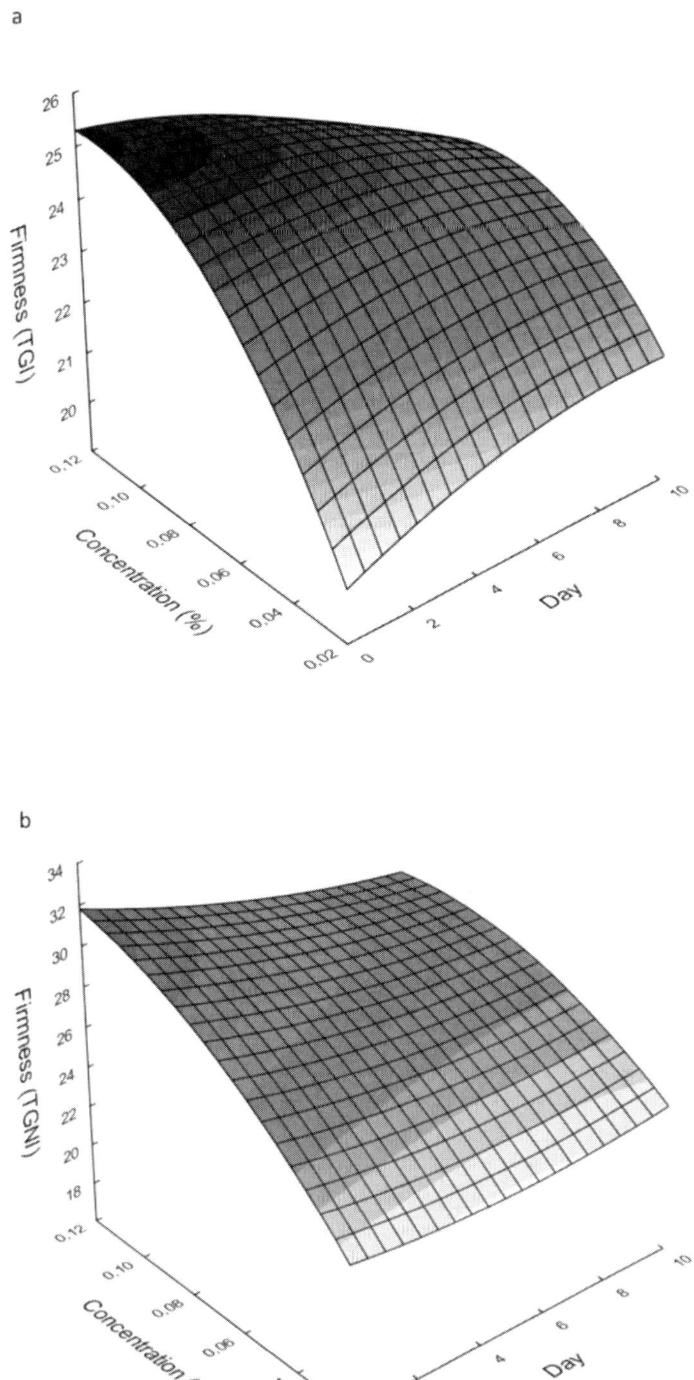

Figure 12. Firmness of yoghurt samples as a function of transglutaminase (TG) concentration and duration of storage: (a) inactivated TG and (b) not inactivated TG [38].

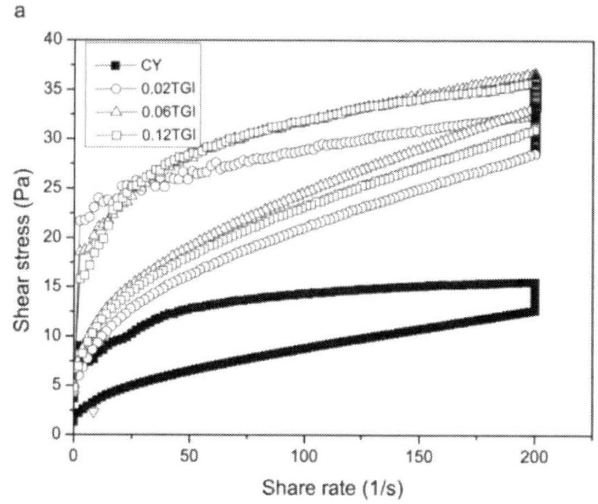

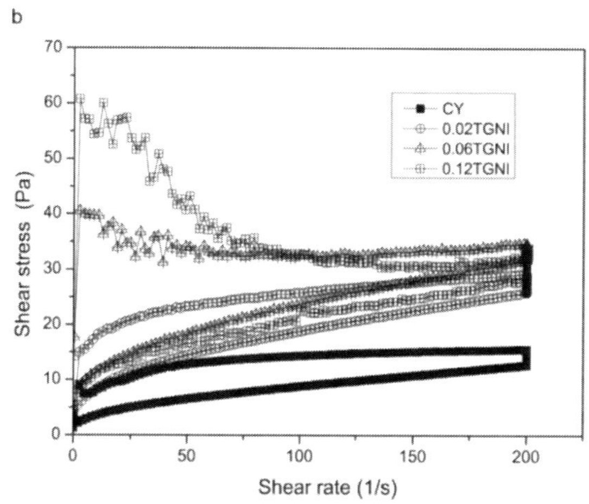

Figure 13. Flow curves of yoghurt samples produced with a) inactivated TG and b) not inactivated TG [38].

Table 3. Textural characteristics of low-fat yoghurt samples [34]

Sample	Firmness (g)	Consistency (gs)	Cohesiveness (g)	Index of viscosity (gs)
Control	19.9±1.7	529±51	−12.6±1.2	−14.1±6.9
0.01%w/w TG, WA[1]	21.7±1.6	577±49	−14.7±1.7	−17.5±2.3
0.02%w/w TG, WA	26.7±1.7	708±51	−18.2±1.5	−30.7±1.8
0.01%w/w TG, A[2]	18.8±2.9	496±63	−13.5±2.6	−10.5±3.2
0.02%w/w TG, A	19.2±1.8	512±44	−13.6±1.7	−10.8±0.87
0.3%w/w WPC	33.8±5.7	853±146	−26.4±7.2	−48.8±26
0.4%w/w WPC	29.6±4.7	754±132	−22.6±4.6	−47.1±19
0.5%w/w WPC	39.7±8.6	995±197	−32.9±9.6	−78.1±25
0.01%w/w TG, WA + 0.3%w/w WPC	16.1±1.7	423±36	−11.9±1.3	−8.9±2.9
0.01%w/w TG, A + 0.3%w/w WPC	13.7±0.3	374±5.8	−8.3±2.5	−4.1±2.0

[1]WA – without activation.

[2]A –activation.

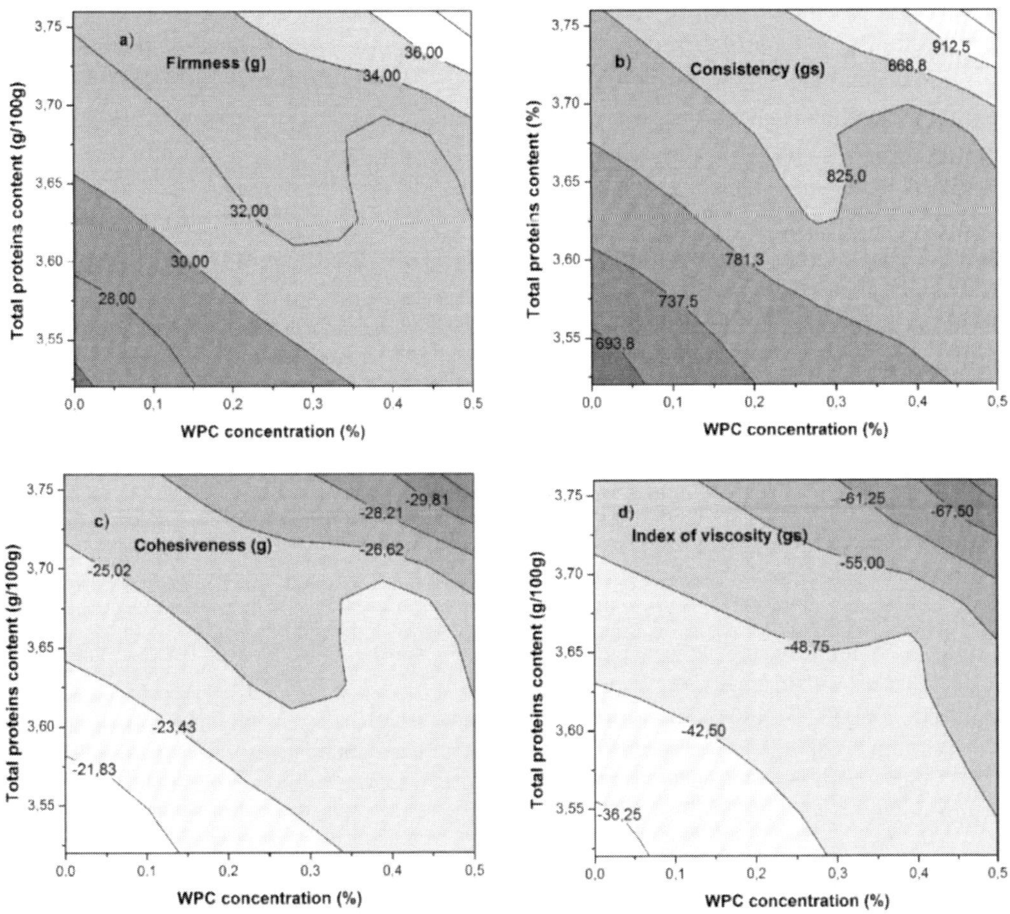

Figure 14. Effect of WPC on textural characteristics of low-fat yoghurt: a) firmness, b) consistency, c) cohesiveness, and d) index of viscosity [34].

Addition of TG to milk at the lowest concentration of 0.01% (w/w) contributed significantly to the formation of stirred yoghurt with improved physical characteristics and better textural properties, higher TG concentrations producing even more favorable effects. Incubation treatment of TG prior to the fermentation (3 h, 25°C) was not efficient, so that non-activated TG proved to be more efficient (by about 30%). When stirred yoghurt was prepared with addition of WPC, its physical characteristics and textural properties were also significantly improved. The effect was proportional to the concentration of the applied ingredient (from 0.3% w/w to 0.5% w/w) and it exceeded the effect achieved by adding TG (by 10–15% with respect to average WHC and by 70–100% regarding average textural properties). However, simultaneous addition of TG and WPC resulted in products with the lowest firmness and consistency, while their cohesiveness and index of viscosity were slightly better than those of the yoghurt prepared with TG alone. It was likely that the increased rate of casein cross-linking in the presence of TG prevailed over the rate of formation of whey proteins aggregates, giving a non-uniform structure of gel matrix and, therefore, a coarse and lumpy product with poor texture (Table 3, Figure 14).

The effect of milk protein polymerization prior to the yoghurt fermentation process with microbial transglutaminase (*Streptoverticillium morbaraense*) was evaluated also by Gauche et al. [33]. They manufactured yoghurt samples with 100% milk or by substituting milk with 20% or 30% of liquid milk whey, aiming at determining the use of natural milk whey in dairy products. Transglutaminase was added at a protein ratio of 0.5 U g^{-1} to all the samples; rheological behavior, syneresis index, and texture profile were monitored.

Table 4. Parameters of the profile texture analysis obtained from the yoghurt samples [33]

Texture parameters					
Sample	F	H	Co	A	G
C100	114.91a ±5.51	142.86a ±6.34	0.49aef ±0.02	–200.34c ±25.30	69.03a ±6.89
TG100	158.50b±7.95	179.19b±9.53	0.45g±0.04	–175.80d±31.18	80.33b±9.03
C80	77.40c ±3.53	101.50c ±4.24	0.51abc ±0.02	–80.46a ±12.31	52.21c ±3.38
TG80	94.05d±1.93	121.42d±4.24	0.50bde ±0.02	–96.61a ±9.23	61.29d±3.54
C70	54.68e ±2.41	74.00e ±3.40	0.55h±0.02	–41.60b±3.12	40.40e ±1.28
TG70	62.00f±7.79	88.18f±2.19	0.52cdf ±0.02	–57.97b±6.01	45.46f±2.16

The values correspond to the mean values ± SD obtained in two repetitions and seven replicates.

Different letters in the same column represent statistically significant differences between values at a significance level of 1% by the Tukey Test.

C 100: 100% of milk; TG100: 100% with transglutaminase (0.5 U g^{-1} protein); C80: 80% milk + 20% milk whey; TG 80: 80% milk + 20% milk whey with transglutaminase (0.5 U g^{-1} protein); C70: 70% milk + 30% milk whey; TG70: 70% milk + 30% milk whey with transglutaminase (0.5 U g^{-1} protein); F: fracturability; H: hardness; Co: cohesiveness; A: adhesiveness; G: gumminess.

The authors found that the addition of transglutaminase (0.5 U g^{-1} proteins) prior to the yoghurt fermentation process led to a significant increase (p <0.01) in almost all the texture parameters analyzed (Table 4). The fracturability, hardness, and gumminess parameters showed a significant increase (p<0.01) in all yoghurt samples subject to enzymatic treatment. These higher values could be due to the additional ε-(γ-glutamyl)lysine bonds created by transglutaminase. Lauber et al. [93] suggested that protein crosslinking predominantly occurred intermolecularly, and the increased stress required to break the yoghurt gel was correlated with the level of casein oligomerization catalyzed by transglutaminase, the latter being increased with the enzymatic reaction time. The introduction of new covalent bonds to milk protein systems using transglutaminase could produce acid milk gels with different textural properties when compared with traditionally manufactured gels, especially when the chemical composition was changed. In particular, when the yoghurt was prepared by substituting 20% of whey and the enzymatic treatment with transglutaminase, the values obtained for the texture profile analysis, despite not being precisely the same, were similar to the values obtained for the reference sample (100% milk) (Table 5).

Improvement of the rheological and physical properties of acid gels from goat milk by combining the dry matter enrichment and the protein cross-linking with microbial transglutaminase was investigated by Ardelean et al. [94]. The incubation with enzyme was performed during 30–420 min, and inactivation was achieved by heat treatment or by addition of N-ethylmaleimide.

Subsequent acidification of enriched and cross-linked goat milk was induced by δ-glucono-lactone at 30°C and resulted in significantly higher stiffness and lower syneresis of the gels compared to gels from untreated goat milk. The highest stiffness values were obtained after 300 min of enzymatic treatment, and an additional impact of the heating step for enzyme inactivation was observed.

This combined procedure of dry matter enrichment and enzyme treatment allowed the production of gels with similar properties as gels from cow milk and appeared to be a promising tool to manufacture goat milk yogurt with satisfying texture properties. The results showed that the introduction of strong covalent isopeptide bonds formed by TG contributed to enhanced textural properties of goat milk acid gels compared with those made without TG, where the weak non-covalent interactions dominated.

Table 5. Rheological properties of set-type yoghurts: fresh and after storage at 5°C for 14 days [95]

Product	Sensory attributes					
	Colour	Taste	Smell	Consistency	Syneresis	Overall acceptability
Fresh yoghurts						
Reference	5.00 (±0.00)	3.96 (±0.17)	3.85 (±0.14)	3.25 (±0.09)	3.33 (±0.12)	3.78 (±0.09)
TG						
(1 U/g protein)	4.94 (±0.03)	3.83 (±0.10)	3.52 (±0.14)	4.01 (±0.06)	4.25 (±0.12)	4.00 (±0.10)
(2 U/g protein)	4.95 (±0.03)	3.78 (±0.09)	3.61 (±0.23)	4.65 (±0.06)	4.78 (±0.11)	4.23 (±0.05)
(3 U/g protein)	4.97 (±0.02)	3.72 (±0.06)	3.57 (±0.11)	4.45 (±0.06)	4.77 (±0.07)	4.25 (±0.05)
Stored yoghurts						
Reference	4.94 (±0.03)	3.57 (±0.08)	3.72 (±0.13)	2.89 (±0.07)	2.50 (±0.26)	3.39 (±0.07)
TG						
(1 U/g protein)	4.95 (±0.03)	3.26 (±0.20)	3.37 (±0.20)	3.60 (±0.09)	3.02 (±0.10)	3.48 (±0.08)
(2 U/g protein)	5.00 (±0.00)	3.30 (±0.20)	3.37 (±0.17)	4.14 (±0.06)	3.96 (±0.10)	3.77 (±0.04)
(3 U/g protein)	4.93 (±0.04)	3.28 (±0.14)	3.30 (±0.20)	4.18 (±0.08)	4.00 (±0.14)	3.77 (±0.07)
Influence of the level of enzyme addition	Ns	Ns	*	**	**	**
Influence of the storage period	Ns	**	*	**	**	**

TG – transglutaminase; Ns – not significant. The tests were performed at 10°C.

* Significant level ($p \leq 0.05$).

** Significant level ($p \leq 0.01$).

Domagała et al. [95] systematically examined rheological properties, textural analysis, and sensory profile of yoghurts produced from goat milk cross-linked by microbial transglutaminase. In general, the higher the amount of added TG used in the milk base the firmer the fresh yoghurt became and the lower syneresis was observed.

The authors found that the texture parameters, such as viscosity, firmness, adhesiveness, and extrusion force of the yoghurts were much higher in the cases of products prepared with added TG as compared with the reference yoghurt (Table 5). It was evident that the TG treatment of milk increased the apparent viscosity and adhesiveness to twice those of the fresh yoghurts. Texture parameters of stored yoghurts were higher than those of the fresh yoghurts. Only the adhesiveness of fresh and stored yoghurts was nearly similar. In addition, yoghurts produced with higher levels of added TG had higher readings of apparent viscosity. For other texture parameters analysis, similar observations were evident at the enzyme addition at the level of 1 and 2 U/g protein. The firmness and extrusion force (i.e. stickiness) readings of the TG-treated yoghurts (fresh and stored containing 2 and 3 U TG /g protein) were similar. However, the adhesiveness and extrusion force of stored yoghurt containing 3 U of TG /g protein were lower than those of products made with 2 U of TG /g protein.

In the case of the set yoghurt, the addition of the TG enzyme at the level of 0.25 U/g protein increased the viscosity of the product by 75% compared to the reference sample, which was comparable to the increase in viscosity achieved by adding 5% of goat milk powder. In the case of stirred yoghurt, pre-treatment of the milk with TG at the level of 2 U/g protein increased the viscosity by more than 3.5 times compared to the untreated sample. However, significant increase in viscosity was observed only for the samples with TG at the levels of 3 and 4 U/g protein. Ardelean et al. [94] concluded that TG treatment of the goat milk base at 1.8 U/g protein combined with fortification of the solids content resulted in a product with good textural properties. Lorenzen et al. [86] reported a strong increase in the gel strength during the first two weeks of storage of set yoghurt produced by simultaneous addition of TG and lactic starter culture. Bonisch et al. [31] found no changes in rheological properties of stirred yoghurt prepared from cross-linked cow milk during cold storage for 6 weeks.

Neve et al. [96] showed that if the enzyme was supplied simultaneously with the yoghurt starter bacteria (*Streptococcus thermophilus* and *Lactobacillus delbrueckii* subsp. *bulgaricus*) without subsequent inactivation, the increased viability of the *L. bulgaricus* culture was observed during cold storage of the yoghurt from cow milk. The authors found that pretreatment of cow milk with TG induced a minor imbalance of the yoghurt starter bacteria during fermentation, whereas a long-term application of TG (without enzyme inactivation) in cow yoghurt resulted in a stabilizing effect on the viability of *L. bulgaricus*.

Farnsworth et al. [91] investigated the effects of enzymatic cross-linking of milk proteins on the functional properties and probiotic culture survivability of goat milk yogurt. They found that the probiotic cultures were relatively stable in the yogurt and their populations remained above 10^6 CFU/g up to 8 weeks at 4°C, suggesting that goat milk yoghurt may be an excellent carrier for the probiotics. No difference ($P > 0.05$) in rate of change of population of all the three probiotic cultures was observed between the reference and the TG-treated samples. When TG was used in goat milk yoghurt preparation, the milk was pretreated for 1 h with enzyme that was subsequently inactivated before addition of the probiotic yogurt starter. This set up mimicked conditions considered to be suitable for TG application for industrial yogurt manufacture. However, the enzyme could also be added to milk together with the starter culture. In that case, the enzymatic reaction proceeded during fermentation.

The effect of transglutaminase on physical, chemical, and sensory properties of set-type yoghurt was investigated by Şanlı et al. [36]. They concluded that the treatment of milk proteins with TG was an efficient method to increase techno-functional properties of set-type yoghurt. It was clearly determined that TG addition to the milk after pasteurization led to increment in viscosity values and decrease of syneresis. Furthermore, TG treatment did not affect the formation of acetaldehyde as a main aroma compound in yoghurt. It also had no unfavorable effect on the sensory properties of the final product. Hence, TG addition following pasteurization step could be suggested for industrial practice.

Wróblewska et al. [97] analyzed immunoreactivity of milk proteins and sensory quality of traditionally prepared kefir enriched by the addition of transglutaminase to milk used for fermentation. With growth of lactic acid bacteria and yeasts, milk proteins were hydrolyzed into short chain peptides. Results of chromatographic analysis demonstrated significant changes in protein absorbance profiles observed at 214 nm and 280 nm. The addition of TG to fresh kefir evoked formation of protein aggregates. Two distinct fractions (at 50–55 mL and 135–155 mL) were observed at 280 nm, suggesting the formation of aromatic acids. The ELISA results obtained during the reaction of kefir proteins with specific antibody against individual milk proteins (β-lactoglobulin, α-lactalbumin, α -casein, β-casein, and κ-casein) indicated differences in the level of proteins as a result of TG addition as well as storage.

The reaction between immunoreactive β-lactoglobulin and specific bodies was not detectable in any of the kefir samples under the adopted conditions of the ELISA assay. The results indicated that the type of kefir could be offered especially to those consumers who are searching for food stuffs with low immunoreactivity.

Application of TG in labneh manufacturing is also very interesting from technological aspects of view. Labneh (concentrated yoghurt) is a semisolid fermented dairy food produced by removing part of the whey from yoghurt to reach total solid levels between 23 and 25 g/100 g, of which 8–11 g/100 g is fat. The texture of labneh is an important parameter that determines the acceptability of the product. The recent increase in the popularity of labneh has led to more interest in its texture, especially in relation to concentration techniques [4, 98]. Increasing the protein level leads to the growing gel strength of yoghurt [88].

Hatice et al. [98] investigated the quality of labneh which was prepared by inoculating goat milk with/without transglutaminase (at ranges of 0–4 U/g protein), followed by heat inactivation. They found that application of TG did not change the chemical composition of the product. However, TG treatment at 2–4 U/g protein increased the firmness of the labneh samples by 14–15 times as compared to the untreated sample.

Scanning electron microscopy pictures revealed that the protein matrices in labneh from TG treated samples appeared to be relatively more compact than in the reference sample. SDS-PAGE scans revealed cross-linking of proteins in labneh samples treated with TG, and conversely, the bands corresponding to casein fractions became weaker as the concentration of TG was increased. Scores of firmness by sensory evaluation indicated that TG treated samples at 2 U/g had a value of 27.4 compared to 12.5 in non-treated samples.

To generalize, addition of TG alone or in combination with different ingredients (milk proteins, whey protein concentrate, glutation) to fermented milk products improves rheological and textural characteristics of samples. The enzyme of TG could be also applied for manufacturing of different fermented milk products via novel technologies - high pressure processing, high pressure homogenization and ultrasonic processing.

NOVEL TECHNOLOGIES FOR FERMENTED MILK PRODUCTION

Numerous studies dealing with application of innovative processing technologies such as high pressure processing (HPP), high pressure homogenization (HPH), and ultrasonic processing (USP) for improvement the texture of fermented milk products [99–103] have been reported.

In convectional homogenization, the pressure of up to 30 MPa is applied, while in HPP and HPH processes the higher pressure is used (the highest operating pressure of 1000 MPa and 400 MPa, respectively).

Figure 15 shows the effect of selected milk treatments and fermentation on casein micelles and gel structure. HPP processing significantly improved the functional properties of milk proteins which was relevant for rheological and textural properties of yoghurt gel. According to Huppertz et al. [104] and Huppertz and de Kruif [105], the following changes occurred during HPP treatment of milk: disruption and/or shrinking of casein micelles, denaturation of whey proteins (especially β-*lg*), and release of colloidal calcium phosphate from casein micelles causing the change of the mineral balance. HPP could be applied at pressures above 300 MPa to yoghurt after fermentation attempting to kill spoilage microorganisms, inhibit further acidification during shelf life, and improve yoghurt texture [99, 106].

Physical properties of stirred yoghurt made from reconstituted skim milk that was pressure-treated at 100, 250, or 400 MPa at 25, 70, or 90°C (during 10 min prior to inoculation with yoghurt cultures) were also studied. Parts of milk HP-treated at 25°C were also heat-treated at 90°C for 10 min before or after pressure treatment. Reference yoghurts were prepared from skim milk heat-treated at 90°C for 10 min. Stirred yoghurts obtained from skim milk HP-treated at 25°C, before or after heat treatment, showed viscosity similar to that made from conventionally heat-treated milk. Lower viscosity was obtained when stirred yoghurts were made with milk HP-treated at elevated temperatures, 70 or 90°C [100].

The published results revealed that the benefits of HPH depended on the fat content of the milk, whereas in the case of HPP the best results were obtained in combination with transglutaminase or heat treatment [99]. It was demonstrated that HPP-treated milk coagulated at higher pH values and the produced yoghurt showed lower amounts of whey separation and increased gel strength [107–110].

Yoghurt samples prepared from HPP in combination with TG-treated milk exhibited the highest values of firmness, lowest values of whey separation, and acidity similar to those of the other samples [101]. All the yoghurt samples made from high pressure processed milk, with or without subsequent TG treatment, exhibited a creamier perception than the samples from heat-treated milk. Yoghurt samples prepared via the conventional procedure were judged as unacceptable after the fifth week of storage due to a strong syneresis and separation of the coagulum. It was suggested that HPP and TG treatment of milk could be useful tools for the dairy industry to achieve products of improved structure and desirable sensory characteristics.

Ultrasonic treatment in food applications involves the transmission of high intensity and high frequency (>18 MHz) sound waves through the liquid food material. Although USP is considered to be a non-thermal technique, considerable heat is generated during its application, and must be controlled during process [111]. Numerous studies revealed that the properties of yoghurts prepared from ultrasound-treated whole milk were improved. Firmness and water-holding capacity increased, and syneresis was weaker as compared with yoghurts prepared from conventionally heated whole milk [112–115]. Other studies reported that the USP of skim milk before fermentation by yoghurt cultures produced gels with higher firmness, shorter gelation time, higher gelation pH, better water-holding capacity, and better sensorial properties than those produced from conventionally heated skim milk [114, 115]. The level of whey protein denaturation in the ultrasonicated milk was reported to be markedly lower than that in the heated milk. That indicated that the improved properties of the yoghurts produced from ultrasound-treated milk could be a consequence of changes in the casein micelles or the interactions between the caseins and denatured whey proteins during ultrasonic treatment. Ultrasonic treatment of reconstituted skim milk (12% w/w) without temperature control could be used to produce gels with markedly increased final G' values and reduced gelation times. It was found [103] that the ultrasonic treatment itself with controlling the temperature during process had only a minor effect on the properties of the skim milk and the subsequent acid gelation.

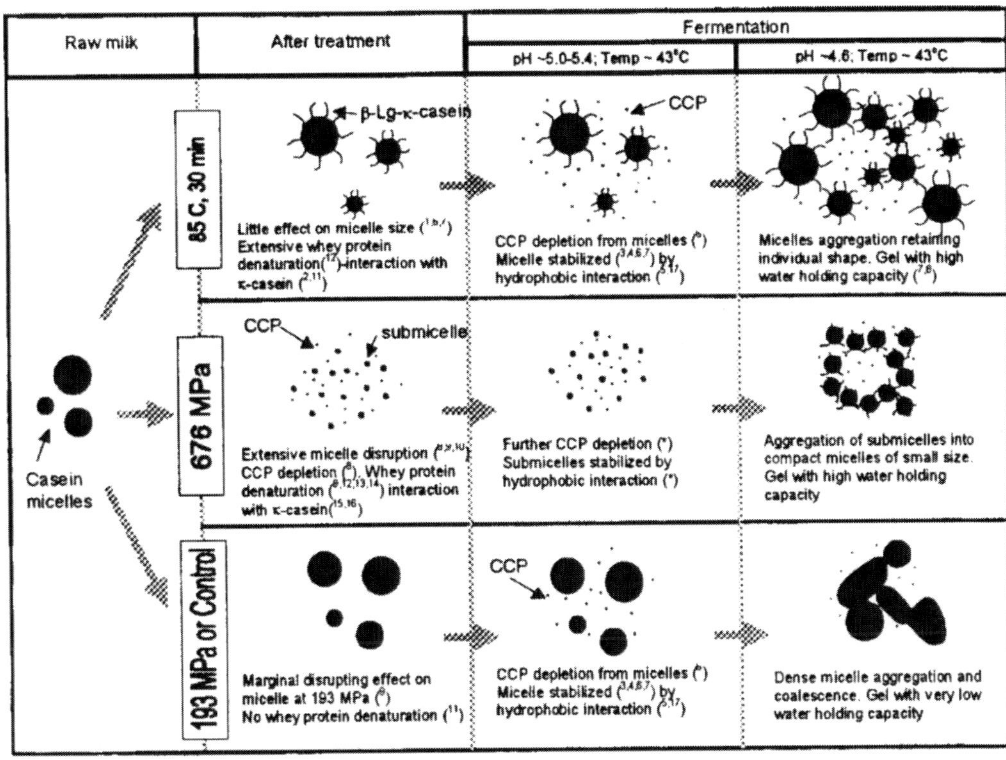

Figure 15. Effect of selected milk treatments and fermentation on casein micelles and gel structure. CCP: Colloidal Calcium Phosphate [108].

Most of the effects of USP of the skim milk were related to the heat generated during the processing, which caused denaturation of the whey proteins, modification of the particle size of the casein micelles and interaction of the denatured whey proteins with the casein micelles. Those effects were very similar to those achieved by conventional heating, and the final G' values and gelation times achieved under optimal USP conditions were very similar to those attained by conventional heating. The obtained results indicated that ultrasonic treatment has a permanent effect on the properties of milk, but the extent of the changes is small. The ultrasonic processing as an emerging technology could be successfully used in dairy systems, due to the ability to preserve food and modulate enzyme activity, to improve processing efficiency, the ability to manufacture products with tailored functionality, and the capability of improving the microstructure through the components interactions. The ultrasound-induced physical effects are used for manufacturing yogurts with superior rheological properties, reduction of yogurt total fermentation time, cutting of cheese blocks, enhancement of whey ultrafiltration, extraction of functional foods, reduction of product viscosity, homogenization of milk fat globules and crystallization of lactose [102, 114–117].

CONCLUSION

In recent years, manufacturers have responded to consumers' demand for low-fat and fat-free dairy products. However, these products somewhat lack the textural and sensory attributes of their full-fat counterparts.

The viscosity and gel structure are influenced by a great number of factors, including milk composition, especially contents of fat and proteins, changes of the ratio of casein to whey proteins, incubation temperature, casein heat treatment of milk, acidity, and type of starter culture.

The selection of the appropriate starter culture type is of great importance in fermented milk products manufacturing. The employed culture may improve the physical attributes of the final products. Ropy and non-ropy starters have been utilized for set and stirred fermented milk product preparation, respectively. Ropy cultures are capable of producing extracellular polysaccharides during the fermentation process, in contrast to non-ropy cultures. Also, milk solids supplementation is a good practice to improve starter culture, especially probiotic bacteria. It is well documented that probiotic bacteria grow slowly in milk because it lacks proteolytic enzymes and contribute poor sensory and rheological characteristics to the product; hence, the practice is to blend these organisms with yoghurt starter culture. Recent studies have been dealing with possibilities of kombucha application as non-conventional starter for milk fermentation. Kombucha is sweetened tea beverage which is fermented by a symbiosis of bacteria and yeast embedded within cellulosic pellicle. The latest results show that the kombucha fermented milk product has the highest angiotensin converting enzyme (ACE) inhibiting activity (63.43%) at the end of storage compared with probiotic and yoghurt products. Based on estimated biological activity, the kombucha product can be considered as an innovative fermented dairy product suitable for human nutrition.

Attempts have been made to improve low-fat yoghurt gel attributes by incorporating different ingredients: inulin, dietary fiber, glutathione, glucono-δ-lactone, low-molecular water–soluble surfactants (monoglycerides or phospholipids), acetylated starch, milk proteins, sodium caseinates, whey protein concentrates, and transglutaminase as well as application of some sophisticated techniques (high pressure homogenization, high pressure processing, ultrasonication).

The fortification of the milk base with certain ingredients or application of transglutaminase results in increase of elastic and viscous modules and improves stability of fermented milk products. In order to produce yoghurt with low caloric value, and good physical and chemical properties, transglutaminase enzyme could be used. Transglutaminase is a potential tool to increase the formation of covalent cross-links in milk products. Activity of transglutaminase depends on several factors: pH value of substrate, incubation temperature, content and ratio of casein and whey proteins, heat treatment, as well as other ingredients addition.

There are two different ways for the application of transglutaminase-either simultaneously or prior to the fermentation. In the former case, the enzyme is gradually inactivated upon acidification. Some authors have concluded that the gel structure of yoghurt samples produced with not inactivated TG is stronger than that of samples produced with inactivated transglutaminase and the reference sample. The application of not inactivated transglutaminase significantly affects water retention properties of the gels and improves the rheological properties and stability of yoghurt samples during storage. Cross-linking of the milk protein can be used as an alternative approach to cow and goat milk solid fortification.

Numerous studies have been dealing with application of innovative processing technologies: high pressure processing, high pressure homogenization, and ultrasonic processing to improve the texture of fermented milk products. The available results show that the benefits of high pressure homogenization depend on the fat content of the milk, whereas in the case of high pressure processing the best results are obtained in combination with transglutaminase or heat treatment.

To conclude, the addition of various starter cultures and ingredients to different fermented milk products (yoghut, kefir, labneh, etc.) and application of sophisticated processing technologies are ways to improved rheology, texture, and sensory characteristics of product, as well as its health benefits.

REFERENCES

[1] Tratnik, Lj. and Božanić, R. 2012, Milk and Dairy Products. *Croatian Dairy Association*, Zagreb, Croatia, p. 510. (in Croatian).

[2] Valli, C.and Traill W. B. 2005, *Food Quality Preference*, 16 (4), 291–304.

[3] Surono I.S. and Hosono, A. 2011, In: Encyclopedia of Dairy Sciences, J.W. Fuquay, P.F. Fox, P.L.H. McSweeney (Eds.), *Elsevier Academic Press*, London, UK, pp. 470–495.

[4] Tamime, A.Y. 2006, Fermented Milks, *Blackwell Science Ltd*, UK, p. 262.

[5] Tamime, A.Y. and Robinson, R.K. 2004, Yoghurt, Science and Technology, *Woodhead Publishing Limited*, Cambridge England, p. 619.

[6] Vélez-Ruiz, J.F. 2008, *In:* Progress in Food Engineering Research and Development, J. M. Cantor (Ed.) ,pp. 223–242, ch. 6, *Nova Science Publishers, Inc.* New York, USA.

[7] Hassan, A.N., Ipsen, R., Janzen, T., and Qvist, K.B. 2003. Microstructure and rheology of yoghurt made with cultures differing only in their ability to produce exopolysaccharides. *J. of Dairy Sci.*, 86, 1632-1638.

[8] Debon, J., Schwinden Prudêncio, E. and Petrus, J.C.C. 2010, *J. Food Eng.*, 99, 28–135.

[9] Stijepić, M., Milanović, S., Djurdjević-Milošević, D., Djurić, M., Glušac, J., Kanurić, K., and Vukić, V. 2012, *Milchwissenchaft/Milk Sci. Int.*, 67 (3), 277–280.

[10] Malbaša, R., Milanović, S., Lončar, E., Djurić, M., Carić, M., Iličić, M., and Kolarov, Lj. 2009, *Food Chem.*,112, 178–184.
[11] Milanović, S., Kanurić, K., Vukić, V., Hrnjez, D., Iličić, M., Ranogajec, M. and Milanović, M. 2012, *African J. Biotechnol.*, 11, 2320–2327.
[12] Iličić, M., Milanović, S, Carić, M., Vukić, V., Kanurić, K, Ranogajec, M. and Hrnjez, D. 2013, *J. Texture Stud.*, 44, 160–168.
[13] Chen, C. and Liu B.Y. 2000, *J. Appl. Microbiol.*, 89, 834–839.
[14] Zhang, H., Yhang, Z., Xin, X. 2011, *J. Beijng Union University*, 2, 11.
[15] Trovatti, E., Serafim, L.S., Freire, C.S.R., Silvestre, A.J.D., and Neto C.P. 2011, *Carbohydr. Polym.*, 86, 1417–1420.
[16] Marsh, A.J., Sullivan, O.O, Hill, C. Ross, R.P. and Cotter P.D. 2014, *Food Microbiol.*, 38, 171–178.
[17] Wu, W., Gai, B. and Ji, B. 2004, *J. Food Sci.*, 4, 10.
[18] Codex Standard for Fermented Milks (CODEX STAN 243-2003).
[19] Bylund,G. 2003, Dairy Processing Handbook, *Tetra Pak Processing System AB, Lund*, p.436.
[20] Tamime, A.Y. 2007, Structure of Dairy Products, *Oxford, England*: Blackwell Publishing, SDT, p. 288.
[21] Kalab, M., and Carić, M. 1990, *In:* Proceeding of the XXIII International Dairy Congress, Montreal, Canada, 1457–1480.
[22] Kalab, M. 1993, *Food Struct.*, 12, 95–114.
[23] Girard, M. and Schaffer-Lequart, C. 2007, *Int. Dairy J.*, 17, 666–673.
[24] Mende, S., Peter M., Bartels, K., Rohm, H. and Jaros, D. 2013, *Food Hydrocolloids*, 32, 178–185.
[25] Prasanna, P.H.P, Grandison, A.S. and Charalampopulous, D. 2013, *Food Res. Int.*, 51, 15–22.
[26] Ramirez-Santiago, C., Ramos Solis, L., Lobato-Calleros, C., Peña-Valdivia, Vernon-Carter, EJ. and Alvarez-Ramirez, J. 2010, *J. Food Eng.*, 101, 229–235.
[27] Cui, B., Lu, Y., Tan, C., Wang, G. and Li, G.H. 2014, *Food Hydrocoll.*,, 35, 576–582.
[28] Aguirre-Mandujano, E., Lobato-Calleros, C., Beristain, C.I., Garcia, H.S., and Vernon-Carter, E.J. 2009, *Food Sci. Technol.*, 42, 938–944.
[29] Krzeminski, A., Prell K.A., Busch-Stockfisch, M., Jochen, W. and Hinrichs, J. 2014, *Int. Dairy J.*, 36, 118–127.
[30] Bönisch , M.P., Huss, M., Weitl, K. and Kulozik, U. 2007, *Int. Dairy J.*, 17, 1360–1371.
[31] Bönisch, M.P, Huss, M., Lauber, S. and Kulozik, U. 2007, *Food Hydrocoll.*, 21, 585–595.
[32] Bönisch, M.P., Lauber, S., and Kulozik, U. 2007, *Int. Dairy J.*, 17, 3–11.
[33] Gauche, C., Tomazi, T., Barreto, P.L.M., Ogliari, P.J. and Bordignon-Luiz, M.T. 2009, *Food Sci. Technol.*, 42, 239–243.
[34] Milanović, S., Iličić, M., Djurić, M., and Carić, M. 2009, *Milchwisenschaft/Milk Sci. Int.*, 4, 388–392.
[35] Guyot, C. and Kulozik, U. 2011, *Int. Dairy J.*, 21, 628–635.
[36] Sanlı, T., Sezgin, E. Deveci, O., Senel E., and Benli, M. 2011, *Food Hydrocoll.*, 25, 1477–1481.

[37] Milanović, S., Carić, M., Đurić, M., Iličić, M., and Duraković, K. 2007, *Acta Periodica Technol.*, 38, 45–52.

[38] Iličić, M., Milanović, S., Carić, M., Dokić, Lj. and Kanurić, K. 2014, *J. Texture Stud.*, 45, 13–19.

[39] Skriver, A., Roemer, H., and Qvist, K.B. 1993, *J. Food Eng.*, 44,175–180.

[40] Milanović, S., Lončar, E., Djurić, M., Malbaša, R., Tekić, M., Iličić, M., and Duraković, K. 2008, *Acta Periodica Technologica*, 39, 37–46.

[41] Iličić, M. 2010, Optimisation of functional fermented dairy beverage technology, PhD Thesis, Faculty of Technology Novi Sad, University of Novi Sad, p. 112.

[42] Dufresne, C. and Farnworth, E. 2000, *Food Res. Int.*, 33, 409–421.

[43] Teoh, A.L., Heard, G. and Cox, J. 2004, *Int. J. Food Microbiol.*, 95, 119–126.

[44] Iličić, M., Kanurić, K., Milanović, S., Lončar, E., Djurić, M. and Malbaša, R. 2012, *Roman. Biotechnol. Lett.*, 17 (1),7013–7021.

[45] Vukić, V., Hrnjez , D., Kanurić, K., Milanović, S., Iličić, M., Torbica, A. and Tomić, J. 2014, *J. Texture Stud.*, 45, 261–273.

[46] Hrnjez, D., Vukić, V., Milanović, S., Iličić, M. and Kanurić, K. 2014, *AgroFOOD Hitech*, 24(5), 24–27.

[47] Hrnjez , D., Vaštag, Ž., Milanović, S.,. Vukić, V., Iličić, M., Popović, Lj. and Kanurić, K. 2014, *J. Funct. Food*, 10, 336–345.

[48] Folkenberg, D.M., Dejmek, P., Skriver, A. and Ipsen, R. 2005, *J. Texture Stud.*, 36, 174–189.

[49] van Marle, M.E., van den Ende, D., de Kruif, C.G., and Mellema, J. 1999, *J. Rheol.*, 43, 1643–1662.

[50] Marshall, V.M., and Rawson, H.L. 1999, *Int. J. Food Sci. Technol.*, 34, 137–143.

[51] Petry, S., Furlan, S., Waghorne, E., Saulnier, L., Cerning, J., and Maguin, E. 2003, *FEMS Microbiol. Lett.*, 221, 285–291.

[52] Ruas-Madiedo, P., Hugenholtz, J. and Zoon, P. 2002, *Int. Dairy J.*, 12, 689-695.

[53] Gibson, G.R. and Roberfroid, M.B. 1995, *J. Nutrit.*, 125 (6), 1401–1412.

[54] Martínez-Villaluenga, C., Frías, J., Gómez, R. and Vidal-Valverde, C. 2005, J. Food Prot., 68, 1246–1252.

[55] Flamm, G., Glinsmann, W., Kritchvsky, D., Prosky, L. and Roberfroid, M. 2001 Critical Reviews in Food

[56] Roberfroid, M.B. 2005, *Brit. J. Nutrit.*, 93 (suppl. 1), 13–25.

[57] Gibson, G.R., Probert, H.M., Van Loo, J., Rastall, R.A. and Roberfroid, M.B. 2004, *Nutrit. Res. Rev.*, 17, 259–275.

[58] Mohameed, H.A., Abu-Jdayil, B. and Al-Shawabkeh, A. 2004, *J. Food Eng.*, 61, 347–352.

[59] Hauly, M.C.O. and Moscatto, J.A. 2002, *Revista Semina: Ciências Exatas e Tecnologia de Alimentos,* 23(1), 23–26.

[60] Tungland, B.C. and Meyer, D. 2006, *Food Sci. Food Safety*, 1, 73–92.

[61] Tziboula, A., Steele, W., West, I.and Muir, D.D. 1998, *Milchwissenschaft/Milk Sci.Int.*, 53, 8–11.

[62] Mistry, V.V. and Moubois, J.L. 2004, *In:* Cheese: Chemistry, Physics and Microbiology, P.F. Fox, P.H.L. McSweeney, T.M. Cogan, and P.T. Guinee (Eds.), pp. 261–284, *Chapman Ltd.*

[63] AACC, 2001, The definition of dietary fiber. *Cereal Foods World*, 46, 112–126.

[64] Rodríguez, R., Jiménez, A., Fernández-Bolaños, J., Guillén, R., and Heredia, A. 2006, *Trends Food Sci. Technol.*,17, 3–15.
[65] Alvarez, A.R. and Peña-Valdivia, C.B. 2009, *J. Prof. Ass. Cactus Dev.*, 11, 26–44.
[66] Marafon, A.P., Sumi, A., Alcântara, M.R., Tamime, A.Y., and Oliveira, M.N. 2011, *Food Sci. Technol.*, 44, 511–519.
[67] Sodini, I., Montella, J.and Tong, P.S. 2005, *J. Sci. Food Agricult.*, 85, 853–859.
[68] Akalin, A.S., Gonc, S., Unal, G. and Fenderya, S. 2007, *J. Food Sci.*, 72(7), M222–M227.
[69] Bhullar, Y.S., Uddin, M.A. and Shah, N.P. 2002, *Milchwissenschaft/Milk Sci.Int.*, 57, 328–332.
[70] Damin, M.R., Minowa, E., Alcantara, M.R., and Oliveira, M.N. 2008, *J. Texture Stud.*, 39, 40–55.
[71] Oliveira, M.N., Sodini, I., Remeuf, F., and Corrieu, G. 2001, *Int. Dairy J.*, 11(11–12), 935–942.
[72] Lucey, J. A. and Singh, H. 1998, *Food Res. Int.*, 30, 529–542.
[73] Torres, I.C., Janhøj, T., Mikkelsen, B.Ø. and Ipsen, R. 2011, Int. Dairy J., 21, 645–655.
[74] Krzeminski, A., Großhable, K., and Hinrichs, J. 2011, *Food Sci. Technol.*, 44, 2134–2140.
[75] Küçükçetin, A. 2008, *Food Res. Int.*, 41, 165–171.
[76] Remeuf, F., Mohammed, S., Sodini, I., and Tissier, J.P. 2003, *Int. Dairy J.*, 13, 773–782.
[77] Gonzalez-Martinez, C., Becerra, M. Chafer, M., Albors, A., Carot, J.M., and Chiralt, A. 2002, *Trends Food Sci. Technol.*, 13, 334–340.
[78] Rodriguez-Nogales, J. 2006, *Int. Dairy J.*, 16, 26–32.
[79] Wang, W., Bao, Y., Hendricks, G.M., and Guo M. 2012, *Int. Dairy J.*, 24, 113–119.
[80] Motoki, M. and Seguro, K. 1998, *Trends Food Sci. Technol.*, 9, 204–210.
[81] Dinnella, C., Gargaro,M., Rossano, R., and Monteleone, E. 2002, *Food Chem.*,78, 363–368.
[82] Yokoyama, K., Nio, N., and Kikuchi, Y. 2004, *Applicat. Microbiol. Biotechnol.*, 64, 447–454.
[83] Bönisch, M.P., Lauber, S., and Kulozik, U. 2004, *J. Food Sci.*, 69, 398–404.
[84] Macierzanka, A., Bordron, F., Rigby, N.M., Mills, C.E.N., Lille, M., Poutanen, K., and Mackie, A.R. 2011, *Food Hydrocoll.*, 25, 843–850.
[85] Færgemand, M., Sørensen, M., Jørgensen, U., Budolsfen, G., and Qvist, K.B. 1999, *Milchwissenschaft*, 54, 563–566.
[86] Lorenzen, P.C., Neve, H., Mautner, A., and Schlimme, E. 2002, *Int. J. Dairy Technol.*, 55, 152–157.
[87] Satoshi, N. K., Nyugyo, K. and Assignee K. (2001). Method for producing yoghurt. *Japanese Patent 2001252011.*
[88] Ozer, B., Kirmaci, H.A., Oztekin, S., Hayaloglu, A., and Atamer, M. 2007, *Int Dairy J.*, 17, 199–207.
[89] Iličić, M., Carić, M., Milanović, S. Dokić, Lj. , Đurić, M., Bošnjak G., and Duraković, K. 2008, *Acta Periodica Technol.*, 39, 11–19.
[90] Wen, Y., Liu, N., and Zhao, X.H. 2012, *Food Technol. Biotechnol.*, 50(4), 473–478.
[91] Farnsworth, J.P., Li, J., Hendricks G.M., and Guo, M.R. 2006, *Small Ruminant Res.*, 65, 113–121.

[92] Ercili-Cura, D., Lille M., Legland D., Gaucel, S., Poutanen K. Partanen, R. and Lantto, R. 2013, *Food Hydrocolloids*, 30, 419–427.
[93] Lauber, S., Henle, T., and Klostermeyer, H. 2000, *Eur. Food Res. Technol.*, 210, 305–309.
[94] Ardelean, A.I., Otto, C. Jaros, D., and Rohm, H. 2012, *Small Ruminant Res.*, 106, 47–53.
[95] Domagała, J., Wszołeka, M., Tamime A.Y., and Kupiec-Teahanc, B. 2013, *Small Ruminant Res.*, 112,154–161.
[96] Neve, V.H., Lorenzen, P.Chr., Mautner, A., Schlimme, E., and Heller, K.J. 2001, *Kieler Milchwirtschaftlichte Forschungsberichte*, 53(4), 347–361.
[97] Wroblewska B., Kolakowski, P., Pawlikowska, K., Troszynska A., and Kaliszewska A. 2009, *Food Hydrocolloids*, 23, 2434–2445.
[98] Hatice S., Alõglua, A., and Oner, Z. 2013, *Small Ruminant Res.*, 109, 31–37.
[99] Loveday, S.M., Sarkar A., and Singh H. 2013, *Trends Food Sci. Technol.*, 33, 5–20.
[100] Udabage, P., Augustin, M., Versteeg C., Puvanenthiran, A., Yoo, J.A., Allen N., McKinnon I., Smiddy M., and Kelly A. L. 2010, *Innovative Food Sci. Emerging Technol.*, 11, 32–38.
[101] Tsevdou, M.S., Eleftheriou, E.G., and Taoukis, P.S. 2013, *Innovative Food Sci. Emerging Technol.*, 17, 144–152.
[102] Chandrapala J., Oliver, C., Kentish, S., and Ashokkumar, M. 2012, *Ultrasonics Sonochem.*, 19, 975–983.
[103] Nguyen N.H.A. and Anema S.G. 2010, *Innovative Food Sci Emerging Technol.*, 11, 616–622.
[104] Huppertz, T., Kelly, A.L., and Fox, P.F. 2002, *Int. Dairy J.*, 12, 561–572.
[105] Huppertz, T. and de Kruif, C.G. 2006, *J. Agricult. Food Chem.*, 54, 5903–5909.
[106] Jankowska, A., Grzekiewicz, A., Winiewska, K., and Reps, A. 2012, *High Pressure Res.*, 32, 339–346.
[107] Ferragut, V., Martinez, V., Trujillo, A., and Guamis, B. 2000, *Milchwissenschaft/Milk Sci.Int.*, 55, 267–269.
[108] Harte, F., Amonte, M., Luedecke, L., Swanson, B., and Barbosa-Cánovas, G.V. 2002, *J. Food Sci.*, 67, 2245–2250.
[109] Needs, E.C., Capellas, M., Bland, A.P., Manoj, P., MacDougal, D., and Paul, G. 2000, *J. Dairy Res.*, 67, 329–348.
[110] Needs, E.C., Stenning, R.A., Gill, A.L., Ferragut, V., and Rich, G.T. 2000, *J. Dairy Res.*, 67, 31–42.
[111] Demirdoven, A. and Baysal, T. 2009, *Food Rev. Int.*, 25, 1−11.
[112] Wu, H., Hulbert, G.J., and Mount, J.R. 2001, *Innovative Food Sci. Emerging Technol.*, 1, 211−218.
[113] Vercet, A., Burgos, J., and Lopez-Buesa, P. 2002, *J. Dairy Res.*, 69, 243–254.
[114] Riener, J., Noci, F., Cronin, D.A., Morgan, D.J., and Lyng, J.G. 2009, *Food Chem.*, 114, 905–911.
[115] Riener, J., Noci, F., Cronin, D.A., Morgan, D.J., and Lyng, J.G. 2010, *Food Chem.*, 119, 1108−1113.
[116] Ashokkumar, M., Kentish, S., Lee, J., Zisu, B., Palmer, M., and Augustin, M.A. 2009, Processing of dairy ingredients by ultra-sonication, PCT Int. Appl., WO2009/079691 A1.

[117] Ashokkumar, M., Bhaskarcharya, R., Kentish, S., Lee, J., Palmer, M., and Zisu, B. 2009, *J. Dairy Sci.*, 92, 5353–5356.

In: Rheology
Editor: Evgeny Karpushkin
ISBN: 978-1-63482-223-7

Chapter 3

KINETICS OF MILK GELATION. PART I. COAGULATION MECHANISM

I. T. Smykov[*]
Department of Physical Chemistry,
All-Russian Scientific Research Institute for Butter- and Cheese-making,
Uglich, Russia

ABSTRACT

In this contribution, the phenomenon of the three-dimensional field of stochastic microflow origin discovered at the stage of coagulation process of milk gelation is discussed. It has been shown that formation of the microflow field results from the Ludwig–Soret and the Marangoni effects at the surface of casein micelle clusters. The appeared microflows cause displacement of the clusters and enhance the probability of interactions involving the clusters and the micelles; the interactions in turn contribute to new gradients properties of the clusters and the microflows intensification. This recursive interaction acts as a feedback accelerating the gelation process.

Keywords: Milk; Gelation mechanism; Microconvection; Microflows; Kinetics; Feedback

INTRODUCTION

Milk gels are by-products of cheese making and the final state of other milk products. Their formation involves a complex system of biochemical reactions and physical interactions between milk components. Therefore, efficient control of milk processing at enterprises and laboratory research should be based on thorough understanding of the phenomena occurring during milk coagulation.

[*] e-mail: i_smykov@mail.ru

Physical and biochemical aspects of milk gelation have been addressed in a number of researches. Modern investigations in the field are concentrated on physical and mathematical models of gelation, and the following concepts have been the most widely accepted: the adhesive hard spheres model, the percolation model, and the fractal model.

The adhesive hard spheres model has been first introduced in the papers dealing with aggregation of destabilized casein micelles in milk [1, 2]. Extensive review of the model development has been given in [3]. The core of the adhesive spheres model is that casein micelles, stabilized by the κ-casein glycomacropeptide corona in the normal state, are destabilized by enzymatic action or other factors enhancing the interaction between the micelles and inducing their primary aggregation [4, 5]. This model has been used to quantify properties of the dispersed system in a variety of states: starting from the aggregates formation up to the gel point [2]. It has been concluded that in case of enzymatic milk gelation the observed viscosity changes can be modeled up to formation of the primary gel structure, whereas further processes taking place in the gelling systems cannot be described with the adhesive hard spheres model.

The analogy between milk gelation and percolation process has been proposed in [6, 7]. According to the percolation theory of casein micelles gelation, the micelles are initially uniformly distributed over the sites of three-dimensional cubic lattice. All the lattice sites are occupied, and therefore a micelle can only interact with its closest neighbors. These interactions occur randomly, leading to sequential formation of dimers, trimers, tetramers, and higher oligomers, thus filling the entire volume with individual micelles clusters of different but finite size. At some point, the clusters growth comes to a threshold characterized by spanning of the whole lattice volume with a single continuous cluster. This threshold is defined as gel point of the system. The percolation theory states that beyond the gel point the number of cross-links in the system is proportional to time [6, 8]. The percolation model is advantageous for computer modeling; however, it has been shown that it is applicable only around the gel point and unable to predict the viscoelastic properties of the formed gel [9].

Several fractal models of milk gel formation [10–13] are based on the concepts of fractal geometry [14]; they aim to give a mathematical description of the gel structure basing on the knowledge of the composition and properties of the constituting particles. In the frame of the fractal gelation models, it is assumed that the particles are moving randomly, collide with each other and prefer to unite into a loose cluster upon collision. These models utilize the concept of the clusters fractal dimension to describe the pattern of the incomplete filling of the considered volume with the bound particles, thus allowing describing of the particles spatial distribution in the cluster with a single number. Further, the clusters are combined into a single three-dimensional structure. The cluster fractal dimension is always less than the geometric one; the fractal dimension can be computer-simulated as well as determined experimentally to explicitly verify the derived models.

The fractal models have been used to relate the gel properties (including the rheological ones) with the gel structure [15, 16]. The general unexplained drawback of the fractal models of gelation is that the clusters fractal decreases as the aggregates grow, the only possibility for this being increase of the spacing between the clusters; however, this has never been experimentally observed. Further open questions related to the fractal models are ambiguity of the gel time (the point at which the fractal clusters are united into a single structure) determination and vagueness of the mechanism of association of large fractal clusters (they are considered immobile in the model, and the free micelles capable of the clusters connection are absent well before the gel point). Hence, complete understanding of the gelation kinetics phenomena seems impossible in the frame of the fractal models.

This work aimed to identify the mechanism of formation of a unite gel structure from the casein micellar clusters within the fractal model of the clusters growth.

1. Material and Methods

1.1. Materials

Whole cow's milk containing 2.6 wt% of casein was incubated at +4°C during 24 h prior to the experiment. Milk rennet coagulation was performed under action of the enzyme preparation with the activity of 100000 units (according to Technical Specification 10.02.851-90, Russia; from local supplier); the enzyme concentration was of 12.5 μg/g, and the initial temperature was of 34°C. Milk acid coagulation was performed using the starter culture of the *Lactobacillus bulgaricus* and *Streptococcus thermophilus* microbial strains (1:1 w/w; Experimental Biofactory, Uglich), the initial temperature being of 40°C.

1.2. Electron Microscopy

Electron microscopy analysis of the disperse phase of the milk samples was carried out using the EM-410 transmission electron microscope (Philips, the Netherlands) using direct detection after heavy metals contrasting.

1.3. Optical Microscopy

Optical microscopy studies (including video recording) of milk gelation kinetics were performed in a thin-layer quartz cell with the optical path length of 5±1 μm. The cell walls were made of 270 μm quartz plates; details of the method are described elsewhere [17].

1.4. Mathematical Analysis

The experimental results were processed taking advantage of the algorithms implemented in Excel (Microsoft), Origin (Origin Lab), and Mathcad 14.0 (Parametric Technology) software. Image analysis was performed using the Image-Pro (Media Cybernetics) software.

1.5. Macroscopic Gelation Monitoring

Milk gelation process at the macroscopic scale was studied taking advantage of the home-made laboratory installation described in [18]. The device was made of two or several thermally insulated Dewar vessels (0.5 L each) equipped with standard temperature sensor

(Pt; accuracy of 0.1°C), standard pH sensor, and the specially designed sensor of viscoelastic gel properties. The latter sensor, based on the Hori method implementation [19], was calibrated using the R-19 rheogoniometer (The Weissenberg Rheogoniometer, England). The installation allowed simultaneous monitoring of the above-mentioned properties in two or more identical measuring cells. The installation was linked to PC station for real-time tracking of the results.

2. Results and Discussion

2.1. Qualitative and Quantitative Analysis of Gelation Kinetics Video Films

This part of the discussion is essentially based on the results commented in [17]. In that work, the gelation was tracked by analyzing sequences of video frames (200 frames during the flocculation phase and 200 frames during the coagulation phase).

The movement of small milk fatty globules and casein micelles at the initial stage of gelation can be described by Brownian dynamics, its characteristic parameter being the average squared displacement over a certain time interval [20]. The average displacement as well as the movement trajectory can be elucidated from the frame-by-frame analysis. Furthermore, the root-mean-squared displacement of fatty globules following the Brownian motion pattern can be calculated using the Einstein equation [20]. Figure 1 displays a combination of the theoretical estimations and experimental evaluations of the mean-squared displacement of fatty globules during the flocculation and early coagulation phases of milk gelation.

The experimentally derived values for the flocculation phase of gelation was highly correlated (r = 0.907, P = 0.95) with the theoretical calculations based on the Einstein equation. Hence, during that stage the fatty globules movement followed the Brownian dynamics rules. That expected result served as validation of accuracy of the proposed approach of fatty globules movement analysis.

The fatty globules motion during the coagulation phase was investigated similarly. The correlation coefficient between the experimentally determined and the theoretically calculated values of the fatty globules mean-squared displacement in the case of early coagulation stage (r = 0.431) was very low; the experimental mean-squared displacement was poorly correlated with the globules radius as well (r = −0.503). Those results contradicted the basics of Brownian dynamics. Hence, our results indicated the previously undiscovered phenomenon of certain processes causing active movement of large objects such as fatty globules during the early coagulation phase of milk gelation.

Further, the analysis of fatty globules movement path [17] concluded that internal microflows with the energy much higher than that of thermodynamic fluctuations arose spontaneously in the sample at the late flocculation and early coagulation phases of milk gelation. Those stochastic microflows appear almost simultaneously over the whole sample volume to induce noticeable active motion of the dispersed particles including the clusters of loosely bound casein micelles.

The above-mentioned stochastic microflows accompanying gelation have not been discussed in the reference literature so far. The effects of these microflows on the gelation process and properties of the products have been unclear. Moreover, the source of the microflows has been unknown as well. It should be emphasized that here we do not mean a well-known gelation in laminar or turbulent flows caused by external force; instead, we discuss stochastic microflows in the dispersed system initiated by internal processes.

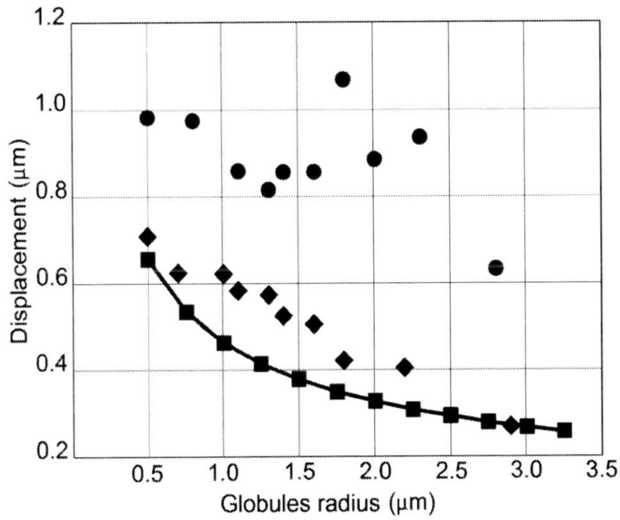

Figure 1. Mean-squared displacement of small fatty globules in milk as function of their size: (*squares*) theoretical calculation based on the Brownian dynamics, (*diamonds*) experimental evaluation during the flocculation phase, and (*circles*) experimental evaluation during the coagulation phase.

As microflows occur over the entire volume of the system simultaneously and are maintained during substantial time period, their existence requires fairly large energy input. These energy sources should be therefore of microscopic size and evenly distributed in the system. The nature of the energy source maintaining the microflows is a key point for understanding of the processes occurring during the gelation and the mechanism of the system structuring.

2.2. Gelation Thermodynamics Research

The observations concerning thermal effects accompanying milk gelation and the related system rheological properties were discussed in detail in [18]. In particular, remarkable changes in the system acidity and temperature were revealed during the late flocculation stage and the early coagulation stage (that is, near the gel point) of the rennet milk gel formation (Figure 2).

Similar results of the system parameters evolution during the acidic coagulation are shown in Figure 3.

From the plots in Figures 2 and 3 it is to be seen that the temperature changes (curves *3*) were significant in the both experiments to reach 1.5–2°C, but the nature of those changes was remarkably different, as reflected in the curves shape. In particular, during the enzymatic coagulation (Figure 2) the onsets of noticeable changes of temperature and acidity coincided with the appearance and steep growth of the measurable storage modulus; hence, they likely reflected the same internal changes in the course of gelation.

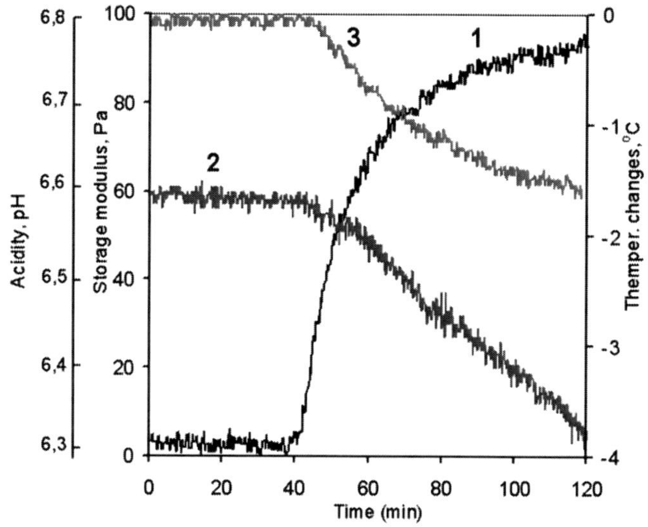

Figure 2. Changes of the storage modulus (*1*), acidity (*2*), and temperature (*3*) during enzymatic gelation of milk.

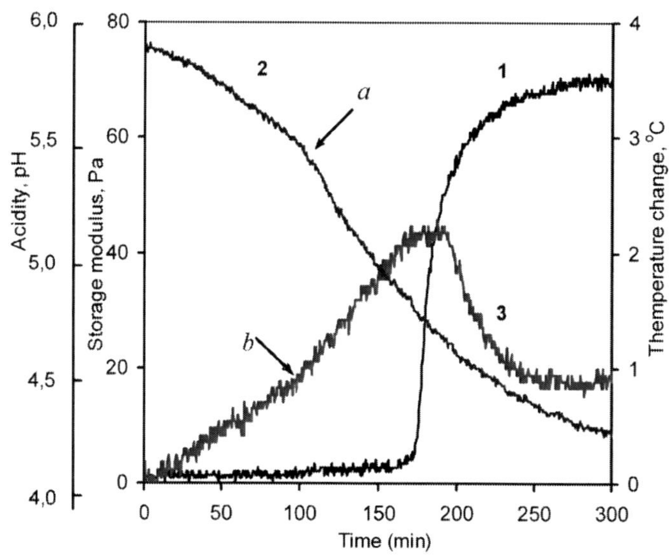

Figure 3. Changes of the storage modulus (*1*), acidity (*2*), and temperature (*3*) during acidic gelation of milk.

However, it was possible that the gel temperature and pH were inherently correlated. In order to check that hypothesis, let us examine the results presented in Figure 3. After introduction of the bacterial starter culture into the milk sample, the microorganisms started propagating, and the observed heating and acidity increase were associated with the occurring metabolic processes. Indeed, curves *2* and *3* in Figure 3 revealed well defined inflections (*a*) and (*b*), coinciding in time. The origin of the inflection was the presence of the two strains in the starter; the initial changes up to the inflection point were mainly due to the *Streptococcus thermophilus* activity, whereas pronounced propagation of *Lactobacillus bulgaricus* started later. The simultaneous slight increase of the storage modulus was probably due to participation of some bacterial metabolism products in the crosslinking, but no gelation was observed at the initial stages. The gel structuring started when the system pH approached the isoelectric point of casein (pH $\approx$ 4.7), leading to steep growth of the storage modulus. After the macroscopic gelation, the system acidity showed up further monotonous increase, whereas the temperature behavior was different. Just after the gelation onset temperature continued increasing due to the ongoing bacteria metabolism, but then temperature decreased because the energy required for the gel structuring exceeded the energy released by the microorganisms.

The above-discussed experimental results suggested definite relationship between the gel structuring and the temperature changes. In particular, the gel structuring caused the temperature decrease, as the energy spent on the casein micelles exceeded the energy gained by the system in the accompanying (bio)chemical processes. The energy required for the casein micelles dispersion to gel has been discussed in detail elsewhere [18]. It has been shown that the most energy-deficient process in the course of gelation are conformational change of casein micelles and formation of new inter- and intramicellar chemical bonds. As seen in Figs. 2 and 3, those processes occurred during the late flocculation and the early coagulation phases (as indicated by significant decrease in temperature). At that point most of the casein micelles were incorporated into the loose fractal clusters.

2.3. Electron Microscopy Studies of Flocculation Phase of Milk Gelation

As demonstrated above, the processes of rennet and acid gelation of milk are accompanied by simultaneous and rather abrupt changes of temperature and other physical properties of the dispersion. In terms of thermodynamics, this possibly means that casein molecules undergo the first-order “all-or-nothing” transition during the late flocculation and the early coagulation phases of gelation.

In our opinion, the phase transition during the late flocculation phase (formation of the micelles from casein molecules) can be treated as the "globule → molten globule" cooperative conformational transition. The features of such transition (including the related kinetic patterns) have been extensively discussed elsewhere [21–28].

Here we address the phenomenon of conformational transition basing on a thorough analysis of the electron microscopy images of micelles clusters taken in the middle to late flocculation phase (see examples in Figure 4). The conformational changes occurring during the gelation were reflected in seemingly distorted, destroyed and/or interconnected micelles.

The images analysis revealed that the phase transition was indeed cooperative, involving most of the casein micelles and the clusters simultaneously. However, cooperativity was not that pronounced in the case of some clusters; therefore, the phase transition of the overall system (from the gel point up to the syneresis onset) occurred within certain finite time, the latter being defined by external conditions as well as by the micelles properties.

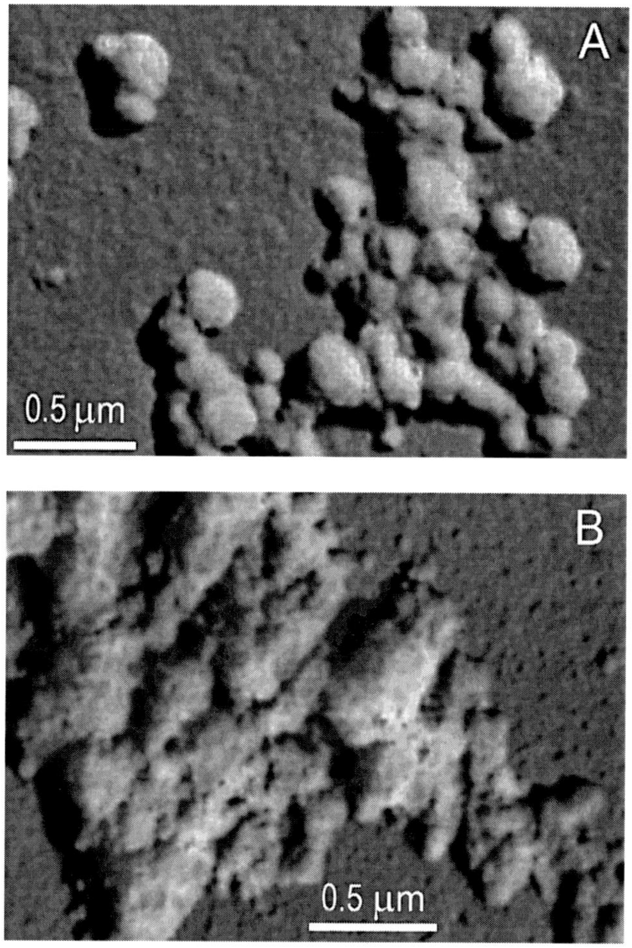

Figure 4. Clusters of casein micelles during (*a*) the middle flocculation phase and (*b*) the late flocculation phase.

2.4. Discussion

The most energy-consuming processes accompanying milk gelation are conformational changes of casein micelles and formation of intra- and intermicellar chemical bonds during the coagulation phase, when micelles clusters are already formed [18]. Hence, the micelles clusters should be regarded as the basic structural elements that cause and limit heat and mass transfer in the system. The rate of conformational changes and the chemical reactions, and, consequently, the observed thermal effects depend on the reactants concentration and temperature. Apart from the chemical processes, the temperature field is determined by heat and mass transfer (i.e. thermal conductivity and diffusion properties of the system). Therefore, the conformational and chemical changes occurring in the system upon gelation are in turn related to macroscopic thermal and diffusion characteristics of the system.

Clusters of micelles in milk serum form a suspension; the clusters interact with the surrounding medium at the interphase boundary and hence should be considered an open system [29]. The processes occurring at the cluster–medium interface are responsible for the dynamics of self-organization in the course of gelation. Cooperative phase transition of the micelles cluster causes the changes of the clusters temperature with respect to that of the environment, which, in turn, gives rise to various thermal heat and mass transfer processes [30, 31]. The surface of an individual cluster as well as the clusters group cannot be considered isothermal.

Overall, in our point of view, milk coagulation is a multi-factor transient ergodic process accompanied by appearance, development, and disappearance of a three-dimensional stochastic temperature field.

2.4.1. Evaluation of Heat and Mass Transfer of Gelation Based on Classical Models

In order to carry out numerical evaluation of heat and mass transfer processes in the system we will define the boundaries as well as initial and boundary conditions with the coefficients taken from [32].

Numerical analysis of thermal processes occurring in the studied system was performed in the frame of n the classical model of Oberbeck–Boussinesq [33], resulting from simplification of the complete Navier-Stokes equations. Our estimates (omitted here for clarity) revealed that the natural convection flow could not occur in the system according the classical Oberbeck–Boussinesq model; that contradicted the obtained experimental results.

That preliminary conclusion called for deeper investigation of heat and mass transfer phenomena occurring at the microscopic level in terms of internal sources of heat using other models.

2.4.2. Microconvection Model for Evaluation of Heat and Mass Transfer

Apparently, a problem similar to the above-discussed was first addressed by researchers from Russia, France, and the USA in the study of convective flows under microgravity conditions at space satellites and stations. It was shown that classical approach of the Oberbeck–Boussinesq approximation was not applicable, and further a number of models to adequately describe the experimental results were developed using a new "microconvection" concept, originally reported in [34].

It has been shown [35, 36] that the Oberbeck–Boussinesq approximation is not suitable to describe thermal gravitational convection if the η parameter is below unity; therefore, a new theoretical microconvection model has been suggested. The parameters of convective flows, especially their velocity, as estimated with the classical Oberbeck–Boussinesq model and the microconvection approach can differ by more than three orders of magnitude [35–38]. On top of that, structure of the occurring flow, its topology and time evolution are different as well. Analytical and numerical study of the effect of small changes in the medium density on the natural convection in a non-isothermal binary mixture at very low Rayleigh numbers and on the microconvection at constant average thermodynamic pressure has been presented in [39].

New methods of numerical analysis of microconvection with various boundary conditions have been proposed in [40, 41]; in particular, thorough comparative analytical and numerical analysis of convective flows in scope of the Oberbeck–Boussinesq approach and the microconvection theory has been performed. Again, these studies have revealed that the Oberbeck–Boussinesq approach is not applicable to analysis of fluid motion at the micrometer scale.

Microconvection model has been modified and used in a variety of other applications, for example, to describe special behavior of nanofluids ("nanofluid" generally states for heterogeneous dispersion with nanoparticles acting as the dispersed). From this point of view milk containing dispersed casein micelles about 100 nm in size can be considered as nanofluid; therefore, the above-described phenomena of heat and mass transfer can be discussed in terms of the microconvection theory.

Authors of the most recognized related studies [42–44] have proposed original models of microconvection in nanofluids. In particular, they have suggested that nano- and microconvection in nanofluids occurs due to the enhanced Brownian motion of nanoparticles, dragging along with the medium. However, other researchers [45, 46] have considered a somewhat opposite mechanism assuming that, nanoconvective fluid flow occurs in the vicinity of the nanoparticles surface due to interphase interactions (e.g., concentrating of impurities in subsurface area), these flow significantly enhancing Brownian motion of nanoparticles. Other semi-empirical microconvection models in liquid polydisperse systems [47] has combined different mechanisms of the flow appearance.

To conclude, recent studies basing on numerous experiments have confirmed existence of microconvective flows in nanofluids, caused by temperature gradient. Microconvective flows act at the range of several micrometers, their velocity being of several micrometers per second. Our estimates coincided with the specified range, and therefore we concluded that microflow occurred in the course of milk gelation due to internal temperature gradient, even though the classical Oberbeck–Boussinesq model did not confirm that. The motive forces and mechanism of the microconvection as well as the related phenomena have remained unclear.

2.4.3. Thermophoresis in the Course of Heat and Mass Transfer during Sol-Gel Transition

Casein micelles, their small clusters, and other components of serum can be affected by the forces emerging from temperature-induced changes of surface energy of the casein micelles in the course of sol-gel transition. The surface energy increases with decreasing temperature, and smaller particles (molecules and ions) move along the decreasing temperature gradient to a cluster surface, i.e. thermophoresis of the micelles [48–51] takes place. The temperature gradient gives rise to uncompensated force acting of the micelles in the same direction. Moreover, other forces may appear during the gel structuring phase, resulting from changes of diffusion coefficient due to the vertical (gravitational) temperature gradient. In real-life systems, the said effects act simultaneously, but their contribution to the overall sol-gel transition process may vary.

Microconvection caused by the concentration gradient of a polymer in the liquid phase due to the lapse rate of temperature has been reported in [52]. The authors have observed islet self-organization of the polymer and suggested that the structuring occurs due to microconvective flows. In the studies of proteins (in particular, lysozyme) thermophoresis [53, 54] it has been pointed that behavior of a protein globules depends on many factors. At least two effects can contribute to behavior of lysozyme preparations: a positive Ludwig–Soret effect (movement of the particles from warm areas to cold ones) and the opposite negative effect. The authors have suggested that the protein particles behavior under temperature gradient is mainly affected by hydrophobic interactions, whereas role of electrostatic interactions is insignificant.

It has been proposed to use thermophoresis as an efficient method to manipulate various nanoparticles during analysis of the forces acting in a colloidal system [55]. Under certain conditions, thermophoresis can be a principal driving force initiating particles movement [56]. It can cause stable microconvective flows and initiate formation of the structures of the Rayleigh–Benard cells in nanofluids. Influence of thermophoresis on the interaction of molecular particles in a colloidal system has been touched upon in [57]; it has been shown that macromolecules can become thermophilic under certain conditions i.e. they obey the negative Soret effect. Possibilities of practical application of the Soret effect for concentration of DNA molecules have been demonstrated in [58, 59]. The experiments have revealed that plasmid DNA can be efficiently concentrated within a volume of about 20 μm in diameter under infrared laser irradiation causing the 1 °C temperature gradient. Furthermore, temperature gradient resulting from local heating with low-power infrared laser irradiation can cause various types of thermophoresis, including toroidal microconvection around the heating zone [59]; hence, DNA concentration takes place due to additive action of thermophoresis and microconvection.

Similarly, concentration of DNA at a substrate within 20 μm radius during the local heating with a focused low-power infrared laser irradiation (local temperature gradient of 1.5°C) has been demonstrated [60]. DNA concentration within even smaller area of 2 μm under the influence of local temperature gradient has been reported in [61]; role of both thermophoresis and microconvection has been found important. The concentration efficiency has been found proportional to molecular mass of the colloid particles. The methods and theoretical models associated with study of heat and mass transfer in nanofluids has been discussed in detail in [62,63]; it has been concluded that phenomena of heat and mass transfer in nanofluids cannot be fully described at the present state of the research in this field, and significant improvement of the theoretical basement is required for better understanding of the phenomena.

To conclude, our analysis has revealed that micrometer-scale thermophoresis of nanoparticles about 100 nm in size can occur under conditions of about 1°C temperature gradient. The reported length and temperature scale of thermophoresis length scale coincides well with typical parameters of the studied casein gelation process. Generally speaking, thermophoretic interaction between the micelle clusters and the individual casein micelles is very likely at the initial stages of coagulation and may be determinative in the formation of the gel spatial structure.

2.4.4. The Marangoni Effect in the Heat and Mass Transfer during the Gelation Process

The Marangoni effect [64, 65] is also important in view of the possibility of initiating of stochastic microflows in disperse systems. The Marangoni effect gives rise to convective motion at the interphase boundary directed to the areas with higher surface tension. There major phenomena originating from the Marangoni effect are as follows:

- if the surface tension gradient is due to the temperature gradient, thermocapillary effect takes place [66, 67];
- if the surface tension gradient is due to the difference in concentration of certain chemicals, the occurring phenomenon is called concentration-capillary effect [68];

- if a chemical reactions occurring at the interface boundary is accompanied by changes in surface tension, the chemocapillary effect is observed [69].

All of the mentioned effects give rise to the tangential capillary forces near the interface boundary leading in turn to appearance and spreading of the Marangoni instability circular cells. When the volumetric forces are relatively small, the surface forces caused by the Marangoni effect majorly contribute to the occurring heat and mass transfer, and influence the shape of the interface boundaries causing convective movement in the whole volume of the dispersion. As surface of the micelles clusters is cooled unevenly in the course of the phase transition, this can significantly affect distribution of the velocity gradient and the components concentration in the subsurface layers of the dispersed material. We further present a detailed analysis of the processes causing heat and mass transfer at the microscopic interface boundary between the micelle clusters and serum as applied to the studied case of milk gelation.

Manifestation of the Marangoni effect in the subsurface liquid layer at the interface boundary can be triggered by various processes. At the surface of the micelles cluster, the effect results from the "molten globule → random coil" cooperative phase transition of casein molecules. The prominent conformational changes of protein molecules are accompanied by the initial reduction of the micelle clusters temperature. As a result, the local surface tension coefficient at the interface boundary changes, and tangential fluid motion directed to the cooled areas emerges, i.e. thermo-capillary (Marangoni) convection occurs at the surface of the micelles cluster.

At the same time protein molecules interact with various components dissolved in the surrounding dispersion medium (primarily with calcium ions), changing their local concentration in the subsurface layer. This results in appearance of the concentration gradients near the surface of the micelles cluster, and gives rise to the concentration-capillary Marangoni convection.

Transition of the protein molecules forming the casein micelles into the random coil conformation facilitates access of water and active components of serum to initially hindered inner parts of the protein globules. Furthermore, intermolecular chemical interactions (formation of disulfide bonds and calcium bridges) become possible. The mentioned chemical interactions are accompanied by the appearance of local temperature and concentration gradients, resulting in the chemocapillary Marangoni convection.

Interface adsorption of molecules with low surface tension is energetically advantageous, decreasing the system free energy. Many macromolecules are composed of hydrophilic (polar) and hydrophobic (non-polar) regions. Such molecules can be adsorbed via interaction of their hydrophobic regions, whereas their hydrophilic are oriented towards water. Rapid adsorption of even a monolayer of such molecules at the active surface sites of the micelles cluster can completely suppress thermocapillary and concentration-capillary convection. Therefore, the crucial condition for Marangoni convection to occur is the absence of macromolecules capable of suppressing the temperature and concentration gradients at the surface of the micelles clusters.

We will further estimate relative significance of the said phenomena [70] using the required parameters as reported in [32].

Intensity of the thermocapillary convection is characterized by the dimensionless Marangoni number:

$$Ma_T = \frac{\xi^2 \cdot \Delta T}{\mu \cdot \alpha} \cdot \frac{d\sigma}{dT} \tag{1}$$

with $\frac{d\sigma}{dT} = \sigma'_T$ being the temperature derivative of the surface tension coefficient.

Intensity of thermogravitational convection is reflected by the dimensionless Rayleigh number:

$$Ra = \frac{g \cdot \rho \cdot \beta \cdot \xi^3}{\mu \cdot \alpha} \cdot \Delta T. \tag{2}$$

These dimensionless numbers characterize the relative significance of thermocapillary or thermogravitational forces with respect to viscous friction forces. The thermocapillary mechanism is predominant over the thermogravitational one when the dynamic Bond number is below unity:

$$Bo_T = \frac{Ra}{Ma_T} = \frac{g \cdot \rho \cdot \beta \cdot \xi^3}{\sigma'_T} \tag{3}$$

This condition limits the characteristic size ξ (correlation radius) of the movement.

Our calculations revealed that the Bond number was well below unity:

$$Bo_T \approx 0.01 < 1, \tag{4}$$

i. e. in the studied case the thermocapillary mechanism of microflow formation dominated over the thermogravitational one.

Mechanism of the concentration-capillary microflows formation is quite similar to that of thermocapillary microflows; however, concentration-capillary microflows are of special interest due to diversity of the operative factors. Moreover, characteristic time of diffusion of dispersed particles, molecules, and ions are longer than the heat diffusion times of heat by 2–4 orders of magnitude. As a result, concentration inhomogeneities in the dispersion medium exist much longer than the heat ones; hence, duration and intensity of the capillary forces at the interface boundary is significantly enhanced.

Intensity of the concentration-capillary convection is determined by the diffusion Marangoni number:

$$Ma_C = \frac{\xi^2 \cdot \Delta C}{\mu \cdot D} \cdot \frac{d\sigma}{dC} \tag{5}$$

with D, diffusion coefficient of Ca^{2+} ions; C, the ions concentration; and $\frac{d\sigma}{dC}$, concentration derivative of the surface tension coefficient.

Diffusion Marangoni and Prandtl numbers ($\Pr_C = \mu / \rho \cdot D$) are usually much larger than the thermal ones. In this regard, concentration-capillary convection should be highly important. The dynamic Bond number for concentration-capillary convection is defined as:

$$Bo_C = \frac{Ra}{Ma_C}. \quad (6)$$

If the Bond number is less than unity, the concentration-capillary mechanism dominates over the thermogravitational one. Our estimates demonstrated that with

$$Bo_C \approx 10^{-5} << 1, \quad (7)$$

the concentration-capillary mechanism was more important than the thermogravitational mechanism in formation if the microflows. In addition, the concentration-capillary convection in the formation of stochastic microflows during milk gelation was much more significant than thermocapillary effects.

To conclude, probability of the formation of concentration-capillary flows in a serum layer near the micelles clusters is very high. After appearance in the microscopic layer, the flow can drag the free casein micelles and their clusters, promoting their interaction and subsequent aggregation into a coherent spatial structure of the gel.

2.4.5. A Generalized Description of the Mechanism of Fermentative Milk Gelation

The micelles interaction first results in formation of di-, tri-, and tetramers of weakly-coupled casein micelles, further evolving into isolated loose fractal clusters. At the late flocculation phase most of the micelles are bound in such loose clusters. At this stage, when approximately 80% of the micelles surface in the cluster is free of glycomacropeptides, the "dense globule → molten globule" cooperative phase transition starts in the casein molecules forming the clusters. This transition is accompanied by a simultaneous, rapid, and abrupt change in the properties of the protein molecules.

The experimental results [17] have showed that stochastic microflows occur almost simultaneously in the whole volume of the dispersed system immediately after the phase transition (late flocculation to early coagulation phase) in the course of gelation. The microflows lead to active movement of the dispersed particles present in the system at this stage, including weakly-coupled casein micelles clusters.

Further studies have demonstrated that the loose clusters of casein micelles (formed by the end of the flocculation phase and evenly dispersed throughout the whole volume of the system) are the energy sinks causing the microflows.

Energy sink into the clusters is caused by the micelles "molten globule → random coil" cooperative phase transition that is in turn due to the external conditions change. The phase transition results in the ingress of water molecules and dissolved ions contributing to formation of new intra- and intermicellar covalent bonds. Endothermic character of the phase transition causes the cluster temperature to be sharply reduced by 1–2°C.

The thermal processes in the course of the phase transition mainly occur at the interface boundary between the dispersion medium and the micelles cluster; therefore, clusters of micelles should be considered the basic elements causing and limiting heat and mass transfer during the gel formation.

Microconvectional flows act over distances of several micrometers, their velocity being of several micrometers per second; they can be modeled in the frame of the microconvection theory. Our estimates confirmed the possibility of such microconvection flows caused by internal gradients of temperature, concentration, density, viscosity, and surface tension at surface layer of casein micelles clusters.

Hence, after the initiation of the gelation process and formation of the fractal clusters of micelles (late flocculation phase) cooperative phase transitions take place in the dispersed system and cause recursive interaction:

- the cooperative phase transitions result in (micro)gradients of temperature, concentration, density, viscosity, and surface tension at the surface of casein micelles, causing thermophoresis, thermocapillary, concentration-capillary, and other types of microflows;
- the microflows cause the dispersed particles movement and strengthen their interactions (both between the clusters and within them), contributing to the appearance of the new gradients, causing in turn new microflows, etc.

This is schematically shown in Figure 5.

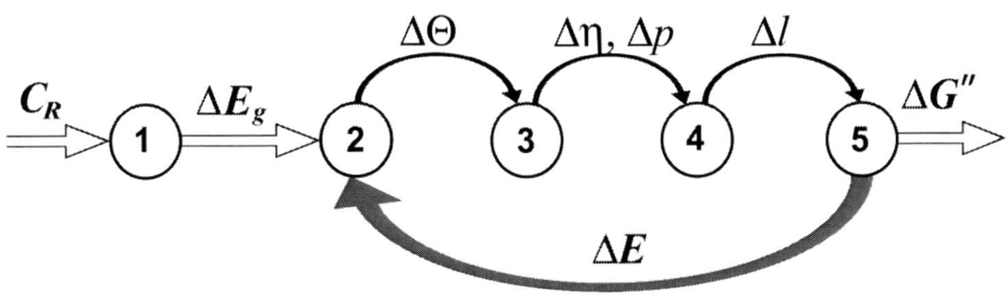

Figure 5. Scheme of the mechanism of recursive gelation: C_R, enzyme concentration; 1 – activation process; ΔE_g, increase of the free energy (flocculation phase); 2, appearance of temperature microgradients (ΔΘ) and decrease in the free energy; 3, appearance of the viscosity and density microgradients ($\Delta\eta$, Δp); 4, formation of the three-dimensional field of stochastic microflows (Δl); 5, formation of new intermicellar bonds and decrease in the free energy (ΔE); G'', increasing storage modulus.

In our opinion, the proposed mechanism accurately describes the processes occurring during the gelation allows formalizing the task of the development of a complete mathematical model of the kinetics of milk gelation.

CONCLUSION

A mechanism of milk gelation kinetics has been developed to explain the observed formation of protein particles clusters and the structure self-organization as described by the fractal model of gelation [10–13]. The recursive mechanism incorporates microconvective (microflows) mass transfer and attachment terms, and is hence applicable to other systems experiencing the sol–gel transition. The derived conditions of microflows occurrence appear reasonable in comparison with the parameters reported elsewhere [60–62, 70].

The flocculation phase of gelation ends up with formation of the micelles clusters. Further, the protein molecules constituting the clusters undergo the cooperative "molten globule → random coil" phase transition. The cooperative phase transition evokes superficial microflows, and the microflows promote three-dimensional clusters movement. The changes accompanying the clusters interaction give rise to new microflows. Hence, internal feedback is established in the disperse system during gelation.

This developed recursive mechanism is expected to aid in future analysis of the sol–gel transitions showing unusual kinetics, and add to understanding of the processes previously reported for fractal gelation kinetics [11–13]. Noteworthily, it may contribute to understanding the challenging gelation processes in medical, biological, and food systems.

Development and verification of a mathematical model of the gelation kinetics based on the above-discussed recursive interaction will be presented in the next contribution.

REFERENCES

[1] De Kruif, C.G., Jeurnink, T.J.M., Zoon, P. 1992, *Neth. Milk and Dairy J.,* 46, 123-137.

[2] De Kruif, C.G. J. 1997, *Colloid Interface Sci.,* 185, 19-25.

[3] De Kruif, C.G. 1999, *Int. Dairy J.,* 9, 183-188.

[4] Holt, C, Horne, D. 1996, *Neth. Milk Dairy J.,* 50, 85-111.

[5] De Kruif C.G., Zhulina E.B. 1996, *Colloids and Surfaces A: Phys. Eng. Aspects.,* 117, 151-159.

[6] Stauffer D. J. 1976, *Chem Society, Faraday Trans II.,* 72, 1354-1364.

[7] De Gennes, P.G. 1979, Scaling concepts in polymer physics. Ithaca, New York: *Cornell University Press.*

[8] Kantor, Y, Webman, I. 1984, *Phys. Rev. Lett.,* 52, 1891-1894.

[9] Horne, D.S. 1999, I*nt. Dairy J.,* 9, 261-268.

[10] Meakin, P. 1983, *Phys. Rev. Lett.,* 51, 1119-1122.

[11] Bremer, L.G.B., van Vliet, T., Walstra, P. 1989, *J. Chem Socie: Faraday Trans.,* 85, 3359-3372.

[12] Chaplain, V, Mills, P, Djabourov, M. 1994, *Colloid Polymer Sci.,* 272, 991-999.

[13] Gisler, T., Ball, R.C., Weitz, D.A. 1999, *Phys. Rev. Lett.,* 82, 1064-1067.

[14] Mandelbrot, B.B. 1977, Fractals: Form, Chance and Dimension. San Francisco. *WH Freeman and Co.*

[15] Shih, W.H., Shih, W.Y., Kim, S.I., Liu, J., Aksay, I.A. 1990, *Phys. Rev. A,* 42, 4772-4779.

[16] Mellema, M., Walstra, P., van Opheusdenm J.H.J., van Vliet, T. 2002, *Adv. in Colloid Interface Sci,.* 98, 25-50.

[17] Smykov, I.T. 2012, *J. Storage and Processing of Farm Products,* 6, 30-37 (in Russian).

[18] Smykov, I.T. 2008, *J. Storage and Processing of Farm Products,* 2, 26-31 (in Russian).

[19] Hori, T. J. 1985, *Food Sci.,* 50, 911-917.

[20] Einstein, A. 1905, *Ann. der Phys.,* 17, 549-560.

[21] Surkov, B.A., Klimovskii, I.I., Krayushkin, V.A. 1982, *Milchwissenschaft,* 37, 393-395.
[22] Privalov, P.L. 1989, *II Annu. Rev. Biophys. Biophys. Chem,.* 81, 47-69.
[23] Myers, C.D. 1990, *In:* Thermal Analysis of Food, Harwalkar, V.R., Ma C-Y., (Eds.), *Elsevier Science Publishers Ltd,* London,
[24] Jagannadham, M.V., Balasubramanian, D. 1985, *FEBS,* 188, 326-330.
[25] Dobson, C.M. 1994, *Curr. Biol.,* 4, 636-640.
[26] Ptitsyn, O.B. 1995, *Adv. Protein Chem.,* 47, 83-229.
[27] Farrell, H.M. Jr., Qi, P.X., Brown, E.M., Cooke, P.H., Tunick, M.H., Wickham, E.D., Unruh, J.J., 2002, *J. Dairy Sci.,* 85, 459-471.
[28] Arai, M., Kuwajima, K., 2000, *Adv. Protein Chem.,* 53, 209-282.
[29] Nicolis, G., Prigogine, I. 1977, Self-Organization in Non-Equilibrium Systems, *Wiley,* New York.
[30] Hu, Y, Larson, R. 2005, *Langmuir,* 21, 3972-3980.
[31] Hu, Y, Larson, R. 2008, *J. Phys. Chem. B.,* 110, 14-17.
[32] Roginski, H., Fuquay, J.W., and Fox, P.F. (Eds.), 2002, Encyclopedia of Dairy Sciences, *Academic Press,* London.
[33] Joserh, D.D. 1976, Stability of fluid motions, *Springer Verlag*, Berlin,
[34] Pukhnachov V.V., Voinov, O.V. 1995, Mathematical model of motion of emulsion under effect of thermocapillary forces and microacceleration. Ninth European Symposium on Gravity Dependent Phenomena in Physical Sciences. Berlin, 32-33.
[35] Pukhnachov V.V., Voinov, O.V. 1996, Thermocapillary motion in an emulsion. *In:* Proceedings of the Third Conference on Fluid Physics in Microgravity, Cleveland, Ohio, 347-352.
[36] Andreev, V.K., Kaptsov, O.V., Pukhnachov, V.V., Rodionov, A.A. 1998, Applications of Group-Theoretical Methods in Hydrodinamics. *Kluwer Academic Publ.*, Dordrecht/Boston/London
[37] Gaponenko, Y., Shevtsova, V.M. 2008, *Microgravity Sci. Techn.,* 20, 307-311.
[38] Ryabitskii, E.A. 2000, *Fluid Dyn.,* 35, 315-320.
[39] Perera, P.S, Sekerka, R.F. 1997, *Phys. Fluids,* 9, 376-391.
[40] Andreev, V.K., Bekezhanova, V.B. 2004, *Fluid Dyn.* 39, 219-229.
[41] Pukhnachov, V.V., Goncharova, O.N. 2004, *XXI ICTAM*, Warsaw.
[42] Jang, S.P., Choi, S.U.S. 2004, *Appl. Phys. Lett.,* 84, 4316-4318.
[43] Prasher, R, Bhattacharya, P., Phelan, P.E. 2005, *Phys. Rev. Lett.,* 94, 025901.
[44] Azizian, M.R., Aybar, H.Ş., Okutucu, T. 2009, *WSEAS Int. Conf.* 53-57.
[45] Hull, J.R., Katti, *Y.* 1988, *Phys. Fluids,* 31, 2368.
[46] Reda, M.R. 2009, *ASME, 7-th Int. Conf.*, 82206, 957-970.
[47] Patel, H.E., Sundrarajan, T., Pradeep, T., Dagupta, A., Dagupta, N., Das, S.K. 2005, *PRAMANA J. Phys.*, 65, 863-869.
[48] Ludwig, C. 1856, *Sitzungsber. Kaiser. Akad. Wiss. (Math.-Naturwiss. Kl.),* Wien, 20, 539.
[49] Soret, C. 1879, *Arch. Sci. Phys. Nat.,* 3, 48.
[50] Talbot, L., Cheng, R.K., Schafer, RW., Willis, D.R. 1980, *J. Fluid Mech.,* 101, 737-758.
[51] Dhont, J., Wiegand, S., Duhr, S., Braun, D. 2007, *Langmuir,* 23, 1674-1683.
[52] Nambu T, Yamauchi Y, Kushiro T Sakurai S. 2005, *Faraday Disc.,* 128, 285-298.
[53] Iacopini, S., Piazza, P. 2003, *Europhys. Lett.,* 63, 247-253.

[54] Piazza, R., Iacopini , S., Triulzi, B. 2004, *Phys. Chem. Chem. Phys.,* 6, 1616-1622.
[55] Piazza, R., Parola, A. 2008, *J. Phys: Condens Matt.,* 20, 153102.
[56] Savino, R., Paterna, D. 2008, *Phys. Fluids,* 20, 017101.
[57] Dhont, J.K.G. 2004, *J. Chem. Phys.,* 120, 1632-1641.
[58] Duhr, S., Arduini, S., Braun, D. 2004, *Eur. Phys. J.,* 15, 277-286.
[59] Braun, D., Libchaber, A. 2002, *Phys. Rev. Lett.,* 89, 188103.
[60] Shin-pin Lin. 2009, Thermophoresis of DNA in micro-channel, China: *Graduate Institute of Phys,* Ph.D. thesis; p. 116.
[61] Hong-Ren, J., Hirofumi, W., Natsuhiko, Y., Masaki, S. 2009, *arXiv:0902.4056v2,* 1-5.
[62] Das, S.K., Choi, S.U.S., Patel, H.E. 2006, *Heat Transfer. Eng.*, 27, 3-19.
[63] Wang, X-Q., Mujumdar, A.S. 2007, *Int. J. of Thermal Sci.,* 46, 1-19.
[64] Marangoni, C. 1865, Sull'espansione delle gocce di un liquido gallegianti sulla superficie di altro liquido. *Tipografia dei fratelli Fusi*, Pavia.
[65] Marangoni, C. 1871, *Ann. der Phys. und Chem.*, 143, 337.
[66] Briskman, V.A., Zuev, A.L., Lyubimova, T.P., Nepomnyashchy, A.A. 1991, *Int. J. Microgravity Sci. Techn.,* 4, 98-99.
[67] Kostarev, K.G., Zuev, A.L., Viviani, A. 2004, *J. Comptes Rendus Mecanique,* 332, 1-7.
[68] Viviani, A., Zuev, A.L. 2008, *Int. J. Energy. Conv. Manag.,* 49, 3232-3236.
[69] Rongy, L., De Wit, A. 2006, *Chem. Phys.,* 124, 164705.
[70] Birikh, R.V, Briskman, V.A, Velarde, M.G., Legros, J-C. 2003, Liquid interfacial systems: oscillations and instability, *Marcel Dekker Inc.,* New York – Basel, p. 367.

In: Rheology
Editor: Evgeny Karpushkin

ISBN: 978-1-63482-223-7

Chapter 4

KINETICS OF MILK GELATION. PART II. MATHEMATICAL MODELING

I. T. Smykov[*]
Department of Physical Chemistry,
All-Russian Scientific Research Institute for Butter- and Cheese-making,
Uglich, Russia

ABSTRACT

This chapter presents the mathematical model of the recursive mechanism of gelation kinetics. Gelation kinetics modeling has been carried out by means of direct and inverse Laplace transformations, thus ensuring consideration of recursive interactions effect in the system and allowing introducing flocculation (induction) phase into the gelation model. The model has been verified using available experimental data. An algorithm of using the gelation kinetics recursive model for prediction of milk gel properties has been introduced. This model of gelation kinetics can also be used for mathematical modeling of various polymerization processes.

Keywords: Milk; Gelation; Mathematical model; Microconvection; Kinetics; Feedback

INTRODUCTION

The first attempt to mathematically describe the process of the gel viscoelastic properties evolution in the course of gelation was introduced by Scott-Blair and Burnett [1] in 1958, and later [2] empirical regression model was refined. The regression model describes the experimental results beyond the gel point, showing an initial rapid increase of the shear modulus, process inflection (at the level of 13.5% of the maximum value) and its subsequent approaching to the equilibrium level at $t \to \infty$. However, the model is based on the use of *a priori* unknown shear modulus limiting value, and cannot predict changes of the gel

[*] e-mail: i_smykov@mail.ru

properties at varied temperature, pH, enzyme concentration, and other factors, and does not characterize the early gelation process (before the gel point). Based on the fact that the shape of the shear modulus change curve is similar to the shape of autocatalytic reactions, another regression model of gelation kinetics was proposed in Ref. [3]. Similarly to the above-discussed model, in the latter case the curve describing the gelation course has a sigmoidal shape, but its bend is at 50% of the maximum (limiting) modulus level; this is nearly impossible in practice. In addition, this model is based on *a priori* knowledge of the shear modulus limiting value, and does not account for the influence of milk properties and external factors on the occurring process as well.

A similar regression model describing gelation kinetics in the dimensionless form as a first-order reaction was proposed by Douillard [4]. Tokita et al. [5] improved the consistent models of this type. In a more recent work [6], Tokita proposed percolation (geometric) model of gelation basing on the analysis of the Douillard model. In the latter Tokita model, an increase of the shear modulus is caused by the consistent strengthening process of the spatial structure by the addition of free system elements and formation of additional bonds. As the basement of percolation model was similar to that of the Douillard model, they suffered from a common basic drawback: a sharp increase in the shear modulus immediately after the gel point, whereas the modulus growth in the experiment is a smooth process.

Further development of the Douillard model was performed in a series of papers by Carlson [7–10], the increasing shear modulus was represented as a sum of two exponents. In essence, this model states the occurrence of "active" sites at the casein micelle that subsequently interact to form the gel structure. Both reactions, activation and deactivation, are presented as first-order processes in the model. The activation occurs via the enzymatic κ-casein hydrolysis, and therefore its rate is proportional to the concentration and activity of the enzyme. The fact that the reaction of the sites deactivation at the micelle must also be of the first rate order is an empirical observation, which is connected with the necessity of small aggregates and free micelles to be incorporated into the gel structure, that remain activated at the moment. At high enzyme concentration hydrolysis of κ-casein is faster than the micelles deactivation, and the Carlson model is transformed into the Douillard model. In general, the Carlson model describes the experimental data better than the Douillard or the Scott-Blair model [11], but it needs further development in order to take into account the gelation induction phase and many more influencing factors. Basing on the dimensionless shape of the Douillard model, a mathematical model of gelation has been developed, taking into account the effect of temperature on the shear modulus [12, 13]. In this model the activation energy of casein micelles has been introduced as a parameter.

The main appeal of the above-discussed works was quantification of milk gel rheological property evolution upon cooling and defining the micelles activation energy in the gelation process. It has been shown that the results of simulation using the proposed mathematical model correlate well with the gel microstructure, and the model allows predictive evaluation of gelation process in similar cases. For example, results of experimental studies of effects of pH, casein concentration, heat treatment, and ionic strength on various properties of the milk clot, including rheology, syneresis, etc have been given in Ref. [14]. According to the experimental results, several regression models have been developed to describe the effect of these factors on the clot properties. Such attention to modeling the gelation and the effect of external and internal factors on the gel properties is due to the urgent need to obtain reliable predictive estimates for application in the dairy products industry as well as because of a real lack of adequate mathematical and physical models that satisfy the requirements of modern researchers.

We aimed to develop a mathematical model of gelation kinetics on the basis of the previously identified recursive mechanism of the gel structure formation from fractal clusters of casein micelles. This mathematical model should account for the flocculation (induction) phase of gelation and should possess predictive capabilities.

1. Mathematical Analysis

Processing of the experimental results was performed taking advantage of the following software: Excel (Microsoft, Redmond, WA, USA), Origin 8 (Origin Lab Co. Northampton, USA), and Mathcad 14.0 (Parametric Technology Co. Needham, MA, USA).

2. Derivation of the Mathematical Model of Enzymatic Gelation Kinetics

2.1. Laplace Transformation and Derivation of the Mathematical Model

To construct a full mathematical model of gelation kinetics we will consider disperse system (milk) as the object of technological control and the kinetics of gelation as a process of the system transition between its states. Such approach allows application of a well-developed mathematical apparatus of the operational calculus [15, 16].

Rennet addition, considered as an input action $x(t)$ on the system (in this case, the major one), influences the control object (milk), along with other parameters such as temperature (Θ), concentration of calcium chloride ($CaCl_2$), rennet concentration (C_R), acidity (pH), etc. All the input effects are conditionally summed up and influence the object simultaneously. In addition, other disturbing factors (Z) can affect the processes in the system.

The input action $x(t)$, initiating transition process in the system, is considered a non-trivial step effect of rennet with concentration C_{Ro} existing at the time $t_0 = 0$. Its representation according to Laplace is the following:

$$L[x(t)] = L[C_{R_O}(t)] = \frac{C_{R_O}}{s} \tag{1}$$

where s is Laplace operator.

Duration of the flocculation phase (primary or induction phase) is characterized by coagulation time (τ), and it can be considered as an element of delay or shift. Such element of the system can be described with the Heaviside function (Φ):

$$\Phi(t) = \begin{cases} 0, \ldots at \ldots t < \tau \\ 1, \ldots at \ldots t \geq \tau \end{cases} \tag{2}$$

where τ is coagulation time.

In terms of operators, the Laplace representation of the Heaviside function is called the shift operator and has the following form:

$$L[\Phi(t)] = e^{-\tau \cdot s} \tag{3}$$

Initially, during the coagulation phase a cooperative phase transition of casein micelles and further activation of the micelles clusters interaction take place. In other words, initial elements of the future structure are formed via the micelles sealing and bonding with each other. This process is characterized by a proportionality factor K_1 and the time constant T_1 determined by the speed of biochemical reactions in the casein micelles and between them. This element of the observed system is of the first order and is aperiodic; representation of its transfer function according to Laplace is the following:

$$W_1(s) = \frac{K_1}{(T_1 \cdot s + 1)} \tag{4}$$

As the micelles clusters start to integrate into a single structure, a certain limit arises: the number of reacting clusters, their elements, and micelles decreases with time due to their limited number in a given volume. Namely, the second aperiodic process occurs in the system, with its aspect ratio K_2 and its time constant T_2, which is determined by the depletion rate of reactive components. Its transfer function is the following:

$$W_2(s) = \frac{K_2}{(T_2 \cdot s + 1)} \tag{5}$$

The first and second aperiodic elements operate consequentially. Their total transfer function has the following form:

$$W_{12}(s) = \frac{K}{(T_1 \cdot s + 1) \cdot (T_2 \cdot s + 1)} \tag{6}$$

where, for simplicity, $K = K_1 \bullet K_2$.

If we draw the inverse Laplace transformation for the resulting function (6), we obtain the equation that mathematically coincides with the Carlson model [9], the differences being only in the terms and notations:

$$L^{-1}[W_{12}(s)] = \frac{K}{T_1 - T_2} \cdot e^{\frac{-t}{T_1}} - \frac{K}{T_1 - T_2} \cdot e^{\frac{-t}{T_2}} \tag{7}$$

This intermediate result confirms the validity of the chosen mathematical approach to model the gelation process.

In our previous studies we have shown that a three-dimensional stochastic microflow field emerges in the system in the course of gelation, contributing to the increase of the number of interacting components in the gel structure per unit of time. This feedback includes only coagulation phase (6) and is characterized by depth (β), which in the considered case shows how much of the energy consumed for gelation is related to initiation of the microflow field in the volume of the object. Then the elements covered by non-unity feedback have the following transfer function:

$$W^*_{12}(s) = \frac{W_{12}(s)}{1+\beta \cdot W_{12}(s)} \tag{8}$$

The model output needs normalizing instantaneous element (K_n) with the transfer function that provides consistency in size and dimension of the input effect and output parameter. This ratio is implied, but further it omitted in the equations for clarity.

To summarize, the functional diagram of the dispersed system (milk) as a control object in a formalized manner used in the operational calculus has the form shown in Figure 1.

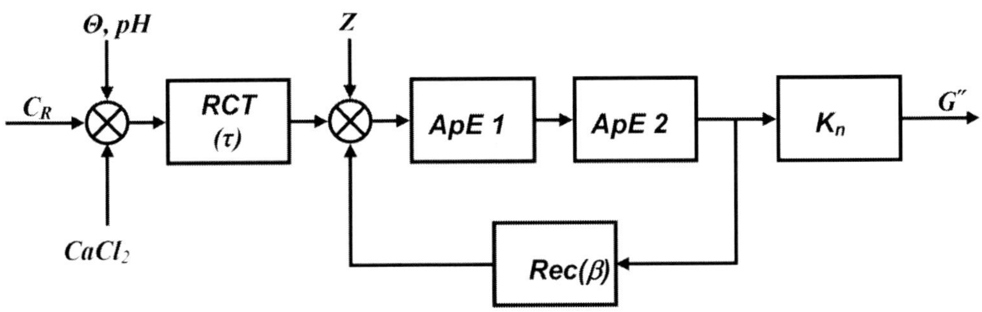

Figure 1. Functional diagram of the model of milk gelation: *RCT (τ)* - the duration of the flocculation phase (coagulation time), *ApE1* - aperiodic element characterizing activation of the gelation, *ApE2* - aperiodic element characterizing the attenuation process, *Rec (β)* - recursive relationship, K_n - normalizing factor.

The total transfer function of the milk model, as an object of technological control, is defined as the product of sequentially bonded elements spanned by feedback, and is as follows:

$$W_m(s) = e^{-\tau \cdot s} \cdot \frac{\dfrac{K}{(T_1 \cdot s+1)\cdot(T_2 \cdot s+1)}}{1+\beta \cdot \left[\dfrac{K}{(T_1 \cdot s+1)\cdot(T_2 \cdot s+1)}\right]} \tag{9}$$

According to Laplace, representation of transient process *h(s)* between the milk system states i.e. its reaction to the introduction of a number of enzyme preparation (gelation kinetics) is as follows:

$$h(s)=\left(\frac{1}{s}\cdot W_m(s)\right)$$

or

$$G(s)=e^{-\tau\cdot s}\cdot\frac{1}{s}\cdot\frac{\dfrac{K}{(T_1\cdot s+1)\cdot(T_2\cdot s+1)}}{1+\beta\cdot\left[\dfrac{K}{(T_1\cdot s+1)\cdot(T_2\cdot s+1)}\right]} \tag{10}$$

In order to transfer into time domain we perform the inverse Laplace transformation of the expression (10) using Mathcad 14.0 software. The resulting expression is extremely cumbersome to write and is not included here; after trivial algebraic manipulations and the introduction of intermediate factors *A* and *B* (note that no functional simplifications have been done) this expression takes the following form:

$$G(t)=\frac{K\cdot\Phi(t-\tau)\cdot\left[2\cdot A\cdot T_1\cdot T_2\cdot\left[1-\cosh[A\cdot(t-\tau)]\cdot e^{B(t-\tau)}\right]-(T_1-T_2)\cdot\sinh[A\cdot(t-\tau)]\cdot e^{B(t-\tau)}\right]}{2\cdot A\cdot T_1\cdot T_2\cdot(1+K\cdot\beta)} \tag{11}$$

where $A=\dfrac{\sqrt{(T_1-T_2)^2-4\cdot K\cdot T_1\cdot T_2\cdot\beta}}{2\cdot T_1\cdot T_2}$ and $B=\dfrac{-(T_1-T_2)}{2\cdot T_1\cdot T_2}$ are intermediate factors,

$\Phi(t-\tau)$ is the Heaviside function, cosh[] and sinh[] are hyperbolic cosine and sine, respectively.

The resulting recursive model of storage modulus evolution in the course of gelation is cumbersome to write, but it is rather easy to be used for the calculations with Mathcad or other similar software, so this recursive model is obtained automatically from the simple expression (10), and all further calculations are fully automated. It should be emphasized that the resulting model is not a speculative construction, but it is the result of mathematical modeling.

2.2. Mathematical Model Verification

We used the experimental results of acidity and enzyme concentration effects on the rheological properties of dairy gel described in [17, Figure 1] to verify the above-derived model. For verification, the available image of the gelation process given in this work was digitized. Figure 2 shows result of verification of the developed recursive model (11) of gelation process.

The correlation between the experimental and calculated data was excellent: $R = 0.99$ at $P=0.95$. The gelation process was characterized by coagulation time $\tau = 8.83$ min, almost identical to the value specified in [17], and a smooth transition from the coagulation phase to flocculation was revealed. Hence, the recursive gelation model not only adequately described the process of coagulation, as do the previously known models, but it also reflected the flocculation stage of the process.

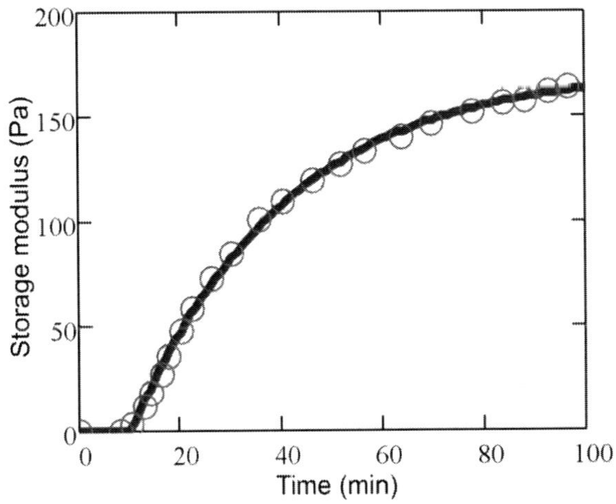

Figure 2. Verification of the milk gelation model. Rings: experimental data [17], solid line: model (11).

Similar verifications of the developed gelation kinetics model performed using a number of published and our own experimental data sets showed that the model high reliably described the results of various experiments by the assessment of external influences on the gel rheological parameters.

To conclude, the derived model is suitable for descriptive purposes. However, the recursive model (11) cannot directly determine the effect of individual input and/or disturbing effects on the gelation kinetics and rheological properties of the gel. However, this task is the key objective the development of a complete mathematical model of gelation, and the regression model (11) can serve as the basis for its solution.

2.3. Development of Complete Recursive Model of Enzymatic Gelation Kinetics

To efficiently use a mathematical model of enzymatic gelation kinetics, it should have the following minimalistic form:

$$Y(t) = F\{\mathrm{C_R}, \mathrm{C_c}, \Theta, \mathrm{pH}\} \tag{12}$$

where $Y(t)$ is an output parameter of the gel (shear modulus, viscosity, etc.), C_R is concentration of the enzyme preparation, C_c is concentration of calcium chloride, Θ is temperature, and pH is acidity.

In fact, parameters of the forming gel depend on many more of the factors, and it is difficult to consider them all at once; in practice no more than three or four factors are commonly varied. Therefore, in the initial version the most important and most frequently used factors have been included in the model. In the final result, the derived model of gelation kinetics should consider the impact of other factors, i.e., it should be scaled by factors.

There have been many reports devoted to the impact of various factors on gelation kinetics and the product rheological characteristics. The results of these studies contribute to the creation of the corresponding mathematical models and to identification of efficient ways of production process management.

Numerous experimental data presented in Refs. [18–21] have shown that the milk coagulation time is strongly dependent on the species of animals, i.e., their genetic traits, an average between-group difference of coagulation time can be up to 50%, and the group spread can be 3–4-fold. Hence, the genetic factor is one of the major factors defining milk properties as well as those of milk clot.

The influence of milk pH on the gelation time has been studied in [22]; it has revealed that this effect is non-linear, and the gelation time decreases sharply with the approach to the isoelectric point.

The consistent effect of acidity and enzyme concentration on the rheological properties of milk clot has been addressed in [17]. Two-factor experiment has revealed that the combined effect of these factors on gelation is not additive. Pair effects of acidity and enzyme concentration are nonlinear.

Milk acidity dramatically affects its ability to rennet coagulation and it also affects its resulting cluster properties. The research of this effect on the coagulation time and the storage modulus has been presented in ref. [23]. Acidity increase reduces the coagulation time, and the closer pH is to the isoelectric point, the faster is the coagulation. Similarly, the storage modulus of the gel decreases with the acidity increase. According to the experimental results, one can conclude that the storage modulus decreases with the decrease in the coagulation time at a constant concentration of the enzyme. The situation will be quite opposite, if the enzyme concentration is varied at constant pH. In the latter case, the storage modulus increases with decreasing coagulation time. Thus, the dependence of the gel storage modulus on the coagulation time is determined by the initial milk acidity.

The influence of calcium and phosphate ions concentration on milk coagulation and storage modulus of the resulting gel has been studied in Ref. [24]. It has been shown that the coagulation time will reach a pronounced minimum at the average values of these ions concentration, the storage modulus variation being less prominent.

Milk-enzymes of different origin are becoming more and more popular in manufacturing practice. The nature of the gelation process depends not only on the concentration of the enzyme insertion, but also on its origin. Comparative studies of the process of gelation in milk using natural rennet, recombinant calf chymosin, microbial coagulant, and recombinant lamb chymosin have been reported in Ref. [25]. The activity has been determined under identical conditions, and the concentration of enzyme preparation samples has been accordingly adjusted in further experiments. The coagulation time dependence on pH is close to exponential one, and the effect of calcium chloride concentration is almost linear.

In general, from the above-mentioned results in follows that qualitative effect of temperature, pH, and concentration of calcium chloride on the milk coagulation time does not depend on the type of enzyme preparation.

Dependence of the maximum value of the gel storage modulus on the acidity at different concentrations of sodium chloride in milk reconstituted from concentrate and obtained by ultrafiltration has been investigated in Ref. [26]. The results have revealed that sodium chloride strongly affects the storage modulus of the formed gel, and its influence is the most pronounced at pH $\approx$ 6.2–6.6. At the milk acidity close to the isoelectric point, the effect of adding salt is extremely suppressed. Hence, the impact of the acidity change on the gel storage modulus is significantly stronger than the influence of sodium chloride addition.

Milk temperature during the coagulation process is one of the most important and most easily applied factors, hence widely used in production environment. Consequently, a number of studies have been devoted to the effect of temperature on the coagulation process. The results have been presented and discussed in Ref. [27]. The effect of temperature effect is nonlinear as well; with heating, the coagulation time decreases exponentially, while the gel storage modulus first increases and then decreases, the dependence being almost quadratic. Therefore, the optimal temperature of gelation exists.

The effects of milk protein and fat concentrations on the coagulation time and the value of the storage modulus have been studied in Ref. [28]. In particular, the increase of milk fat concentration almost linearly decreases the coagulation time and increases the storage modulus value. This result confirms the importance of milk fat globules as the natural centers of the gel structural elements formation.

In order to determine the main parameters determining the features of gelation process (gelation time, clot cut, storage modulus, viscosity, clot density rate, etc.), multiple devices are used, based on a variety of physical principles. About twenty devices produced by different companies and used in the laboratory practice as well as in manufacturing have been reviewed and discussed in Ref. [29]. Each of the devices has own advantages and drawbacks. For laboratory studies devices based on the use of torsional vibrations at low frequencies and at low amplitudes, spectroscopic devices, and optical devices operating in the near-infrared region have found the widest application. In industry, devices operating on hot wire method and reflector optic devices are most popular. Construction and operation of modern fiber optic device to monitor the coagulation of milk have been discussed in detail in Ref. [30]. Advantages and disadvantages of the method have been analyzed, some applications in the dairy industry have been given, and the results of experimental studies have been introduced.

All the mentioned devices and equipment have certain methodological and instrumental measurement error of gelation parameters. Therefore, the use of several different devices helps to clarify and supplement with the characteristic parameters of gelation.

However, usually a researcher has access to the only or to a pair of devices, and therefore the results reported by different research groups are not always coinciding, even being performed under identical process conditions. This seriously complicates the task of mathematical modeling of gelation process and requires a thorough comparative analysis of the results given in literature.

Ref. [31] is devoted to comparative studies of measurement of milk coagulation time, of clot time cut, and other parameters obtained using six devices based on different principles. From the discussion it follows that the measurement results coming from the devices differ by up to 50%, being sometimes even qualitatively different. Whereas the quantitative differences can be compensated by the elimination of systematic errors, the qualitative differences are almost unavoidable. Further, the discussion illustrates the complexity of mathematical modeling of gelation kinetics; on the other hand, by imposing certain contingencies, these results suggest possible solutions to complex problems.

Concentration of casein in milk also strongly influences on the formation process and properties of rennet gel. The model experiments with casein concentrates [32] have revealed that the increase of casein concentration in reconstructed milk generally decreases its coagulation time; however, the storage modulus of model samples has shown a different behavior as compared with conventional low-fat milk. Despite the decreased coagulation time, the storage modulus goes down with increasing casein content. The behavior of reconstituted milk has been explained by the high content of insoluble calcium compounds in casein micelles of the used concentrates.

Results of modeling the relationship of coagulation time and the density of the milk clot have been shown in Ref. [33]. The impact of some important factors such as acidity has been, however, ignored; as shown above it can dramatically changes the observed relation.

The influence of several external factors on the milk coagulation time cannot be represented as a simple algebraic sum of individual factors effects. Instead, binary, triple, and multiple interactions exist between the factors and the gelation kinetics. Many studies have been dedicated to this aspect. In particular, the combined influence of pH, calcium chloride concentration, temperature, and the enzyme concentration has been analyzed in Ref. [34]. The dependences of the coagulation time on the listed individual factors have been elucidated, and multiple regression equation relating the coagulation time with all the considered external factors has been derived. The multi-factor study has showed that acidity has the most prominent effect on the coagulation time, while temperature makes the highest contribution to the prediction of the clot density parameters. Except for the clot density, the coagulation parameters are also determined by the interaction of acidity, temperature, and calcium concentration. Coagulation time depends on bi- and three-factor interactions. The interaction of acidity – calcium concentration is the only combination of factors that affects the clot density. Linear regression model obtained in this work is of interest; however, the definitely important nonlinearity of the effects has been ignored in the model, raising questions concerning its broad applicability.

Comprehensive study of the influence of various factors on the parameters that somehow characterize the coagulation process has been reported in ref. [35]. Similarly to the above-discussed Ref. 34, the latter report includes the experimental results and a series of mathematical models allowing determination of the coagulation time and the limiting value of milk clot storage modulus. However, the accuracy of these equations is doubtful, because at pH <5 (below the isoelectric point) the milk rapidly (immediately) curdle, and according to the above equations, the coagulation time should be pretty long. The same applies to the formulas to determine the maximum modulus of milk clot: if the concentration of casein equals to zero, how comes that any storage modulus is determined?

Coagulation time and storage modulus are the basic and reliably measured parameters of the gelation process. The influence of various factors on these parameters has been considered in Ref. [36], and some mathematical models of the effects have been shown. According to our verification study, the relations presented in [36] can accurately predict the milk coagulation in the desired conditions.

However, the influence of the calcium chloride concentration is absent in these relations. In addition, the nature of the considered factors influence should also be specified on the basis of data published by other authors.

After careful comparison of theoretical and experimental research, we have amended and supplemented the empirical formula given in Ref. [36]. As a result, the above model of the effect of external factors on the coagulation time (τ) has been transformed as follows.

$$\log\tau = 0.0176\cdot(\log C_r)^2 - 0.446\cdot\log C_r + 0.084\cdot(pH)^2 - 0.84\cdot pH$$
$$+0.0053\cdot\Theta^2 - 0.414\cdot\Theta + \frac{0.0272\cdot\Theta}{\log C_r} + 0.25\cdot C_c^{1.5} - 0.3\cdot C_c^{0.5} + 11.6 \tag{13}$$

In Eq. (13) there is a constant term (= 11.6), which is used for the valuation and can be changed in whatever way due to the exclusion of any factor influence, or the introduction of additional factors, for example, concentration of casein, fat, salt, etc.

The milk coagulation time, calculated according to the given values of external influences using Eq. (13), is one of the main factors determining the values of the parameters of the gelation kinetics recursive model (11).

The gelation kinetics recursive model contains four variable parameters: K, T_1, T_2, and β.

The K parameter (static transfer factor) in this model is a normalizing one; it is the constant number and does not depend on the properties of milk and external influences, $K \equiv 5000$. It can be changed in this or that way; consequently, other parameters will also change in relations to the external factors, but the simulation result will remain the same.

The T_1 parameter is time constant that characterizes flow intense of intra- and intermicellar biochemical reactions.

The T_2 parameter is time constant that characterizes the decay of the coagulation process, which is the result of the dispersed system reactive elements depletion.

The β parameter is a feedback factor that determines the efficiency of recursive interactions between micelles clusters, free micelles, and between clusters and micelles (stochastic microflow intensity).

The time constant T_1 in the model (11) depends, along with the coagulation time, on internal and external factors, being related via a simple correlation:

$$T_1 = k_1\cdot\tau \tag{14}$$

where τ is measured time in minutes and k_1 is the first factor depending on the intrinsic properties of milk, the latter value is of $0.3 < k_1 < 0.4$.

Similarly to T_1, the time constant T_2 is related to the influencing factors via the coagulation time:

$$T_2 = k_2\cdot(\tau)^{1/2} \tag{15}$$

where k_2 is the second factor depending on the intrinsic properties of milk, it is usually in the range of 300 to 1100.

The acidity and the static transmission K strongly affect the value of the feedback factor β:

$$\beta = \frac{k_r\cdot\tau^{1/3}}{K\cdot(pH-5.44)} \tag{16}$$

where k_r is a factor depending on a combination of the intrinsic milk properties, and $10 < k_r < 16$.

Hence, milk temperature, acidity, and concentrations of introduced rennet clot and calcium chloride being set, the coagulation time is unambiguously determined using Eq. 13. Further, according to Eqs. (14)–(16), the time constant and the feedback factor are determined. Substituting these values in (11), we obtain a recursive numerical model of storage modulus change in the gelation process for a particular kind of milk.

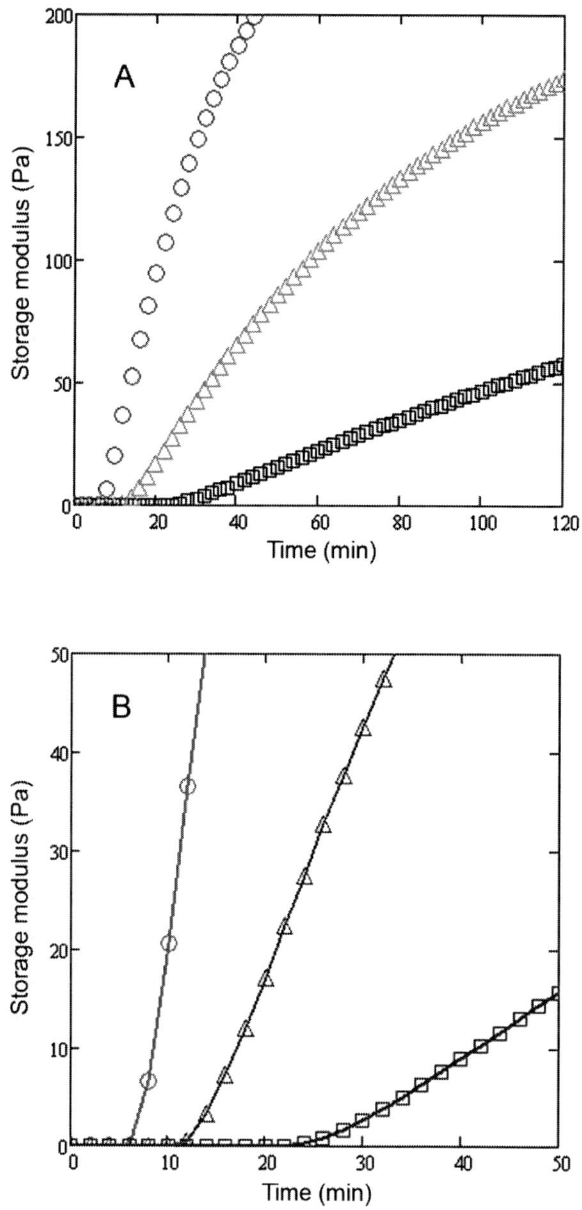

Figure 3. Effect of rennet concentration on the milk coagulation time and the storage modulus of the gel: Δ – 147; ○ – 400; □ – 52 (10^3.units/kg-milk).

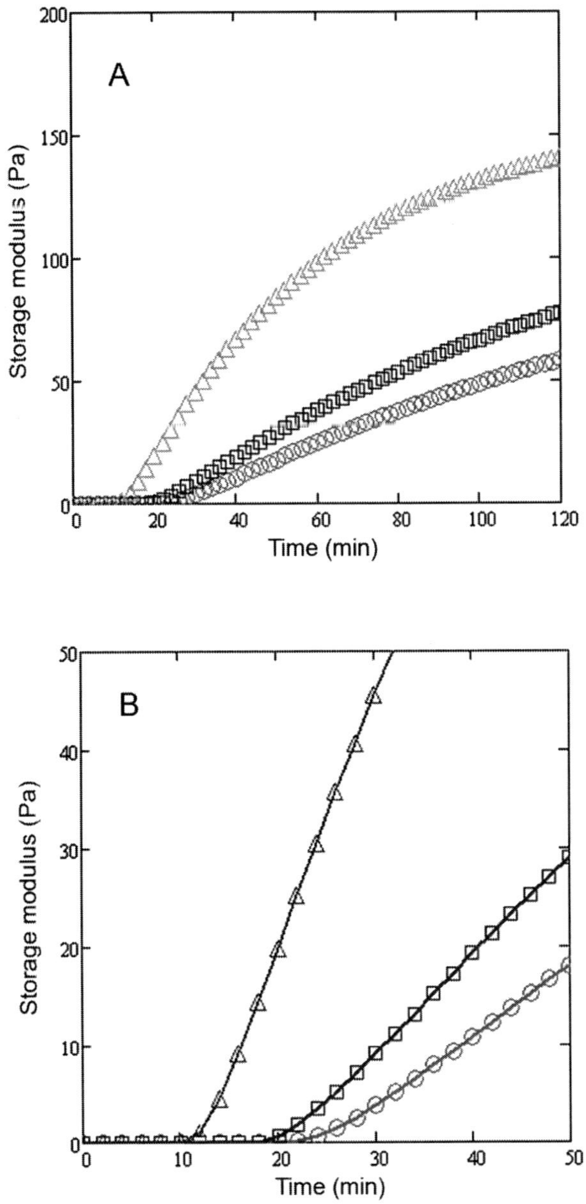

Figure 4. The influence of temperature on the rennet milk coagulation time and the storage modulus of the gel: Δ – 35°C; ○ – 30°C; □ – 45°C.

The simulation results are illustrated by Figures 3A–6A showing the effects of individual parameters on the coagulation time and kinetics of storage modulus evolution. The simulations were performed using the lowest, the highest, and the average value of each of the factors. The enlarged views illustrating the initial phase of coagulation are given in Figures 3B–6B.

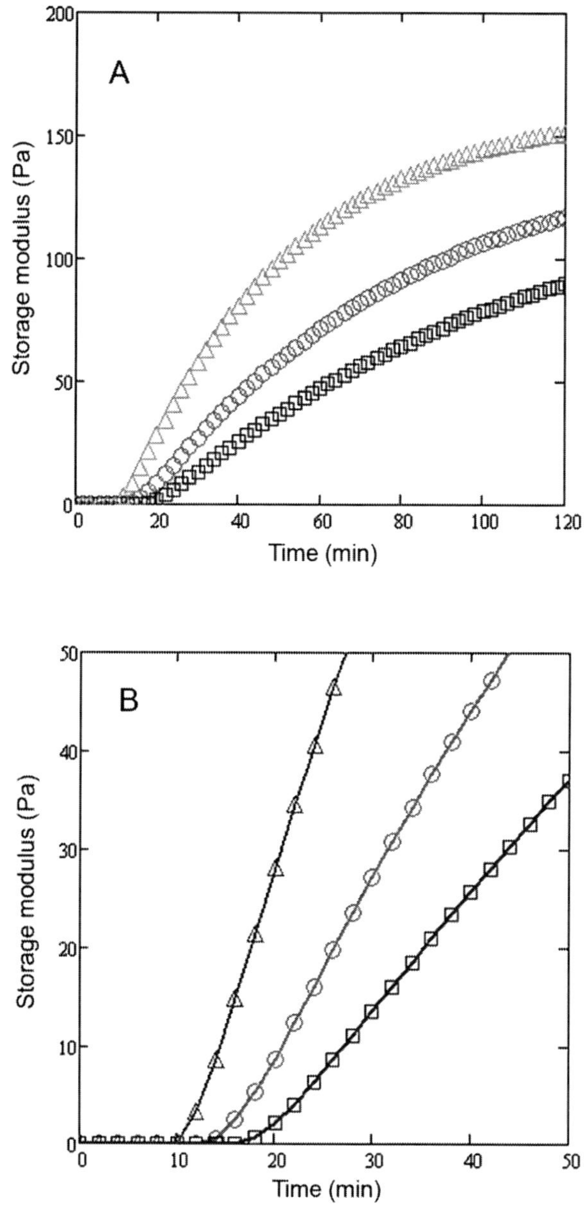

Figure 5. Effect of the calcium chloride concentration on rennet milk coagulation time and the storage modulus of the gel: Δ – 0.4; ○ – 0; □ – 1.5 g/kg.

In general, the obtained model relations are close to the results of experimental studies, obtained by different authors (see discussion above). The closeness of the fit is due to the fact that the gelation model has been developed basing on the analysis and synthesis of the results of these studies; however, the relations may differ from the results obtained in other studies, where other milk or other pretreatment of milk was used.

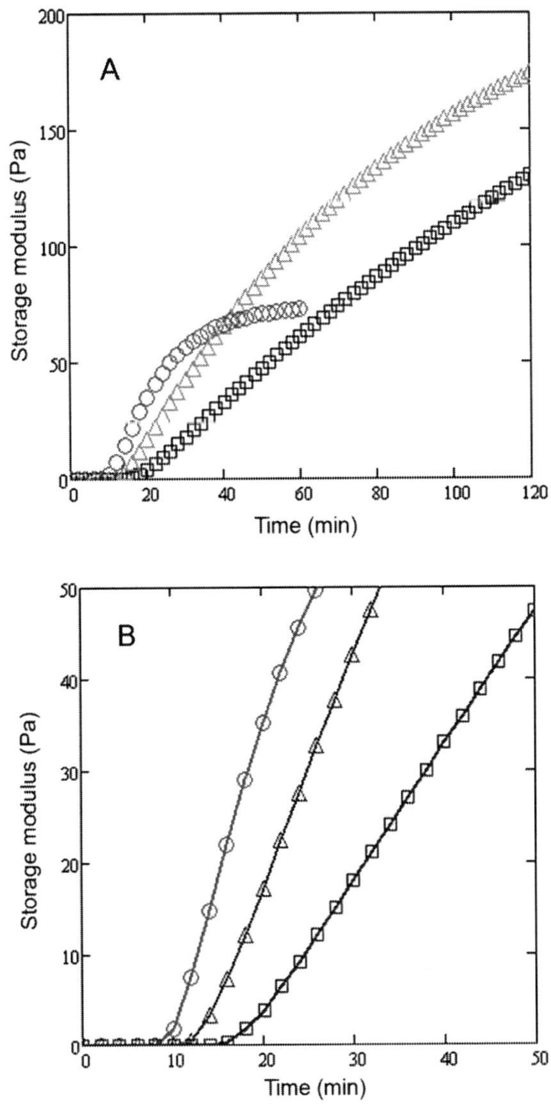

Figure 6. Effect of acidity during the milk coagulation time and the storage modulus of the gel: Δ – pH 6.2; ○ – pH 5.7; □ – pH 6.6.

2.4. Verification of Complete Mathematical Model Milk Gelation Kinetics

We have performed a numerical verification of a fully recursive model of gelation using the results of the experimental research published in Ref. [24]. That study shows the influence of pH values on the coagulation time and the gels elastic modulus, the milk source and other external conditions being the same. The purpose of verification is the definition of the adequacy of a recursive model of gelation in the setting of fixed internal gelation properties of milk, i.e. at constant values of the coefficients k_1, k_2 and k_r and the significantly varied acidity value.

Figure 7 shows a graphic representation of the modeling results obtained using the recursive model of gelation (11) and the empirical model of the effect of external influences (13).

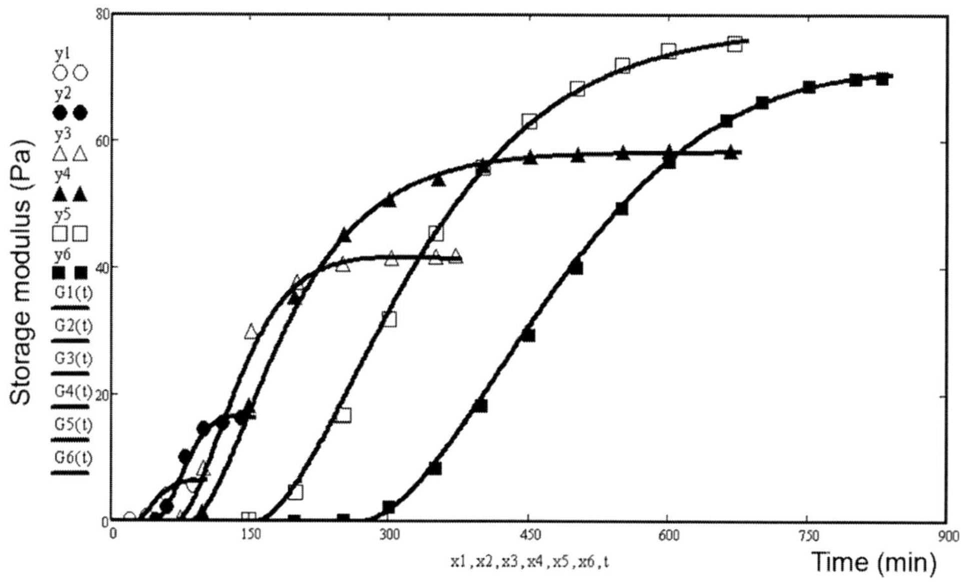

Figure 7. Modeling the impact of acidity during the rennet coagulation time of milk and the storage modulus of the gel (solid lines) according to experimental data [24]: pH 6,7 (■); 6,4 (□); 6,0 (▲); 5,8 (△); 5,6 (●) and pH 5,4 (○).

The coefficients of correlation between the modeling results and the experimental data are close to unity. This leads to conclusion that the description of the processes occurring during a rennet gel formation with a recursive model is adequate. The process of gelation is deteriorating rapidly as acidity approaches the isoelectric point, confirming the existence of its acid instability in enzymatic coagulation of milk.

It should be noticed that the obtained result is a consistent, efficient and unbiased one only for the studied case. The usage of a recursive model for predicting estimates is only possible in strict compliance of the certain conditions of the experiment. This includes ensuring the constant initial properties of the source of milk, including its storage and initial treatment, contamination etc., which affect the types of dependences (13)–(16) and, finally, the predicted results. At the same time, the basic recursive model of the gelation (11) remains in any case the same.

Algorithm of the application of the recursive gelation model to get a prognostic evaluation of the obtained results is as follows:

- judging from the point of the posed task, select the type of source of milk (modeled, recovered from the dried milk; from the individual animal or mixed, etc.) which is a test and subsequently, the reference sample;
- select the specific conditions of gelation in the test sample of milk (enzyme and calcium chloride concentrations, temperature, and acidity), the values of which will be used later in (13);

- perform the elastic modulus measurements during the milk gel formation and determine gelation time using the available equipment;
- determine the theoretical value of the coagulation time, putting it in (13) the specific conditions of gelation in the test sample of milk. If necessary, adjust it to the magnitude of the experimental one by changing the value of the free term of the formula;
- the iteration method improves coefficients k_1, k_2 and k_r in Eqs. (14)–(16) within the range of ± 10%, so that the experimental points (curves) and theoretical values are very close to each other. These coefficients remain unchanged in the future for this kind of milk;
- after testing the properties of a reference sample of milk one can go to modeling the effect of external influences on the process of gelation;
- if it is necessary to estimate the evaluations of the impact of other external influences on the process of gelling on experimental or literature data, additional terms and their dependencies will be added to the formula (13);

Gelation processes are widespread not only in nature, but also in many of modern biochemical and chemical production technologies of various materials and products.

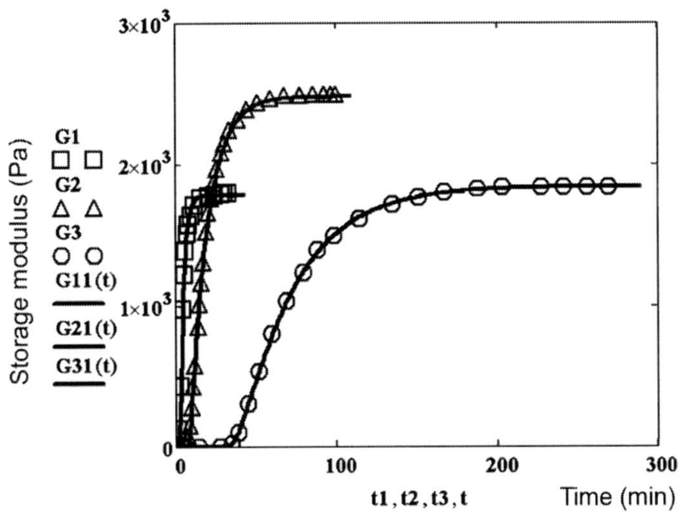

Figure 8. Effect of temperature on the process of formation of a polyacrylamide gel: points [39]; the result of simulation of process using a recursive model (11): solid lines.

First of all it concerns the polymerization processes, theory and appropriate mathematical models of which were first proposed in Flory [37, 38] and are still widely used. For example, Ref. [39] depicts the effect of temperature on the formation of polyacrylamide gel from aqueous solution of acrylamide. Comparison of the given plots with our results on the enzymatic gelation in milk reveals their definite similarity; thus, it is possible to describe the process of acrylamide polymerization using the proposed recursive model of gelation. Figure 8 shows a simulated result obtained using a recursive model (11), in which the values of the time constant of activation of the gelation process, coagulation time, and the feedback coefficient are selected appropriately. Time was expressed in minutes.

CONCLUSION

Mathematical models of gelation kinetics have been developed to fit the observed kinetics of rennet milk gelation [1, 4, 8, 9, 11]. The first of the models (basic) is of the regression type and hence is applicable to disperse systems where gelation/coagulation or polymerization may occur. The second model (complete) has prognostic ability and intends exclusively to describe the kinetics of gelation in milk, based on the values of concentrations of the enzyme and calcium chloride, temperature, and acidity.

The derived evolution of the storage modulus during gelation appears to be reasonable when compared to the available published data [13, 17, 33, 34, 36].

Gelation kinetics modeling was carried out by means of direct and inverse Laplace transformations, allowing, for the first time ever, to take into account the effect of recursive interactions in the system and to introduce a flocculation (induction) phase into the gelation mathematical model.

It is interesting to note that the same kinetics of structuring is observed for other disperse systems, perhaps suggesting that underlying processes is governed by the feedback. If the model describes the kinetics of gelation/coagulation or polymerization in disperse systems, we can assume that recursion may be present in these processes, i.e. the feedback between the images of appearing product elements and source components that affect the intensity of their interactions in solution.

We have developed a recursive model, which can also be used for mathematical modeling of gelation/coagulation or polymerization kinetics of various chemical compounds [39]. It may contribute to our understanding of challenging areas where gelation or coagulation processes occur, such as medical and biological systems and food industry. It is also realized, that the roles of an internal feedback in gelation/coagulation and polymerization/crystallization need to be explored.

We expect that, with further development, the internal feedbacks discussed herein will play an important role in the analysis of different applications, including gelation, coagulation, polymerization, and crystallization.

REFERENCES

[1] Scott-Blair, G.W., Burnett, J. 1958, *J. Dairy Res.,* 25, 297-303.
[2] Scott-Blair, G.W., Oosthuizen, J.C. 1961, *J. Dairy Res.,* 28, 165-173.
[3] Tuszynski, W.B. 1971, *J. Dairy Res.,* 38, 113-125.
[4] Douillard, R. 1973, *J. Texture Stud.,* 4, 158-165.
[5] Tokita, M.K., Kikichi, R., Niki, R., Arima, S. 1982, *Biorheology,* 19, 209-219.
[6] Tokita, M.K.1989, *Food Hydrocolloids*, 3, 263-264.
[7] Carlson, A., Hill, Jr., C.G., Olson, N.F. 1987, *Biotechnol. Bioeng.*, 29, 560-572.
[8] Carlson, A., Hill, Jr., C.G., Olson, N.F. 1987, *Biotechnol. Bioeng.*, 29, 582-589.

[9] Carlson, A., Hill, Jr., C.G., Olson, N.F. 1987, *Biotechnol. Bioeng.*, 29, 601-611.
[10] Carlson, A., Hill, Jr., C.G., Olson, N.F. 1987, *Biotechnol. Bioeng.*, 29, 612-624.
[11] Esteves, C.L.C., Lucey, J.A., Pires, E.M.V. 2001, *J. Dairy Res.,* 68, 499-510.
[12] Zhong, Q., Daubert, C.R. 2003, 3-rd Int. Symp. Food Rheo. Str. 273-277.
[13] Zhong, Q. 2002, Int. Forum Mat. Char. Karlsruhe, Germany.
[14] Daviau, C., Famelart, M-H., Pierre, A., Goudedranche, H., Maubois, J-L. 2000, *Lait,* 80, 397-415.
[15] Dyke P.P.G. 2001, An Introduction to Laplace Transforms and Fourier Series. Springer Undergraduate Mathematics Series.
[16] Riley, K.F., Hobson, M.P. Bence, S.J. 2010, Mathematical methods for physics and engineering (3rd ed.). *Cambridge University Press.*
[17] Lopez, M.B., Lomholt, S.B., Qvist, K.B. 1998, *Int. Dairy J.*, 8, 289-293.
[18] Frederiksen, P.D., Hammershøj, M., Bakman, M., Andersen, P.N., Andersen, J.B., Qvist, K.B., Larsen, L.B. 2011, *Dairy Sci. & Tech.,* 91, 309–321.
[19] Cecchinato A, Penasa M, De Marchi M, Gallo L, Bittante G, Carnier P. 2011, *J. Dairy Sci.,* 94, 4205-4213.
[20] Frederiksen, P.D., Andersen, K.K., Hammershøj, M., Poulsen, H.D., Sorensen, J., Bakman, M., Qvist, K.B., Larsen, L.B. 2011, *J. Dairy Sci.,* 94, 4787–4799.
[21] Cecchinato, A., Ribeca, C., Maurmayr, A., Penasa, M., De Marchi, M., Macciotta, N.P.P., Mele, M., Secchiari, P., Pagnacco, G., Bittante, G. 2012, *J. Dairy Sci.,* 95, 450-454.
[22] Ezeh, V.N., Lewis, M.J. 2011, *Int. J. Dairy. Tech.,* 64, 179-187.
[23] Choi, J., Horne, D.S., Lucey, J.A. 2007, *J. Dairy Sci.*, 90, 2612-2623.
[24] Udabage, P., McKinnon, I.R., Augustin, M.A. 2001, *J. Dairy Sci.,* 84, 1569-1575.
[25] Rogelj, I., Perko, B., Francky, A., Penca, V., Pungercar, J. 2001, *J. Dairy Sci.,* 84, 1020-1026.
[26] Karlsson, A.O., Ipsen, R., Schrader, K., Ardo, Y. 2005, *J. Dairy Sci.,* 88, 3784-3797.
[27] Esteves, C.L.C., Lucey, J.A., Hyslop, D.B., Pires, E.M.V. 2003, *Int. Dairy J.,* 13, 877-885.
[28] Guinee, T.P, Gorry, C.B., O'Callaghan, D.J., O'Kennedy, B.T., O'Brien, N., Fenelon, M.A. 1997, *Int. J. Dairy Tech.,* 50, 99-106.
[29] O'Callaghan, D.J., O'Donnell, C.P., Payne, F.A. 2002, *Int. J. Dairy Tech.,* 55, 65-74.
[30] Fagan, C.C., Castillo, M., Leedy, M., Payne, F.A., O'Donnell, C.P., O'Callaghan, D.J. 2007, *J. Agric. Food Chem.,* 55, 8836-8844.
[31] O'Callaghan, D.J., O'Donnell, C.P., Payne, F.A. 2000, *J. Food Eng.,* 43, 155-165.
[32] Ferrer, M.A., Hill, A.R., Corredig, M. 2008, *J. Dairy Sci.,* 91, 959-969.
[33] Bittante, G. 2011, *J. Dairy Sci.,* 94, 5821-5832.
[34] Najera, A.I., de Renobales, M., Barron, L.J.R. 2003, *Food Chem.,* 80, 345-352.
[35] Davlau, C., Famelart, M-H., Pierre, A., Goudedranche, H., Maubois, J-L. 2000, *Lait,* 80, 397-415.
[36] Hori, T. 1985, *J. Food Sci.,* 50, 911-917.
[37] Flory, P.J. 1941, *J. Am. Chem. Soc.*, 63, 3083-3090, 3091-3096, 3096-3100.
[38] Flory, P.J. 1953, Principles of Polymer Chemistry, *Cornell University Press,* NY, Ithaca.
[39] Calvet, D., Wong, J.Y., Giasson, S. 2004, *Macromolecules,* 37, 7762-7771.

In: Rheology
Editor: Evgeny Karpushkin

ISBN: 978-1-63482-223-7

Chapter 5

RHEOLOGY AND TEXTURE OF CRESS SEED GUM

Seyed M. A. Razavi* and Sara Naji-Tabasi
Food Hydrocolloids Research Center,
Department of Food Science and Technology,
Ferdowsi University of Mashhad (FUM), Mashhad, Iran

ABSTRACT

Hydrocolloids have been widely used in food industry to modify texture, improve moisture retention, control water mobility, and maintain overall product quality during storage. Numerous studies have been carried out on rheological and functional properties of hydrocolloids. Various food formulations such as soups, gravies, salad dressings, sauces and toppings contain hydrocolloids as additives to attain the desired viscosity and mouth feel. Recently, research in this area has received increasing attention in view of seeking for new hydrocolloid sources with appropriate properties as consumer demands have changed throughout the years, and understanding of gum functionality within the industry has increased. The seeds of hydrocolloid for many consumers are regarded as totally natural, therefore researches have been concentrated more on seed gums. *Lepidiumsativum* seed is one of a novel source of gum.

Lepidiumsativum of the Cruciferae family, commonly called "Garden Cress", is a polymorphic species and have valuable medical effects. Cress seeds adsorb water quickly when soaked in water and produce a large amount of mucilaginous substances. Garden cress is claimed to be useful in treatment of various diseases including asthma, hypertension, hepatotoxicity, hyperglycaemia, anuresis, and fractures. Macromolecular component of the cress seed gum has molecular weight of 540 kDa and radius of gyration of 75 nm, its chain being nearly as rigid as that of xanthan. As far as the rheological properties are concerned, cress seed gum exhibits pronounced shear thinning in steady shear rheology and a weak gel type behavior in dynamic rheology when the polymer concentration is high (> 1 wt%). The functionality of seed gums depends on their sugar composition; cress seed gum is a galactomannan with a high mannose to galactose ratio (M/G of 8.2). The high fraction of unsubstitutedmannan of cress seed gum causes enhanced chain association and leads to gelling properties.

* e-mail: s.razavi@um.ac.ir

The foremost reason behind the ample use of hydrocolloids in foods is their ability to modify the rheology of food systems. The rheology of hydrocolloids is strongly influenced by different factors such as temperature, pH, stress, and presence of other components. For many purposes, it is important to be aware of the factors determining and limiting stability of hydrocolloid solutions and of the reasons responsible for their degradation. Therefore, this contribution is devoted to rheology and texture of cress seed gum under different processing conditions (shear rate, time of shearing, thermal treatments, concentration, pH, salts, and sugars) and its applications.

1. INTRODUCTION

Garden cress (*Lepidiumsativum* L.) is a traditional medicinal plant loaded with nutrients and hydrocolloids (Figure 1). It is a member of the Brassicaceae family which is indigenous to southwest Asia and referred to over many centuries ago in Western Europe [1, 2]. Its common names are: garden cress, pepper grass, garden pepper cress, water cress, and pepper wort [3]. Garden cress is a cool season annual and is an erect herbaceous plant which grows up to 50 cm rapidly. Its seeds can be harvested within 70–90 days to give 800–1000 kg/ha [4]. Cress leaves are opposite, ovate or ovate-lanceolate, glabrous, 1.5–8.0 cm or larger and thin (Figure 1a). The flowers are brownish purple, violet or straw-colored, unscented and are borne in terminal and axillary paniculate cymes (Figure 1b) [5]. The knowledge pertaining to the medicinal significance of garden cress have passed from previous to the present generation and was consumed in Persia even before bread was known [6]. The seeds, leaves and roots of garden cress are of some economic value but its crop is mainly cultivated for its seeds [7].

(a) (b)

Figure 1. Garden cress (*Lepidiumsativum* L.) plant: (*a*) leaves, (*b*) flowers.

The cress seeds are brownish-red and oval in shape, with mean dimensions of 2.692±0.102 mm in length, 1.243±0.066 mm in width, and 0.947±0.060 mm in thickness (Table 1, Figure 2a). Its bulk density is 742.60±1.52 (kg/m^3) and 100 seeds equal 0.196±0.008 g (with moisture content of 9.8±0.09, wt.%) [8, 9]. Cress seed has a high nutritional value and can thus be utilized as a functional food ingredient. It has been found to contain much of proteins and iron, and significant amount of calcium [6]. As its bran is asource of pentosans (11.0%) with a swelling index of 18–19 mL, it has a potential to be used as a dietary fiber [9]. Alkaloids, flavonoids, cardiotonic glycosides, glucosinolates, sterols, tannins, and triterpenes are important phytochemical constituents imparting pharmacological characteristics to garden cress seed [10]. The seeds also contain seven imidazole alkaloids: lepidines B, C, D, E, and F (dimeric) as well as lipidenes A and B (monomeric) [11]. Alkaloids from *L. sativum* seeds contain lepidine and semilepidine, which have been shown to suppress blood glucose, cholesterol, triglyceride, and urea level in alloxan-induced diabetic rats upon administration at a dose of 250 mg/kg during 21 days [12]. The major secondary compounds from this plant are the glucosinolates: glucotropaeolin, gluconapin, gluconasturtin, and glucobrassicanapin [13]. Therefore, it is used in treating and curing certain diseases. In the Tafilalet region of Morocco, *L. sativum* seeds have been used for the management of diabetes with an average success rate of 65 per 320 diabetic persons [14]. The cress seeds are known for aperient, diuretic, demulcent, aphrodisiac, carminative, galactagogue, emmenagogue and rubefacient effects [15]. Furthermore, they can be applied as a poultice for hurts and sprains [16] and to improve the clinical symptoms and severity of asthmatic attacks [17].Cress seeds contain approximately 21.54 wt% of solvent-extractable oil and are rich in ω-3 fatty acids. The major fatty acids in cress seed oil are α-linolenic (34%) and oleic (22%) acids. The total tocopherol and carotenoid content of cress seed oil are 327.42 and 1.0 mmol/100 g oil, respectively [19]. The total phytosterol (mainly sitosterol, campesterol, and avenasterol) concentration in cress oil is 14.41 mg/g. Cress seed oil is considered to be fairly stable, since its natural antioxidants (tocopherol, phytosterol, and carotenoids) protect the oil from rancidity [19]. Cress seed oil supplementation to the human diet is efficient in depressing cholesterol, triglycerides, α-linolenic acid and arachidonic acid levels in serum and liver tissues as well as for converting linoleic acid into the long-chain fatty acids (eicosapentaenoic and docosahexaenoic acids) in serum, liver, heart, and brain tissues [4].

Table 1. Some physico-mechanical properties of cress seeds [18]

Property	Mean±SD	Property	Mean±SD
Length, mm	2.692±0.102	Bulk density, kg/m^3	742.60±1.52
Width, mm	1.243±0.066	Static coefficient of friction on	
Thickness, mm	0.947±0.060	Fiber Glass	0.362±0.004
GMD*, mm	1.476 ±0.500	Galvanized iron	0.317±0.005
Surface area, mm^2	6.769±0.461	Glass	0.240 ±0.010
Sphericity	0.546±0.019	Plywood	0.444 ±0.013
Mass of 1000 seeds, g	1.958±0.085	Rubber	0.446 ±0.023
True volume, mm^3	1.83±0.00	Filling contact angle, deg	17.3±1.7
Apparent volume, mm^3	1.04 ±0.11	Emptying contact angle, deg	19.3±1.1
True density, kg/m^3	1070.70±2.47	Terminal velocity, m/s	4.20±0.39

*Geometric mean diameter.

Figure 2. (a) Cress seeds, (b) the seeds soaked in water, (c) mucilaginous mass, and (d) cress seed gum powder [20].

When soaked in water, cress seeds swell and get covered with transparent colorless mucilage. Cress seed contains significant amount of gum (cress seed gum CSG) (6.5-15 wt.%) with outstanding functional properties comparable to these of xanthan gum (XG). Recently researchers tried to apply CSG as a new source of hydrocolloids in order to satisfy the increasing demand for hydrocolloids with specific functionality. Among commercial gums, seed hydrocolloids are still among the most used in food formulations because of their appropriate price, easy availability, and proper functionality [20]. CSG is an anionic polysaccharide with zeta potential of −10.78±0.19 mV [21]. CSG is mostly of carbohydrate nature, the major sugars being mannose, arabinose, galacturonic acid, fructose, glucuronic acid, galactose, rhamnose, and glucose. The macromolecular component of the CSG has molecular weight of 540 kDa, and its chains are nearly as rigid as those of xanthan [22]. In view of rheological properties, it shows a strong shear-thinning behavior, even more pronounced than in the case of xanthan. The rheological characteristics confirmed possibility of CSG application as a novel thickener and a texture modifier [23]. The intrinsic viscosity of CSG in water is 1.33 mL/mg, and hydrodynamic behavior of the hydrocolloid extract is typical of the polyelectrolytes, with intrinsic viscosity decreasing with increasing NaCl concentrations. Dynamic viscosity measurements of dilute to concentrated CSG solutions show liquid-like behavior at lower concentration and viscoelastic gel-like at higher concentration [1]. Cress seed gum offers a wide variety of food applications because of its yield stress, pseudoplasticity and thixotropy properties. It is stable in the presence of salts (NaCl and $CaCl_2$) and over a wide range of pH, with a synergic effect of added sugars (sucrose and lactose) on its flow behavior [24]. CSG gel possesses elastic, adhesive and cohesive properties, which could maintain product integrity during heating, refrigeration and freeze-thawing operations. CSG offers applications in emulsifying and foaming, being stable under different thermal conditions [23, 25, 26]. Various functional attributes of cress seed gum provide essential potency for employing it as a novel food hydrocolloid source.

2. Cress Seed Gum

Mucilage is a complex carbohydrate with a great capacity towards water absorption (6.5–15 wt.% of water is found in cress seeds) [3, 27]. As shown in Figure 2 (b & c), the outer pericarp (or outer epidermis) of seeds, when soaked in water, rapidly swells to form a gelatinous mass [28] of cellulose and uronic acid-containing polysaccharides. Water can penetrate between the polyuronide chains, allowing the latter to hydrate and swell, thus dispersing the cellulose micelles. The gel is a network of hydrated cellulose micelles, scattered with more heavily hydrated uronide chains [2]. The optimized conditions (temperature, water-to-seed ratio, and extraction time) to maximize yield and viscosity of aqueous extract from cress seed at lowest protein content have been found in [3].

2.1. Extraction Optimization

When cress seed is allowed to soak in water, the mucilaginous layer of the seed swells but is not dissolved and cannot be squeezed away. Therefore, the major problem in isolation of cress seed mucilage is that it cannot be easily separated from the seeds [5, 29]. Hence, general methods of separation of mucilage are not applicable, and different procedures have been tried: vacuum pumping and filtering through the Buckner funnel without any filter [5] or passing the seeds through an extractor with a rotor to scrape the gum layer on the seed surface [3]. The extract was further passed through filters and dried (Figure 2d). Also, the mucilage could be purified by ethanol [5, 29] or acetone [5] precipitation.

CSG extraction conditions allowing for the highest yield and viscosity at the lowest protein content have been optimized using response surface methodology (RSM). A central composite face design (CCFD) with four independent variables: temperature (25–85°C), pH (3–10), extraction time (10–25 min), and water-to-seed ratio (10:1–80:1) was used to study the response variables (yield, viscosity, and protein content) [3].

The effect of extraction time and temperature on the yield is presented in Figure 3a; increasing temperature or extraction time has little effect on extraction yield. The effect of water-to-seed ratio and pH on the yield is given in Figure 3b. Increasing the dilution up to a certain value (45:1) increases the yield which is further decreased at the higher water-to-seed ratios. Increasing pH at the optimal dilution increases the yield as well. In the alkaline medium, the yield could be increased due to hydrolysis of insoluble constituents into soluble ones. It was found that the extraction yield of CSG significantly increased from 5.24 to 6.8% as the ratio of water to raw material was increased from 10 to 30, likely due to an increased driving force for the mass transfer of the polysaccharides [30]. Therefore, the 30:1 ratio of water to raw material was chosen as the center in the RSM experiments [3].

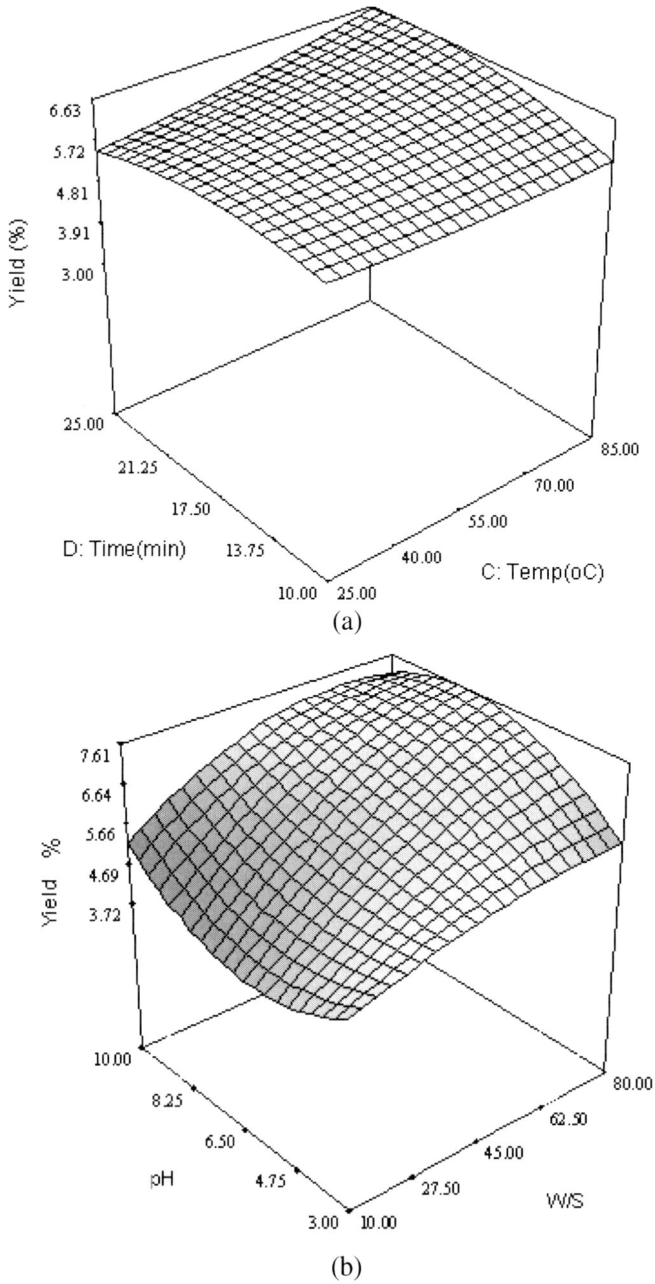

Figure 3. Response surface for the effect of independent variables on extraction yield of cress seed gum: (a) time and temperature (pH 6.5, water-to-seed 45:1), and (b) pH and water-to-seed ratio (temperature 55°C, time 17.5 min) [3].

The effect of pH/water-to-seed ratio and water-to-seed ratio/temperature on apparent viscosity of CSG is shown in Figs. 4a and 4b, respectively. Decreasing water-to-seed ratio and increasing pH increased the extract viscosity. The viscosity of cress seed mucilage is a decreasing function of water-to-seed ratio. Increase of pH at low water-to-seed ratio increases the viscosity (Figure 4a). Viscosity as function of extraction temperature shows a maximum at certain point. Therefore, high pH, low water-to-seed ratio, and mild extraction temperature should result in high viscosity of CSG [3].

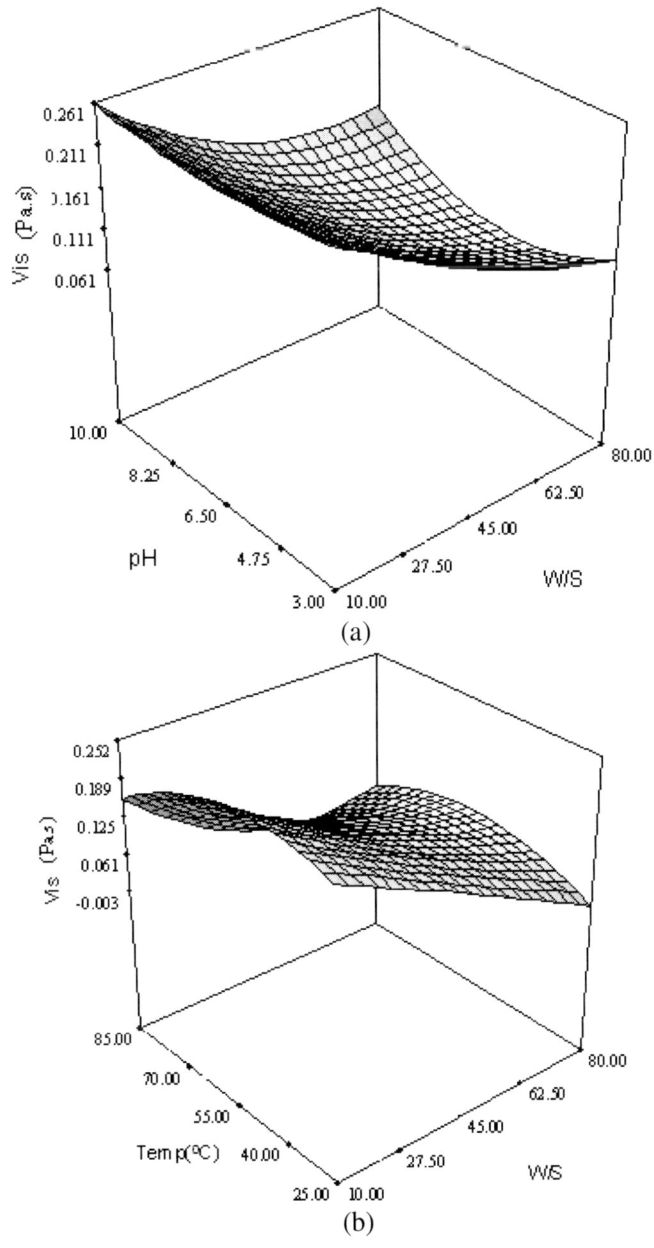

Figure 4. Response surface for the effect of independent variables on apparent viscosity of cress seed gum (as determined at 25°C, concentration of 2 wt.%, and shear rate of 300 s^{-1}: (a) pH and water-to-seed ratio (temperature 55°C and time 17.5 min) and (b) water-to-seed ratio and temperature (pH 6.5 and time17.5 min) [3].

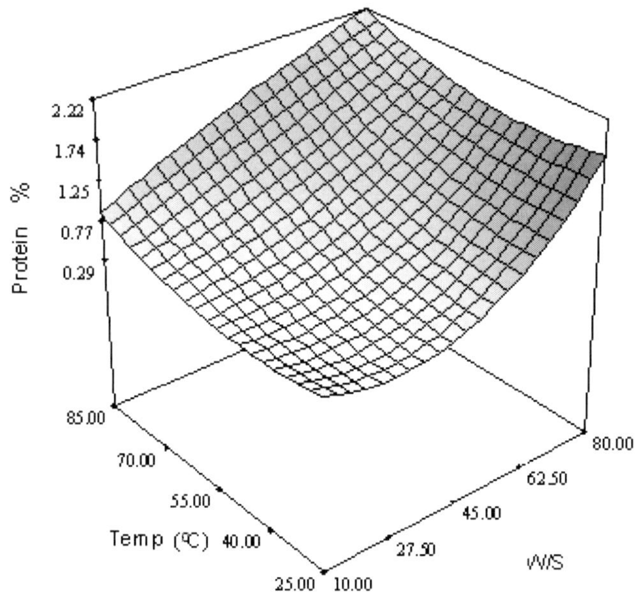

(a)

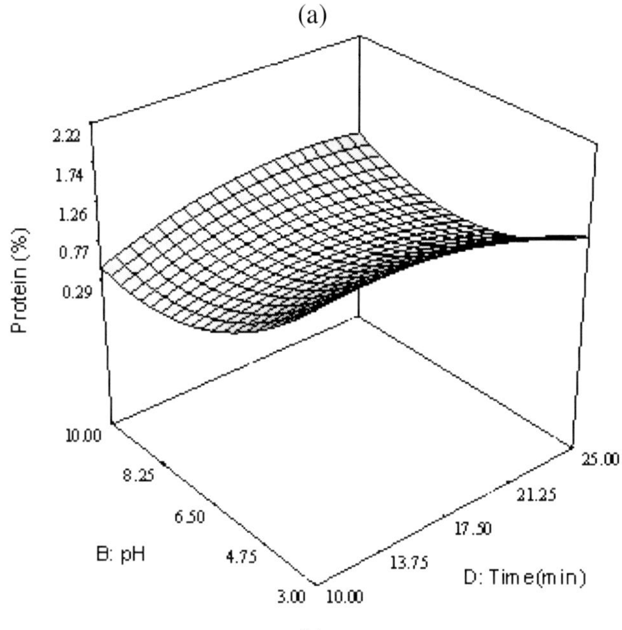

(b)

Figure 5. Response surface for the effect of extraction variables on protein content of cress seed gum: (a) temperature and water-to-seed ratio (pH 6.5 and time 17.5 min) and (b) pH and time (water-to-seed 45:1 and temperature 55°C) [3].

The effect of independent variables on protein content of CSG is shown in Figure 5. Decreasing temperature and water-to-seed ratio decreases the protein content of CSG. At higher temperature, the amount of protein is up due to superior mass transfer rates. On the other hand, increasing pH decreases protein content. The increase in protein content with decrease in pH might be due to the enhanced protein solubilization at lower pH region [31].

The extraction time has a linear effect on protein content. The highest protein content is observed at highest temperature (85°C), the lowest pH (3), and the highest water-to-seed ratio, reaching 5.4 wt.% of the extracted gum from cress seed. Numerical optimization has revealed the optimum extraction conditions as follows: 35°C, pH 10, extraction time 15 min, and water-to-seed ratio of 30:1 [3].

2.2. Physicochemical Properties

The morphological and physical evaluation of CSG shows (Table 2) that it is a lustrous brownish-white powder (Figure 2d) with characteristic odor.

Table 2. Physicochemical parameters of cress seed mucilage

Item	[5]	[32]	[33]
Description	brownish white powder	brownish white powder	n/a[c]
Odor	characteristic	Characteristic	n/a
Appearance	lustrous	lustrous	n/a
Solubility	n/a	forms colloidal solution, soluble in warm water, practically insoluble in ethanol, acetone, diethyl ether and chloroform	n/a
Particle size (μm)	n/a	n/a	189.57
pH	7.8 (0.5%)	neutral (1%)	6.2 (0.5 %)
Mass loss on drying	19.3%	7%	n/a
Swelling index	2.8	18%	3.7
Contact angle (deg)	40	n/a	36.14
Bulk density (g/mL)	0.673	n/a	0.37
Trapped density (g/mL)	0.7	n/a	0.62
Carr's index[a]	3.8%	n/a	40.32%
Hausner'sratio[b]	1.040	n/a	1.67
Compressibility index	0.03857	n/a	n/a
Viscosity (cP)	319.93 (0.5%)	n/a	8.05 (0.5%)
Melting point (°C)	228	>200 (decomp.)	n/a
Total ash	n/a	4.5% (w/w)	n/a
Acid insoluble ash	n/a	0 (absence of sandy matter)	n/a
Sulphated ash	n/a	0.95% (w/w)	n/a

[a] Carr's index=(tapped density–bulk density)×100%/tapped density.
[b]Hausner's ratio=tapped density/bulk density.
[c] n/a – no data.

CSG is soluble in warm water, practically insoluble in ethanol, acetone, diethyl ether, and chloroform. Its swelling index is 18%. Mucilage decomposes above 200°C, like other common polysaccharides. The heavy metal content is of 20–25 ppm (lead) and below 1 ppm (arsenic) [32].

The main chemical components of the cress seed and CSG and their composition are listed in Tables 3 and 4, respectively. CSG is mainly composed of carbohydrates, with sugars content of nearly 77%: mannose (38.9%), arabinose (19.4%), galacturonic acid (8.0%), fructose (6.8%), glucuronic acid (6.7%), galactose (4.7%), rhamnose (1.9%), and glucose (1.0%). The presence of the two uronic acids (~15%) leads to polyelectrolyte nature of the cress seed gum. CSG is poorly soluble in cold water and forms a considerable amount of hydrogel (76%), due to the high mannose content [1]. The amount of the uronic acids is comparable to that found in gum arabic (≈16%) [34]. In addition, the FT-IR spectra of the polymer contain the C–O, C–O–C, O=C–O, C–H, and O–H stretching bands (Figure 6), confirming the presence of carboxyl groups and polyelectrolyte nature of CSG. Acid equivalent weight of CSG is of 1025.64 g/mol [22].

Table 3.Chemical composition of cress seeds [3]

Component	wt.%[a]
Moisture	5.29±0.15
Ash	5.05±0.11
Proteins	22.75±0.88
Fats	23.40±1.50
Carbohydrates[b]	43.51±2.64

[a]mean±SD of three determinations.
[b] by difference.

From Table 4 it is to be seen that CSG contains only a small fraction of proteins and fats (2.45% and 1.85%, respectively). The elution peaks in light scattering and RI profiles (Figure 7) coincide with the UV peak, indicating that the macromolecular component of CSG is somehow associated with protein. Hence, the macromolecular component exists in the forms of glycoprotein or protein–polysaccharide complex, similarly to the case of gum arabic [35]. Ca^{+2} is the major ion present in CSG with a content of 0.17%. Trace amounts of K^+, Na^+, and Mg^+ are found as well, as shown in Table 4.

Table 4. Chemical composition, constituting sugars and molecular parameters of CSG [1]

Composition (wt.%)		Sugars (wt.%)		Molecular parameters	
Moisture	7.17	Man	38.9	M_w (kDa)	540 (4800, 480)[c]
Ash	11.5	Ar	19.4	M_w/M_n	1.44
Protein	2.45	GalA	8.0	R_g (nm)	75
Fat	1.85	Fru	6.8	Mass (%)	91.4 (1.2, 90.2)[c]
Sugar	77.03[a]	GluA	6.7		
Ca	0.17	Gal	4.7		
K	0.062	Rha	1.9		
Na	0.039	Glu	1.0		
Mg	0.0076	Total	87.4[b]		

[a] by difference.
[b] total sugar content recovered in HPAEC–PAD measurement.
[c] the values in parentheses are the parameters derived assuming bimodal distribution of the CSG peaks.

Weight-average molecular weight of the extract is of 540 kDa, and its radius of gyration (R_g) is of 75 nm (Table 4). Under assumption of bimodal distribution, the fraction corresponding to the shoulder shows an extremely high molecular weight of 4800 kDa, and the fraction corresponding to the major peak has molecular weight of 480 kDa. The higher molecular weight fraction accounts for only 1.2 wt% of the whole sample, and is assigned to the aggregated form of CSG molecules.

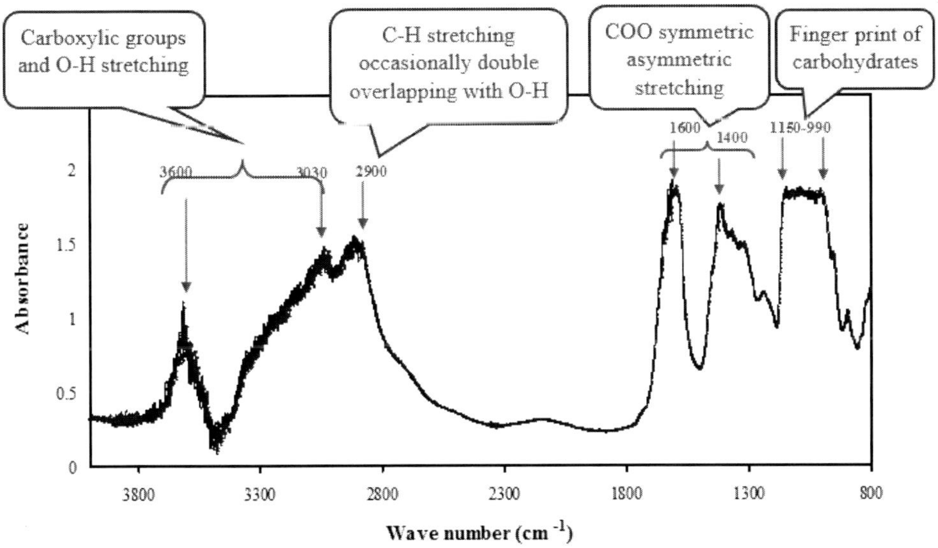

Figure 6. FT-IR spectrum of CSG [22].

The macromolecular component in CSG has a semi-rigid chain conformation with flexibility in between that of random coil and rigid rod [1]; this is similar to the case of xanthan gum [36, 37]. The semi-rigid chain conformation is favorable for entanglements formation, which can explain the relatively high viscosity and pseudoplasticity of the CSG described in the following sections.

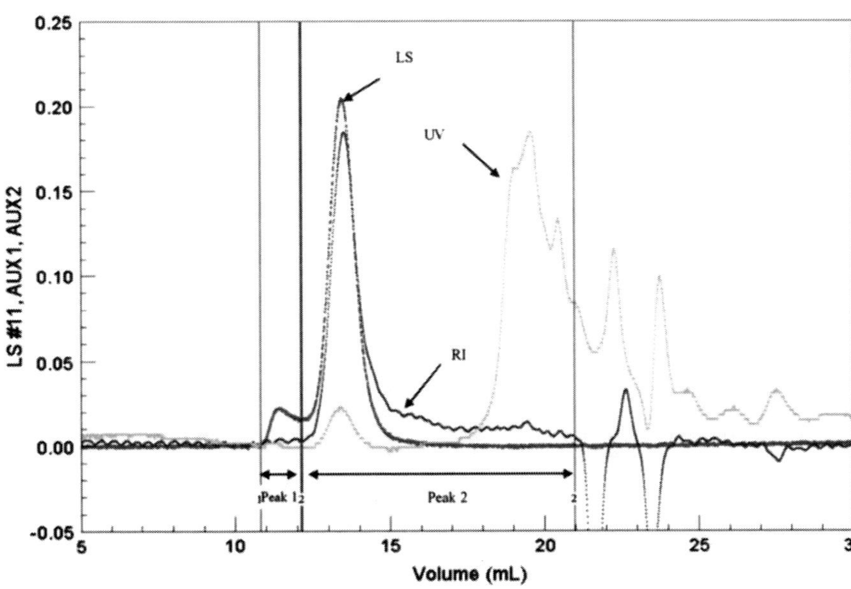

Figure 7. GPC traces (refractive index, light scattering, and UV detection) of CSG [1].

3. Rheological Properties

3.1. Dilute Solutions

One of the most significant features of the biopolymers is high viscosity of their aqueous solutions, even at low concentration. Viscosity measurements of dilute solution (where individual molecules are isolated and move independently) provide important information the solute molecules behavior and on relationship between their structure and properties. Dilute solution studies, in particular, determination of intrinsic viscosity, provide deep insight into the fundamental molecular characteristics of macromolecules in solution [38].

CSG solution has intrinsic viscosity of 1.33 mL/mg (in water) and 0.546 mL/mg (in 0.2 mol/L NaCl) at 25°C, as estimated using the Huggins (Eq. 1) and Kraemer (Eq. 2) empirical expressions.

$$\frac{\eta_{sp}}{C} = [\eta] + k_1[\eta]^2 C \quad (1)$$

$$\frac{ln\ \eta_{rel}}{C} = [\eta] + k_2[\eta]^2 C \quad (2)$$

In Eqs. 1 and 2, k_1, k_2 and C are the Huggins constant, Kraemer constant, and the solute concentration, respectively.

These values are much higher than the reported for many other natural hydrocolloids [22]. Galactomannans such as locust bean gum and Tara gum have intrinsic viscosity of 1.103 and 1.496 mL/mg, respectively, close to that of CSG [39].

By using Mark-Houwink-Sakurada equation, Karazhiyan et al. [1] estimated the exponent α=0.84, suggesting that the macromolecular component of CSG has a semi-rigid chain conformation with flexibility somewhat in between random coil and rigid rod.

As seen in Figure 8, the best model for intrinsic viscosity estimation of cress seed gum solution is the Tanglertpaibul and Rao model (Eq. 3), showing better linear fit with higher correlation (R^2) in comparison with Huggins (Eq. 1), Kraemer (Eq. 2) and Higiro's equations (Eqs. 4 and 5) [40].

$$\eta_{rel} = 1 + [\eta]C \quad (3)$$

$$\eta_{rel} = e^{[\eta]C} \quad (4)$$

$$\eta_{rel} = \frac{1}{1-[\eta]C} \quad (5)$$

In Eqs. 2–5, η_{rel}, $[\eta]$, and C are relative viscosity, intrinsic viscosity, and concentration, respectively.

The plot of log (η_{sp}) versus lg$C[\eta]$, is known as the ''master curve'', is used to determine the coil overlap parameter (critical concentration) and the dilute Newtonian domain [41, 42]. Critical overlap parameter $C^*[\eta]$ at which macromolecular chains begin to overlap is 4.53 in water and 2.36 in 0.2 mol/L NaCl for cress seed gum (Table 5 and Figure 9). For a wide range of hydrocolloids with flexible coil conformation, the overlap usually occurs at $C^*[\eta]$ of 4.0.

The comparatively lower value of $C^*[\eta]$ in the case of CSG polymer can be due to a strong tendency of chain association promoting macromolecules entanglement. Some of the stiff hydrocolloids like xanthan also show lower critical overlap values owing to interchain interactions. The critical overlap parameters of CSG correspond to critical concentrations of 0.34% and 0.43% (w/w) in water and 0.2 mol/L NaCl, respectively.

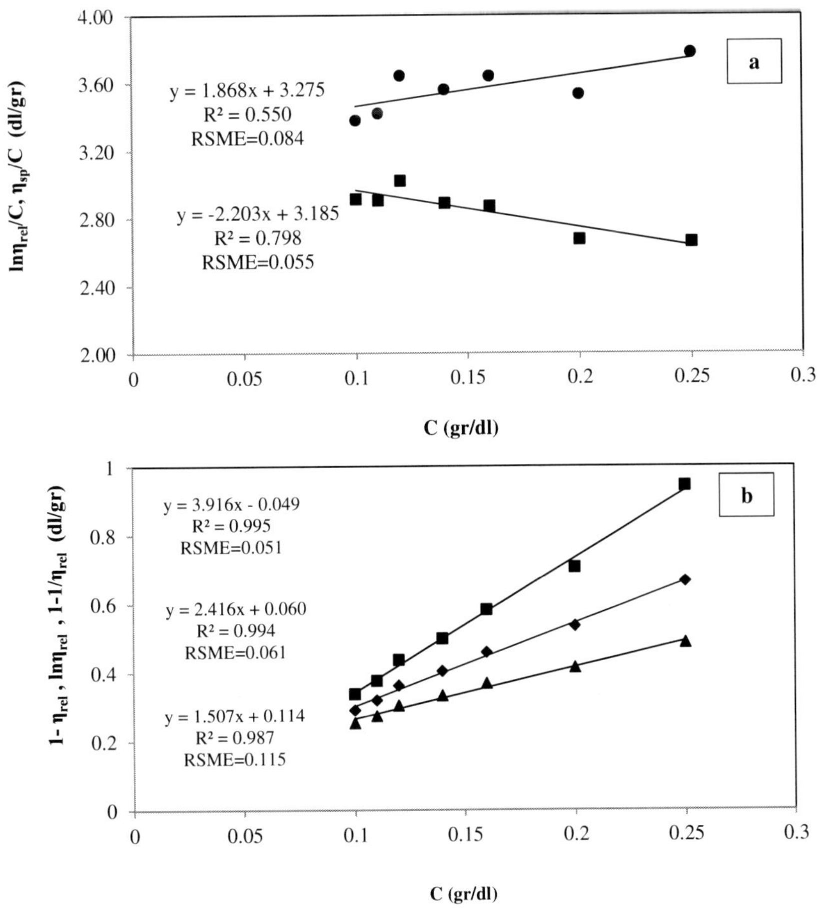

Figure 8. Typical viscosity plots of CSG in deionized water at 25°C: (a) dual Huggins (●) and Kraemer (■) plots; (b) triple Tanglertpaibul–Rao (■), Higiro 1(♦), and Higiro 2 (▲) plots [40].

Table 5. Analysis of the $\lg\eta_{sp,0}$ versus log $C[\eta]$ plot for CSG in water and 0.2 mol/L NaCl [22]

Parameter	Water	0.2 mol/L NaCL
$[\eta]$ dL g^{-1}	13.3	5.46
$\lg C^*[\eta]$	0.656	0.373
$C^*[\eta]$	4.53	2.36
C^* (g dL^{-1})	0.341	0.433
$\lg\eta_{sp,0}$ at C^*	1.21	0.97
slope below C^*	0.96	1.048
slope above C^*	1.78	1.99

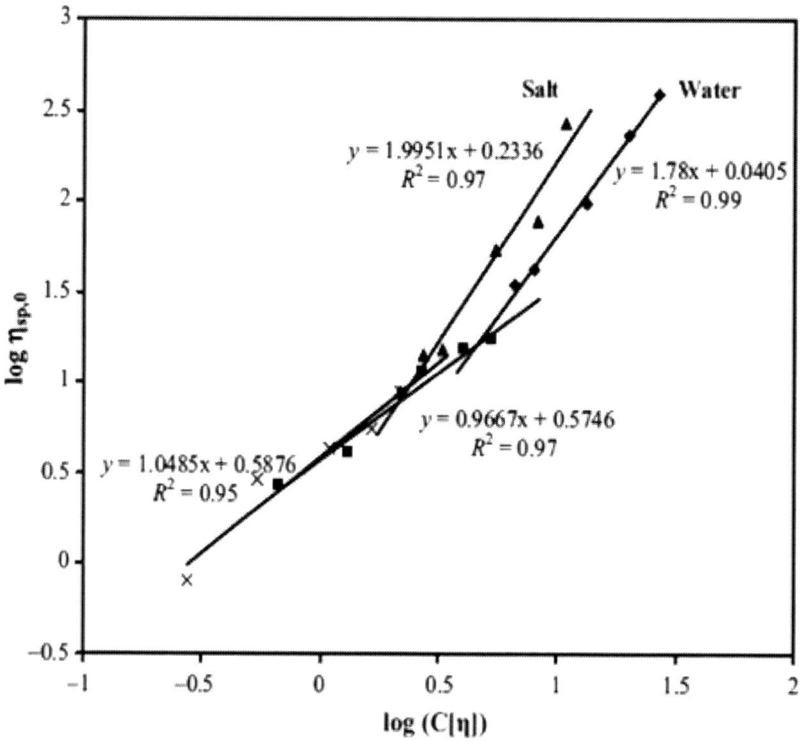

Figure 9. Double logarithmic plot of zero shear specific viscosity versus coil overlap parameter ($C[\eta]$) of CSG in water and 0.2 mol/L NaCl [22].

The relatively lower concentration required for the chains overlap in water is likely due to the expanded chain conformation arising from intramolecular electrostatic repulsion without salt screening [22].

The slope of the discussed plot ranges from 1.1 to 1.4 in dilute solutions, being of 2.7 to 5.1 in concentrated solutions [42, 43]. In the cases of concentrated CSG solutions, the slope is significantly lower due to strong tendency of the chains association at high concentration leading to formation of hydrogel [22].

3.1.1. Influence of Sugars

As shown in Table 6, no entanglement occurs in the dilute solutions upon addition of sucrose or lactose [24]. Addition of 10–20 wt.% of sucrose to CSG solutions decreases the observed intrinsic viscosity, further addition of this sugar up to 40 wt.% results in intrinsic viscosity growth. This trend is observed in the case of lactose as well, with a minimum intrinsic viscosity at 5 wt.% of the sugar.

The initial decrease of the intrinsic viscosity with the sugars addition is not completely clear. From the point of view of the solvent quality, the reduction in $[\eta]$ suggests that 20% sucrose and 5% lactose are poorer solvents for CSG than water.

Due to the sugar composition (relatively high mannose/galactose ratio M/G of 8.2), polymer chains association at very dilute CSG solutions are favorable. Hence, the reduction in intrinsic viscosity at 5 wt.% of lactose and 20 wt.% of sucrose should be attributed to shrinkage of CSG macromolecules due to the suppressed macromolecules association. The intrinsic viscosity of polysaccharides like guar and locust bean gum was also reduced by addition of 5 wt.% of sucrose [44], attributed to similar changes of the macromolecules environment.

The highest intrinsic viscosity reached at approximately 40 wt.% of sucrose and 15 wt.% of lactose indicates that these media are the best solvents for CSG.

Several factors may contribute to this effect, such as decrease in dielectric constant of the solvent, dehydration of sucrose or lactose, and aggregation of the hydrocolloid molecules [24, 44, 45]. To generalize, the observed minimum at the intrinsic viscosity dependences on the sugar concentration results from the two competing factors: reduction of the interchain association of CSG resulting in decrease of the intrinsic viscosity and the CSG coils expansion leading to increase of the intrinsic viscosity [8, 40, 44], both indicating the improving solvent quality.

Table 6. Intrinsic viscosity determined for CSG upon addition of two sugars (25°C) [40]

Additive	Concentration wt.%	Tanglertpaibul and Rao model		Higiro 1 model		Higiro 2 model	
		[η] (dL/g)	R^2	[η] (dL/g)	R^2	[η] (dL/g)	R^2
None	0	3.916 (±0.142)bB	0.995	2.416 (±0.077)aA	0.994	1.507 (±0.542)aA	0.987
Sucrose	10	3.796(±0.010)cAB	0.995	2.383 (±0.005)bA	0.994	1.510 (±0.003)aA	0.987
	20	3.409 (±0.027)cA	0.978	2.363 (±0.019)bA	0.968	1.644 (±0.012)aB	0.955
	30	3.876 (±0.039)cB	0.995	2.433 (±0.016)bA	0.996	1.541(±0.013)aAB	0.993
	40	4.434 (±0.012)cC	0.998	2.894 (±0.154)bB	0.995	1.902 (±0.004)aC	0.989
Lactose	5	3.890 (±0.024)aA	0.997	2.519 (±0.015)aA	0.998	1.648 (±0.010)aA	0.993
	10	4.109 (±0.011)cB	0.999	2.617 (±0.008)bB	0.997	1.648 (±0.005)aA	0.988
	15	4.658 (±0.005)cD	0.998	2.827 (±0.004)bD	0.997	1.738 (±0.000)aA	0.988

Results are expressed as mean±SD for three replications.

A–D: Mean values followed by the same letters in the same column for each sugar are not significantly different (P>0.05).

a–c: Mean values followed by the same letters in the same raw for each sugar are not significantly different (P>0.05).

3.1.2. Influence of Salts

Addition of salts (NaCl or $CaCl_2$) reduces intrinsic viscosity of CSG solution due to the conformational changes of CSG random coils [24]. At 25 mmol/L of NaCl, an initial abrupt decrease of [η] occurs, followed by slight increase at 50 mmol/L of NaCl and further by a moderate decrease in [η] at higher NaCl concentrations (Table 7). Adding $CaCl_2$ causes a monotonous decrease in intrinsic viscosity up to 10 mmol/L of $CaCl_2$, the [η] value remaining almost constant at 15 mmol/L of $CaCl_2$. Polyelectrolyte nature of CSG is mainly responsible for this intrinsic viscosity reduction. The polyelectrolyte coils are expanded due to electrostatic repulsion between the chain segments in the salt-free solution; addition of ionic salt screens the intramolecular electrostatic repulsion, suppressing molecular chain expansion and decreasing the coil hydrodynamic volume. Furthermore, at a given salt concentration, the intrinsic viscosity decreases with increasing temperature, possibly indicating shrinkage of the polymer coil, e.g. due to hydrophobic interactions [22, 40]. When in a good solvent, the polymer segments favor to interact with the solvent rather than with other polymer segments. Addition of a salt deteriorates the solvent quality and hence decreases the volume occupied by the macromolecules [46]. Precipitation of CSG occurs at 0.1 mol/L $CaCl_2$ and 0.5 mol/L NaCl, thus confirming that excess of the both salts results the macromolecules repulsion and favors the macromolecules aggregation [40]. Intrinsic viscosity of some other hydrocolloids like levan [47], red microalga exopolysacharide [48], pectin [49], hydroxypropyl methyl cellulose [50], xanthan [51], *k*-carrageenan [52], sage seed gum [53] and Balangu seed gum [54] behaves similarly at increasing ionic strength of the solution.

Table 7. Intrinsic viscosity values for CSG with addition of two salts at selected concentrations (25 °C) [40]

Additive	Concentration mmol/L	Tanglertpaibul and Rao model		Higiro 1 model		Higiro 2 model	
		[η] (dL/g)	R^2	[η] (dL/g)	R^2	[η] (dl/gr)	[η] (dL/g)
None	0	3.916 (±0.142)bD	0.995	2.416(±0.077)aD	0.994	1.507 (±0.542)aB	0.987
NaCl	25	2.335 (±0.015)cA	0.99	1.597 (±0.011)bA	0.998	1.102 (±0.008)aA	0.995
	50	3.184 (±0.012)cC	0.991	1.987 (±0.006)bC	0.967	1.257 (±0.003)aA	0.994
	75	3.115 (±0.017)cC	0.989	1.963 (±0.008)bC	0.997	1.254 (±0.004)aA	0.997
	100	2.918 (±0.021)cB	0.993	1.881 (±0.019)bB	0.999	1.245 (±0.009)aA	0.995
$CaCl_2$	5	3.065 (±0.013)cB	0.984	1.963 (±0.009)bB	0.995	1.275 (±0.007)aA	0.998
	10	2.212(±0.042)cA	0.998	1.545(±0.026)bA	0.996	1.088 (±0.017)aA	0.989
	15	2.219(±0.058)cA	0.991	1.542 (±0.037)bA	0.996	1.081 (±0.024)aA	0.995

Results are expressed as mean±SD for three replications.

A–D: Mean values followed by the same letters in the same column for each sugar are not significantly different (P> 0.05).

a–c: Mean values followed by the same letters in the same raw for each sugar are not significantly different (P>0.05).

3.2. Steady Shear

3.2.1. Flow Behavior

Typical flow curves for CSG solutions at different concentrations (25°C) are shown in Figure 10. CSG solutions exhibit shear-thinning behavior at all concentrations (the viscosity decreases with increasing shear rate). Similar behavior is observed at other temperatures as well. The explanation of the effect is as follows. With growing shear rate, the intermolecular interactions (in particular, the entanglements) may be disrupted due to anisotropy of the shear deformation, and the observed shear stress is decreased [55]. The conspicuously strong shear-thinning behavior observed for cress seed gum is attributable to its rather rigid chain conformation that gives rise to a highly entangled macromolecular solution [1] and formation of gel-like structure via the intermolecular association (the hydrogel content in CSG is of 76%) [22, 56].

Szczesniak and Farkas [56] showed that gum solutions with high value of flow behavior index (n) tend to feel slimy in the mouth. When less slimy mouth feel characteristics are desired, the choice should be a gum system having a lower n value. Therefore, CSG is suitable for the applications that require less slimy texture or mouth feel, e.g. a thickener for dysphagia treatment.

As seen in Figure 10, the initial reduction of shear stress was steep, becoming less prominent at higher shear rates. Such behaviour is advantageous for processing applications, as the markedly different viscosity behavior is required over different ranges of shear rates associated with when the hydrocolloid is used as stabilizer or thickener for various processing operations (mixing, pumping, and pouring) [57]. At comparable concentration and temperature, the magnitude of flow index of CSG is lower and its consistency coefficient is higher than those of carboxymethyl cellulose, pectin, palmate-tuber salep, and monoi leaves gum, whereas CSG behaves less shear-thinning with lower consistency coefficient than do xanthan, carrageenan, tragacanth, yellow mustard gum, basil seed gum, *Lepidiumperfoliatum* seed gum, sage seed gum, and *Alyssum homolocarpum* seed mucilage. Yield stress of CSG is lower than that of xanthan and carrageenan [24].

Several time-independent rheological models (power-law, Eq. 6; Herschel–Bulkley, Eq. 7; Casson, Eq. 8; Heinz–Casson, Eq. 9; and Bingham, Eq. 10 ones) can be employed to quantitatively evaluate the steady shear flow behavior of cress seed gum [24].

$$\tau = k_p \dot{\gamma}^{n_P} \tag{6}$$

withk_P, the consistency coefficient (Pa.s^n) and n_P, the flow behavior index (dimensionless).

$$\tau = k_H (\dot{\gamma})^{n_H} + \tau_{0H} \tag{7}$$

with k_H, consistency coefficient (Pa.s^n); n_H, flow behavior index (dimensionless); and τ_{0H}, yield stress (Pa).

$$\tau^{0.5} = k_{0C} + k_C \dot{\gamma}^{0.5} \tag{8}$$

with k_{0C} (Pa$^{0.5}$) and k_C (Pa$^{0.5}$.s$^{0.5}$), intercept and slope, respectively, of the $\tau^{0.5}$ versus$(\dot{\gamma})^{0.5}$ plot. Further, the values of k_{0C}^2 and k_C^2 can be used as the Casson yield stress (τ_{0C}, Pa) and Casson plastic viscosity (η_c,Pa.s), respectively.

$$\tau^n = k_{0HC} + k_{HC} (\dot{\gamma})^{n_{HC}} \tag{9}$$

with $(k_{HC})^{1/n}$, the Heinz–Casson plastic viscosity (η_{HC}, Pa.s); n_{HC}, the Heinz–Casson flow behaviour index (dimensionless); and $(k_{0HC})^{1/n}$, the Heinz–Casson yield stress (τ_{0HC}, Pa).

The Herschel–Bulkley model was found to be the most suitable model to fit the rheological properties of CSG as a function of concentration, thermal treatment, salts (NaCl and $CaCl_2$) or sugars (sucrose or lactose) concentration, and pH (3–9) [23, 24, 26]. CSG reveals yield stress (τ_0) according to the Herschel–Bulkley model. The existence of τ_0 indicates that there is a cross-linked or other associated structure in a material that should be broken down before a flow can occur at an appreciable rate [24].

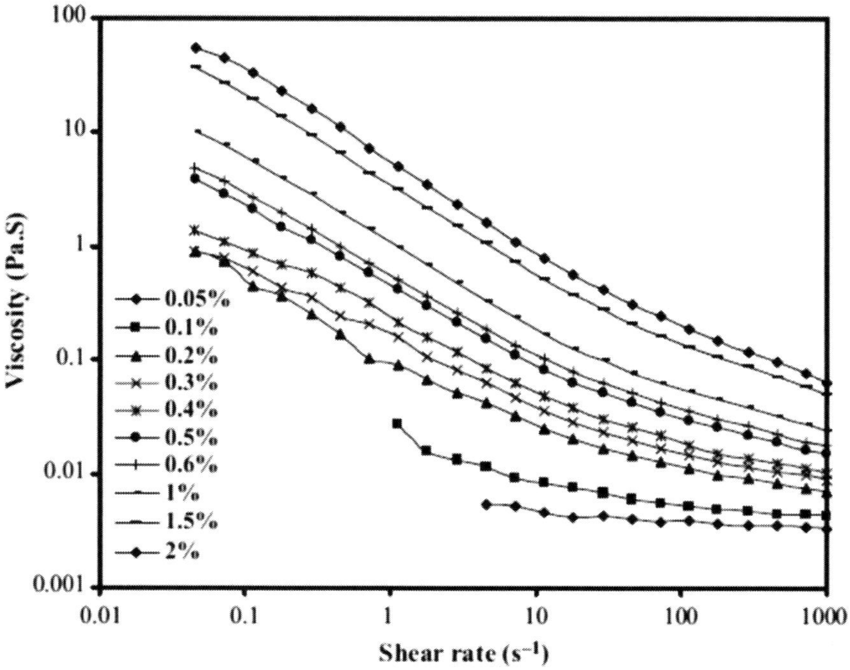

Figure 10. Flow curves for different concentrations of CSG in water at 25°C [22].

3.2.1.1. Influence of CSG Concentration

Viscosity of CSG solutions at extremely low concentration, below 0.1 wt.%, is constant with shear rate, being typical of Newtonian fluid and hence indicating the absence of any intermolecular entanglements. With increasing concentration, non-Newtonian shear-thinning behavior becomes more prominent. As mentioned above, the strong shear-thinning behavior of CSG is mainly due to: (i) enhanced macromolecular entanglements of the relatively rigid chains of cress seed mucilage [1] and (ii) the presence of gel-like structure due to intermolecular association [24]. CSG reveals high M/G ratio (8.2); galactomannans with higher M/G ratios (like locust bean gum) show even stronger tendency to form associates, while galactomannans with lower M/G ratio (like guar gum) are free of chain associations in solutions. Considering this fact and the high-hydrogel content in the CSG extract (76%), the second mechanism leading to the shear-thinning phenomenon should be predominant [22]. As expected, the consistency coefficient increases with the concentration of CSG. Higher solid content generally causes an increase in the viscosity due to enhanced entangling of the polymer chains [1]. As a result, increasing the CSG concentration leads to growth of the apparent viscosity (Figure 10). The yield stress increases by 70% with changing its concentration from 1 wt.% to 2 wt.%; the corresponding increase is of 38% and 146% for xanthan and sage seed gum, respectively [58, 59].

Concentration dependence of viscosity at each temperature (at 50 s^{-1}) has been evaluated using a power-law model (Eq. 10) and an exponential model (Eq. 11) in [1]:

$$\eta = aC^b \tag{10}$$

$$\eta = a\, exp(bC) \tag{11}$$

In Eqs. 10 and 11, *C* is concentration of CSG solution (wt.%); *a* and *b* are constants. The analysis results are summarized in Table 8. It can be seen that parameter *a* decreases and parameter *b* increases with increasing temperature. The increase of *b* with temperature means that the concentration dependence becomes stronger at heating due to enhanced intermolecular interaction [1].

Table 8. Concentration dependence coefficients for apparent viscosity of CSG measured at shear rate of 50 s^{-1} [1]

Temperature (°C)	Power law $\eta = aC^b$			Exponential $\eta = a\ exp\ (bc)$		
	a	*b*	R^2	*A*	*b*	R^2
10	0.26	1.64	0.99	0.28	0.057	0.97
25	0.19	1.75	0.96	0.21	0.611	0.98
40	0.16	1.83	0.97	0.16	0.662	0.98
55	0.064	2.47	0.99	0.073	0.846	0.99
70	0.036	2.75	0.99	0.056	0.853	0.97

3.2.1.2. Influence of Temperature

The effect of temperature on viscosity of CSG solution is shown in Figure 11. The flow behaviour index decreases trend with temperature, indicating that CSG solutions tend to become more shear-thinning upon heating. Theconsistency coefficient and the viscosity decreased with temperature as well. Such behavior is attributed to enhanced mobility of the CSG macromolecules, revealing less resistance to flow. Temperature dependency of the CSG solution viscosity at 50 s^{-1} was fitted with the Arrhenius model (Eq. 12). Figure 12 illustrates that the change of apparent viscosity with temperature generally obeyed the Arrhenius law.

$$\eta = \eta_0 \exp(E_a/RT) \tag{12}$$

withη_0, frequency factor or viscosity coefficient at a reference temperature (Pa.s); E_a, activation energy (kJ/mol); *R*, gas constant (kJ·mol^{-1}·K^{-1}); and *T*, absolute temperature (K).

The activation energy indicates the energy barrier of an elementary flow process. The magnitude of the temperature effect varies with concentration as shown by the $E_a$values in Table 9.

In general, higher activation energy causes greater temperature effect on the viscosity [60]. As seen from Table 9, the activation energy for CSG decreases from 15.59 kJ/mol (at 2 wt.%) to 7.82 kJ/mol (at 3.5 wt.%). Hence, the effect of temperature on viscosity is more pronounced at lower CSG concentrations.

Table 9. Arrhenius model fit of apparent viscosity (at 50 s^{-1}) for CSG at different concentrations [1]

Concentration (wt.%)	η_0 (Pa·s)	E_a (kJ/mol)	R^2
2.0	0.001	15.59	0.96
2.5	0.012	10.83	0.98
3.0	0.031	9.23	0.99
3.5	0.075	7.82	0.99

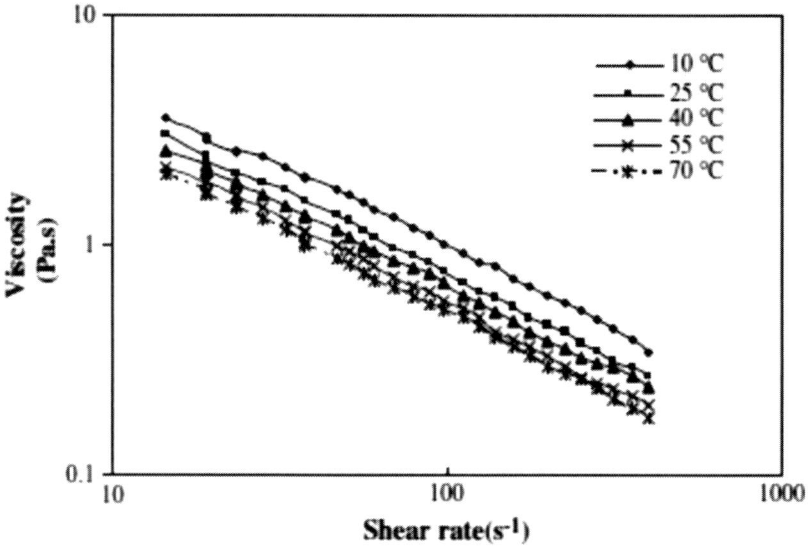

Figure 11. Apparent viscosity of cress seed gum (3 wt.%) at selected temperatures [1].

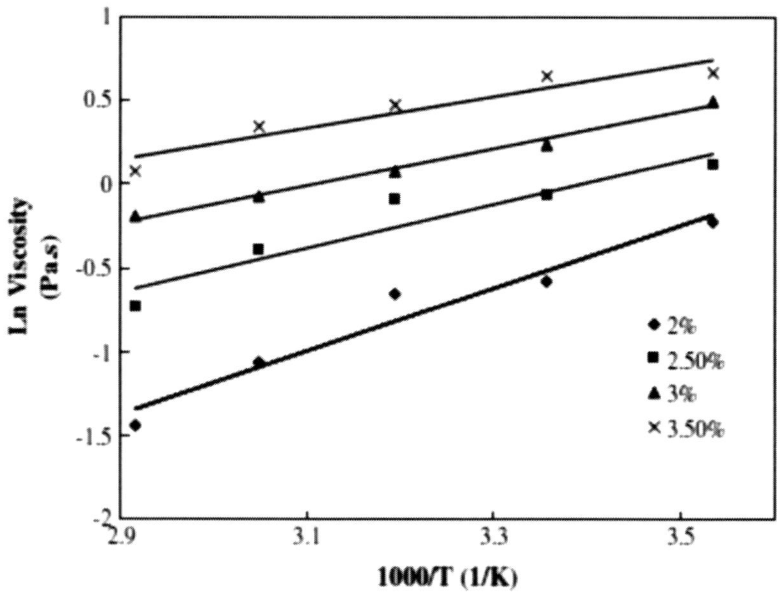

Figure 12. Arrhenius plots for viscosity at 50 s^{-1}. Experimental data: (♦) 2 wt.%; (■) 2.5 wt.%; (▲) 3 wt.%; (×) 3.5 wt.%; lines show the Arrhenius model fit [1].

3.2.1.3. Influence of pH

Properties of many food biopolymers depend on their chemical structure and are sensitive to solution conditions (pH, ionic strength, specific ions, etc.). In addition, their functionality in foods products is affected by the interactions with other food components (proteins, polysaccharides, lipids, sugars, and salts) [61].pH is among key factors altering the hydrocolloids properties. As shown in Table 10, effect of pH (3–9) on rheological properties of CSG show that neither flow behaviour index nor yield stress are significantly affected by pH at different CSG concentrations (1–2 wt.%). However, the consistency coefficient changes significantly at 1.5 wt.% of the gum concentration [24]. These results demonstrate that the acidity has little effect on apparent the flow characteristics of CSG solutions at pH of 3–9. In particular, consistency coefficient is the highest at pH 5, changing by only 18–21% with pH increasing from 3 to 5 or decreasing from 9 to 5. Similarly, the apparent viscosity amplitude as function of pH is of 7–17% depending on CSG concentration. Close values of CSG solution viscosity over a wide range of pH is very similar to the case of gum Tragacanth dispersions [62]. Increase in apparent viscosity with pH changing from 3 to 5 is more prominent at higher CSG concentrations. This may be induced by the combined effect of the changing Huggins coefficient and the expanding hydrodynamic volume of the mucilage macromolecules. When a hydrocolloid chain close to the rod-like conformation, its consistency coefficient is the highest [63], and this probably occurs around pH 5 in the case of CSG solutions.The stability under acidic and alkaline conditions of CSG is a good reason for its application in food and pharmaceutical formulations [24].

Table 10. Regression parameters of the Herschel–Bulkley model and apparent viscosity ($50\ s^{-1}$) for aqueous CSG solutions at different pH (25°C) [24]

Parameter	Concentration (wt. %)	pH			
		3	5	7	9
k (Pa s^n)	1.0	0.015 (±0.006)^{ab}	0.022 (±0.000)^{b}	0.009 (±0.002)^{a}	0.016 (±0.003)^{ab}
	1.5	0.021 (±0.003)^{a}	0.049 (±0.002)^{b}	0.034 (±0.009)^{ab}	0.026 (±0.007)^{a}
	2.0	0.133 (±0.002)^{a}	0.098 (±0.009)^{a}	0.142 (±0.063)^{a}	0.170 (±0.116)^{a}
n	1.0	0.935 (±0.079)^{a}	0.877 (±0.001)^{a}	1.026 (±0.062)^{a}	1.026 (±0.041)^{a}
	1.5	0.950 (±0.036)^{b}	0.834 (±0.002)^{a}	0.857 (±0.045)^{ab}	0.885 (±0.040)^{ab}
	2.0	0.696 (±0.001)^{a}	0.766 (±0.014)^{a}	0.673 (±0.073)^{a}	0.653 (±0.137)^{a}
τ_0(Pa)	1.0	1.258 (±0.142)^{a}	1.365 (±0.135)^{a}	1.448 (±0.158)^{a}	1.313 (±0.151)^{a}
	1.5	2.278 (±0.062)^{a}	2.472 (±0.014)^{b}	2.352 (±0.082)^{ab}	2.215 (±0.008)^{a}
	2.0	2.727 (±0.036)^{a}	3.091 (±0.018)^{a}	2.610 (±0.158)^{a}	2.328 (±0.590)^{a}
η_a(Pa s)	1.0	0.037 (±0.001)^{ab}	0.040 (±0.002)^{b}	0.038 (±0.002)^{ab}	0.034 (±0.001)^{a}
	1.5	0.062 (±0.000)^{ab}	0.075 (±0.000)^{c}	0.065 (±0.000)^{b}	0.061 (±0.002)^{a}
	2.0	0.094 (±0.000)^{ab}	0.101 (±0.001)^{b}	0.090 (±0.002)^{a}	0.085 (±0.005)^{a}

Results are expressed as mean ± SD for two replications.

$^{a–c}$: Mean values followed by the same letters in the same row are not significantly different (P> 0.05).

3.2.1.4. Influence of Salts

Study of the effect of salt type and concentration on flow behavior of CSG solutions is important to elucidate polyelectrolyte nature of the macromolecule and to assess their functional properties.

As seen in Table 11, salts (NaCl and $CaCl_2$) have prominent influence on flow behavior (flow behavior index, consistency coefficient, and yield stress) of CSG solutions [24].

Addition of 100 mmol/L of NaCl to 1, 1.5, and 2 wt.% CSG solutions decreases the consistency coefficient by 38%, 53%, and 83%, respectively; further increase of NaCl concentration does not changing the said parameter. Hence, whereas viscosity of salt-free CSG solutions significantly grows with the polymer concentration, the effect is reduced at 100 mmol/L of NaCl. $CaCl_2$ has similar effect on flow properties of CSG; addition of as low as 5 mmol/L of $CaCl_2$ reducing the *k* values.

The effect of salts on yield stress of CSG solutions depends on concentration of the gum: addition of 5 mmol/L of $CaCl_2$ to 1.5 and 2 wt.%% CSG solution causes 5% and 26% increase in yield stress, respectively, no changes being observed in the case of 1 wt.% solution (Table 11). In the cases of 1.5 and 2 wt.% solutions, the presence of NaCl induces higher plasticity, without any definite specified trend with increasing the salt concentration. Above 100 mmol/L of NaCl and 5 mmol/L of $CaCl_2$, increase of the salts concentration has little effect on *k*, *n,* and τ_0 parameters of the CSG solutions.

Apparent viscosity of CSG solutions (50 s^{-1}) is slightly decreasing upon addition of $CaCl_2$ and is almost constant with NaCl concentration. As mentioned above, major polysaccharides of CSG are anionic uronic acids; addition of salts suppressed ionization of the carboxyl groups and, in turn, decreases the third electro-viscous effect. The salt effects are more remarkable at higher concentration of CSG in solution, when the suppressed polymer–polymer interactions [64] and decreased hydrodynamic domain [65] have more pronounced influence on the flow properties. On top of that, the salts can enhance flexibility of the macromolecules, inducing their conformational changes that results in the viscosity drop [24].

3.2.1.5. Influence of Sugars

Most of food formulations include sugar additives that affect rheological properties of hydrocolloids. Sucrose and lactose are sugars widely utilized in food industry. Changes in the Herschel–Bulkley model parameters and apparent viscosity (50 s^{-1}) of CSG as function of sucrose and lactose concentration are shown in Table 12. All the rheological parameters reported there are significantly influenced by CSG and sucrose concentrations, except for flow behavior index. Increase in sucrose concentration is accompanied by increasing *k*, $\tau_{0,}$ and apparent viscosity (η_a) and decreasing *n* value. The lower flow behavior index indicates that at higher concentration of sucrose the CSG solutions are more pseudoplastic. On the contrary, with increasing lactose concentration, the *k* value and apparent viscosity increase, whereas the *n* value and yield stress are almost constant. Furthermore, the apparent viscosity as function of CSG concentration (1–2 wt.%) is practically identical in the sugar-free solution and in the presence of sucrose. The influence of sucrose presence on apparent viscosity is more pronounced at higher gum concentration.

Table 11. Regression parameters of the Herschel–Bulkley model and apparent viscosity (50 s^{-1}) for aqueous CSG solutions as affected by salt type and concentration at 25°C [24]

Additive, mmol/L		k (Pa s^n)			n(-)			τ_0(Pa)			η_a(Pa s)		
		1 wt.% CSG	1.5 wt.% CSG	2 wt.% CSG	1 wt.% CSG	1.5 wt.% CSG	2 wt.% CSG	1 wt.% CSG	1.5 wt.% CSG	2 wt.% CSG	1 wt.% CSG	1.5 wt.% CSG	2 wt.% CSG
NaCl	100	0.019 (±0.001)^{a}	0.045 (±0.012)^{a}	0.063 (±0.044)^{a}	0.950 (±0.005)^{b}	0.894 (±0.054)^{b}	0.942 (±0.174)^{b}	1.750 (±0.049)^{ab}	3.503 (±0.140)^{c}	5.115 (±1.231)^{bc}	0.051 (±0.002)^{a}	0.098 (±0.002)^{b}	0.136 (±0.004)^{a}
	150	0.017 (±0.002)^{a}	0.049 (±0.005)^{a}	0.167 (±0.082)^{a}	0.963 (±0.030)^{b}	0.877 (±0.016)^{b}	0.763 (±0.057)^{ab}	1.833 (±0.079)^{ab}	4.361 (±0.178)^{d}	3.934 (±0.393)^{ab}	0.052 (±0.001)^{a}	0.116 (±0.004)^{d}	0.139 (±0.006)^{ab}
	200	0.022 (±0.005)^{a}	0.052 (±0.028)^{a}	0.125 (±0.041)^{a}	0.939 (±0.033)^{ab}	0.881 (±0.081)^{b}	0.800 (±0.065)^{ab}	1.704 (±0.044)^{a}	3.656 (±0.212)^{c}	5.663 (±0.441)^{c}	0.050 (±0.001)^{a}	0.102 (±0.001)^{c}	0.167 (±0.006)^{c}
	250	0.021 (±0.003)^{a}	0.059 (±0.005)^{a}	0.176 (±0.034)^{a}	0.948 (±0.028)^{ab}	0.852 (±0.016)^{ab}	0.732 (±0.037)^{a}	1.868 (±0.033)^{b}	4.340 (±0.102)^{d}	3.887 (±0.289)^{ab}	0.055 (±0.001)^{b}	0.118 (±0.001)^{d}	0.137 (±0.003)^{a}
	300	0.016 (±0.002)^{a}	0.053 (±0.013)^{a}	0.164 (±0.084)^{a}	0.986 (±0.027)^{b}	0.860 (±0.043)^{b}	0.775 (±0.098)^{ab}	1.804 (±0.097)^{ab}	3.161 (±0.099)^{b}	4.342 (±0.623)^{b}	0.051 (±0.001)^{a}	0.092 (±0.000)^{a}	0.148 (±0.007)^{b}
$CaCl_2$	5	0.014 (±0.002)^{a}	0.032 (±0.003)^{a}	0.108 (±0.026)^{a}	0.960 (±0.030)^{b}	0.911 (±0.015)^{c}	0.796 (±0.034)^{b}	1.361 (±0.054)^{a}	2.796 (±0.023)^{b}	3.854 (±0.014)^{b}	0.039 (±0.001)^{a}	0.079 (±0.002)^{a}	0.124 (±0.005)^{a}
	20	0.015 (±0.001)^{a}	0.033 (±0.001)^{a}	0.091 (±0.034)^{a}	0.943 (±0.029)^{ab}	0.900 (±0.006)^{bc}	0.812 (±0.053)^{b}	1.403 (±0.055)^{a}	2.820 (±0.096)^{b}	3.777 (±0.235)^{b}	0.039 (±0.001)^{a}	0.075 (±0.002)^{a}	0.117 (±0.003)^{a}
	35	0.014 (±0.001)^{a}	0.035 (±0.003)^{a}	0.097 (±0.063)^{a}	0.972 (±0.029)^{b}	0.880 (±0.004)^{b}	0.821 (±0.089)^{b}	1.453 (±0.031)^{a}	2.853 (±0.067)^{b}	3.642 (±0.331)^{ab}	0.041 (±0.001)^{a}	0.078 (±0.001)^{b}	0.116 (±0.007)^{a}
	50	0.017 (±0.003)^{a}	0.034 (±0.003)^{a}	0.126 (±0.053)^{a}	0.949 (±0.027)^{ab}	0.900 (±0.016)^{bc}	0.766 (±0.087)^{b}	1.455 (±0.036)^{a}	2.780 (±0.072)^{ab}	3.559 (±0.412)^{ab}	0.042 (±0.000)^{a}	0.077 (±0.001)^{b}	0.118 (±0.002)^{a}

Results are expressed as mean ± SD for three replications. $^{a-c}$: Mean valuess followed by the same letters in the same column for each salt are not significantly different (P>0.05).

Table 12. Regression parameters of the Herschel–Bulkley model and apparent viscosity (50 s^{-1}) for aqueous CSG solutions as affected by sugar type and concentration at 25°C [24]

Additive (wt.%)		k (Pa s^n)			n(-)			τ_0(Pa)			η_a(Pa s)		
		1 wt.% CSG	1.5 wt.% CSG	2 wt.% CSG	1 wt.% CSG	1.5 wt.% CSG	2 wt.% CSG	1 wt.% CSG	1.5 wt.% CSG	2 wt.% CSG	1 wt.% CSG	1.5 wt.% CSG	2 wt.% CSG
Sucrose	0	0.031 (±0.009)^a	0.097 (±0.006)^a	0.370 (±0.167)^a	0.890 (±0.047)^a	0.780 (±0.011)^{bc}	0.610 (±0.080)^{ab}	1.782 (±0.114)^a	2.657 (±0.070)^{ab}	3.041 (±0.510)^a	0.055 (±0.003)^a	0.090 (±0.001)^a	0.140 (±0.003)^a
	10	0.044 (±0.008)^a	0.099 (±0.019)^a	0.645 (±0.037)^b	0.880 (±0.029)^a	0.825 (±0.027)^c	0.578 (±0.020)^a	2.009 (±0.091)^{ab}	3.195 (±0.085)^{abc}	3.041 (±0.155)^a	0.067 (±0.002)^b	0.113 (±0.003)^b	0.177 (±0.007)^b
	20	0.066 (±0.016)^a	0.329 (±0.108)^b	0.663 (±0.056)^b	0.759 (±0.156)^a	0.673 (±0.054)^a	0.612 (±0.018)^{ab}	2.009 (±0.126)^{ab}	2.444 (±0.480)^a	3.451 (±0.280)^{ab}	0.077 (±0.000)^b	0.137 (±0.003)^c	0.208 (±0.006)^c
	30	0.107 (±0.021)^b	0.362 (±0.140)^b	0.679 (±0.041)^b	0.813 (±0.037)^a	0.697 (±0.080)^{ab}	0.644 (±0.011)^{ab}	2.161 (±0.246)^{ab}	2.850 (±0.738)^{abc}	4.330 (±0.429)^{bc}	0.094 (±0.002)^c	0.160 (±0.004)^d	0.249 (±0.007)^d
	35	0.135 (±0.039)^{bc}	0.318 (±0.072)^b	0.808 (±0.178)^b	0.803 (±0.050)^a	0.741 (±0.040)^{abc}	0.656 (±0.041)^b	2.289 (±0.351)^b	3.498 (±0.257)^{bc}	4.247 (±1.008)^{bc}	0.105 (±0.004)^d	0.183 (±0.006)^e	0.285 (±0.008)^e
	40	0.165 (±0.022)^c	0.377 (±0.130)^b	0.786 (±0.120)^b	0.797 (±0.003)^a	0.738 (±0.060)^{abc}	0.673 (±0.021)^b	2.294 (±0.231)^b	3.620 (±0.721)^c	4.850 (±0.817)^c	0.111 (±0.000)^e	0.201 (±0.010)^f	0.311 (±0.005)^f
Lactose	5	0.028 (±0.004)^a	0.099 (±0.015)^a	0.563 (±0.040)^a	0.915 (±0.028)^a	0.787 (±0.032)^a	0.497 (±0.090)^a	1.858 (±0.111)^{ab}	2.650 (±0.280)^a	2.758 (±0.290)^a	0.056 (±0.001)^a	0.095 (±0.006)^a	0.142 (±0.001)^a
	10	0.031 (±0.005)^a	0.153 (±0.059)^{ab}	0.501 (±0.141)^a	0.926 (±0.026)^a	0.744 (±0.072)^a	0.605 (±0.049)^a	2.103 (±0.107)^{bc}	2.729 (±0.449)^a	3.373 (±0.512)^a	0.064 (±0.001)^b	0.107 (±0.001)^b	0.168 (±0.001)^b
	15	0.054 (±0.015)^b	0.221 (±0.094)^b	0.583 (±0.117)^a	0.847 (±0.053)^a	0.696 (±0.063)^a	0.587 (±0.034)^a	2.119 (±0.180)^c	2.495 (±0.489)^a	3.084 (±0.374)^a	0.071 (±0.001)^c	0.112 (±0.001)^b	0.173 (±0.001)^b

Results are expressed as mean ± SD for three replications. $^{a-c}$: Mean values followed by the same letters in the same column for each sugar are not significantly different (P>0.05).

Nishinari et al. [66] stated that sucrose promoted hydrogen bonding in κ-carrageenan systems, increasing the number of junction zones and stabilizing their structure. Elsewhere, explanation for similar effect of lactose addition was presented by Mazurkiewicz et al. [67], who found that increase in the viscosity of dextran solutions with increasing concentration of some low-molecular additives resulted from changes in water activity rather than the direct interactions of the additives with the macromolecules. As the apparent viscosity of CSG solutions containing 10 wt.% of sucrose or 10 wt.% of lactose are close, the ordering of lactose and sucrose in CSG solutions is nearly the same. Therefore, our results are consistent with reports for other hydrocolloids. A similar finding was reported in the case of guar gum [68]; it was assumed that some interactions between the polymer and sugars took place. Increase of apparent viscosity with increasing sucrose concentration was also reported for yellow mustard mucilage [69], oat gum [70], carrageenan [71], wheat starch [72], carboxymethylcellulose [73], *Alyssum homolocarpum* seed gum [74], and waxy maize starch [75].

3.2.1.6. Influence of Heat Treatment

Processing of food products presumes different types of thermal treatment such as pasteurization, sterilization, evaporation, cooking, baking, freezing, and drying. Hydrocolloids are used to confer stability to the products undergoing heating/cooling. Therefore, study of the functional properties related to stability at high/low temperatures is essential to evaluate capability of hydrocolloid to be used in food formulations. Evaluations of functional properties related to stability of hydrocolloids against thermal treatments have been scarce so far [76–79].

Table 13. Steady-shear rheological parameters of CSG solutions (1 wt.%, 25°C) under different heating, refrigerating, and freezing treatment regimes*

Model	Witness	Heating				Cooling		Freezing	
	(25°C)	60°C/ 30 min	80°C/ 23 min	100°C/ 18 min	121°C/ 15 min	4°C/ 96 h	10°C/ 40 h	−18°C/ 24 h	−30°C/ 15 h
Power law model									
K_p (Pa·s^n)	0.65^c	0.89^b	0.88^b	0.90^b	1.01^a	1.00^a	0.89^b	0.67^a	0.66^a
n_p	0.44^a	0.42^a	0.43^a	0.43^a	0.42^a	0.40^a	0.42^a	0.46^a	0.46^a
R^2	0.985	0.981	0.980	0.998	0.987	0.983	0.983	0.977	0.980
Herschel–Bulkley model									
τ_{0H} (Pa)	2.38^b	2.87^b	2.87^b	2.89^b	3.31^a	3.16^a	3.00^a	2.92^a	2.68a^b
K_H (Pa·s^n)	0.06^a	0.10^a	0.09^a	0.10^a	0.10^a	0.08^a	0.07^a	0.05^a	0.04^a
n_H	0.83^a	0.76^a	0.78^a	0.75^a	0.76^a	0.77^a	0.80^a	0.86^a	0.86^a
R^2	0.999	0.999	0.999	0.998	0.998	0.999	0.999	0.998	0.999

* Different letter superscripts indicate significant differences at $P<0.05$ in the Duncan test.

Effect of different heating conditions (60°C/30 min, 80°C/23 min, 100°C/18 min, and 121°C/15 min) on properties of 1 wt.% CSG solution is presented in Figure 13 and Table 13. The apparent viscosity decreases with increasing shear rate in all the cases, indicating non-Newtonian shear-thinning behavior; the *n* values are below unity for all the heating regimes.

The consistency coefficient and viscosity increases significantly after heating; the effect of even shorter heating at higher temperature is more pronounced, but no significant difference has been found between the other treatments. Yamazaki et al. [79] reported that heating of *Corchorusolitorius* L. solution at 60°C during 30 min led to increasing of the solution viscosity. However, lower viscosity at higher temperature was observed in that study. Seemingly, irreversible intermolecular arrangement occurs in CSG solution upon heating, contributing to the enhanced viscosity. Semi-rigid chain conformation is again responsible for prominent shear-thinning and aggregation properties, and association leads to higher viscosity at elevated temperature. The flow behavior index is nearly constant under all heating conditions, and so is pseudoplasticity feature. The almost unchanged flow behavior index of CSG solutions is especially important since increasing in n value leads to slimy feeling in mouth. When appropriate mouth feel properties are desired for heat-treated food, low n value gums with high stability of the property in the course of thermal process are an excellent choice [56]. In addition, the heating treatments do not lead to CSG destruction; on the contrary, xanthan gum solution underwent pronounced viscosity drop, deteriorating its functional properties [25].

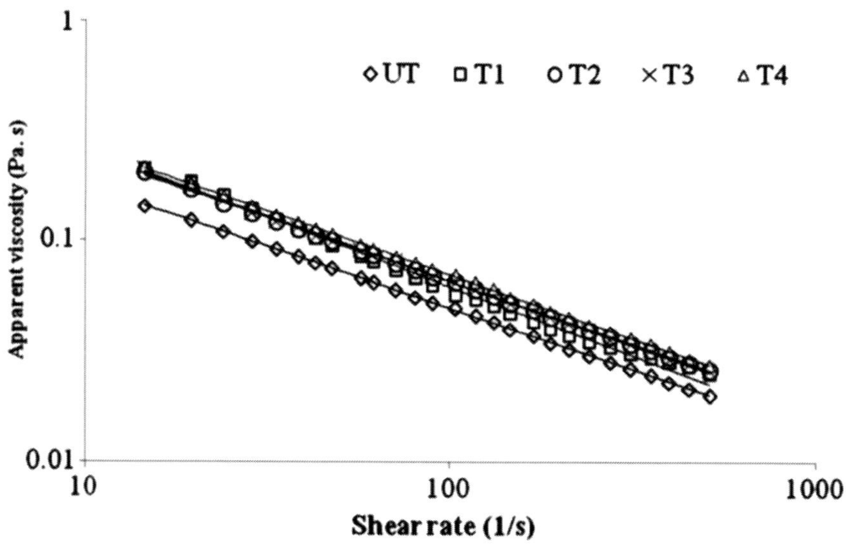

Figure 13. Influence of thermal treatment on apparent viscosity of CSG solution (1 wt.%). UT: untreated, T1: 60°C/30 min, T2: 80°C/23 min, T3: 100°C/18 min, T4: 121°C/15 min [25].

3.2.1.7. Influence of Refrigeration Condition

Refrigerating or cold storage of food products is a common way to extend their shelf life. However, the final quality of the so stored products may be changed due to the phase separation of liquid during storage. The related problems can be solved by incorporating appropriate hydrocolloid into a formulation. The influence of cooling treatment (4°C/96 h and 10°C/40 h) on rheological properties of CSG solutions (1 wt.%) are summarized in Table 13 [23]. Shear-thinning behavior has been retained, and the flow behavior index of the CSG solution has been nearly unchanged after storage in the refrigerator. Stability of shear-thinning behavior of the hydrocolloid is especially important in view of its potential application as food texture modifier and the wide use of low temperature treatment in food industry.

Refrigeration significantly increases $k_P$values of the CSG solution. Longer cooling at lower temperature (4°C/96 h) produces more prominent effect on viscosity. On cooling, the polymer network builds up with double helices forming junction points of the polymer chains [80]; the junctions limit polymer chain arrangement and stretching [81] and consequently lead to increase of the solution viscosity. This effect is significant in the case of CSG due to a strong tendency for molecular association between the unsubstitutedmannan units.

According to the Herschel–Bulkley model analysis, the yield stress of CSG solution increases significantly after exposition to low temperature, but the increase is not influenced by temperature and incubation time (Table 13). Since yield stress helps to keep a food formulation in the shape, refrigerating can improve the related CSG properties [23].

3.2.1.8. Influence of Freezing/Thawing Process

The market of frozen foods is growing and has become a larger proportion of the total food market. Nevertheless, texture instability still remains the most significant challenge for these food products, especially taking into account possible post-production temperature fluctuations. When a food product is frozen, formation of ice crystals within its matrix causes physical stress; upon thawing, melting of the ice crystals leads to loss of moisture and integrity, deteriorating the overall quality of frozen foods [82]. If these ice structures cannot appear, the frozen food will regain its flavor, texture, and color after thawing. Due to high water holding capacity of hydrocolloids, they can be used to minimize the negative effects of freezing food products, by preventing ice crystal growth and controlling moisture level. Increased viscosity is a peculiar feature of some hydrocolloid solutions that restrict diffusion in the system, but it does not have limiting impact on the rate of ice crystal growth because of poor correlation between viscosity and growth rate. Hindrance of the expanding crystal due to entrapment in polymer chains is far more important, and suppression of water crystallization is correlated with the polymer degree of cross-linking and its tendency towards gel formation. In the gelled systems, small ice crystals are usually formed, leading to beneficial effects on texture [83–86]. Consequently, the ideal polysaccharide for frozen foods application should possess gel or gel-like properties, forming a three-dimensional network to restrict water mobility either through junction zone formation or polyelectrolyte complexes [86, 87]. CSG adsorbs water and increase viscosity of foods, but can it tolerate the negative effect of freezing? To answer this question, the effect of freezing treatment (–18ºC/24 h and –30ºC/15 h) on cress seed gum solution (1 wt.%) was investigated, and the results are presented in Table 13 [26].

The consistency coefficient (k) values of CSG solution increases only slightly after freezing. The slight growth of viscosity of CSG solutions is attributed to the polymer chains intermolecular association during freezing. Indeed, the local polymer concentration increases due to partial water crystallization, leading to formation of the chains aggregates that are retained after freezing. In addition, semi-rigid chain conformation of CSG can be retained during under the both freezing regimes to account for pseudoplastic behavior. The temperature and time of freezing do not have any significant effect on the k values as well. Therefore, the rheological properties of CSG solution remains unchanged after freezing [26].

3.2.2. Thixotropy

Thixotropic behavior of a hydrocolloid is defined as the breakdown of network crosslinks formed by associated polymer chains; it is a common feature of many hydrocolloid solutions with the polymer chains interacting to form three-dimensional structures [88]. When a material is sheared at a constant rate, viscosity of a thixotropic material will decrease over time, implying a progressive breakdown of the structure [89].

Modeling of thixotropic behavior of food products has been based on certain equations including the first-order stress decay with a zero (Eq.13) or non-zero (Eq. 14) equilibrium stress value, Weltman model (Eq. 15), and structural kinetic model (Eq. 16):

$$\tau = \tau_0 e^{-kt} \tag{13}$$

with τ_0, initial shear stress and k, breakdown rate constant.

$$\tau - \tau_{eq} = (\tau_0 - \tau_{eq})\,e^{-kt} \tag{14}$$

with τ_0, τ_{eq}, and k being initial shear stress, equilibrium stress, and breakdown rate constant, respectively.

$$\tau = A + B(\ln t) \tag{15}$$

with A and B being initial shear stress and degree of breakdown, respectively.

$$\left[\left(\frac{\eta - \eta_\infty}{\eta_0 - \eta_\infty}\right)\right]^{1-n} = (n-1)kt + 1 \tag{16}$$

withη_0, initial apparent viscosity at $t = 0$ (structured state); η_∞, steady-state apparent viscosity at $t \to \infty$ (non-structured state); n and k, order of the structure breakdown reaction and its rate constant, respectively.

Experimentally, the flow kinetics is easily revealed by running a hysteresis cycle. The hysteresis area (Eq. 17) gives an estimate of the magnitude of the product thixotropy [90]:

$$\text{Hysteresis loop area } S = \int_{\dot{\gamma}_1}^{\dot{\gamma}_2} k\dot{\gamma}^{n} - \int_{\dot{\gamma}_1}^{\dot{\gamma}_2} k'\dot{\gamma}^{n'} \tag{17}$$

with$k$, k' and n, n' being consistency coefficients and flow indices for the increasing and the decreasing shear rate measurements, respectively.

The hysteresis area is a measure of energy per unit of time and per unit of volume needed to eliminate the influence of time in flow behavior [91]. A difference between upward and downward curves are observed at CSG concentration of 6 wt.% and higher, indicating the time dependence of rheological properties [1]. The downward (decelerating) curve is located under the upward (accelerating) curve under a variety of conditions, reflecting thixotropic behavior of the solutions (Figure 14). As shown in Table 14, the hysteresis area increases with increasing extract concentration and with increasing temperature, due to enhanced intermolecular interactions (see discussion of the viscosity behavior above). Indeed, CSG is a galactomannan with a high M/G ratio of 8.2. The intermolecular association of galactomannans occurs between the unsubstituted regions of mannan backbone via formation of the side-by-side linkage [92]. Therefore, the high fraction of unsubstitutedmannan of CSG enhances the association and explains the thixotropy behavior of CSG solutions. Another galactomannan with a high M/G ratio of 3.28 (locust bean gum) is known for high tendency to form associates as well, but guar gum with higher galactose fraction (M/G ratio of approximately 1.59) does not reveal any significant thixotropy as galactose side chains prevent the linkage between neighboring molecules [93]. Thixotropic properties of CSG impart a variety of utilitarian characteristics desired for applications in products such as salad dressing and mayonnaise [23, 26].

Table 14. Hysteresis area at different concentrations of CSG at different temperatures [1]

Concentration (wt.%)	Temperature (°C)	Hysteresis area ($\times 10^3$ Pa s^{-1})
6.0	5	50
	25	106
	45	129
	65	150
6.25	5	95
	25	117
	45	131
	65	161
6.50	5	125
	25	137
	45	159
	65	172
6.75	5	132
	25	152
	45	161
	65	175

At a constant shear rate, the apparent viscosity decreases with time, confirming the thixotropic behavior of CSG. The transient viscosity profile of CSG solutions was modeled using the above-mentioned kinetic models (Eqs. 13–16). Karazhiyan et al. [1] reported the first-order stress decay model with a non-zero equilibrium stress (Eq. 14) as the best-fit model (R^2> 0.9). However, the Weltman (Eq. 15) and the structural kinetic (Eq. 16) models have recently been successfully applied to describe the experimental data (R^2>0.9) [23, 26]. According to the first-order stress decay model with a non-zero equilibrium stress, the initial shear stress (τ_0) and equilibrium stress (τ_{eq}) decrease with increasing temperature and increase with increasing CSG concentration [1].

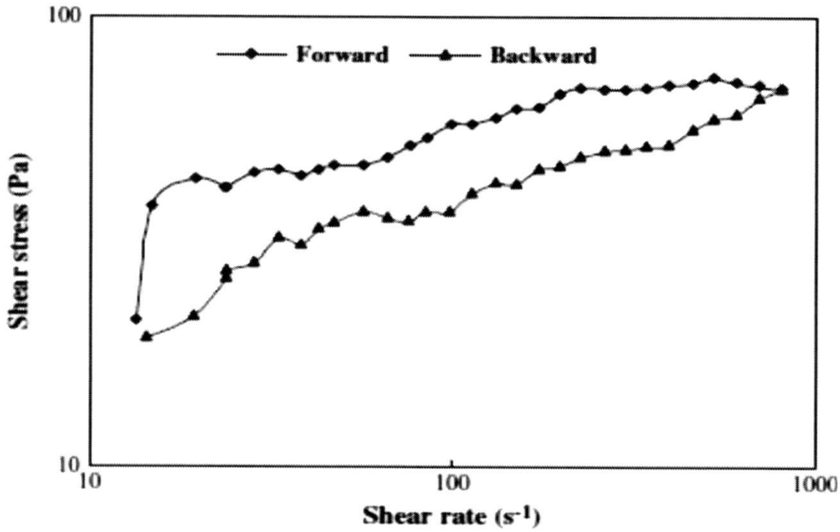

Figure 14. A typical hysteresis loop of the flow curves of CSG solution at (6.75% wt.% and 45°C) [1].

The breakdown rate constant (*k*) does not reveal any clear dependence on temperature, concentration, or shear rate. Viscosity (at 50 s^{-1}) of CSG solutions decreases rapidly with time during the first 400 s of shearing and approaches a constant value after approximately 1000 s [25]. From the Weltman model, the extent of thixotropy or structural breakdown (*B* value) of CSG is noticeably higher than these of xanthan. Thixotropic behavior was also reported for some other hydrocolloids such as Salep, Balangu seed gum [94], *Alyssum homolocarpum* seed gum [95], methylcellulose, hydroxypropylmethyl cellulose [88], and oat starch [96].

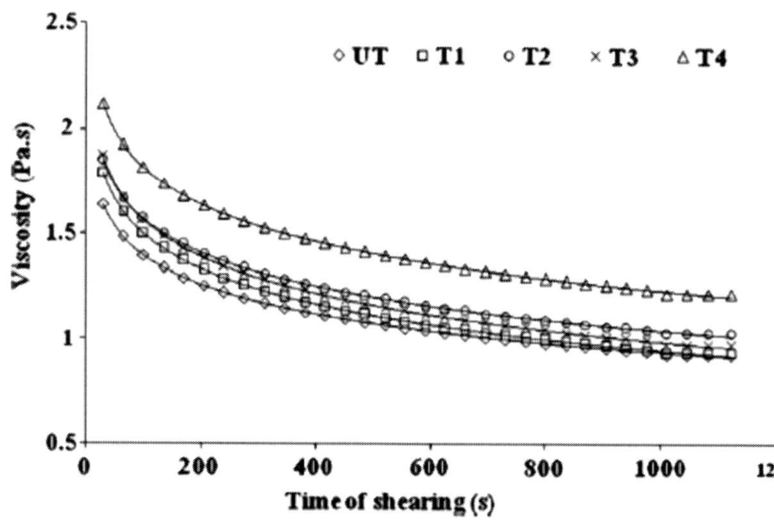

Figure 15. Viscosity of cress seed gum solution (6 wt.%) as function of shearing time at different thermal conditions (shear rate of 50 s^{-1}). UT: untreated, T1: 60ºC/30 min, T2: 80ºC/23 min, T3: 100ºC/18 min, T4: 121ºC/15 min [25].

3.2.2.1. Influence of Heating Treatment

Time-dependent rheological behavior of CSG solutions under different heating conditions (untreated, 60°C/30 min, 80°C/23 min, 100°C/18 min, and 121°C/15 min) is illustrated in Figure 15. All the CSG solutions exhibit thixotropic behavior. According to the first-order stress decay model, heating treatment of CSG solutions does not influence the decay rate constant (*k*), which is an indication of how fast the gum solutions reaches the equilibrium stress under the shear. Thermal treatment increases the initial stress (τ_0) and equilibrium stress (τ_{eq}) values of CSG solutions.

However, the increase is significant only in the case of autoclaving condition (121°C/15 min). In general, even shorter treatment at higher temperature causes stronger association of CSG polymeric chain leading to more prominent solution structuring and the related increase in the initial stress and equilibrium stress values. Hysteresis area for CSG increases with the treatment temperature (Table 15), the increase being more prominent under autoclaving conditions. However, the change of hysteresis area is not that significant [25].

Table 15. Power law model parameters and hysteresis loop area for cress seed gum solutions (6 wt.%) under different thermal conditions [25]

Treatments	Upward curve			Downward curve			Hysteresis area
	k (Pa.s^n)	*n*	R^2	*k* (Pa.s^n)	*n*	R^2	(Pa. s^{-1})
UT	24.28^c	0.28^a	0.991	9.58^a	0.42^a	0.993	13618^a
T1	30.22^b	0.26^a	0.978	9.60^a	0.42^a	0.984	12814^a
T2	32.92^{ab}	0.25^a	0.990	9.68^a	0.43^a	0.995	13716^a
T3	34.11^a	0.25^a	0.991	10.06^a	0.43^a	0.992	13267^a
T4	33.45^a	0.26^a	0.989	10.03^a	0.43^a	0.985	16235^a

Mean values in a column followed by different superscripts are significantly different at $P \leq 0.05$ (Duncan test).UT: untreated, T1: 60ºC/30 min, T2: 80ºC/23 min, T3: 100ºC/18 min, T4: 121ºC/15 min.

The structural kinetics model postulates that change in time-dependent flow properties is associated with shear-induced breakdown of the internal fluid structure, and the rate of this breakdown in the course of shear depends on the kinetics of the structured state → non-structured state transition [91, 97]. For all the studied CSG solutions, it is found that their transient viscosity data can be fitted with the second-order structural kinetic model (n=2). The difference in kinetic order may be related to the complicated binding pattern in different polymers [92, 98]. It has been reported that the structure breakdown of locust bean gum and starch pastes obey the second- and third-order kinetics behavior, respectively. Initial apparent viscosity (η_0) of cress seed gum solution increases significantly under heating, but there is no significant difference between the different thermal treatment conditions. The ratio of initial viscosity to the steady one $\eta_0/\eta_\infty$is a measure of degree of structure breakdown [94]. Heating treatment causes slight increase in η_0/η_∞values of CSG solutions. Therefore, the extent of thixotropy remains constant after thermal treatments. The rate constant (*k*), which indicates the rate of the thixotropic breakdown, does not change after heating as well.

Table 16. Influence of cooling on the time-dependent rheological behavior of CSG gels (6 wt.%) [23]

Model/parameters	Witness (25ºC)	Cooling 1 (4ºC/96 h)	Cooling 2 (10ºC/40 h)
First-order stress decay model, with a non-zero equilibrium stress			
τ_0 (Pa)	92.09[a]	96.63[a]	99.31[a]
τ_{eq} (Pa)	49.72[a]	57.72[a]	57.70[a]
K (s^{-1})	0.007[a]	0.004[a]	0.006[a]
R^2	0.911	0.970	0.984
Weltman model			
A (Pa)	116.06[a]	126.38[a]	125.70[a]
–B (Pa)	10.07[a]	9.96[a]	10.14[a]
R^2	0.951	0.953	0.920
Second-ordered structural kinetics model			
η_0 (Pa·s)	2.27[a]	2.09[a]	2.30[a]
η_0/η_∞	2.13[a]	2.05[a]	2.26[a]
K × 10^3 (s^{-1})	12.00[a]	12.30[a]	12.17[a]
R^2	0.992	0.967	0.990
Hysteresis area	13618[a]	17282[a]	15197[a]

Mean values in a column followed by different superscripts are significantly different at $P \leq 0.05$ (Duncan test).

3.2.2.2. Influence of Refrigeration Condition

Refrigeration treatment has no significant impact on thixotropic properties of CSG solution. The parameters of all time-dependent models show that the CSG solutions remain intact after refrigerating (Table 16). The conditions of cooling (4ºC/96 h and 10ºC/40 h) do not have any influence on time-dependent properties of CSG solutions. It is found that the hysteresis loop area of CSG increases only slightly after refrigeration (Table 16). This clearly demonstrates the potential of using CSG in stabilizing of refrigerated foods [23].

3.2.2.3. Influence of Freezing/Thawing Treatment

Properties of 6 wt.% CSG solutions after freezing treatment (–18ºC/24 h and –30ºC/15 h) show that CSG solution are tolerant to the freezing-thawing process (Table 17). The time-dependent flow of the CSG solutions after freeze-thaw treatment obeys the first-order stress decay model (Eq. 14), and no significant changes in the initial stress (τ_0) and equilibrium stress (τ_{eq}) values was found. The effect of freezing on thixotropy of CSG is similar to that in the case of xanthan. On top of that, freeze-thaw treatment slightly increases the extent of structure breakdown of CSG solutions.

CSG solution exhibits significant increase in the relative hysteresis area after freezing reflecting the enhanced thixotropic behavior (Table 17). The increment is more noticeable after freezing at –30°C during 15 h, which demonstrate more pronounced crystallization of water at the lower temperature, enhancing alignment and association of polymer chains in the frozen state. Increment of the relative hysteresis area is in good agreement with the corresponding increase in η_0/η_∞ values. However, increase of hysteresis loop area after freezing more significant. This may be related to the reason that η_0/η_∞ values are obtained at a constant shear rate, and the associates are only partially ruptured; whereas in the hysteresis loop area experiments, under action of the varied shear rate more of the interchain associates can be broken down. Similar behavior was observed in the case of locust bean gum solutions and was attributed to certain structural heterogeneity [92].

Table 17. Influence of freezing process on the time-dependent rheological behavior CSG gels (6 wt.%) [26]

Model/parameters	Witness (25°C)	Freezing 1 (–18°C/24 h)	Freezing 2 (–30°C/15 h)
First-order stress decay model, with a non-zero equilibrium stress			
τ_0 (Pa)	92.09±11.23^a	104.52±7.98^a	102.89±9.57^a
τ_{eq} (Pa)	49.72±9.45^a	52.82±10.23^a	56.75±5.77^a
k (s^{-1})	0.007±0.00^a	0.005±0.01^a	0.007±0.01^a
R^2	0.911	0.974	0.986
Second-ordered structural kinetics model			
η_0 (Pa·s)	2.27±0.10^{ab}	2.33±0.55^a	2.38±0.56^a
η_0/η_∞	2.13±0.46^a	2.23±0.31^a	2.14±0.29^a
k × 10^3 (s^{-1})	12±0.01^a	17±0.01^a	20±0.01^a
R^2	0.992	0.995	0.987
Hysteresis area	13618^b	23051^a	26815^a

Mean values in a column followed by different superscripts are significantly different at $P \leq 0.05$ (Duncan test).

Freezing of galactomannans like CSG, is accompanied with progressive increase in local concentration of the polymer as liquid water turns into ice, thus promoting intermolecular association retained after subsequent thawing in the form of junction zones [99]. This effect is more pronounced in the cases of galactomannans with higher M/G ratio, as the crosslinking likely occurs via association of unsubstitutedmannan units [99, 100].

3.3. Dynamic Properties

3.3.1. Effect of Concentration

As CSG concentration increases from 0.1 to 2.0 wt.%, its dynamic rheological properties change from liquid-like through viscoelastic to these of weak gel. As shown in Figure 16a, at very low concentrations (normally below C^*), the loss modulus $G''$is higher than the storage modulus G'at all frequencies. This type of behaviour is indicative of a dilute liquid-like solution in the absence of macromolecules entanglements [101, 102] At higher concentrations, as shown in Figure 16b, both $G'$and $G''$are dependent on frequency and they crossover at a certain frequency. The crossover frequency decreases with increasing concentration. This is typical rheological behavior of a concentrated solution of entangled macromolecules [103]. Dynamic rheological behavior at even higher concentrations (Figure 16c) indicates the presence of weak gel with G' being higher than G''over a wide frequency range, the both modules being slightly frequency-dependent [101].

Such behavior further confirms the ability of CSG solutions to become structured. This ability arise from the interchain association owing to the high M/G ratio [100, 104], and it is in line with the observations of strong shear-thinning and high hydrogel content as commented above [22].

3.3.2. Effect of Temperature

Effect of temperature dependence on the dynamic rheological properties is shown in Figure 17 using 2 wt.% CSG solution as a representative example. It can be seen that both storage modulus (G') and loss modulus (G'') decrease with increasing temperature from 25 to 80°C. Interestingly, G' remains larger than G'' over the whole temperature range, and the modules dependence on frequency is not altered significantly. This implies that the CSG solution can retain weak-gel type behavior up to a quite high temperature, and the structuring and associating ability is not susceptible to temperature variation [22].

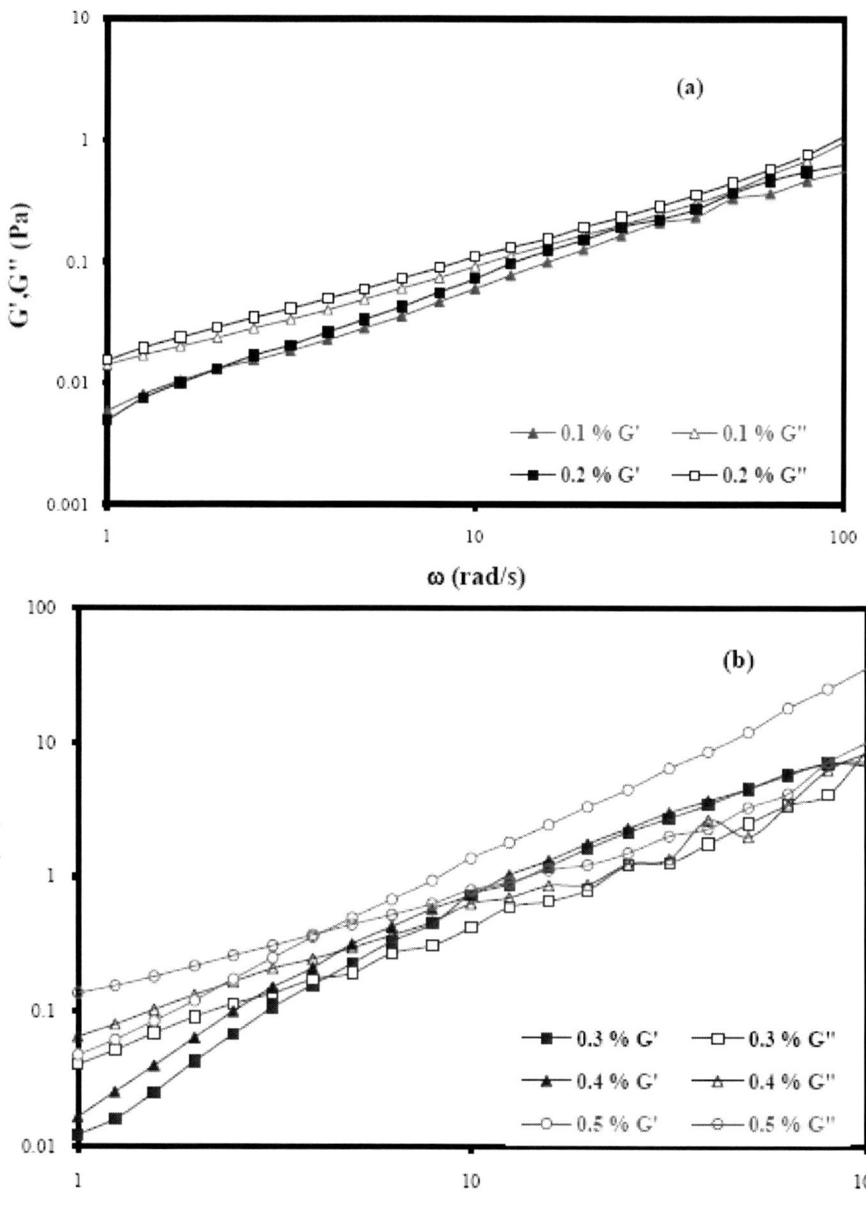

Figure 16. (Continued).

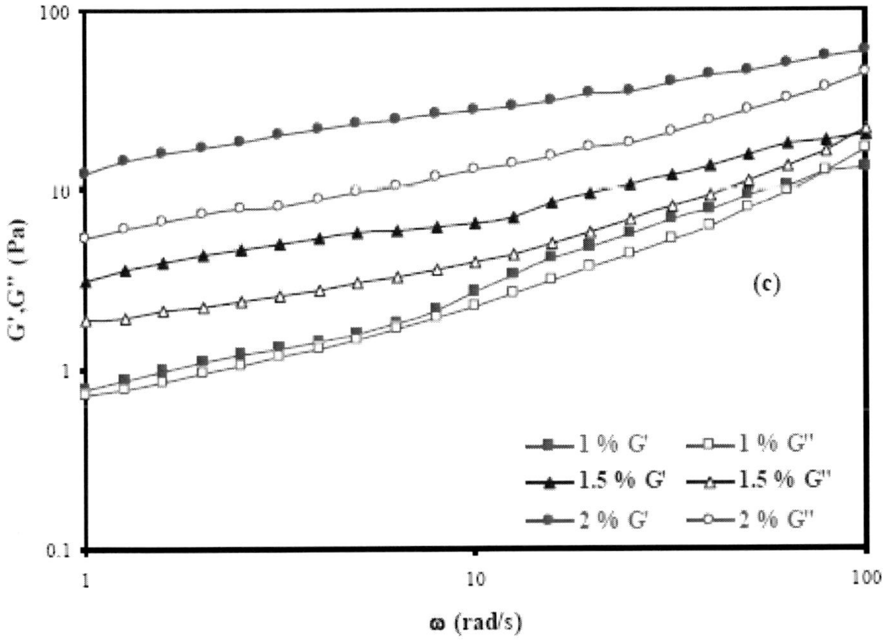

Figure 16. Frequency sweep of G' and G'' of CSG: (a) dilute solutions, (b) semi-dilute solutions, and (c) concentrated solutions [22].

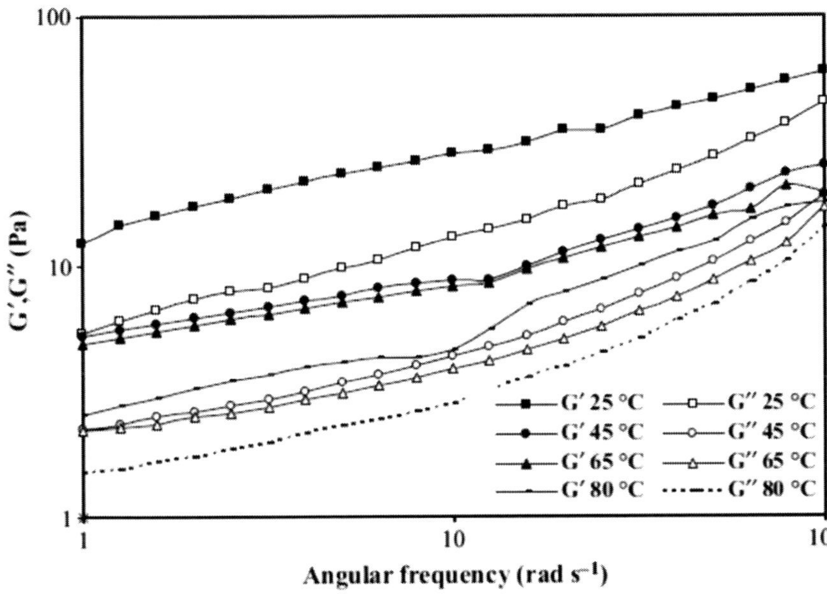

Figure 17. Frequency sweep test of 2 wt.% CSG solution at varied temperatures [22].

3.3.3. Comparison of Steady Shear and Dynamic Rheology: The Cox–Merz Rule

Empirically, the variation of dynamic viscosity (η^* or η') as a function of angular frequency can be compared to steady shear viscosity (η) [105].

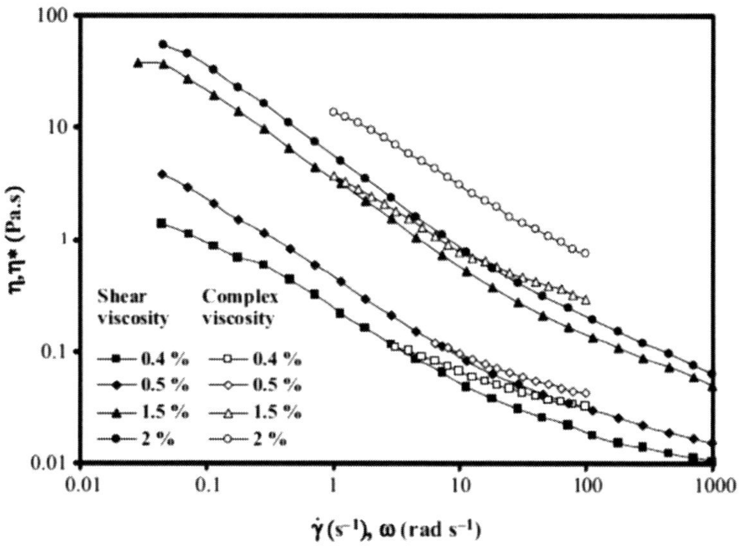

Figure 18. Cox–Merz plot of dynamic viscosity (η^*) and steady shear viscosity (η) at different concentrations of CSG solution [22].

According to the Cox–Merz rule, the two mentioned parameters should be identical if polymer solution is devoid of energetic interactions. This rule has been confirmed experimentally for most of the random-coil biopolymer solutions. In contrast, solutions of polysaccharides (e.g. xanthan, gellan, and schizophyllan) with rigid and ordered chain conformation, behaving like structured liquids, violate the Cox–Merz's rule. In particular, at small deformation η^* is substantially higher than the steady-shear viscosity (η) at equal deformation rates, indicating a tenuous network which remains intact under low-amplitude oscillation but is disrupted under continuous shear [106–109].Figure 18 compares the steady shear viscosity and dynamic viscosity of CSG at varied concentrations. It can be seen that at all the concentrations, the η^* is larger than η, deviating from the Cox–Merz rule. This again manifests the CSG solution being a structured liquid [22].

4. Textural Properties

Texture of a food product is one of the key components of the overall food quality acceptability. In order to keep and improve food texture, hydrocolloids are widely employed as thickening, stabilizing, and gelling agents in food systems due to several advantages including ease in control of the final product utilitarian properties and its acceptability for the consumer [110].Gel formation is a phenomenon involving association or cross-linking of the polymeric chains to form a three-dimensional network that traps or immobilizes solvent (water). Junction zones in hydrocolloid gels may be formed via hydrogen bonding, hydrophobic interactions, electrostatic bonding, and other forces generally weaker as compared to covalent bonds [34, 111]. Although all hydrocolloids thicken aqueous dispersions, only few of gums can potentially form gels.

Textural characteristics of CSG gel as potentially applicable hydrocolloid additive have been investigated by penetration and Texture Profile Analysis (TPA) tests. CSG gel shows reasonable hardness in penetration and TPA tests (11 and 102.67 g, respectively). Cohesiveness, consistency, elasticity, springiness, and gumminess are further important textural properties of CSG gel, but its most prominent characteristic is adhesiveness. Adhesiveness is ability of a gel specimen to become sticky and hence refers to amount of work needed to overcome the attractive forces between the surface of a food product and the surface of the tested material in contact [112]. It is a surface characteristic that depends on a combined effect of adhesive and cohesive forces, viscosity, and viscoelasticity of the sample [113]. This parameter is important, because it may influence overall quality, appearance, and shelf life of a food product. Adhesiveness can be observed in two different ways: adhesion to manufacturing equipment or sticking to fingers and parts of the mouth, and may be considered a positive characteristic (salad dressing, puddings, confectionary products, bakery goods) or a negative one (spaghetti and other pasta products, bread crumbs, some meat products) [34, 114, 115]. Substantial adhesiveness is found for CSG gel when dissolved in water that makes it interesting for binding and agglomerating powders. In general, gels are not physically steady systems, and they tend to turn into unstable (or metastable) state during under a number of conditions. Hence, sufficient knowledge on textural characteristics of hydrocolloid gels in different conditions will assist appropriate selection of a texture modifier for processed foods; especially the food products subjected to different thermal treatment.

4.1. Effect of Heat Treatment

It is of practical importance that CSG gel can maintain its integrity under different thermal conditions. CSG gel can be heated without melting, and different thermal treatments (60°C/30 min, 80°C/23 min, 100°C/18 min, and 121°C/15 min) can improve its textural parameters.Hardness, consistency, cohesiveness, springiness and, adhesiveness of CSG gel increase slightly after the listed thermal treatments, except for heating at 121°C, the latter causing significant increment of the properties. High temperature treatment leads to the gel network with relatively high strength as observed in both penetration and TPA tests, and the highest texture values belong to the 121°C/15 min regime. Chewiness, depending on hardness, dramatically increases after all the listed thermal treatments. At the same time, thermal treatment has no effect on the CSG gel elasticity. The results suggest that heating temperature is more important for textural characteristics of cress seed gum gel than the heating time. The thermal behavior of textural properties of various gum gels is different due to the features of junction zones, and one of the major factors influencing the strength of junction zones is their length [110, 116]. Hydrocolloids like gelatin melt at much lower temperature because the junction zones are merely bound by weak hydrogen bonds. On the other hand, some gels like alginate ones do not melt in boiling water because of the strong calcium ionic bridges in the junction zones [117].The increase of CSG gel hardness and consistency after thermal treatment is attributed to stronger junction zones between cress seed gum molecules. As mentioned above, the CSG molecules exist in semi-rigid conformation with flexibility in between those of random coil and of rigid rod. Such chain conformation leads to formation of numerous macromolecular entanglements accounting for the high stability of hardness. The strength of network after heating increases, so the work needed to overcome the internal junction arises significantly [118].

4.2. Effect of Refrigeration Conditions

Cooling increases the hardness of CSG gel. However, this increase is not significant at 4°C/96 h or 10°C/40 h) [23]. The hardness increase is due to formation of clusters of ordered chains along the polysaccharide molecule upon cooling. Further cooling leads to aggregation of the junction points, building up a rigid gel structure [36]. Similarly, cooling of gelatin gel leads to ordering of additional fragments of the polymer chains, either via growth of the existing junctions or via formation of new ones [119]. It is generally believed that the growth of existing junctions gives the major contribution to the increased rigidity upon cooling [119].

CSG gel consistency increases after cooling (the effect is significant at 4°C/96 h). The same behavior is observed for elasticity, springiness, gumminess, and chewiness [23]. Adhesiveness value decreases; again, the reduction is significant only in the case of cooling at 4°C/96 h. Cohesiveness of CSG gel does not change significantly upon cooling storage. This is attributed to the strong internal bonds and appears to be effective in improving the texture and structure of CSG gel. The same effect of improving cohesiveness upon cooling was reported for xanthan/potato starch [120]. To conclude, when the product is cooled, a cohesive CSG gel is formed, maintaining product integrity during refrigeration operations and reducing moisture losses throughout the product shelf life to efficiently preserve flavor, texture, and color of the product [23].

4.3. Effect of Freeze/Thaw Treatment

Freezing treatment (–18°C/24 h or –30°C/15 h) strengthen the three-dimensional network and enhances hardness, consistency, and gumminess of CSG gel. The gel gets stiffer when the freezing temperature is decreased from –18°C to –30°C [26]. Formation of stronger gels at lower freezing temperature means higher conversion of water into ice. Due to ability of CSG to form strong gels, it can be used to control the shape of frozen foods. Increase in hardness after freezing-thawing process was also reported for the locust bean gum [97], Tamarind seed kernels-sugar gel [121], tapioca/xanthan gel [122], curdlan/locust bean gum mixture [82], and xanthan gel [86].

Cohesiveness remains unchanged after freezing at –18°C during 24 h, whereas its slight increase is observed upon freezing at –30°C during 15 h. According to the importance of the shape stability after thawing, it is highly desirable to employ a gel that can remain stable over freezing and thus stabilize texture of food products [86].

As reported in [26], there is a significant increase in the springiness of CSG gel after freezing treatments ($P>0.05$). This is related to the strong intermolecular association of CSG typical of galactomannans with high M/G ratio, and to the enhanced macromolecular entanglement of relatively rigid macromolecules.

It was reported the freeze-thaw process makes Tamarind seed kernels and locust bean gum gels more elastic [66, 92]. As reported in [92], the polymers are excluded from the growing ice crystals during freezing, thus enriching the unfrozen phase with polymer and promoting the interchain association. Successive freeze-thaw process possibly pushes the already formed network together, as revealed by increase in hydrocolloid pore size, and eventually leads to an increase of elasticity.

5. Applications

In view of diverse properties of CSG described above, there are several interesting applications for this material as liquids slow properties modifier; fat substitute and gelling agent in food industry; and suspending agent, super-disintegrant, and controlled release scaffold in pharmacy.

5.1. Super-Disintegrant

Mucilage isolated from cress seeds can be used as a disintegrant for tablet formulations due to its outstanding disintegrating properties [123, 124]. Enhanced dissolution of Nimesulide tablet can be obtained when cress seed mucilage is used in formulation. The disintegration and mean dissolution time of optimized formulation sample were of 17 s and 5.27 s, respectively, which is better than other tablets prepared from sodium starch glycolate (SSG) and Ac-Di-Sol as synthetic disintegrating agent [123].

5.2. Suspending Agent

Cress seed mucilage can be applied as stabilizer and thickener of choice when high viscosity is desired, especially in cosmetic, pharmaceutical, and food industries. Pharmaceutical manufactures should give ample attention to dissolution characteristics of suspensions similar to those given for solid and semisolid dosage forms. CSG suspensions are stable during 6 month without any change in color, odor, and taste [125].

5.3. Drug Delivery Systems

CSG alone or in combination with other synthetic polymer can be applied as a novel drug delivery system. For example, diffusion of venlafaxine (an antipsychotic drug) from buccalmucoadhesive gel formulations prepared by combining cress seed mucilage and carbopol 934P (a high molecular weight polymer of acrylic acid cross-linked with alkyl ethers of sucrose) was studied. Measurement of mean plasma concentration of venlafaxine indicated that a combination of cress seed mucilage and carbopol 934P allowed prolonged and efficient venlafaxine delivery as compared with oral and intravenous solution administration. Mucoadhesive nature and improved applicability to buccal mucosa revealed some interesting differences that could be relevant to the application [21].

5.4. Controlled Release Tablet

Mucilage of cress seed was successfully used to prepare matrix for sustained release of ibuprofen. The drug release from such tablet usually occurred via a diffusion-controlled mechanism. Ibuprofen sustained release tablet was prepared successfully using cress seed mucilage to retard the drug release and achieve required dissolution profile. Drug release kinetics of that formulation corresponded to the Korsemeyer–Peppas model. However, the release kinetics could not be predicted unambiguously as it appeared to be a complex process involving swelling, diffusion, and erosion. The tablet formulation containing 24% of cress seed mucilage revealed the non-Fickian release kinetics of ibuprofen with maximum release of ibuprofen after 12 h. CSG exhibited an appreciable physicochemical properties and suited best for the development of sustained release tablets as indicated by the drug release studies. The retarding system of CSG with drug provided prolonged duration of treatment and reduced adverse effect in patients [126].

5.5. Emulsion Stabilizer

One of the main food applications of hydrocolloids is stabilizing emulsion system like mayonnaise, salad dressings, sauces, and spreads. However, only a few of hydrocolloids can act as emulsifying agents (gum Arabic, hydrophobic modified starches, modified celluloses, some kinds of pectin, and some galactomannans: guar, locust bean gum, and fenugreek seed gum) [61, 127]. Naji et al. [26] showed that CSG possessed good emulsifying properties at 1% w/v concentration (emulsifying capacity >90%) (Figure 19). Moreover, mucilage of cress seed (0.4% w/v) exhibited excellent emulsifying properties as compared with acacia gum [128]. The stabilizing action of CSG is majorly through modification of aqueous phase viscosity and due to the yield stress imparted to the product. On a molecular level, the large polysaccharides form a thick stabilizing layer and protect oil droplets against aggregation. The emulsion stabilization may also occur via the steric hindrance due to the electrostatic repulsion of glucuronic and galactoronic acids of CSG located at the dispersed oil droplets. On top of that, glycoprotein structure of CSG increases surface activity of CSG solution, imparting enhanced emulsifying properties [26]. Interestingly, thermal processing has beneficial effect on emulsifying behavior of CSG. For example, after 15 min of autoclaving at 121°C the emulsifying capacity of CSG reached 96% (Figure 19).Seemingly, heating caused unfolding of the protein moiety, so that all non-polar amino acid units were directed towards the oil phase. This would improve both emulsifying capacity and long term stability. Under certain thermal treatment conditions, the protein unfolding can be only partial [76]. Of note, heat treatment of CSG solution increased its viscosity which also affected the emulsifying properties [25].

A minor decrease of emulsifying capacity of CSG was observed after freezing. Upon (fast) freezing, small ice crystals were formed with a relatively large surface area of ice–liquid interface; exposure of the protein molecules to the ice–liquid interface led to the proteins damage and, consequently, to decrease of the surface activity of CSG. No significant difference was found between the effects of freezing conditions (temperature and time) on emulsifying behavior of CSG and xanthan [26].

Emulsifying properties of CSG remained unchanged after refrigeration. Since the emulsion stabilization by hydrocolloids results mainly from the increased viscosity of the continuous phase [129], and cooling has no negative effect on viscosity of CSG, the gum solutions retained its high ability to stabilize emulsions [23].

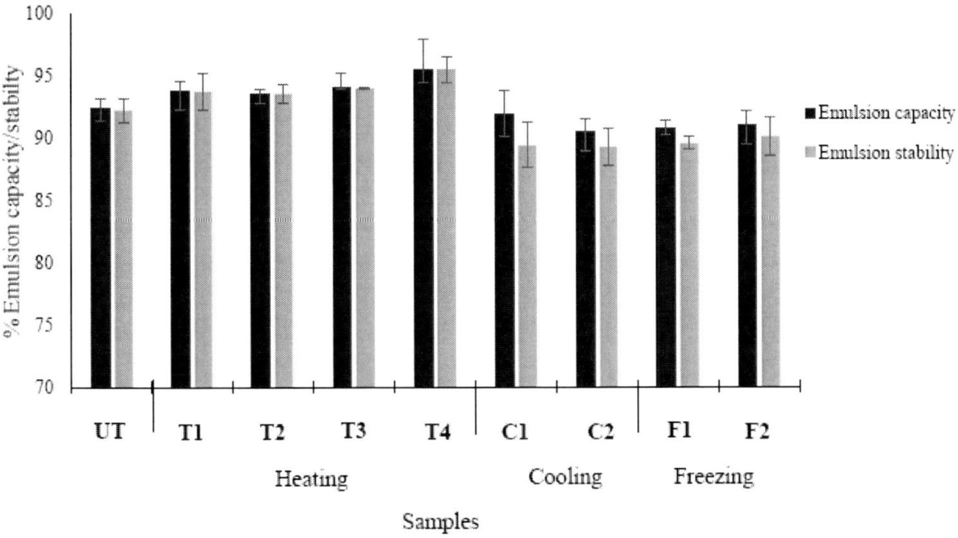

Figure 19. Emulsion properties of CSG after different thermal treatment: UT: untreated; T1: 60ºC/30 min, T2: 80ºC/23 min, T3: 100ºC/18 min, and T4: 121ºC/15 min); C1: 10ºC/40 h and C2: 4ºC/96 h); F1: –18ºC/24 h and F2: –30ºC/15 h). Emulsion stability was expressed in the emulsion volume after heating at 80°C.

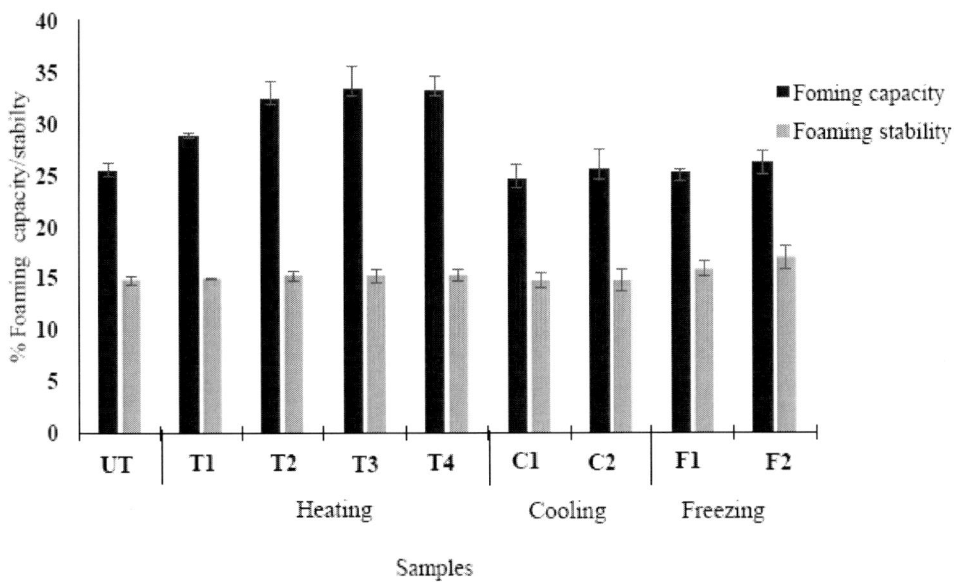

Figure 20. Foaming properties of CSG after different thermal treatment: UT: untreated; T1: 60ºC/30 min, T2: 80ºC/23 min, T3: 100ºC/18 min, and T4: 121ºC/15 min; C1: 10ºC/40 h and C2: 4ºC/96 h; F1: –18ºC/24 h and F2: –30ºC/15 h).

5.6. Foaming Agent

Foams are complex systems, products containing a gaseous phase stabilized in a continuous matrix in order to confer an aerated structure to the material [130]. In food products, incorporation of finely dispersed gas is of great practical importance as it has a pronounced effect on parameters of appearance, texture, consistency, digestibility, size, and overall organoleptic acceptability. The presence of a foaming agent affects the properties of the gas–liquid interface, in particular, the interface viscosity and its permeability. The most common polysaccharides used for foaming are guar gum, xanthan gum, and κ-carrageenan [131–133].

CSG also imparts positive effects on foaming properties (capacity and stability) in food systems, even more pronounced than those in the cases of xanthan gum, guar gum, and flaxseed gum (Figure 20). The foam capacity is attributed to the ability of a molecule to decrease the surface tension at the air–liquid interface. CSG, existing in the form of glycoprotein complex, reduces the surface tension and stabilizes the interfacial film, thus preventing the leakage of air and collapse of the structure. Of note, the foaming capacity of CSG is retained after thermal treatment (heating, freezing, and cooling); in certain cases, the foaming ability is even enhanced after the treatment [25].

5.7. Gluten Substitute

Gluten matrix is a major determinant of the important rheological characteristics of dough (elasticity, extensibility, resistance to stretching, mixing tolerance, and gas holding ability) [134]. The gluten-free bread dough can only keep gas if gluten is replaced by another gel. The gums enhance cohesion of starch granules, and the bread produced can compete with wheat bread, being free of the detrimental effects of gluten [135]. The most commonly used gluten substitutes are hydrocolloids: xanthan, pectin, guar gum, Arabic gum, galactomannans, and methylcellulose [136]. Application of CSG as a gluten substitute in composite rice-wheat flour resulted in improved product parameters such as water absorption, dough development time (time to reach maximum consistency), dough stability, and viscosity. The extensibility value of dough was increased with the cress seed gum concentration growing from 0.3% to 0.6%, and then decreased at 1% of the gum. The crumb firmness decreased with increasing CSG concentration, whereas the water activity was almost unchanged. The effect of CSG on crumb firmness reduction is more pronounced than that of guar gum [137].

5.8. Edible Film Production

Various polysaccharides have been used for preparation of edible films, including starch [138], psyllium seed gum [139], tapioca [140], corn starch [141], cellulose and cellulose derivatives (hydroxypropyl-/methylcellulose, carboxymethylcellulose, and methylcellulose) [142, 143].

Recent studies have demonstrated that CSG is promising for preparation of edible films, and glycerol as plasticizer improves the product quality [144]. All the so produced films were smooth and homogeneous. However, increasing the glycerol concentration enhanced water vapor permeability through CSG films. Glycerol significantly affected the mechanical properties of CSG films to prevent their cracking, increase tensile strength, and decrease elongation at break. However, with 50 wt% of glycerol the tensile strength decreases down to 42%, while the elongation at break increased up to 115% of the original values. As a result, increasing of glycerol concentration improved CSG film extensibility and reduced its resistance. The effects could be due to small amount of plasticizer inserted between the polymer chains to produce a "cross-linker" effect and decrease the polymer the free volume. No marked differences were found between the surface microstructure of the reference and glycerol-plasticized films (Figure 21). The cross section images revealed that the surface of glycerol-free CSG films was rough, whereas glycerol addition led to smooth and uniform surface morphology without noticeable cracks or openings. The low production cost, hydrophilicity, biocompatibility, and biodegradability of the films along with their good mechanical properties make CSG-based materials an excellent alternative to common synthetic plastics [144].

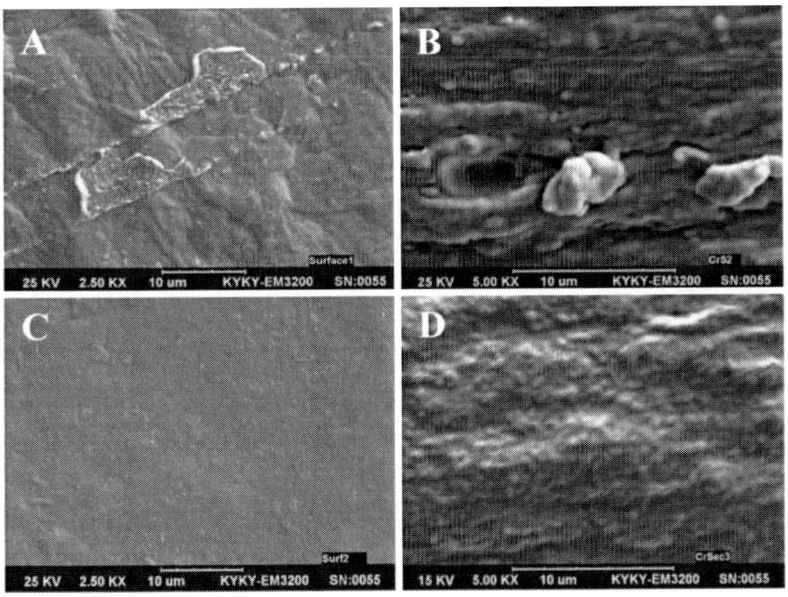

Figure 21. Scanning electron microscopic images of the CSG-based films: (A and B) glycerol-free, (C and D) plasticized with 35 wt% of glycerol. The images display the films surface at 2500× magnification (A and C) and cross-sections at 5000× magnification (B and D) [144].

CONCLUSION

Nowadays hydrocolloids are important in preparation of various formulations. Cress seed gum is a plant-derived hydrocolloid with high molecular weight (540 kDa), taking a conformation is in between random coil and rigid rod. The rheological characteristics demonstrate high potential of CSG as a novel thickener. In addition possible food applications arising from the rheological benefits and low cost of CSG, it exhibits a variety of other possible applications including texture modification, emulsification, foaming, fat substitution, suspending, and edible film production. Food products are subject to different types of thermal treatment during processing operations, and various additives like salts, sugars, acids, etc. may be used. The formulations containing cress seed gum are stable against heating, refrigeration, freezing, pH changes, salts or sugar addition. Consequently, the ranges of application of cress seed gum should further expand because of its stability under different conditions. Hence, the demand for cress seed gum is expected to be enhanced.

REFERENCES

[1] Karazhiyan, H., Razavi, S., Phillips, G.O., Fang, Y., Al-Assaf, S., Nishinari, K., and Farhoosh, R. 2009, *Food hydrocolloids*, 23, 2062–2068.

[2] Sharma, S. and Agarwal, N. 2011, *Indian J. Nat. Prod. Resour*, 2, 292–297.

[3] Karazhiyan, H., Razavi, S., and Phillips, G.O. 2011, *Food Hydrocolloids*, 25, 915–920.

[4] Diwakar, B., Dutta, P., Lokesh, B., and Naidu, K. 2008, *Prostaglandins, LeukotrienesEssent. Fatty Acids*, 78, 123–130.

[5] Bhatia, N.M., Salunkhe, S.S., Mali, S.S., Gadkari, S.S., Hajare, A.A., Gaikwad, S.V., and Karade, R.S. 2014, *Pharm. Lett.*, 6(1), 65–70.

[6] Agarwal, N. and Sharma, S. 2013, *Indian J. Traditional Knowl.*, 12, 699–706.

[7] Datta, P., Diwakar, B., Viswanatha, S., Murthy, K., and Naidu, K. 2011, *Int. J. Appl. Res. Nat. Prod.*, 4, 37–42.

[8] Behrouzian, F., Razavi, S., and Phillips, G.O. 2014, *Bioact. Carbohydrates Dietary Fibre*, 3, 17–28.

[9] Mathews, S., Singhal, R., and Kulkarni, P. 1993, *Food/Nahrung*, 37, 69–71.

[10] Ghante, M., Badole, S., and Bodhankar, S. 2011, *In:* Nuts and seeds in health and disease prevention, Preedy, V., Watson, R., and Patel, V. (Eds.), *ElsevierPress*, London.

[11] Nayak, P.S., Upadhyaya, S.D., and Upadhyaya, A. 2008, *J. Sci. Res.*, 1, 121–127.

[12] Shukla, A., Bigoniya, P., and Srivastava, B. 2012, *Res. J. Med. Plant*, 6.

[13] Macleod, A.J. and Islam, R. 1976, *J. Sci. Food Agric.*, 27, 909–912.

[14] Eddouks, M., Maghrani, M., Lemhadri, A., Ouahidi, M.-L., and Jouad, H. 2002, *J. Ethnopharmacol.*, 82, 97–103.

[15] Nadkami, K. 1954, Indian materiamedica, *Dhootapa-peshwarPrakashan Ltd*, Panvel, India.

[16] Chopra, R.N., Chopra, I.C., and Varma, B. 1986, Glossary of Indian Medicinal Plants: Suppl. [1], *Council of Scientific & Industrial Research, Publications &Informations Directorate.*

[17] Paranjape, A.N. and Mehta, A.A. 2007, *Iran. J. Pharmacol. Ther.*, 5, 55–59.

[18] Niknia, S. and Razavi, S.M.A. 2008, The 5th National Conference on Agricultural Machinery Engineering and Mechanization, 27–28 Aug., *Ferdowsi University of Mashhad*, Iran.

[19] Diwakar, B.T., Dutta, P.K., Lokesh, B.R., and Naidu, K.A. 2010, *J. Am. Oil Chem. Soc.*, 87, 539–548.

[20] Razavi, S.M.A., Farhoosh, R., and Bostan, A. 2007, Research project grant no.1475. *Ferdowsi University of Mashhad*, Iran. Unpublished report.

[21] Nerkar, P.P. and Gattani, S.G. 2012, *J. Mater. Sci. Mater. Med.*, 23, 771–779.

[22] Karazhiyan, H., Razavi, S., Phillips, G.O., Fang, Y., Al-Assaf, S., and Nishinari, K. 2011b, *Int. J. Food Sci. Technol.*, 46, 1066–1072.

[23] Naji, S. and Razavi, S. 2014, *Food Biosci.*, 5, 1–8.

[24] Behrouzian, F., Razavi, S., and Karazhiyan, H. 2013, *Int. J. Food Sci. Technol.*, 48, 2506–2513.
[25] Naji, S., Razavi, S., and Karazhiyan, H. 2012, *Food Hydrocolloids*, 28, 75–81.
[26] Naji, S., Razavi, S.M., and Karazhiyan, H. 2013, *Food Bioproc. Technol.*, 1–10.
[27] Basheer, A.G. 2013, *J. Pharm. Biol. Sci.*, 1, 66–73.
[28] Azuma, J.-i. and Sakamoto, M. 2003, *Trends Glycosci. Glycotechnol.*, 15, 1–14.
[29] Bailey, K. 1935, *Biochem. J.*, 29, 2477.
[30] Bendahou, A., Dufresne, A., Kaddami, H., and Habibi, Y. 2007, *Carbohydr. Polym.*, 68, 601–608.
[31] Cui, W., Mazza, G., Oomah, B., and Biliaderis, C. 1994, *LWT-Food Sci. Technol.*, 27, 363–369.
[32] DivekarVarsha, B. 2010, *Int. J. Pharma Res. Dev.*, IJPRD/2010/PUB/ARTI/VOV-2/ISSUE-1/MAR/006
[33] Patel, H., Kardile, D., Puvar, A., Prajapati, R., and Patel, M. 2010, *Int. J. Pharm. Appl. Sci.*, 1(3), 56–62.
[34] Phillips, G. and Williams, P. 2000, Handbook of hydrocolloids. *Woodhead Publishing Ltd.*, Abington, Cambridge.
[35] Randall, R., Phillips, G., and Williams, P. 1989, *Food Hydrocolloids*, 3, 65–75.
[36] Lee, H.-C. and Brant, D. A. 2002, *Macromolecules*, 35, 2212–2222.
[37] Sato, T., Norisuye, T., and Fujita, H. 1984, *Macromolecules*, 17, 2696–2700.
[38] Pamies, R., Cifre, J.G.H., Martínez, M.d.C.L., and de la Torre, J.G. 2008, *Colloid Polym. Sci.*, 286, 1223–1231.
[39] Sittikijyothin, W., Torres, D., and Gonçalves, M. 2005, *Carbohydr. Polym.*, 59, 339–350.
[40] Behrouzian, F., Razavi, S., and Karazhiyan, H. 2014, *Food Hydrocolloids*, 35, 100–105.
[41] Launay, B., Cuvelier, G., and Martinez-Reyes, S. 1997, *Carbohydr. Polym.*, 34, 385–395.
[42] Morris, E. R., Cutler, A., Ross-Murphy, S., Rees, D., and Price, J. 1981, *Carbohydr. Polym.*, 1, 5–21.
[43] Lapasin, R. and Pricl, S. (1999). Rheology of industrial polysaccharides: Theory and applications, *An Aspen Publ*, USA.
[44] Richardson, P.H., Willmer, J., and Foster, T.J. 1998, *Food Hydrocolloids*, 12, 339–348.
[45] Chen, T.-S., and Joslyn, M. 1967, *J. Colloid Interface Sci.*, 25, 346–352.
[46] Harrison, G., Franks, G., Tirtaatmadja, V., and Boger, D. 1999, *Korea-Australia Rheol. J.*, 11, 197–218.
[47] Stivala, S.S. and Bahary, W.S. 1978, *Carbohydr. Res.*, 67, 17–21.
[48] Eteshola, E., Gottlieb, M., and Arad, S.M. 1996, Chem. Eng. Sci., 51, 1487–1494.
[49] Yoo, S.-H., Fishman, M.L., Hotchkiss Jr, A.T., and Lee, H.G. 2006, *Food Hydrocolloids*, 20, 62–67.
[50] Jumel, K., Harding, S.E., Mitchell, J.R., To, K.-M., Hayter, I., O'Mullane, J.E., and Ward-Smith, S. 1996, *Carbohydr. Polym.*, 29, 105–109.
[51] Higiro, J., Herald, T., Alavi, S., and Bean, S. 2007, *Food Res. Int.*, 40, 435–447.
[52] Viturawong, Y., Achayuthakan, P., and Suphantharika, M. 2008, *Food Chem.*, 111, 106–114.

[53] Razavi, S., Moghaddam, T.M., Emadzadeh, B., and Salehi, F. 2012, *Food Hydrocolloids* 29, 205–210.
[54] Mohammad Amini, A. and Razavi, S. 2012, *Int. J. Biol. Macromol.*, 51, 235–243.
[55] Song, K.-W., Kim, Y.-S., and Chang, G.-S. 2006, *Fibers Polym.*, 7, 129–138.
[56] Szczesniak, A.S. and Farkas, E. 1962, *J. Food Sci.*, 27, 381–385.
[57] Moreno, J., Vargas, M., Madiedo, J.M., Muñoz, J., Rivas, J., and Guerrero, M.G. 2000, *Biotechnol. Bioeng.*, 67, 283–290.
[58] Marcotte, M., TaherianHoshahili, A.R., and Ramaswamy, H. 2001, *Food Res. Int.*, 34, 695–703.
[59] Razavi, S., Taheri, H., and Quinchia, L.A. 2011, *Food Hydrocolloids*, 25, 451–458.
[60] MohdNurul, I., MohdAzemi, B.M.N., and Manan, D.M.A. 1999, *Food Chem.*, 64, 501–505.
[61] Dickinson, E. 2003, *Food Hydrocolloids*, 17, 25–39.
[62] Phillips, P.A., Stephen, G.O., Stephen, A.M., and Churms, S.C. 2006, *In*: Food Polysaccharides and Their Applications, Stephen, A.M. and Phillips, G.O. (Eds.), pp. 455–495, *Taylor and Francis*, London.
[63] Feng, T., Gu, Z., and Jin, Z. 2007, *Food Sci. Technol. Int.*, 13, 55–61.
[64] Launay, B., Doublier, J., and Cuvelier, G. 1985, *In:* Functional properties of food macromolecules, Mitchell J.R. and Ledward D.A., *Springer*, Berlin.
[65] Chen, R.H. and Chen, W.Y. 2001, *J. Appl. Phycol.*, 13, 481–488.
[66] Nishinari, K., Watase, M., Williams, P.A., and Phillips, G.O. 1990, *J. Agric. Food Chem.*, 38, 1188–1193.
[67] Mazurkiewicz, J., Rębilas, K., and Tomasik, P. 2006 *Food Hydrocolloids*, 20, 21–23.
[68] Chenlo, F., Moreira, R., and Silva, C. 2011, *J. Food Eng.*, 107, 234–240.
[69] Cui, W., Eskin, N., and Biliaderis, C. 1993, *Food Chem.*, 46, 169–176.
[70] Dawkins, N. and Nnanna, I. 1995, *Food Hydrocolloids*, 9, 1–7.
[71] Ozdemir, M. and Sadikoglu, H. 1998, *Int. J. Food Sci. Technol.*, 33, 439–444.
[72] Al-Malah, K., Azzam, M., and Abu-Jdayil, B. 2000, *Food Hydrocolloids*, 14, 491–496.
[73] Cancela, M., Alvarez, E., and Maceiras, R. 2005, *J. Food Eng.*, 71, 419–424.
[74] Koocheki, A., Taherian, A. R., and Bostan, A. 2013, *Food Res. Int.*, 50, 446–456.
[75] Wang, B., Wang, L.-J., Li, D., Özkan, N., Li, S.-J., and Mao, Z.-H. 2009, *Carbohydr. Polym.*, 77, 472–481.
[76] Buffo, R.A., Reineccius, G.A., and Oehlert, G.W. 2002, *J. Food Eng.*, 51, 341–345.
[77] Kök, M.S., Hill, S., and Mitchell, J. 1999, *Carbohydr. Polym.*, 38, 261–265.
[78] Yamamoto, F. and Cunha, R. 2007, *Carbohydr. Polym.*, 68, 517–527.
[79] Yamazaki, E., Kurita, O., and Matsumura, Y. 2009, *Food Hydrocolloids*, 23, 655–660.
[80] Glicksman, M. 1982, Food hydrocolloids, *CRC Press Boca Raton*, FL.
[81] Bae, I.Y., Oh, I.-K., Lee, S., Yoo, S.-H., and Lee, H. G. 2008, *Int. J. Biol. Macromol.*, 42, 10–13.
[82] Williams, P.D., Sadar, L.N., and Lo, Y.M. 2009, *J. Food Proc. Preserv.*, 33, 126–139.
[83] Lo, C. and Ramsden, L. 2000, *Food/Nahrung*, 44, 211–214.
[84] Pai, V.B. and Khan, S.A. 2002, *Carbohydr. Polym.*, 49, 207–216.
[85] Sanderson, G.R. 1996, Gums and their use in food systems, *Food technology*, USA.
[86] Soma, P.K., Williams, P.D., and Lo, Y.M. 2009, *Front. Chem. Eng. China*, 3, 413–426.
[87] Saha, D. and Bhattacharya, S. 2010, *J. Food Sci. Technol.*, 47, 587–597.
[88] Imeson, A. 2011, Food stabilisers, thickeners and gelling agents, *John Wiley & Sons.*

[89] Abu-Jdayil, B. and Mohameed, H.A. 2004, *Eur. Food Res. Technol.*, 218, 123–127.
[90] Halmos, A. and Tiu, C. 1981, *J. Texture Stud.*, 12, 39–46.
[91] Tárrega, A., Durán, L., and Costell, E. 2004, *Int. Dairy J.*, 14, 345–353.
[92] Mao, C.-F. and Chen, J.-C. 2006, *Food Hydrocolloids*, 20, 730–739.
[93] Tanaka, R., Hatakeyama, T., and Hatakeyama, H. 1998, *Polym. Int.*, 45, 118–126.
[94] Razavi, S. and Karazhiyan, H. 2009, *Food Hydrocolloids*, 23, 908–912.
[95] Koocheki, A., Mortazavi, S.A., Shahidi, F., Razavi, S.M.A., and Taherian, A. 2009, *J. Food Eng.*, 91, 490–496.
[96] Berski, W., Ptaszek, A., Ptaszek, P., Ziobro, R., Kowalski, G., Grzesik, M., and Achremowicz, B. 2011, Carbohydr. Polym., 83, 665–671.
[97] Abu-Jdayil, B. 2003, *J. Food Eng.*, 57, 97–102.
[98] Dzuy Nguyen, Q., Jensen, C.T., and Kristensen, P.G. 1998, *Chem. Eng. J.*, 70, 165–171.
[99] Giannouli, P. and Morris, E. 2003, *Food Hydrocolloids*, 17, 495–501.
[100] Dea, I., Morris, E.R., Rees, D.A., Welsh, E.J., Barnes, H.A., and Price, J. 1977, *Carbohydr. Res.*, 57, 249–272.
[101] Clark, A.H. and Ross-Murphy, S.B. 1987, *In:* Biopolymers, pp. 57–192, *Springer.*
[102] Steffe, J.F. 1996, Rheological methods in food process engineering, *Freeman press.*
[103] Ross-Murphy, S. 1984. Rheological methods. *In*: Biophysical methods in food research. Chan, H. W. S. (Ed.). Palo Alto, CA, *Blackwell.*pp: 138-199.
[104] Richardson, P., Clark, A., Russell, A., Aymard, P., and Norton, I. 1999, *Macromolecules*, 32, 1519–1527.
[105] Cox, W. and Merz, E. 1958, *J. Polym. Sci.*, 28, 619–622.
[106] Bot, A., Smorenburg, H.E., Vreeker, R., Pâques, M., and Clark, A.H. 2001, *Carbohydr. Polym.* 45, 363–372.
[107] Chronakis, I. S. and Kasapis, S. 1995, *Dev. Food Sci.*, 37, 75–109.
[108] Fang, Y., Takemasa, M., Katsuta, K., and Nishinari, K. 2004, *J. Rheol.*, 48, 1147–1166.
[109] Miyoshi, E. and Nishinari, K. 1999, *Colloid Polym. Sci.*, 277, 727–734.
[110] Saha, D. and Bhattacharya, S. 2010, *J. Texture Stud.*, 41, 459–471.
[111] Roopa, B. and Bhattacharya, S. 2010, *J. Food Proc. Eng.*, 33, 466–480.
[112] Szczesniak, A.S. 1975, *J. Texture Stud.*, 6, 139–156.
[113] Huang, M., Kennedy, J., Li, B., Xu, X., and Xie, B. 2007, *Carbohydr. Polym.*, 69, 411–418.
[114] Fiszman, S. and Damásio, M. 2000, *J. Texture Stud.*, 31, 55–68.
[115] Hoefler, A. 2001, Carrageenan: chemistry, functionality, and applications, *Oxford PA.*
[116] deVries, J. 2004, *In:Gums and stabilizers for the food industry*, Philips G. and Williams P. (Eds.), Vol. 12., *RSC.*
[117] Draget, K., Phillips, G., and Williams, P. 2009, *In:* Handbook of hydrocolloids, Philips G. and Williams P. (Eds.), *Elsevier*, pp. 807–828.
[118] Naji, S., Razavi, S. M., Karazhiyan, H., and Koocheki, A. 2012, *Electron. J. Environ., Agric. Food Chem.*, 11, 222–237.
[119] Ledward, D. 2009, *In:* Handbook of hydrocolloids, Philips G. and Williams P. (Eds.), *Elsevier*, pp. 67-86.
[120] Mandala, I., Palogou, E., and Kostaropoulos, A. 2002. *J. Food Eng.*, 53, 27-38
[121] Nishinari, K., Yamatoya, K., and Shirakawa, M. 2009, *In:* Handbook of hydrocolloids, Philips G. and Williams P. (Eds.), *Elsevier*, pp. 247–267.

[122] Sae-kang, V. and Suphantharika, M. 2006, *Carbohydr. Polym.*, 65, 371–380.
[123] Mehta, K.K., Patel, H.H., Patel, N.D., Vora, C.N., and Patel, N. 2010, *Int. J. Pharm. Pharm. Sci.*, 2.
[124] Sharma, V. and Arora, V. 2012, *Int. J. Pharm. Sci. Res.*, 3, 3947–3954.
[125] Ramling, P. and Sachinkumar, P. 2011, *Indo Am. J. Pharm. Res.*, 1, 181–186.
[126] Salunkhe, S. S., Mali, S. S., Bhatia, N. M., Gadkari, S. S., Bhatia, M. S., and Hajare, A. A. 2014. *Journal drug discovery and therapeutics.* 2, 01-06.
[127] Garti, N., Slavin, Y., and Aserin, A. 1999, *Food Hydrocolloids.*, 13, 145–155.
[128] Patel, M., Chauhan, G., and Patel, L. 1987, *Indian J. Hosp. Pharm.*, 24, 200–202.
[129] Lepine, A., Mezdour, S., Erazo-Majewicz, P., and Michon, C. 2008, *In:* Gums and Stabilisers for the Food Industry 14, Phillips, G. and Williams, P., (Eds.), pp. 245–256, *RSC*, Cambridge.
[130] Marinova, K.G., Basheva, E.S., Nenova, B., Temelska, M., Mirarefi, A.Y., Campbell, B., and Ivanov, I.B. 2009, *Food Hydrocolloids*, 23, 1864–1876.
[131] Miquelim, J.N., Lannes, S., and Mezzenga, R. 2010, *Food Hydrocolloids*, 24, 398–405.
[132] Sciarini, L., Maldonado, F., Ribotta, P., Pérez, G., and León, A. 2009, *Food Hydrocolloids*, 23, 306–313.
[133] Wang, X., Cai, X., Hu, J., Shao, N., Wang, F., Zhang, Q., Xiao, J., and Cheng, Y. 2013, *J. Am. Chem. Soc.*, 135, 9805–9810.
[134] Lazaridou, A., Duta, D., Papageorgiou, M., Belc, N., and Biliaderis, C. 2007, *J. Food Eng.*, 79, 1033–1047.
[135] Mohammadi, M., Sadeghnia, N., Azizi, M.-H., Neyestani, T.-R., and Mortazavian, A.M. 2013, *J. Ind. Eng. Chem.*, 20, 1812–1818.
[136] Mahmoud, R., Yousif, E., Gadallah, M., and Alawneh, A. 2013, *Ann. Agric. Sci.*, 58, 19–25.
[137] Sahraiyan, B., Naghipour, F., Karimi, M., and Davoodi, M. G. 2013, *Food Hydrocolloids*, 30, 698–703.
[138] Osés, J., Niza, S., Ziani, K., and Maté, J.I. 2009, *Int. J. Food Sci. Technol.*, 44, 1360–1366.
[139] Ahmadi, R., Kalbasi-Ashtari, A., Oromiehie, A., Yarmand, M.-S., and Jahandideh, F. 2012, *J. Food Eng.*, 109, 745–751.
[140] Vásconez, M.B., Flores, S.K., Campos, C.A., Alvarado, J., and Gerschenson, L.N. 2009, *Food Res. Int.*, 42, 762–769.
[141] Psomiadou, E., Arvanitoyannis, I., and Yamamoto, N. 1996, *Carbohydr. Polym.*, 31, 193–204.
[142] Pérez, C.D., Flores, S.K., Marangoni, A.G., Gerschenson, L.a.N., and Rojas, A.M. 2009, *J. Agricul. Food Chem.*, 57, 6844–6855.
[143] Sánchez-González, L., Vargas, M., González-Martínez, C., Chiralt, A., and Cháfer, M. 2009, *Food Hydrocolloids*, 23, 2102–2109.
[144] Jouki, M., Khazaei, N., Ghasemlou, M., and HadiNezhad, M. 2013, *Carbohydr. Polym.*, 96, 39–46.

In: Rheology
Editor: Evgeny Karpushkin
ISBN: 978-1-63482-223-7

Chapter 6

VISCOELASTIC AND THIXOTROPIC BEHAVIORS OF WAXY CRUDE OIL AND ITS W/O EMULSION

Houxing Teng*[1,2]*, Guangyu Sun*[1] *and Jinjun Zhang*[1]
[1]College of Mechanical and Transportation Engineering,
China University of Petroleum, Beijing, China
[2]College of Pipeline and Civil Engineering,
China University of Petroleum, Qingdao, Shandong Province, China

Crude oil is a primary and essential energy source throughout the world. The declining resources of conventional light oil and increasing demand for petroleum have resulted in ever-increasing production of waxy crude oil, which represents a significant amount of the petroleum reserves and productions in the world [1]. In China more than 80% of crude oil produced is waxy crude oil. Waxy crude oil shows complex temperature-dependent and shear-and-thermal history-dependent rheological behaviors [2]. Below the wax appearance temperature (WAT), molecules of wax precipitate out and form solid wax crystals in the crude oil. As temperature further decreases, the ever-increasing wax crystals will interlock with each other and form a three-dimensional sponge-like network structure, completing the transition of crude oil from sol to colloidal gel. The gelled waxy crude oil shows complex rheological behaviors, such as viscoelasticity, yield stress, and thixotropy. The poor flowability of waxy crude oils at lower temperatures may result in severe flow assurance problems of oil pipelines [3–5], such as restart failure after the extended shutdown of a pipeline. Therefore, the rheological properties of waxy crude oil have always been the crucial concern for flow assurance. In this chapter, these complex rheological behaviors of waxy crude oils are reviewed with emphasis on the new advances in modeling the viscoelasto-thixotropic behaviors of waxy crude oil.

The rheological behavior will be dramatically changed after the waxy crude oil is emulsified. The yielding mechanism of concentrated W/O emulsion gel of waxy crude oil is different from that of the water-free waxy crude oil gel due to the participation of flocculated water droplets, and the viscoelasticity is enhanced with the increase of water cut and with the

* tel: +86-10-89734627; e-mail: zhangjj@cup.edu.cn

decrease of droplet size. Moreover, owing to the combined effects of wax crystal structure and dispersed droplets, the thixotropy of the emulsion gel is enhanced with the increase of water fraction.

1. Rheological Behavior of Waxy Crude Oil Varying with Temperature

Waxy crude oil shows complex temperature-dependent rheological behaviors, owning to the crystallization of wax with the decreasing of temperature. Waxy crude oil contains mixtures of light and heavy hydrocarbons, resins, and asphaltenes. The wax basically consists of normal alkanes, and some other alkanes such as isoalkane, cycloalkanes and aromatics, usually ranging from C_{18} to C_{65}. At temperatures above the wax appearance temperature (WAT), waxy crude oils behave as a Newtonian fluid as they form homogeneous solutions with wax dissolved in the liquid oil. As the temperature of crude oil decreases, molecules of wax begin to precipitate at the WAT and form solid crystals in the crude oil. As the temperature further decreases, the amount of precipitated solids increases and the crude oil becomes a pseudo-plastic fluid at a temperature called the abnormal temperature [2], and then begins to exhibit thixotropic behavior at some temperature slightly below the abnormal temperature. When the temperature decreases further to the gelation temperature, the precipitated wax crystals form a three dimensional sponge-like network, completing the transition of waxy crude oil from sol to colloidal gel [6], and consequently the waxy crude oil shows complex rheological behaviors, such as viscoelasticity, yielding behavior and thixotropy [6, 7]. As an example, Figure 1 shows the microscopy image of the wax crystals of a typical waxy crude oil, produced in Daqing oilfield that is the largest oilfield in China, with a wax content of 25%. It is surprising that only about 2–3% of precipitated wax, accounted on the basis of total mass of the crude oil, is enough to let gelation occur, with the other 97–98% of liquid oil being trapped within the spongy-like structure of wax crystals [8]!

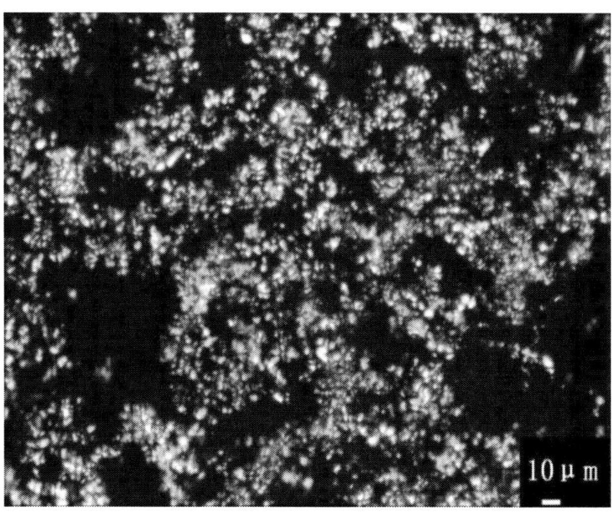

Figure 1. Microscopy image of wax crystals of Daqing waxy crude oil at 33°C.

The rheological properties of waxy crude oil are affected by its thermal and shear history [9–13], i.e. heating a waxy crude oil up to different temperatures and then cooling it down to a temperature below the abnormal temperature will result in different rheological properties, and the waxy crude oil subjected to different shear during the cooling process may also exhibit different rheological properties. For the waxy crude oils treated with the pour point depressant (PPD), a kind of chemical additive for reducing the pour point and apparent viscosity below the abnormal temperature, the effects of thermal and shear history are more significant [14], which is related to the change in the PPD-modified morphology and structure of waxy crystals induced in the shear and thermal process [15]. The change of morphology and structure of wax crystals can be characterized by fractal analysis on microscopic images of wax crystals [16, 17]. Thus the relationship between the shear history effect and the change in the morphology of waxy crystals could be developed [15].

2. Viscoelastic and Thixotropic Behavior of Waxy Crude Oil

Due to the existence of the wax crystal structure, the mechanical response of a gelled waxy crude oil is nearly linearly elastic at the beginning of deformation, and with the increase of shear strain the mechanical response shifts from linear to non-linear viscoelasticity [18]. As the shear strain further increases, the three-dimensional network composed of loose clusters is disrupted into a small number of relatively large flocs or aggregates, losing the connectivity of network structure [6, 18]. Consequently, the mechanical response transfers from the viscoelastic regime to the viscoelastic-and-thixotropic coexistence regime, accompanying with the yielding process. In this process, the elasticity decays, and the mechanical response gradually transfers to be dominated by the viscous effect. After further shearing for a period of time, the internal network structure is nearly totally destroyed, thus the viscoelastic behavior is very weak and even can be ignored [18]. In summary, the mechanical response of a gelled waxy crude oil can be divided into the three regions: a pre-yield (viscoelastic) region, a yielding (viscoelastic–thixotropic) region and a post-yield (thixotropic and/or viscous) region [18–20].

Figure 2 shows the variations of the storage modulus G' and loss modulus G'' during a quiescent cooling and isothermal process for Daqing waxy crude oil at 33°C. It can be seen that at temperatures higher than the WAT (39–50°C), G'' is relatively small (<1 Pa) and the value of G' is so small that it is verified to zero at the beginning of experiments, which indicates that the oil is purely viscous. When the oil is cooled to temperatures below the WAT, both G' and G'' increase sharply with decreasing temperature, but G' increases more strongly than G''. Below the gelation temperature (at which $G' = G''$, that is 34.5°C in this case), G' becomes higher than G'', meaning that the oil shows more elastic behavior than viscous behavior. At the end of the isothermal process, G' is well above G'' and the loss angle is within 25° nearby the gelation temperature, which indicates that the oil has already formed a gel structure.

The test result of stepwise increases and decreases in shear rate for Daqing waxy crude oil at 33°C is shown in Figure 3. It can be seen that the flow curve at the first shear rate (1 s^{-1}) assumes the bell-shape before the shear stress gradually comes to its equilibrium value. This stress overshoot characterizes the viscoelastic behavior. From Figure 3, it can also be observed that the transient flow curves after the maximum shear rate (128 s^{-1}) are nearly parallel to the *X*-axis (i.e. the shear stress almost does not change with time), indicating that the structure rebuild-up rate is very small compared with the structure breakdown rate.

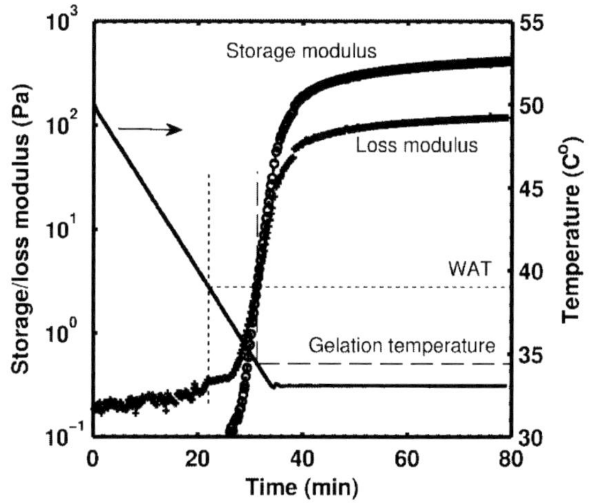

Figure 2. Viscoelastic parameters versus time during the cooling and isothermal processes for Daqing waxy crude oil at 33°C [1].

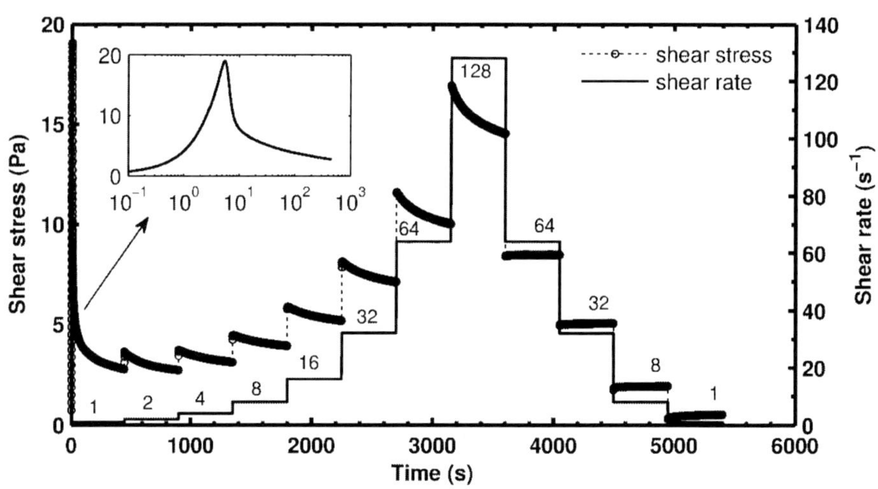

Figure 3. Stress response to stepwise increase and decrease in shear rate for waxy crude oil at 33°C. (The inset shows the stress variation with time for the first shear rate 1 s^{-1}) [2].

Moreover, it can also be observed that after the highest shear rate (128 s^{-1}) the shear stresses at the end of each shear rates are less than those under the same shear rates before the maximum shear rate. These all suggest that the thixotropy of waxy crude oil shows partial irreversibility, which was also demonstrated by other researchers [6, 22].

[1] Reprinted with permission from [21]. Copyright (2013) American Chemical Society.
[2] Reprinted with permission from [21]. Copyright (2013) American Chemical Society.

3. Rheological Behavior of the W/O Emulsion Gel of Waxy Crude Oil

3.1. Viscoelastic Behavior

3.1.1. Enhancement of Viscoelasticity by Dispersed Droplets

The viscoelasticity of the W/O emulsion gel of waxy crude oil is enhanced by dispersed water droplets.

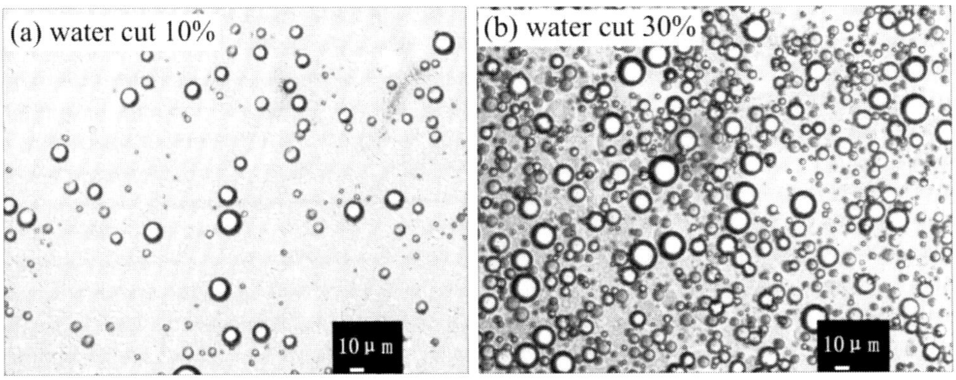

Figure 4. Examples of photocopied microscopy images of crude oil emulsion.

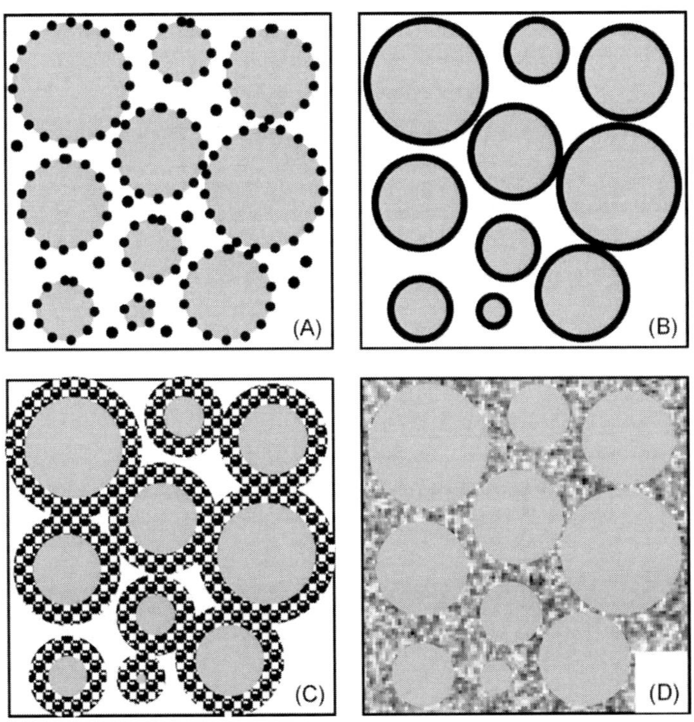

Figure 5. Scheme of the gelation of waxy crude oil emulsion [3].

[3] Reprinted from [25], Copyright (2008), with permission from Elsevier.

As stated in Section 1, for waxy crude oil below the WAT, the precipitated wax particles will interconnect with each other and form a three-dimensional network structure that entraps the liquid oil within it, resulting in the gelation of the crude oil and the emergence of non-Newtonian rheological features such as viscoelasticity, yielding behavior, and thixotropy. As to the water-in-waxy crude oil emulsion, the paraffinic wax will not only precipitate out in the continuous oil phase but also nucleate or absorb at the oil-water interface simultaneously [23, 24]. The microscopy images of crude oil emulsions are shown in Figure 4. When the temperature is just below the WAT, the precipitated wax crystals will adsorb on the droplet surface (Figure 5a). With the temperature continuously lowered, the wax crystals will cover the whole interface and stabilize the emulsion (Figure 5b). Then the precipitated solid wax crystals will continuously grow up on water droplets or between them (Figure 5c), and finally the dispersed water is entrapped by wax crystal network (Figure 5d). As a result, the interface of dispersed water droplets provides more binding sites for the paraffin crystalline particles, making the emulsion easier to gel and the gel structure enhanced [3, 23].

The viscoelasticity of emulsion is enhanced by dispersed droplets owing to the existence of oil-water interfacial tension [26]. When an external shear stress is applied to the emulsion, the droplets will deform and thus additional interfacial area will be produced. This process will consume work due to the resistance of interfacial tension and interfacial energy [27–30]. Once the external stress is reduced or removed, the deformed droplets will return to its original shape. By this way, the elasticity of emulsion is enhanced. Besides, the viscoelastic enhancement of emulsion is also believed to be resulting from the formation of some internal structures of water droplets, such as the flocculation and bridgings of droplets [31].

3.1.2. Effects of Dispersed Phase on the Viscoelasticity of Emulsions

(1) Effect of the volume fraction of dispersed phase

In general, the viscoelasticity of disperse system is enhanced with the increase of the volume fraction. The early researches on the relationship between the storage modulus and the volume fraction of dispersed phase have been conducted by means of theoretical derivations based on rigorous physical assumptions [32–38]. However, these theoretical models are not suitable for the systems where the dispersed droplets are not monodisperse or the volume fraction of dispersed phase is high [39], which makes them hard to be used in engineering applications. On top of that, a variety of semi-empirical models were thereupon brought out to adjust the interactions between droplets [39–41].

As to the viscoelastic behavior of waxy crude oil emulsion gel, it has been shown that during the small amplitude oscillatory measurement in the cooling process, with the increase of the dispersed water the storage modulus grows up and the loss angle decreases (Figure 6). The gelation temperature of emulsion gels increases with the increase of water cut (Figure 7) [42]. When the structure equilibrium is reached after some quiescent duration, there exists a power law relation between the storage modulus and the volume fraction of water cut for the water cut above 35% [25].

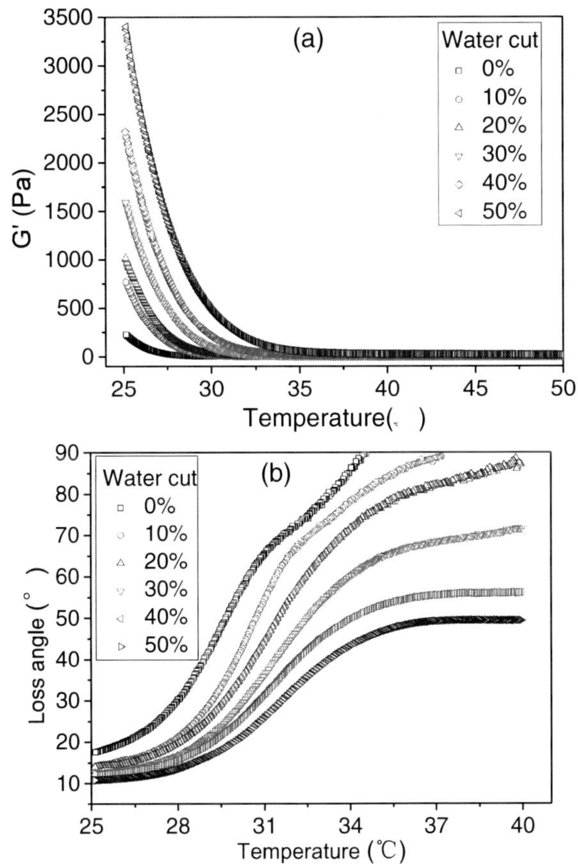

Figure 6. Evolution of storage modulus and loss angle of waxy crude oil emulsion with temperature during the cooling process [4].

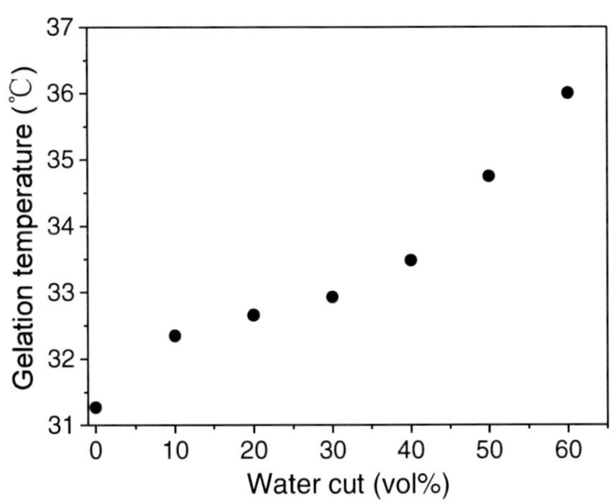

Figure 7. Gelation temperature of waxy crude oil emulsions versus water cut. [5]

[4] Reprinted with permission from [42]. Copyright (2014) American Chemical Society.
[5] Reprinted with permission from [42]. Copyright (2014) American Chemical Society.

Besides the oscillatory measurement, the creep-recovery behavior within the linear viscoelastic region was also investigated to obtain further information about the viscoelasticity of waxy crude oil emulsion gel [42]. It was demonstrated that both the creep and recovery behavior may well be described by a mechanical analogy model with one Maxwell model and two Kelvin–Voigt models associated in series. However, the structural recovery capability of emulsion gels is significantly stronger than that of crude oil. This may be ascribed to the recoverable nature of water droplets. To be specific, the deformed water droplets possess a spontaneous tendency to restore to their original spherical shape once the stress is removed, while the wax crystal structure is more brittle and breaks irreversibly upon application of stress. Actually when the water cut is higher than 40%, the deformation produced in the creep process can be all recovered, as shown in Figure 8 [42].

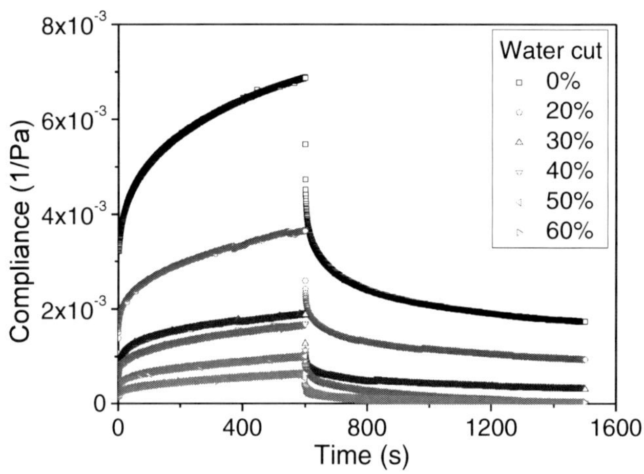

Figure 8. Creep-recovery behavior of waxy crude oil and its emulsion gel [6].

(2) Effect of the droplet size of dispersed phase

The rheological behavior of concentrated emulsion is dependent on oil droplet size [43, 44]. According to the study on the effect of droplet size on the viscoelasticity of petroleum emulsions, Pal maintained that fine concentrated emulsions, i.e. emulsions with a smaller average droplet size, have much higher storage moduli than the corresponding coarse emulsions [45].

In order to clear up the role of droplet size on the storage modulus of concentrated emulsions quantitatively, some further studies have been conducted. Based on model calculations and careful experiments, most research results have indicated that there is a linear relationship between the storage modulus and the reciprocal of droplet diameter [46–50], as shown in the well-known semi-empirical correlation proposed by Princen [46].

$$G = a\left(\sigma/R_{32}\right)\phi^{1/3}\left(\phi - b\right) \tag{1}$$

where G is the elastic modulus of the concentrated emulsion, σ is the interfacial tension, R_{32} is the Sauter mean droplet radius, ϕ is the volume fraction of the dispersed phase, a and

[6] Reprinted with permission from [42]. Copyright (2014) American Chemical Society.

b are constants. Princen maintained that this equation is universally valid for different kinds of highly concentrated emulsions and foams.

Some others correlations are also exist, such as the linear relationship between the storage modulus and the reciprocal of the square of droplet diameter, implying that elasticity is determined by the surface of droplets [44], which may be ascribed to the type of the emulsion used and to the difference in the procedure of averaging droplet size for polydisperse samples.

3.1.3. Topics Needed to Be Further Studied

Given the diverse interactions and polydispersity of the droplets in emulsions, no universal theoretical correlation has been built up between the viscoelasticity and the dispersed droplets. As we know, the crude oil emulsion above the WAT is a Newtonian fluid when it is very dilute. In this case, the emulsion is inelastic and its rheological behavior is very simple. However, when the concentration of internal phase becomes high, there exists a critical volume fraction beyond which the rheological properties are considerably modified [51]. In fact, the dispersed droplets come into contact with each other at high concentration and thus the non-rigid interface of the droplets becomes deformed, which makes it hard to study the viscoelasticity theoretically. Furthermore, the viscoelastic property is also affected by the polydispersity of droplets [29], making the theoretical research even harder.

Studies on the viscoelasticity of waxy (or heavy) crude oil emulsion and/or emulsion gel are in urgent need to be complemented compared with the achievements obtained on model oil emulsion. So far, most researches focus on the concentrated model oil emulsion, while the rheological properties of crude oil emulsion may be different from those of model oil emulsion for that crude oil contains complex components such as paraffins, asphaltenes, and resins. For water-in-waxy crude oil emulsions below the WAT, at the present time it has only known that the precipitated wax crystals may interact strongly with the dispersed droplets, making the viscoelasticity remarkably enhanced by the dispersed water. More rheological studies in combination with micro-mechanism investigation are expected for further research.

3.2. Yield Behavior

Waxy crude oil shows yield behavior at temperatures nearby pour point because of the precipitation of wax crystals and the gelation of the oil [52]. Yielding is an important time-dependent rheological characteristic and an indication of the structural strength. So far, the yield behaviors of waxy crude oil under different loading conditions have been intensively studied. As to waxy crude oil emulsion, most researches focus on the yield behaviors of crude oil emulsion gel after the precipitation of wax. Nevertheless for high water cut emulsions with the absence of wax crystals, there still exists yield stress originating from the weak flocculated water droplet network [23].

3.2.1. Yield Stress

Experimental results show that the yield stress of waxy crude oil emulsion gel increases monotonically with the increase of dispersed water [25, 42, 53–57]. Guo [58] explained this phenomenon as follows. For an emulsion gel, the applied stress is shared by both the wax crystal structure and the water droplets, thus the yield stress of an emulsion gel is higher than that of the crude oil, and the yield stress increases with increasing water cut for the number of droplets increases. Nevertheless, divergence appeared when the increase extent of yield stress with water cut was involved. The results of Visintin et al. [25], Guo et al. [57], and Li et al. [56] have demonstrated that when water cut is below 30% the yield stress of waxy crude oil emulsion gel increases slowly with increasing water cut, while for water cut above 30% the yield stress increases dramatically. This is different from the results obtained by Rønningsen [53], which have showed that the yield stress rises markedly with increasing water cut in the whole water cut range.

However, Oh and Deo [59] have found that the yield stress of the emulsion gel with a small amount of wax content decreases with increasing water cut, and they ascribed this phenomenon to the limited amount of wax that is insufficient to entrap all of the water droplets. As a result, the incomplete wax crystal structure may allow dispersed droplets to move freely and not act like wax particles. Similarly, Paso et al. [23] have found that there exists an optimal water cut at which the structural strength of the emulsion gel reaches its maximum. This optimal water cut imparts a large specific interfacial area to the gel while retaining a sufficient volumetric density of crystal-crystal and crystal-interface linkages to strongly bind the water-oil interfaces to the paraffin crystal network. From the above researches, it can be concluded that the wax precipitation and wax content is crucial to the structural strength of waxy crude oil emulsion gel.

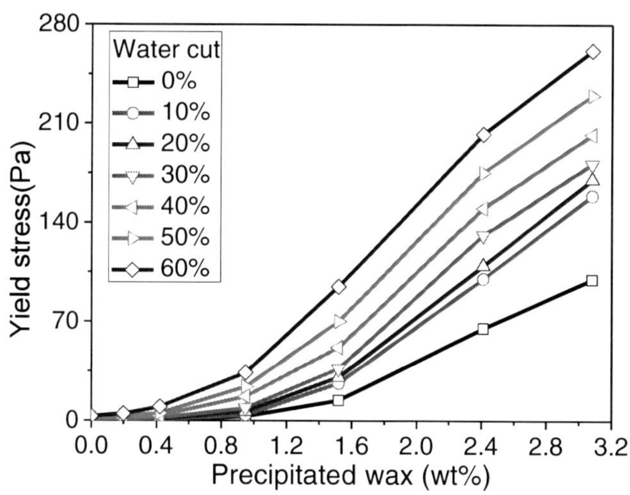

Figure 9. Yield stress of waxy crude oil and its emulsion versus precipitated wax .[7]

The yield stress of waxy crude oil emulsion gel increases with decreasing temperature, which is similar to the situation of water-free waxy crude oil. As can be easily understood, the paraffinic wax in the oil precipitates out continuously with the lowering of temperature, thus the strength of the wax crystal network structure enhances, and thus the yield stress increases (Figure 9). It has been found that the yield stresses of crude oil and its emulsion gel are exponential to the relative temperature, which is defined as the temperature difference between the test temperature and the gelation temperature [42]:

[7] Reprinted with permission from [42]. Copyright (2014) American Chemical Society.

$$\tau_y = \alpha \exp\left(-\beta\left(T_m - T_g\right)\right) \tag{2}$$

where τ_y is the yield stress; T_m is the test temperature; T_g is the gelation temperature; α represents the yield stress at the gelation temperature; β characterizes the increasing rate of structural strength with decreasing temperature.

3.2.2. Yield Strain

With the experimental methods continually advanced and the knowledge of the yield phenomenon gradually deepened, whether the yield stress is an inherent nature of a material is suspected by some researchers [60–62], while the strain at yielding is thought to be a parameter to characterize the yielding process of a material more reasonably [63, 64]. In contrast to the yield stress, the yield strain is basically independent of the loading conditions. That is to say, the gel structure will yield and flow as long as the shear strain reaches a critical value, as shown in Figure 10.

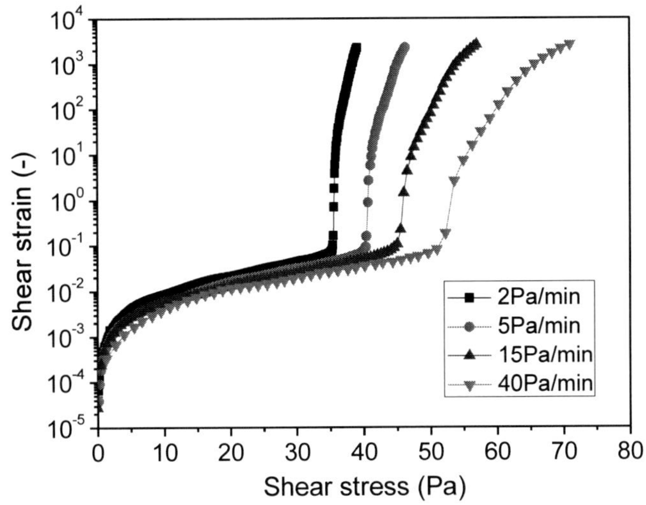

Figure 10. Evolution of the shear strain of waxy crude oil at different loading rates of stress [64].

The yield strain of waxy crude oil emulsion gel increases remarkably with the increase of dispersed water, indicating that significant deformation of water droplets occurs before the maximum yield stress attained [23, 42]. Paso et al. [23] have mentioned that the yield strain increases from about 1.5% in the absence of emulsified water to nearly 20% at the water cut of 70%.

Further investigations on the evolution of the yield strain of waxy crude oil and its emulsion gel with precipitated wax have been carried out to investigate the yielding mechanism of emulsion gels [42]. The results showed that the yield strain of concentrated emulsions (above 30%) first increases and then decreases with decreasing temperature (i.e., with increasing precipitated wax) (see Figure 11). Further inquiries have revealed that the yielding mechanism of waxy crude oil emulsions is changed by the precipitation of wax crystals. When the water cut is above 30%, as the temperature decreases, the yield mechanism changes from the domination of the damage of droplet clusters to the domination of the fracture of wax crystal network, due to the increasing amount of precipitated wax and the formation of wax crystal network that may entrapping the dispersed water.

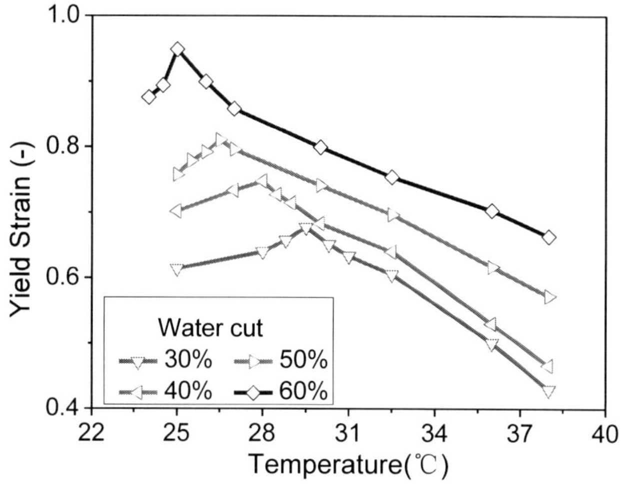

Figure 11. Evolution of the yield strain of the concentrated emulsions with temperature below the WAT. [8]

3.3. Thixotropic Behavior

Thixotropy is another important time-dependent characteristic for water-in-waxy crude oil emulsion gel. The thixotropic behavior of emulsion gel resembles that of the water-free crude oil. The difference is that the thixotropic behavior of waxy crude oil emulsion gel is not only influenced by the wax crystal structure but also influenced by the dispersed water droplets.

As the water fraction increases and the temperature decreases, the thixotropy of emulsion gel enhances [56, 58]. This is attributed to the combined effects of wax crystal structure and dispersed droplets. During shearing, the coalescence of dispersed droplets occurs, and the increased droplet size may reduce the viscosity and thus the shear stress of emulsions decreases. Moreover, the coalescence makes the space between droplets increase, which results in weaker interactions between droplets and thus the shear stress reduces consequently.

It has demonstrated that after shearing the structure of water-free waxy crude oil gel cannot be fully recovered [6, 11, 22, 65–69]. So is the situation for the water-in-waxy crude oil emulsion gel. According to the traditional definition of thixotropy, the broken-down structure should be able to fully recover on rest. There remains the controversy all long about whether or not the not-fully-recovered structured fluids could be called thixotropic fluids [70–73]. Nevertheless, the term "thixotropy" is still used in many works for this kind of fluid [21, 66, 68, 71, 74–78].

[8] Reprinted with permission from [42]. Copyright (2014) American Chemical Society.

4. MODELING THE VISCOELASTIC AND THIXOTROPIC BEHAVIORS OF WAXY CRUDE OIL

4.1. Viscoplastic Models for Waxy Crude Oil

Most of the existing thixotropic models belong to the structural kinetics model, which uses a scalar structural parameter λ to characterize the instantaneous level of microstructure, and its value varies between the values of 0 for the totally broken-down structure and 1 for the fully developed structure. The inelastic structural thixotropic model is composed of an equation of state and a kinetic equation, and their general formats are shown as follows [79].

$$\begin{aligned} &\tau = \tau(\lambda, \dot{\gamma}) \\ &\frac{\mathrm{D}\lambda}{\mathrm{D}t} = g(\lambda, \dot{\gamma}) \end{aligned} \tag{3}$$

The thixotropic behavior of materials including waxy crude oil has been extensively studied and various thixotropic models of this type have been put forward [79–84]. Some were introduced in the reviews of Cheng [85] and Barnes [86], and extensive lists of available models have been compiled by Mujumdar et al. [87] and Mewis and Wagner, respectively [70].

Currently most used thixotropic model for waxy crude oil is that developed by Houska for slurry of solid materials [81], which was later widely used for restarting flow computations of waxy crude oil pipelines. The model is as follows.

$$\begin{aligned} &\tau = \tau_{y0} + \tau_{y1}\lambda + (k + \Delta k\lambda)\dot{\gamma}^{n} \\ &\frac{\mathrm{d}\lambda}{\mathrm{d}t} = a(1-\lambda) - b\dot{\gamma}^{m}\lambda \end{aligned} \tag{4}$$

In this model, the yield stress comprises a residual, permanent component τ_{y0} when the thixotropic structure is completely broken down $(\lambda = 0)$, and a structural contribution $\lambda\tau_{y1}$. As for the apparent viscosity, it is characterized by a structure-parameter-dependent consistency Δk (the so-called structural consistency) and a completely unstructured consistency k. The shear thinning behavior is described by a kinetic index n. In this equation, a stands for the structure build-up rate, and b for the breakdown rate, and m as well as n are fitting parameters.

However, for the Houska model and some other thixotropic models such as the Moore model [80] and Alexandrou model [88], it has been demonstrated that after a sudden change in shear rate, the structural parameter changes exponentially to its equilibrium value [89], thus the shear stress of waxy crude oil predicted by those models decreases to its equilibrium value more quickly than the measured data [90]. To address this problem, Teng and Zhang [91] proposed a new viscoplastic thixotropic model for waxy crude oil. In the kinetic equation for the structural parameter, a new pre-factor, which contains shear strain as variable, is introduced for the structure buildup and breakdown term. And the kinetic equation's breakdown term is assumed to be a function of energy dissipation rate rather than on shear rate, which will be discussed later in section 4.4. The new model is as follows [91].

$$\begin{aligned} \tau &= \lambda\tau_y + (k + \Delta k\lambda)\dot{\gamma}^{n_1} \\ \frac{d\lambda}{dt} &= \frac{1}{1+\gamma^{n_2}}\left[a(1-\lambda) - b\lambda\phi^m\right] \end{aligned} \tag{5}$$

Compared with the Houska model, it can be seen that the permanent part τ_{y0} in new model is neglected. This is because that for waxy crude oil, after shearing under a high shear rate for a period of time, the waxy crude oil behaves as a shear thinning fluid with the residual yield stress negligible. In this equation, n_2 and m are positive dimensionless constants.

Figure 12 shows the model fitting results of stepwise shear rate test for Daqing waxy crude oil at 33°C. The fitting and predictive capability of the proposed model is shown to be excellent for the stepwise shear rate test and hysteresis loop test. And it has demonstrated that for the breakdown term of structural kinetic equation, the assumption that it is a function of energy dissipation rate is proved to be better than the assumption that the breakdown term is related with shear rate, for it takes account of not only the influence of shear rate but also the influence of shear stress on the breakdown rate of structural parameter. After a sudden change in shear rate, the proposed kinetic equation predicts the structural parameter with a varying, non-exponential shape [91].

As to the determination of the unknown model parameter, here take Teng and Zhang model for example, the constants of the equation of state (τ_y, k, Δk and n_1), the ratio b/a and m can be determined via substituting the equilibrium structural parameter obtained from the kinetic equation into the equation of state through a least square fitting to the steady state flow curve. And the values of the remaining parameters (and/or all of the model parameters) can be determined via fittings to transient data. Another method to obtain the model parameter is that the values of all model parameters are obtained via fittings to the data pertaining to transient flow through a computer program with the nonlinear least-square method. The kinetic equation can be solved numerically by the fourth order Runge–Kutta discretization method.

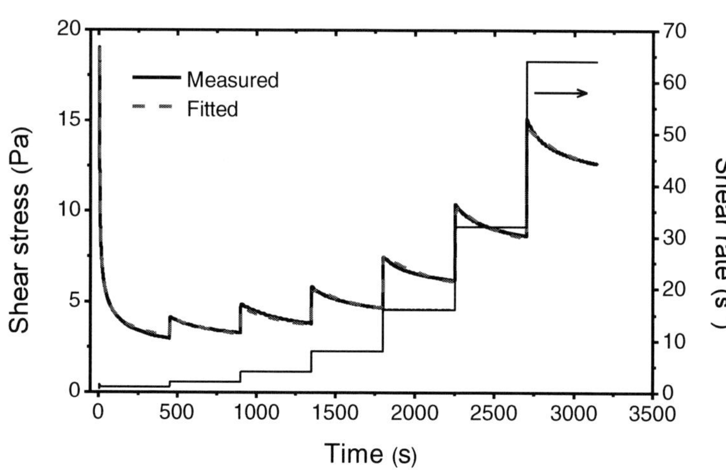

Figure 12. Model fittings of stepwise shear rate test for Daqing waxy crude oil at 33°C.

It has been found that the hysteresis loop test alone is somewhat not suitable to be used to determine the values of model parameter [21, 91]. The reason is that the transient data of stepwise shear rate test contains the steady state information, for the steady-state values of stress at different shear rates can be extracted from the test of the stepwise change of shear rate, while the transient data of hysteresis loop doesn't contain steady state information. For the hysteresis loop test, due to the simultaneous change of two variables of shear rate and time during the experiment, the structural parameter, apparent viscosity and other parameters don't arrive at their equilibrium values, which leads to the fact that the fitted values of model parameter may deviate from their true values, especially for the model parameters (a, b and m) related with the structural kinetic equation. However, for the stepwise shear rate test, the apparent viscosity at each shear rate nearly reaches its equilibrium value, so are the structural parameter and other parameters. Thus, the model parameters should be fitted from the stepwise shear rate test, and should not be fitted from the hysteresis loop test.

4.2. Viscoelasto-Plastic Models for Waxy Crude Oil

As stated above, most of the existing thixotropic models used for waxy crude oil are viscoplastic model, utilizing a yield stress to characterize the yielding behavior [86, 87]. The viscoplastic models assume that the material begins to flow only if the shear stress is greater than yield stress. For this reason, they can only depict the thixotropic behavior after yielding, but cannot account for the initial viscoelastic behavior before the yield point [92]. Some of the recent studies attempted to describe both the viscoelastic and thixotropic effects of thixotropic materials with yield stress [86, 87, 93–95]. There are generally two types of structural kinetic thixotropic model, Type I and Type II, to take account of both the viscoelasticity and the thixotropy of materials [94]. Type I model assumes that the total shear stress is comprised of an elastic and a viscous stress, using the elastic stress to represent the elastic behaviors [87, 94, 96, 97]; while Type II model, also named Maxwell-type model, utilizes the Maxwell [92, 93] or Jefferys [95] mechanical analog to account for the viscoelastic property.

Type I model uses the elastic stress to account for the viscoelastic behavior. The stress equation, which is usually algebraic, is not established on a mechanical analog, but originates from the Bingham model. And to describe the evolution of elastic strain pertaining to the microstructure, an additional kinetic equation is needed. Doraiswamy et al. [96] proposed a model for concentrated suspensions, which assumes that the material has an initial elastic behavior before yielding and a purely viscous behavior afterward. Mujumdar et al. [87] extended this model to describe the smooth transition of thixotropic materials with yield stress from an elastic dominated phase to a viscous dominated phase. Dullaert and Mewis [94] developed a general structural kinetics thixotropic model for thixotropic systems based on inelastic suspending fluid to describe both the structure-dependent elastic and viscous contributions of particles and medium. At the very beginning of deformation, the shear stress predicted by the Dullaert–Mewis model [94] has elastic and viscous components; while for the Mujumdar model [87], the shear stress only has the elastic component without the viscous component. This is one distinction between them. Nevertheless, many papers show that for waxy crude oil [6, 7, 18-20] and other thixotropic materials with yield stress [87, 98, 99], the initial mechanical response is elastic. From this aspect, the initial mechanical response predicted by the Mujumdar model seems more realistic and plausible for these materials.

Type II model is usually established on a mechanical analog composed of a series of ideal spring and dashpot, and incorporates structural kinetic theory to describe the thixotropic behavior [100]. In this case, thixotropy can be considered as a special case of more general non-linear viscoelasticity [92]. For the viscoelastic information is already embedded in the stress equation, an additional evolution equation for elastic strain is not needed. The mechanical analogs of different models are usually different, and many of the published models have been reviewed in literature [87, 92]. More recently, de Souza Mendes [93] proposed a novel Maxwell model for elasto-viscoplastic thixotropic fluids. For the kinetic equation for the structural parameter, the steady-state flow curve is used as an input and the breakdown term is assumed to be dependent on stress rather than on shear rate. Subsequently, another similar model is developed by replacing the Maxwell mechanical analog with the Jeffreys mechanical analog [95].

Of Type I and Type II methods, Mewis and Wagner [70] recommends models of Type I while the recent review of de Souza Mendes and Thompson [101] argues that Type II method is more robust. One main argument that de Souza Mendes and Thompson are against Mewis and Wagner as well as the Type I method is that "yielding in the classical sense does not depend on shear history". Indeed, there is some controversy among rheologists regarding whether yielding is dependent on shear histories or not. Some scholars [84, 88, 102], besides Mewis and Wagner, argue that "yield stress depends on the history of the sample and there is a close link between thixotropy and yield stress, because both phenomena are believed to be caused by the same underlying physics - the internal microstructure". More recently, Moller et al. showed that this controversy disappears when a clear distinction is made between two types of yield stress fluids: thixotropic and non-thixotropic fluids; the yield stress of thixotropic fluids depends on the shear history of the sample, while for the non-thixotropic ones the yield stress depends only on the shear rate [103]. As to waxy crude oil, it shows yield stress and thixotropy, and the experimental data show that the yield stress does depend on the shear history of the sample [6, 7, 18, 104]. The viscoelasticity of waxy crude oil, very sensitive and fragile, is different from that of polymer-like materials (e.g. polymer solution, polymer melt), for the internal structure is destroyed very sharply nearby the yield point. The waxy crude oil's viscoelasticity mainly refers to the viscoelastic behavior before (and/or nearby) the yield point; the most dominant and important rheological behavior of waxy crude oil is still thixotropy. The Maxwell-type model seems inappropriate to describe this viscoelasto-plastic behavior.

To fully describe both the viscoelasticity and thixotropy of waxy crude oil, Teng and Zhang [21] proposed a new viscoelasto-plastic thixotropic model, without using a mathematical yield criterion (a critical yield strain or a critical yield stress) to tackle the yielding behavior. In this model, the shear stress is resolved into an elastic component and a viscous component. The elastic response at different strains is described by a non-linear damping function together with a structural parameter. A damping function, Soskey and Winter equation, often used to describe the nonlinear viscoelasticity of polymers, is adopted here to describe the shear modulus decay of waxy crude oil as a function of elastic strain. The evolution of elastic strain is thought to be continuous and described by a specially proposed kinetic equation. The variation of structural parameter is described by a new kinetic equation in which a pre-factor with shear strain as variable is introduced, and the breakdown term is assumed to be dependent on the rate of energy dissipation rather than on shear rate and/or shear stress. The mechanical response predicted by the model is viscoelastic when the structural parameter is close to one, and as the structural parameter decreases the mechanical response gradually shifts to be thixotropic.

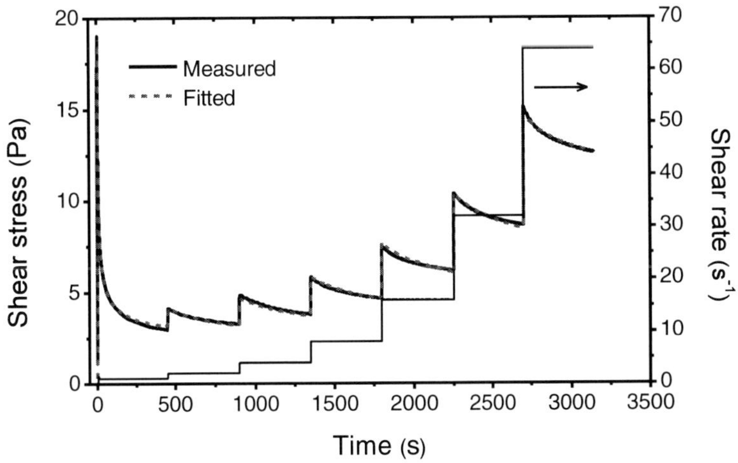

(a) shear stress vs. time

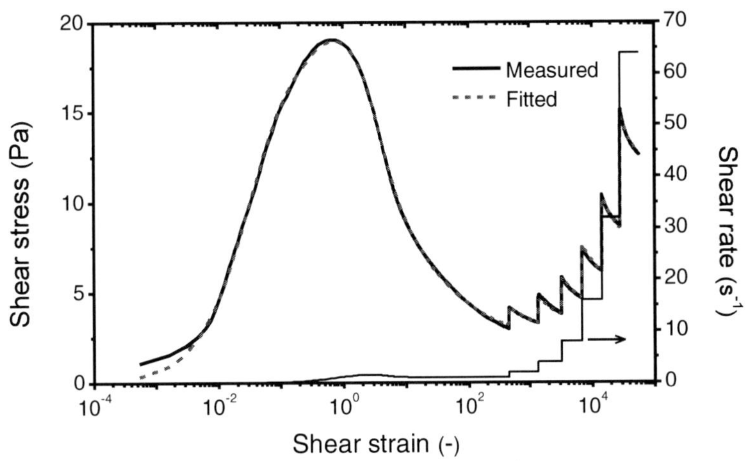

(b) shear stress vs. shear strain

Figure 13. Model fittings of stepwise shear rate test data for Daqing waxy crude oil at 33°C .[9]

The model is as follows:

The equation of state:

$$\tau = \lambda G_0 h(\gamma_e)\gamma_e + (1-\lambda)\cdot(\lambda\Delta k + k)\dot{\gamma}^{n_1}$$
$$h(\gamma_e) = \frac{1}{1 + p_1 \cdot \gamma_e^{\,p_2}} \tag{6}$$

[9] Reprinted with permission from [21]. Copyright (2013) American Chemical Society.

The kinetic equation for the structural parameter λ:

$$\frac{d\lambda}{dt} = \frac{1}{1+\gamma^{n_2}}\left[a(1-\lambda) - b\lambda\phi^{m}\right] \tag{7}$$

The kinetic equation for the elastic strain γ_e:

$$\frac{d\gamma_e}{dt} = \left[g_1 - (1-g_1)s\dot{\lambda}\right]\dot{\gamma}$$
$$\text{with} \quad g_1 = e^{-(p_1\cdot\gamma^{p_2})} \quad \gamma = \int_0^t \dot{\gamma}(t')dt' \tag{8}$$

In summary, the proposed model contains eleven adjustable parameters, i.e. G_0, k, Δk, n_1, n_2, a, b, m, s, p_1, p_2. And these model parameters can be categorized into the three groups: (1) three parameters related with the viscous property: k, Δk, n_1; (2) another three parameters corresponding to the elastic property: G_0, p_1 and p_2; (3) and the rest four parameters, n_2, a, b and m, which are used to describe the time evolution of structural parameter.

Figure 13 shows the model's fitting results of stepwise shear rate test for Daqing waxy crude oil at 33°C. The proposed model well depicts the complete transient flow curve without a discontinuity, from the initial elastically-dominated region, through the viscoelasticity and thixotropy coexistence region accompanying with yielding process, and finally to the viscously-dominated region. The new model can offer predictions of elastic and viscous property evolution with time. And for waxy crude oil with the increase of the shear rate in the start up flow, the peak shear stress predicted by the model is shifted toward to a larger value, while the time for the shear stress reaching its peak value reduces.

4.3. Thixotropic Models for the W/O Emulsion Gel of Waxy Crude Oil

Usually these thixotropic models proposed for water-free waxy crude oils can be directly applied to waxy crude oil emulsion gels [23, 58], and the fitting results are shown to be good due to the similarity in rheological behaviors between waxy crude oil and its emulsion gel. For instance, Figure 14 shows the fitting results of hysteresis loops of waxy crude oil emulsion gel by the thixotropic model proposed by Teng and Zhang [21].

Although the thixotropic models proposed for water-free waxy crude oil can describe well the thixotropy of crude oil emulsion gel, they cannot manifest the influence of water cut. The water cut of an emulsion can vary broadly. Guo [58] has found that under constant shear rate, the relative shear stress, which is defined as the ratio of the shear stress of emulsion to that of the continuous phase, i.e. the water-free oil, sheared at the same shear rate and the same time, is almost independent on time; and the average relative shear stress is observed to vary exponentially with water cut. Based on this phenomenon, Guo proposed a three-parameter model that can accurately estimate the thixotropic behavior of emulsion gel [58].

$$\tau_r = (A - B\dot{\gamma})\exp(K\phi) \tag{9}$$

where τ_r is the average relative shear stress; ϕ is the water cut; $\dot{\gamma}$ is the shear rate, s^{-1}; A, B, and K are fitting parameters. By using this method and the thixotropic measurement data of crude oil and its emulsion at three water cuts, the shear stress decay of waxy crude oil emulsion at arbitrary water cuts can be predicted with acceptable accuracy, which is of great significance in engineering applications.

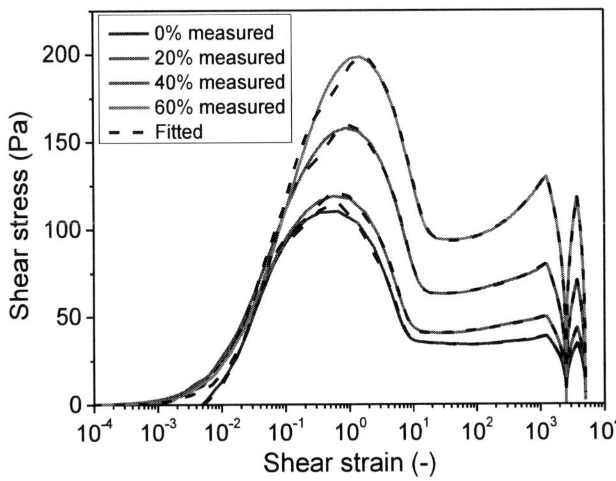

Figure 14. Model fitting results of hysteresis loops for waxy crude oil and its emulsion gel.

4.4. Kinetic Equations for the Structural Parameter

As for the kinetic equation describing the evolution of structural parameter λ, in analogy with chemical reaction kinetics, it is usually assumed that its time evolution is controlled by the combined result of structure buildup and breakdown rates. So far many kinetic equations have been put forward. Recently Teng and Zhang [21, 91] developed a new kinetic equation for the structural parameter as follows.

$$\frac{d\lambda}{dt} = \frac{1}{1+\gamma^{n_2}}\left[a(1-\lambda) - b\lambda\phi^m\right] \quad (10)$$

In this equation, the first term on the right-hand side exhibits the structure buildup, while the second term expresses the structure breakdown.

Some kinetic equations introduce a common pre-factor for the buildup and breakdown terms [70]. Phenomenologically, the pre-factor generates stretched exponentials for the evolution of λ. The pre-factor doesn't change the equilibrium value of λ at each shear rates. It only changes the descending rate of λ coming to its equilibrium value. By this way, it substantially improves the model's fitting capability. For example, the model proposed by Dullaert and Mewis [94] introduces a pre-factor with time as variable, that is $1/t^{n_2}$. As a consequence, the value of the pre-factor at the same time point is the same no matter what the shear rate of start-up flow is. Nevertheless, the descending rate of λ at a higher shear rate of start-up flow is obviously different from that at a lower shear rate. So it seems somewhat inconsistent with the actual evolution of λ. To solve this problem, Teng and Zhang [21, 91] introduce a new pre-factor $1/(1+\gamma^{n_2})$, which takes shear strain as variable. This pre-factor follows the same functionality of the Perkins-Turner correlation [105]. At the same time point, as the shear rate of start-up flow increases, the shear strain γ gets larger value, hence the value of pre-factor becomes smaller. Therefore, the descending rate of λ at a higher shear rate is different from that at a lower shear rate. Thus the pre-factor that takes γ as variable seems more plausible than that takes time as variable.

For the thixotropic materials with yield stress, it is physically reasonable to expect that for the structural parameter, (i) the breakdown rate is low at early times when the microstructure is nearly un-deformed, and as time elapses and the shear strain increase, the breakdown rate increases; (ii) nearby the yield point the breakdown rate comes to its maximum value; (iii) after that it begins to decay. Generally, the kinetic equation for structural parameter assumes the following form:

$$\frac{\mathrm{d}\lambda}{\mathrm{d}t} = g(\lambda,\dot{\gamma}) - \lambda f(\tau,\dot{\gamma}) \tag{11}$$

where $g(\lambda,\dot{\gamma})$ expresses the structure buildup rate, while $f(\tau,\dot{\gamma})$ represents the structure breakdown rate. Yziquel et al. [106] as well as Mewis and Wagner [70] have summarized that the evolution of λ can be categorized into the following three types of possible kinetic equation. The mostly used one assumes that the structure change is due to the shear rate [81, 87, 94], and some suppose that the structure evolution is associated with the shear stress [93, 95], while only a few kinetic equations assume that the variation of λ is related to a combination of the two, e.g. the rate of energy dissipation [106, 107]. As analyzed by de Souza Mendes [93, 95], if the function $f(\tau,\dot{\gamma})$ is taken to be dependent on shear rate (e.g. $f(\tau,\dot{\gamma}) \propto \dot{\gamma}^{m}$), $f(\tau,\dot{\gamma})$ is zero at $\dot{\gamma} = 0$, and increases monotonically with the increase of $\dot{\gamma}$. For a completely structured material initially at rest, if it is suddenly subjected to a constant shear rate (artificially assumed here), the stress is zero at t =0 and increases with time during the elastic regime until it reaches the yield stress. However, the function $f(\tau,\dot{\gamma})$ does not change with time. As a consequence, the breakdown term $\lambda f(\tau,\dot{\gamma})$ is maximum at the onset of applied shear rate (because $\lambda = 1$, and $f(\tau,\dot{\gamma})$ is constant), and decreases as λ decreases. This is not consistent with the actual evolution of λ. Therefore, de Souza Mendes [93, 95] came to the conclusion that it is non-physical to assume that the destruction of microstructure is a function of shear rate. One thing to note is that the non-physical responses mentioned above lie in the viscoelastic part before the yield point. Therefore, it doesn't show up in the thixotropic models (e.g. visco-plastic models) without considering the viscoelastic part before the yield point.

To overcome this shortcoming, the two models proposed by de Souza Mendes [93, 95] assume that the breakdown term of the structural kinetic equation is a function of the instantaneous stress. This is physical because for the fully structured gelled waxy crude oil, it is the bond between structural units that determines the gel strength, and the external force is the agent that causes structure breaking. However, for the thixotropic dominated regime (away from the yield point) when the network structure is broken down into isolated flocs suspending in the liquid, it is the surface shear stress experienced by an isolated floc that causes flocs breaking down further. And it has shown that the surface shear stress is proportional to the shear rate raised to a power [86]. Therefore, in this regime it is physical to assume that the breakdown term of the structural kinetic equation is dependent on shear rate. This is also the reason why most of the viscoplastic models' kinetic equations assume that the breakdown term is dependent on shear rate.

To take consideration of the initial elastic dominated regime and the thixotropic (and/or viscous) dominated regime (away from the yield point), Teng and Zhang [21, 91] assumed that $f(\tau,\dot{\gamma})$ is a function of a combination of shear stress and shear rate, e.g. the rate of energy dissipation ϕ, which is defined as $\phi = \tau\dot{\gamma}$ in the simple shear flow. For the constant shear rate condition, the shear stress will increase with time during the elastic regime until it reaches its peak value, thus the energy dissipation rate increases with time, so are the values of $f(\tau,\dot{\gamma})$ and $\lambda f(\tau,\dot{\gamma})$. As a result, the breakdown term $\lambda f(\tau,\dot{\gamma})$ will arrive at its maximum value nearby the peak shear stress point. This is in line with the actual evolution of λ. It has been demonstrated that by using this kinetic equation for the structural parameter the model we proposed can better predict the thixotropic behaviors of waxy crude oil than the kinetic equations using either shear stress or shear rate alone in the breakdown term [21, 91].

ACKNOWLEDGMENTS

The authors gratefully acknowledge the financial support from the National Natural Science Foundation of China (No. 51134006).

REFERENCES

[1] Rønningsen, H.P., Bjoerndal, B., Baltzer Hansen, A., Batsberg Pedersen, W. 1991, *Energy Fuels*, *5*, 895–908.
[2] Yan, D. and Luo, Z. 1987, *SPE Production Engineering*, Nov., 267–275.
[3] de Oliveira, M.C.K., Carvalho, R.M., Carvalho, A.B., Couto, B.C., Faria, F.R., Cardoso, R.L. 2010, *Energy Fuels*, *24*, 2287–2293.
[4] Zhang, J., Yu, B., Li, H., Huang, Q. 2013, *Pet. Sci.*, *10*, 538–547.
[5] de Oliveira, M.C.K.; Teixeira, A.; Vieira, L.C.; de Carvalho, R.M.; de Carvalho, A.B.M.; do Couto, B.C. 2011, *Energy Fuels*, *26*, 2688–2695.
[6] Visintin, R.F., Lapasin, R., Vignati, E., D'Antona, P., Lockhart, T.P. 2005, *Langmuir*, *21*, 6240–6249.
[7] Magda, J.J., El-Gendy, H., Oh, K., Deo, M.D., Montesi, A., Venkatesan, R. 2009, *Energy Fuels*, *23*, 1311–1315.
[8] Li, H., Zhang, J., Yan, D. 2005, *Pet. Sci. Technol.*, *23*, 1313–1322.
[9] Wardhaugh, L.T., Boger, D.V. 1991, *AIChE J.*, *37*, 871–885.
[10] Ding, J., Zhang, J., Li, H., Zhang, F., Yang, X. 2006, *Energy Fuels*, *20*, 2531–2536.
[11] Jia, B., Zhang, J. 2012, *Ind. Eng. Chem. Res.*, *51*, 10977–10982.
[12] Hou, L., Zhang, J., Sun, L. 2009, *Pet. Sci. Technol.*, *27*, 2168–2176.

[13] Li, H., Zhang, J. 2014, *Pet. Sci. Technol*, *32*, 521–526.
[14] Zhang, J., Zhou, S., Li, H. Proceedings of the XIV International Congress on Rheology; Seoul, Korea, Aug 22–27, 2004.
[15] Yi, S., Zhang, J. 2011, *Energy Fuels*, *25*, 5660–5671.
[16] Gao, P., Zhang, J., Ma, G. 2006, *J. Phys.: Condens. Matter.*, *18*, 11487–11506.
[17] Yi, S., Zhang, J. 2011, *Energy Fuels*, *25*, 1686–1696.
[18] Chang, C., Boger, D.V., Nguyen, Q.D. 1998, *Ind. Eng. Chem. Res.*, *37*, 1551–1559.
[19] Oh, K., Deo, M.D. 2008, *Energy Fuels*, *23*, 1289–1293.
[20] Oh, K., Jemmett, M., Deo, M. D. 2009, *Ind. Eng. Chem. Res.*, *48*, 8950–8953.
[21] Teng, H., Zhang, J. 2013, *Ind. Eng. Chem. Res.*, *52*, 8079–8089.
[22] Rønningsen, H.P. 1992, *J. Petrol. Sci. Eng.*, *7*, 177–213.
[23] Paso, K., Silset, A., Sørland, G., Gonçalves, M.D.A., Sjöblom, J. 2009, *Energy Fuels*, *23*, 471–480.
[24] Haj-shafiei, S., Ghosh, S., Rousseau, D. 2013, *J. Colloid Interface Sci.*, *410*, 11–20.
[25] Visintin, R.F., Lockhart, T.P., Lapasin, R., D'Antona, P. 2008, *J. Non-Newtonian Fluid Mech.*, *149*, 34–39.
[26] Lee, H.M., Lee, J.W., Park, O.O. 1997, *J. Colloid Interface Sci.*, *185*, 297–305.
[27] Gladwell, N., Rahalkar, R.R., Richmond, P. 1986, *Rheol. Acta*, *25*, 55–61.
[28] Dolz, M., Hernández, M.J., Delegido, J. 2008, *Food Hydrocolloid*, *22*, 421–427.
[29] Otsubo, Y., Prud'homme, R.K. 1994, *Rheol. Acta*, *33*, 29–37.
[30] Leal-Calderon, F., Thivilliers, F., Schmitt, V. 2007, *Curr. Opin. Colloid Interface Sci.*, *12*, 206–212.
[31] Schramm, L.L. 1992, *Adv. Chem. Ser.*, *231*, 428.
[32] Batchelor, G. K., Green, J.T. 1972, *J. Fluid Mech.*, *56*, 401–427.
[33] Kerner, E.H. 1956, *Proc. Phys. Soc. B*, *69*, 808–813.
[34] van der Poel, C. 1958, *Rheol. Acta*, *1*, 198–205.
[35] Uemura, S., Takayanagi, M. 1966, *J. Appl. Polym. Sci.*, *10*, 113–125.
[36] Schowalter, W.R., Chaffey, C.E., Brenner, H. 1968, *J. Colloid Interface Sci.*, *26*, 152–160.
[37] Palierne, J.F. 1990, *Rheol. Acta*, *29*, 204–214.
[38] Heinrich, G., Klüppel, M., Vilgis, T.A. 2002, *Curr. Opin. Solid State Mater. Sci.*, *6*, 195–203.
[39] Pal, R. 2002, *J. Colloid Interface Sci.*, *245*, 171–177.
[40] Lewis, T.B., Nielsen, L.E 1970,. *J. Appl. Polym. Sci.*, *14*, 1449–1471.
[41] Gašperlin, M., Tušar, L., Tušar, M., Kristl, J., Šmid-Korbar, J. 1998, *Int. J. Pharm.*, *168*, 243–254.
[42] Sun, G., Zhang, J., Li, H. 2014, *Energy Fuels*, *28*, 3718–3729.
[43] Clark, P.E., Pilehvari, A. 1993, *J. Petrol. Sci. Eng.*, *9*, 165–181.
[44] Malkin, A.Y., Masalova, I., Slatter, P., Wilson, K. 2004, *Rheol. Acta*, *43*, 584–591.
[45] Pal, R. 1996, *AIChE J.*, *42*, 3181–3190.
[46] Princen, H.M., Kiss, A.D. 1986, *J. Colloid Interface Sci.*, *112*, 427–437.
[47] Aronson, M.P., Petko, M.F. 1993, *J. Colloid Interface Sci.*, *159*, 134–149.
[48] Langenfeld, A., Schmitt, V., Stébé, M.J. 1999, *J. Colloid Interface Sci.*, *218*, 522–528.
[49] Babak, V.G., Langenfeld, A., Fa, N., Stébé, M.J. 2001, *Progr. Colloid Polym. Sci*, *118*, 216–220.
[50] Pal, R. 2002, *J. Non-Newtonian Fluid Mech.*, *105*, 21–33.

[51] Jager-Lézer, N., Tranchant, J.F., Alard, V., Vu, C., Tchoreloff, P.C., Grossiord, J.L. 1998, *Rheol. Acta*, *37*, 129–138.
[52] Stokes, J.R., Telford, J.H. 2004, *J. Non-Newtonian Fluid Mech.*, *124*, 137–146.
[53] Rønningsen, H.P. 2012, *Energy Fuels*, *26*, 4124–4136.
[54] Rønningsen, H.P., Karan, K. *Proc. 10th BHR Int. Conf. Multiphase*, Cannes, France, June 13–15, 2001
[55] Haj-shafiei S. Rheological properties and flow behaviour of wax-stabilized water-in-oil emulsions. M.Sc. Thesis, Ryerson University, Toronto, Canada, 2011.
[56] Li, S., Huang, Q., He, M., Wang, W. 2014, *J. Dispersion Sci. Technol.*, *35*, 1114–1125.
[57] Guo, L., Wang, L., Song, Y. 2011, *Sci. Technol. Eng.*, *11*, 4372–4376. (In Chinese)
[58] Guo L. Study on thixotropic behaviors of waxy crude oil and its emulsions. Ph.D. Thesis, China University of Petroleum (Beijing), Beijing, 2010.
[59] Oh, K., Deo, M.D. 2011, *Fuel*, *90*, 2113–2117.
[60] Barnes, H. 1999, *J. Non-Newtonian Fluid Mech.*, *81*, 133–178.
[61] Evans, I.D. 1992, *J. Rheol.*, *36*, 1313–1316.
[62] Astarita, G. 1992, *J. Rheol.*, *36*, 1317–1318.
[63] Li, C. Research on the structural characteristics of gelled waxy crude oil and the mechanism of its pour point reduction. Ph.D. Thesis, Shandong University, Jinan, China, 2003.
[64] Hou, L. Study on rheological properties of gelled waxy crudes. Ph.D. Thesis, China University of Petroleum (Beijing), Beijing, 2005.
[65] Kané, M., Djabourov, M., Volle, J.L., Lechaire, J.P., Frebourg, G. 2003, *Fuel*, *82*, 127–135.
[66] Teng, H., Zhang, J. 2013, *Petrol. Sci.*, *10*, 395–401.
[67] Webber, R.M. 2001, *Ind. Eng. Chem. Res.*, *40*, 195–203.
[68] Kané, M., Djabourov, M., Volle, J.L. 2004, *Fuel*, *83*, 1591–1605.
[69] Wardhaugh, L.T., Boger, D.V. 1987, *Chem. Eng. Res. Des.*, *65*, 74–83.
[70] Mewis, J., Wagner, N.J. 2009, *Adv. Colloid Interface Sci.*, *147*, 214–227.
[71] Nguyen, Q.D., Boger, D. V. 1985, *Rheol. Acta*, *24*, 427–437.
[72] de Kretser, R.G., Boger, D.V. 2001, *Rheol. Acta*, *40*, 582–590.
[73] Roussel, N. 2006, *Cem. Concr. Res.*, *36*, 1797–1806.
[74] Guo, L., Zhang, J., Han, S., Teng, H. 2013, *Petrol. Sci. Technol.*, *31*, 895–901.
[75] Chang, C., Nguyen, Q.D., Rønningsen, H.P. 1999, *J. Non-Newtonian Fluid Mech.*, *87*, 127–154.
[76] Vinay, G., Wachs, A., Agassant, J.F. 2005, *J. Non-Newtonian Fluid Mech.*, *128*, 144–162.
[77] Davidson, M.R., Nguyen, Q.D., Rønningsen, H.P. 2007, *J. Petrol. Sci. Eng.*, *59*, 1–16.
[78] Vinay, G., Wachs, A., Frigaard, I. 2007, *J. Non-Newtonian Fluid Mech.*, *143*, 141–156.
[79] Cheng, D.C.; Evans, F. 1965, *Br. J. Appl. Phys.*, *16*, 1599–1617.
[80] Moore, F. 1959, *Trans. Br. Ceram. Soc.*, *58*, 470–494.
[81] Houska, M. Engineering aspects of the rheology of thixotropic liquids. Ph.D. Thesis, Czech Technical University of Prague, Prague, 1981.
[82] Lacuna, J.L., Rudé, E., Mans, C. 1996, *Progr. Colloid Polym. Sci.*, *100*, 252–258.
[83] Toorman, E.A. 1997, *Rheol. Acta*, *36*, 56–65.
[84] Coussot, P., Nguyen, Q.D., Huynh, H.T., Bonn, D. 2002, *J. Rheol.*, *46*, 573–589.
[85] Cheng, D.H. 1987, *Int. J. Cosmet. Sci.*, *9*, 151–191.

[86] Barnes, H.A. 1997, *J. Non-Newtonian Fluid Mech.*, *70*, 1–33.

[87] Mujumdar, A., Beris, A.N., Metzner, A.B. 2002, *J. Non-Newtonian Fluid Mech.*, *102*, 157–178.

[88] Alexandrou, A.N., Constantinou, N., Georgiou, G. 2009. *J. Non-Newtonian Fluid Mech.*, *158*, 6–17.

[89] Kirkwood, D., Ward, P. 2008, *Mater. Lett.*, *62*, 3981–3983.

[90] Zhang, J., Guo, L., Teng, H. 2010, *Appl. Rheol.*, *20*, 53944–53950.

[91] Teng, H., Zhang, J. 2013, *Rheol. Acta*, *52*, 903–911.

[92] Labanda, J., Llorens J. 2006, *Rheol. Acta*, *45*, 305–314.

[93] de Souza Mendes P.R. 2009, *J. Non-Newtonian Fluid Mech.*, *164*, 66–75.

[94] Dullaert, K., Mewis, J. 2006, *J. Non-Newtonian Fluid Mech.*, *139*, 21–30.

[95] de Souza Mendes, P.R. 2011, *Soft Matter*, *7*, 2471–2483.

[96] Doraiswamy, D., Mujumdar, A.N., Tsao, I., Beris, A.N., Danforth, S.C., Metzner, A.B. 1991, *J. Rheol.*, *35*, 647–685.

[97] Huang S., Lu C. 2005, *Acta Mech. Sin.*, *21*, 330–335.

[98] Liddel, P.V., Boger, D.V. 1996, *J. Non-Newtonian Fluid Mech.*, *63*, 235–261.

[99] Putz, A.M., Burghelea, T.I. 2009, *Rheol. Acta*, *48*, 673–689.

[100] Zhu, C., Smay, J.E. 2011, *J. Rheol.*, *55*, 655–672.

[101] de Souza Mendes, P.R., Thompson R.L. 2012, *J. Non-Newtonian Fluid Mech.*, *187–188*, 8–15.

[102] Møller, P., Mewis, J., Bonn, D. 2006, *Soft Matter*, *2*, 274–283.

[103] Møller, P., Fall, A., Chikkadi, V., Derks, D., Bonn, D. 2009, *Phil. Trans. R. Soc. A*, *367*, 5139–5155.

[104] Hou, L. 2012, *Rheol. Acta*, *51*, 603–607.

[105] Perkins, T., Turner, J. 1971, *J. Pet. Technol.*, *23*, 301–308.

[106] Yziquel, F., Carreau, P.J., Moan, M., Tanguy, P.A. 1999, *J. Non-Newtonian Fluid Mech.*, *86*, 133–155.

[107] Fredrickson, A. 1970, *AlChE J.*, *16*, 436–441.

In: Rheology
Editor: Evgeny Karpushkin

ISBN: 978-1-63482-223-7

Chapter 7

RHEOLOGY OF POLYSACCHARIDE ION-GELS PREPARED IN IONIC LIQUIDS

***Kamalesh Prasad*[1,2*] *and Jitkumar Bhatt*[1,2]**

[1]Marine Biotechnology and Ecology Discipline,
CSIR-Central Salt and Marine Chemicals Research Institute (CSIR-CSMCRI),
Council of Scientific & Industrial Research (CSIR)
Gijubhai Badheka Marg, Bhavnagar, (Gujarat), India,
[2]AcSIR- Central Salt & Marine Chemicals Research Institute
G. B Marg, Bhavnagar, (Gujarat), India

ABSTRACT

Various polysaccharides such as cellulose, starch, guar gum, agar, carrageenan, xanthan gum, tamarind gum, etc. have been reported to form strong ion-gels in ionic liquids. Composite ion-gels of the polysaccharides in various ionic liquids have been reported as well. The so obtained ion-gels are viscoelastic in nature, and various rheological properties have been thoroughly evaluated for such gel systems. In this chapter, the characteristic rheological behavior of the polysaccharide ion-gels prepared in different ionic liquids is discussed. Thixotropy and self-healing behavior of a couple of polysaccharide ion-gels based on guar gum and tamarind gum are discussed in detail.

Keywords: Polysaccharide, ion-gel, viscoelasticity, self-healing ability, thixotropy

INTRODUCTION

Polysaccharides are widespread in nature and have been regarded as structural materials and as sources of water and energy for animals and plants [1]. They have become increasingly important because of the unique structure and properties, much different from those of typical

* Phone +91-278-2567760. Fax +91-278-2567562; e-mail: kamlesh@csmcri.org / drkamaleshp@gmail.com

synthetic polymers. Many of the polysaccharides have been applied in food and beverage industries. Some of them are used as thickening and gelling agents to improve the texture of food products as well to minimize syneresis (water release) from water-based food products [2]. The so applied polysaccharides are referred to as texture modifiers or structure stabilizers.

The feature of some polysaccharides to form gel in water at low concentration along with their favorable interactions with sugar yielding gel make them promising substrates for applications in food industry. Their non-toxicity, low processing cost, and abundance are further factors that stimulate their emerging applications. Among the available polysaccharides, a few can form gel without any additives, whereas there are many non-gelling polysaccharides which are used as food thickeners [3]. Modified starch, agar, κ/ι-carrageenan, low/high-methoxy pectin, gellan gum, alginate, guar gum, and methyl/hydroxypropyl cellulose are among common polysaccharides used as gelling agents in food applications.

Polysaccharide-Based Ion-Gels

Ion-gels are semi-solid (soft) gel materials prepared by incorporating of macromolecules in an ion-conducting liquid. Polymerization of appropriate monomers in such liquids can also lead to formation of ion-gels.

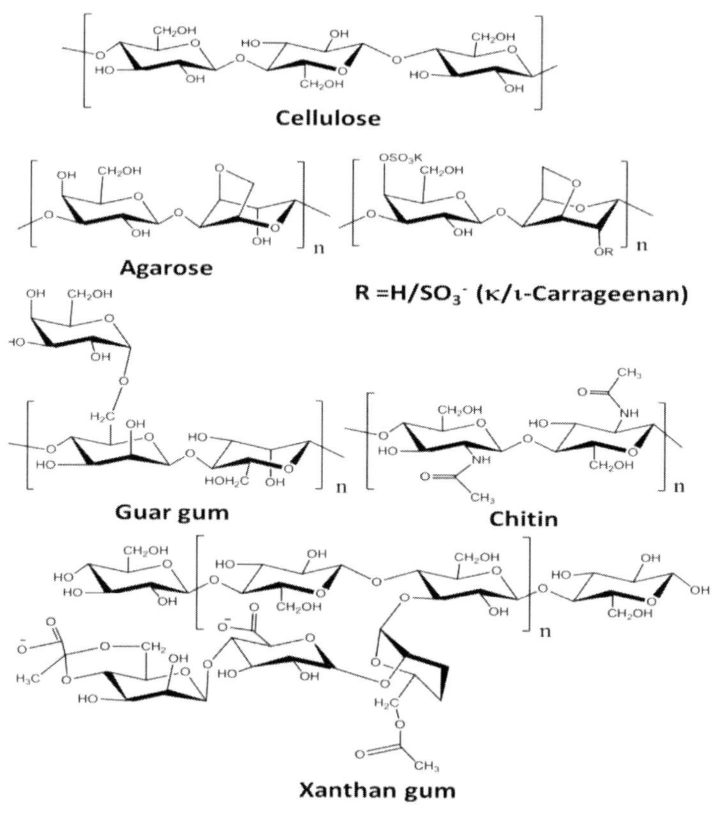

Figure 1. (Continued).

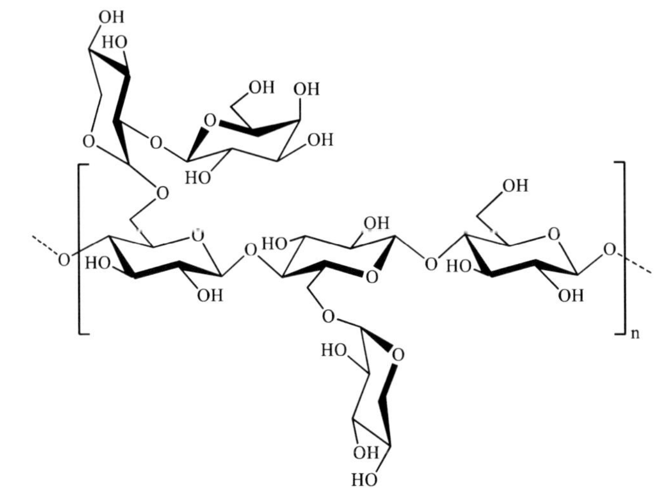

Tamarind gum

Figure 1. Repeating units of some selected polysaccharides used to prepare ion-gels.

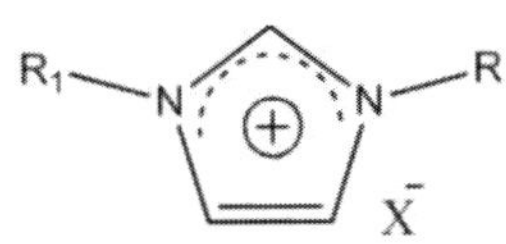

R= Allyl/Butyl ; R_1= Methyl/H
X= Cl/Br

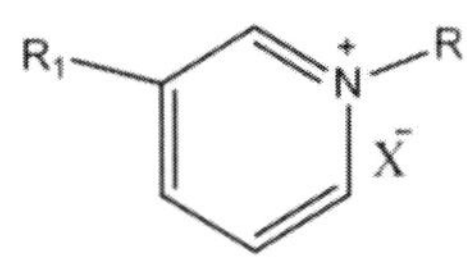

R= Butyl ; R_1= Methyl
X= Cl

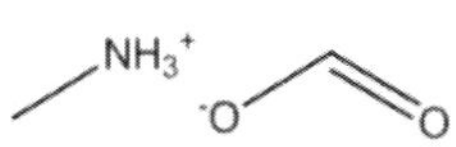

Methylammonium formate

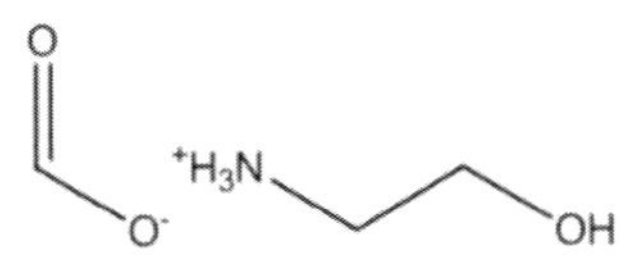

N-(2-hydroxyethyl)ammonium formate

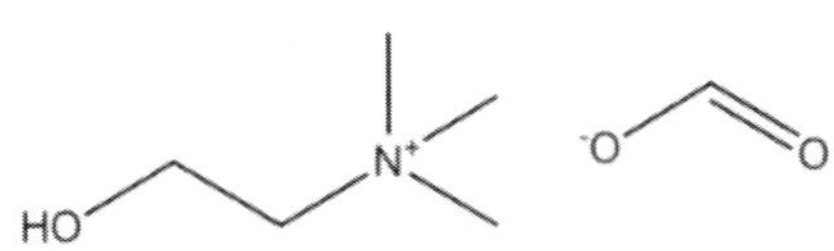

Choline formate

Figure 2. Structure of some selected ionic liquids used to prepare polysaccharide ion-gels.

Ionic liquids (ILs) are salts with melting point below room temperature and possess unique properties such as high ionic conductivity, excellent thermal stability of the structure, low vapor pressure, etc. making them a suitable media for the preparation of ion-gels [4]. The ion-gels are useful in development of electromechanical actuators due to the feature of differential ion migration leading to differential gel swelling in response to an applied electric field [5].

The above-described relatively new solvent systems have recently become popular in carbohydrate chemistry since the discovery of cellulose dissolution in 1-butyl-3-methylimidazolium chloride (BmimCl) [6] affording preparation of various cellulose derivatives [7]. Besides conventional ILs, those bearing polymerizable groups have been reported to show very good ionic conductivity [8]. Polysaccharides such as cellulose, agarose, carrageenan, guar gum, and xanthan gum have been used to prepare ion-gels in ILs. Chemical structures of selected polysaccharides and ionic liquids having been reported to form ion-gels are shown in Figures 1 and 2.

The polysaccharide ion-gels in ionic liquids are generally prepared via the heating-cooling process. In particular, the appropriate amount of polysaccharide capable of the gel formation in the IL is first heated up to complete dissolution, and the so obtained concentrated solution turns to a gel upon cooling to room temperature. The gels are isolated by adding some polar poor solvent (or precipitant) for the polysaccharides such as acetone, alcohols, etc. Representative images of some polysaccharide ion-gels are collected in Figure 3.

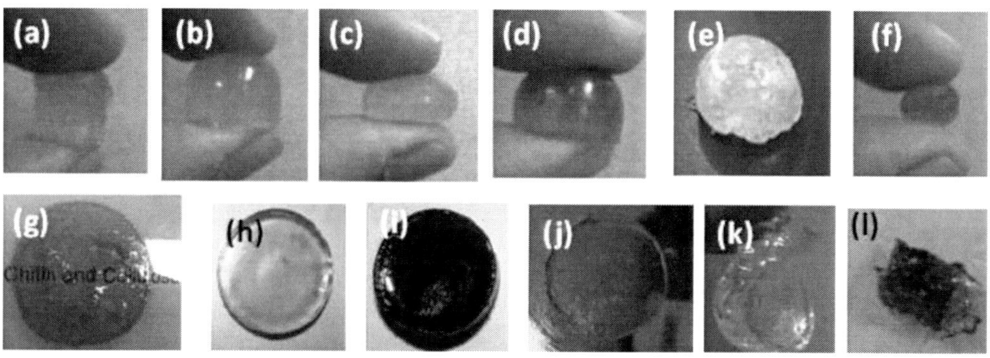

Figure 3. Images of some polysaccharide-based ion-gels prepared in various ionic liquids: (*a*) cellulose in BmimCl, (*b*) κ-carrageenan in BmimCl, (*c*) ι-carrageenan in BmimCl, (*d*) κ-carrageenan/cellulose composite in BmimCl, (*e*) xanthan gum in BmimCl, (*f*) agarose in BmimCl, (*g*) chitin/cellulose composite in the mixture of AmimBr and BmimCl, (*h*) guar gum in BmimCl, (*i*) guar gum/MWCNT nanocomposite in BmimCl, (*j*) guar gum in choline formate, (*k*) tamarind gum in BmimCl, and (*l*) chitin in AmimBr.

RHEOLOGY OF ION GELS

The term 'rheology' is derived from the Greek word *rheos* meaning stream or flow. In general, rheology studies the materials flow behaviour. Many of the materials possess viscous and solid properties simultaneously and are thus known as "viscoelastic materials". Biopolymers can form liquid solutions at low concentration range of the polymer, whereas at higher concentration they tend to behave as soft solids (gels).

Viscoelastic gels response to the applied stress or strain with structure deformation and flow. In rheological terms, when a force F is applied to a unit area A_0, certain deformation (strain) of the material surface occurs. Deformation properties of soft gels are generally studied using two geometries of the rheometer cell: cone/plate and plate/plate geometries (Figure 4).

For fluids and ion gels the shear viscosity is defined by

$$\eta = \frac{\tau}{{}^{d\gamma}/_{dt}} = \frac{\tau}{\dot{\gamma}}$$

where τ is standing for shear stress and γ being shear strain (deformation with respect to the gap size).

As far as fluids and ion-gels testing is concerned, the most usual trace of shear stress against strain rate for polymer solutions is that of decreasing slope indicating pseudoplastic behavior. Some low-molecular weight systems or dispersion of swollen particles reveal Newtonian behavior (viscosity η independent of the strain rate) or dilatancy ("shear thickening") respectively (Figure 5).

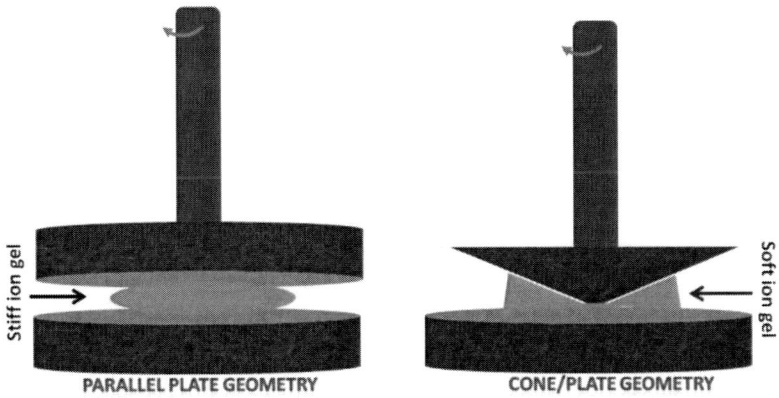

Figure 4. Parallel plate and cone plate geometry used in rheometry for characterization of fluids and gels.

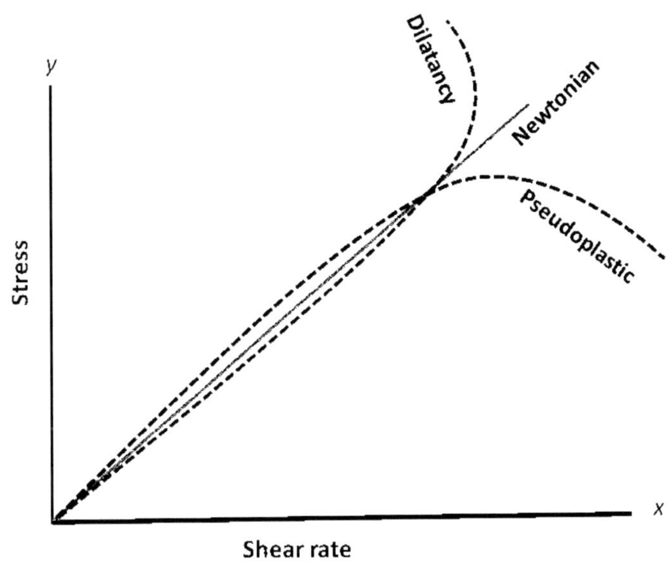

Figure 5. Flow curves for different states of matter.

In order to determine the viscoelastic properties of ion-gels, they may be subjected to periodic oscillation shear (dynamic testing). Sinusoidal shear strain is most often applied:

$$\gamma = \gamma_0 \sin(\varpi t)$$

Where γ_0 is shear strain amplitude; ω is oscillation frequency (may be expressed as $2\pi f$, f being frequency in Hz); and t is time. The mechanical response of viscoelastic materials (expressed in terms of shear stress τ) is somewhat intermediate between an ideal pure elastic solid (obeying the Hooke's law) and an ideal pure viscous fluid (obeying the Newton's law); the shear stress output signal is therefore phase-shifted with respect to the imposed deformation:

$$\tau = \tau_0 \sin(\varpi t + \varphi) = G'(\varpi)\gamma_0 \sin(\varpi t) + G''(\varpi)\gamma_0 \cos(\varpi t)$$

Where $G'(\omega)$ and $G''(\omega)$ are being the frequency-dependent shear storage modulus and the shear loss modulus, respectively.

G' gives information about the elasticity (the energy stored in the material during deformation, the in-phase response), whereas G'' describes the viscous character (the energy dissipated in the form of heat during the deformation, the out-of-phase response).

The ratio between the viscous and the elastic modules is expressed as the loss factor being equal to tangent of the phase shift angle δ:

$$\tan\delta = \frac{G''}{G'}$$

The loss tangent is a measure of the ratio of the energy lost and stored during the oscillatory deformation. The phase angle δ is equal to 90° in the case of a purely viscous material, to 0° for a purely elastic material, and is of $0° < \delta < 90°$ for viscoelastic materials.

WEAK ION-GELS

Several important biopolymer ion-gel systems appear to give a gel-type mechanical spectrum with $G' > G''$ and both the modulus being almost frequency-independent. However, the shear modulus strongly depend on strain (the linear viscoelastic region usually extends only up to strains of <0.1), unlike the stiff ion-gel systems. With the biopolymer gel systems, steady shear measurements can be performed, although the response is much different to that produced by stiff gels. In particular, almost no approach to the Newtonian plateau is observed; instead, the values of log η vs. log $\dot{\gamma}$ are nicely fitted with a power law over a wide range of shear rate, with a slope of ≈0.8–0.9. The nature of the employed ionic liquids is an important factor in determining the rheological properties of the polysaccharide ion-gels. For instance, in the case of microcrystalline cellulose solutions (0.1–15 wt%) prepared in two different ionic liquids (1-butyl-3-methylimidazolium chloride, BmimCl and 1-ethyl-3-methylimidazolium chloride, EmimCl), the shear viscosity of the solutions in EmimCl was lower than that measured in BmimCl [9]. The Newtonian plateau steady flow behaviour was reported for cellulose, microcrystalline cellulose, spruce sulfite pulp, and bacterial cellulose dissolved in EmimAc and BmimCl for dilute to semi-dilute solutions at temperatures from 0 to 100 °C [9].

Another ionic liquid, 1-allyl-3-methylimidazolium chloride AmimCl, was found to solubilize cellulose (0.1–3.0 wt%); the obtained cellulose solutions showed shear thinning behaviour at high shear rates in the steady shear measurements. However, the dilute cellulose solutions revealed another shear thinning region at low shear rates, likely reflecting the properties of the ionic liquid. The values of frequency-dependent G' and G'' modulus were intermediate between the theoretical predictions using the Zimm model and the elongated rod-like particle models, whereas a hybrid model combining the two theories agreed well with the experiment. Further, the system viscoelasticity was dependent on the molecular mass of cellulose, and the empirical time temperature superposition principle held true at the experiment temperature, while the Cox–Merz rule failed for the studied solutions [10].

For many polymer solutions, dependence of complex viscosity η^* on frequency ω and dependence of the steady-shear viscosity η on the shear rate $\dot{\gamma}$ are close at the overlapping ranges of numerical values of ω and $\dot{\gamma}$. This empirical correlation, often called the Cox–Merz rule, has already been observed for a number of biopolymer solutions, but a few exceptions have been reported as well. Even in the cases of viscosities mismatch useful information can be extracted from the comparison of (η, $\dot{\gamma}$) and (η^*, ω) profiles. If the viscosity/shear rate profile is typical of the entanglement network (showing up) a Newtonian plateau, the superimposition is always observed. However, if the weak gel "power law" behavior is observed, then η^* is almost always greater than η, the difference becoming increasingly large with at lower frequency (and shear rate). Application of the Cox–Merz rule has proved a valuable mean to characterize the nature of structured liquids or weak ion-gels.

Chitin could be solubilized in AmimBr up to the concentration of 5 wt%, whereas the 7 wt% chitin in that IL was a soft gel as shown in Figure 3*l* [11]. The so obtained gel was studied by means of shear viscometry and thixotropy profiling. Following the positive outcome in the cases of the non-Newtonian flow behavior of many polymer solutions, the Ostwald–de Waele two-parametric power-law model was applied to analyze the measurement results [12]:

$$\tau = K\dot{\gamma}^n$$

Where τ is shear stress in Pa; $\dot{\gamma}$ is shear rate in s^{-1}; K, consistency index in $Pa{\cdot}s^n$; and n, flow behavior index. Generally, the viscosity behavior index is of $0< n <1$ for shear-thinning or pseudoplastic fluids and $n >1$ for shear-thickening or dilatant fluids.

The non-Newtonian or shear-thinning flow behavior for both the 5 wt% solution and the 7 wt% soft gel was evident from the value of the flow index ($n < 1$). The extracted parameters were as follows: for 5 wt% chitin solution $n = 0.31$ and $K = 2.02$, for 7 wt% chitin gel $n = 0.076$ and $K = 172.24$. The much lower flow behavior index and the much higher consistency index of the gel indicated stronger and more stable structure of the gel compared with that of the solution. The highest apparent viscosity of the 7 wt% chitin gel was of 219.6 Pa·s and the lowest strain was of 0.0034, whereas those parameters of the 5 wt% chitin solution were of 6.19 Pa·s and 11.73, respectively. Those measured values further confirmed that the 7 wt% chitin in AmimBr is a gel state, and the 5 wt% chitin in AmimBr was a fluid. It should be mentioned that for AmimBr $K = 1$ and $n = 0.40$, reflecting its fluid behavior.

Flow curves of 5 wt% chitin solution exhibited time-independent shear-thinning behavior during a thixotropic loop experiment over the tested shear rate range in which the upward and downward flow curves fully coincided (Figure 6a).

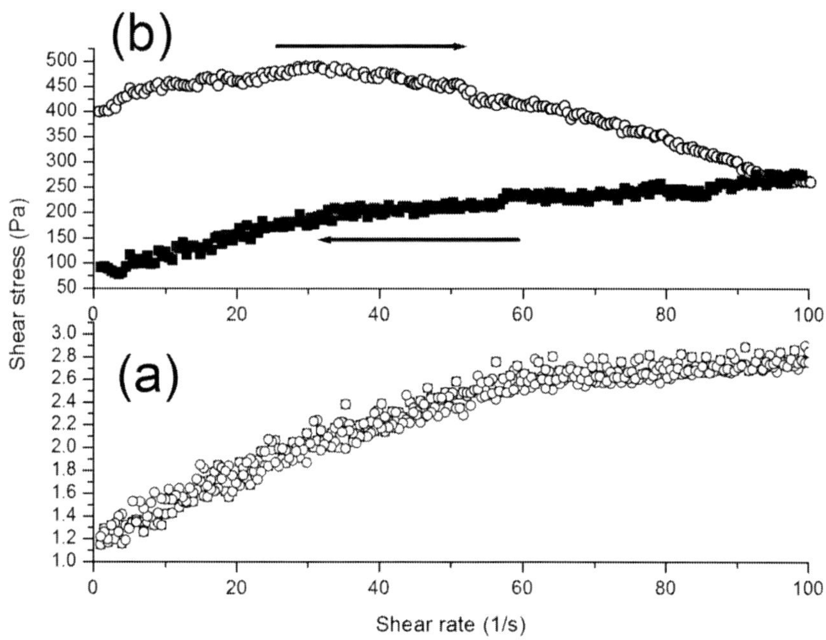

Figure 6. Hysteresis loop test for (*a*) 5 wt% chitin solution and (*b*) 7 wt% chitin gel.

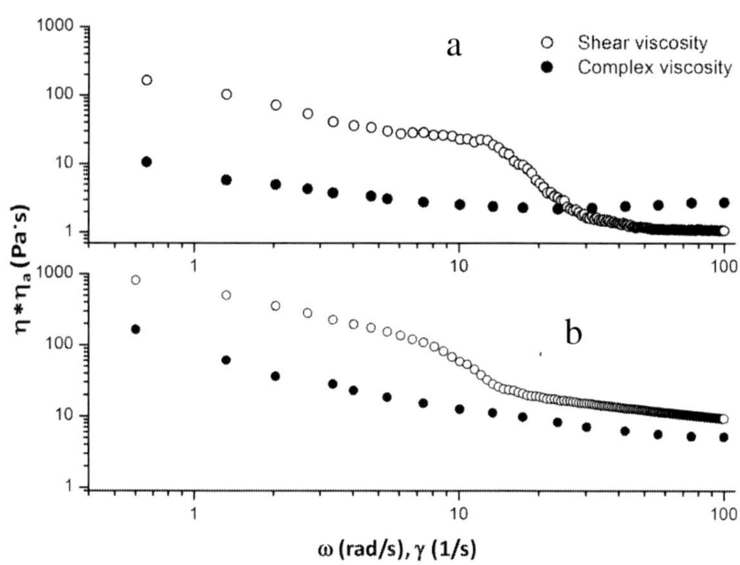

Figure 7. Cox–Merz plot for (*a*) 5 wt% chitin solution and (*b*) 7 wt% chitin weak gel.

The shear-thinning flow behavior was caused by disruption of chitin macromolecules entanglements under the applied shear. Both the upward and downward curves were well fitted with the Ostwald–de Waele model ($R^2 = 0.99$) showing consistency index of 1 $Pa{\cdot}s^{0.22}$ and flow behavior index of 0.22. The 7 wt% chitin gel exhibited time-dependent shear-thinning (thixotropic) behavior (Figure 6b). Both the upward ($R^2 = 0.96$) and the downward ($R^2 = 0.99$) curves were well described by the power law model. The consistency indices for upward and downward curves were of 622.32 Pa·s$^{0.13}$ and 71.23 Pa·s$^{0.28}$, respectively (hence, the flow behavior indices were of 0.13 and 0.28, respectively). The hysteresis loop area between the two curves was of 18357.18 Pa·s^{-1}. The imposed shear caused the chitin network structure to break, leading to the shear-thinning behavior. After decreasing the shear rate, the structure was built up only partially during the experiment time as confirmed by the lower consistency index value of the downward curve in comparison with that of the upward curve. The thixotropy of the 7 wt% chitin gel further confirmed its network structures. Although the 5 wt% chitin system demonstrated the presence of network structures as well, it did not possess sufficient stability to show thixotropy under the applied shear conditions, typical of the solution (fluid) state. When the complex viscosity of both 5 wt% (Figure 7a) and 7 wt% (Figure 7b) chitin in the ionic liquid was compared with the shear viscosity under the same shear rate conditions, violation of the Cox–Merz rule was clearly indicated. This confirmed the absence of macromolecules entanglements in the solution and the gel; instead, the presence of aggregated structure was more likely.

STIFF POLYSACCHARIDE ION-GEL

The biopolymers such as cellulose (Figure 3a) [13], κ/ι-carrageenan (Figs. 3b and 3c) and their composite with cellulose (Figure 3d) [14], xanthan gum (Figure 3e) [15], chitin/cellulose composite (Figure 3f) [16], guar gum and its nanocomposite with multiwalled carbon nanotubes (MWCNT) (Figs. 3i and 3j) [17, 18], and tamarind gum (Figure 3k) [19] have been reported to form strong ion-gels in different ionic liquids. Suitable concentrations (5–10 wt%) of cellulose, κ/ι-carrageenan, xanthan, guar, and tamarind gum form strong gels in BmimCl [13–15, 17–19], agarose forms strong gel in methylammonium formate and N-(2-hydroxyethyl) ammonium formate [20], and chitin/cellulose composite gel can be prepared in the mixture of BmimCl and AmimBr [16]. In the cases of all the mentioned systems, $G' \gg G''$, the both modules being nearly frequency-independent, typical of the strong gel-like structure.

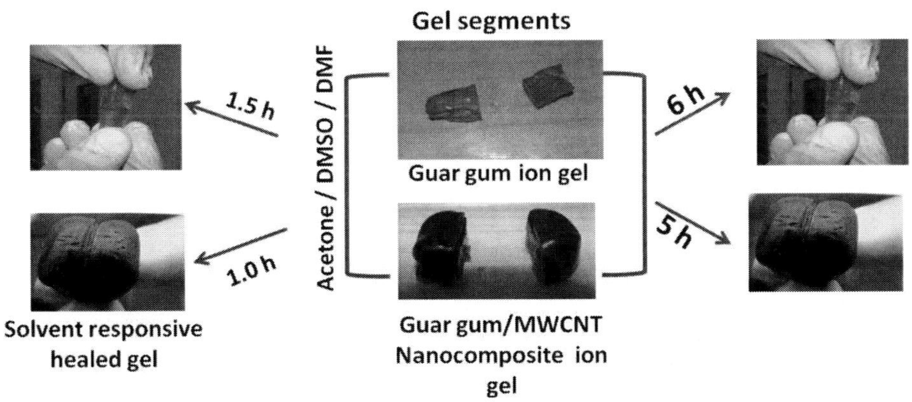

Figure 8. Self- and solvent-responsive healing of guar gum-BmimCl ion-gel and its nanocomposite gel with MWCNT.

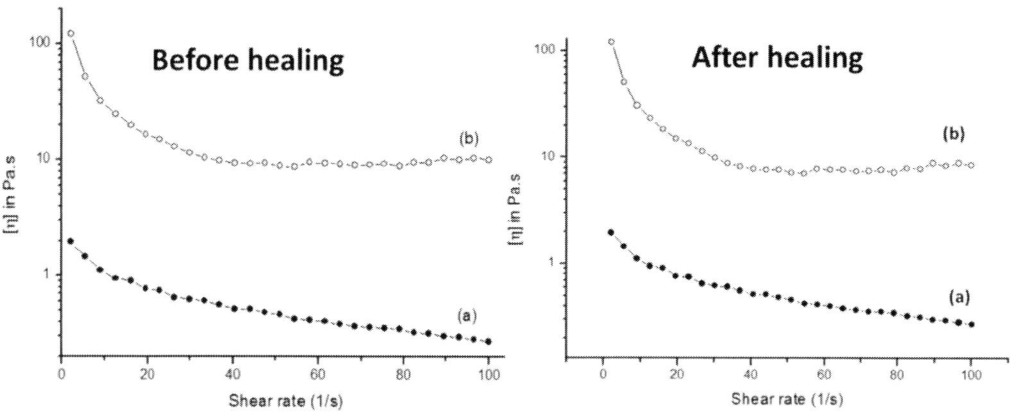

Figure 9. Steady-flow behavior of guar gum/BmimCl and its nanocomposite gel with MWCNT before and after healing.

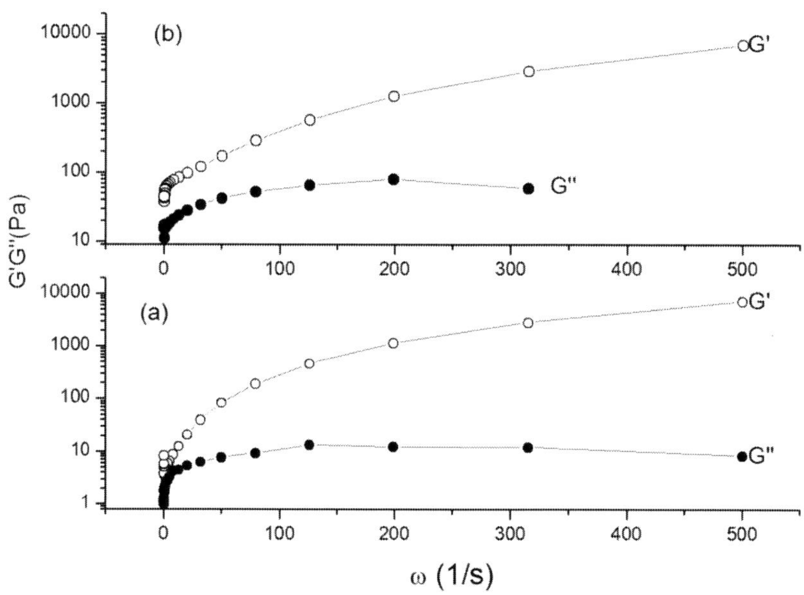

Figure 10. Frequency-dependent viscoelastic properties of (*a*) guar gum ion-gel and (*b*) guar gum/MWCNT nanocomposite ion-gel.

Guar gum is a galactomannan extracted from the seed of the leguminous shrub *Cyamopsis tetragonoloba*. It was demonstrated to form a viscoelastic gel in BmimCl, the optimal concentration being of 10% w/v [17]. A nanocomposite gel of the gum with enhanced strength could be prepared with 0.2% w/v of MWCNT in the ionic liquid. When the so prepared gels were completely separated into two segments, they could self-heal at room temperature without any external stimulus [17]. The gel segments also got healed in the presence of aprotic polar solvents such as acetone, DMSO, or DMF (Figure 8) [18], whereas no healing was observed in the presence of acetonitrile. The mentioned viscoelastic gel systems were thixotropic, and recovery of the storage modulus with time in the course of several deformation cycles was observed in the experiment.

The steady-flow behavior of the guar gum ion-gel prepared in the presence of MWCNT exhibited much higher shear viscosity in comparison to the MWCNT-free guar gum ion-gel, indicating formation of stronger gel network in the presence of carbon nanotubes. Likely, the cation-π interactions between BmimCl and MWCNT and the hydrogen bond formation between guar gum and BmimCl promoted the formation of solid-like structures through favorable chain entanglements (Figure 9).

As seen from data in Figure 10, in the cases of the both guar gum-based gels, the flow behavior of the healed gels was identical to that of the respective gels before damage, suggesting complete restoration of the gels structure. Measurement of the frequency-dependent viscoelastic properties of the GG-BmimCl and GG-BmimCl/MWCNT nanocomposite (NC) gel revealed predominance of the storage modulus over the whole accessed frequency range, typical of true gel systems (Figure 10). Of note, the NC gel was more elastic than the MWCNT-free guar gum gel, as evidenced by the loss factor values.

Ionotropic gelation involving BmimCl/BmimCl interactions (both Van der Waals and Coulombic) and cooperative guar gum/BmimCl interactions and guar gum intermolecular hydrogen bonding contributed to formation of a stable associative system having longer interactions lifetime. The structure formation was further enhanced in the presence of MWCNT. The time-dependent viscoelastic response over up to 15 min revealed the dominance of storage modulus over the loss modulus for the both gels, confirming their solid-like behaviour. Again, both the modules of the NC gel were higher as compared with those of the parent ion-gel reflecting formation of stronger network in the NC. The ionic liquid used, BmimCl, possessed a dual functionality since it behaved as a solvent and as a junction promoter binding the guar gum chains in a physical gel network. Those cross-links were irregular, forming extended junction zones due to the cooperative effect. The three-dimensional network was disturbed under mechanical strain (higher shear rate), but was restored during the relaxation phase (zero shear stress), the latter process being responsible for the self-healing of the gels (Figure 11). The area of the hysteresis loop for the guar gum–BmimCl gel was of $2.33 \cdot 10^5$ $Pa \cdot s^{-1}$, lower than that of guar gum–BmimCl/MWCNT NC gel ($7.9 \cdot 10^5$ $Pa \cdot s^{-1}$). That difference explained the faster response of the NC gel to mechanical strain towards self-healing.

In order to investigate recovery of storage modulus of the gels upon relaxation, the gels were fractured employing high strain, and then the relaxation stage was monitored. In particular, the gels were initially subject to 1% strain at 1 Hz for 300 s (the shear modulus were monitored during the process), and then the gels were fractured employing very high strain at 1 Hz during 100 s. It was observed that G' of the guar gum–BmimCl gel was recovered to the original value during the relaxation phase (Figure 11). The fracture-recovery cycles were repeated ten times. The guar gum–BmimCl/MWCNT NC gel showed similar restoration of the storage modulus during twelve consecutive cycles. The storage modulus of the nanocomposite ion-gel was about twice higher as compared to the MWCNT-free gel; indicating stronger gel network formed in the presence of carbon nanotubes, that was seemingly a reason of the higher stability of the nanocomposite gel under shear. To summarize, the discussed experiments confirmed the self-healing nature of the studied gels.

The guar gum–BmimCl and its NC gel healed in the presence of polar aprotic solvents showed different relaxation profiles [18]. When the gels were fractured in a same as described above, the storage modulus of guar gum–BmimCl gel was recovered to the original value upon relaxation (Figure 12). The cycle could be repeated three times only (cf. ten cycles of self-healing discussed above). The guar gum–BmimCl/MWCNT nanocomposite gel behaved similarly in the present of the solvent (Figure 12), the corresponding storage modulus values being about 20% higher in comparison MWCNT-free gel.

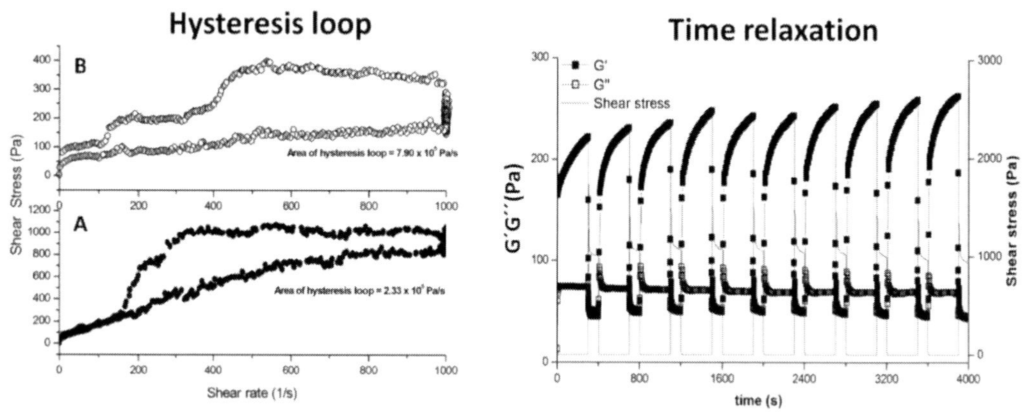

Figure 11. Left: investigation into the self-healing ability of guar gum–BmimCl ion-gel (*a*) and the MWCNT nanocomposite gel (*b*) via the hysteresis loop studies. Right: relaxation studies of guar gum–BmimCl ion-gel [17]. Reproduced with permission from Elsevier.

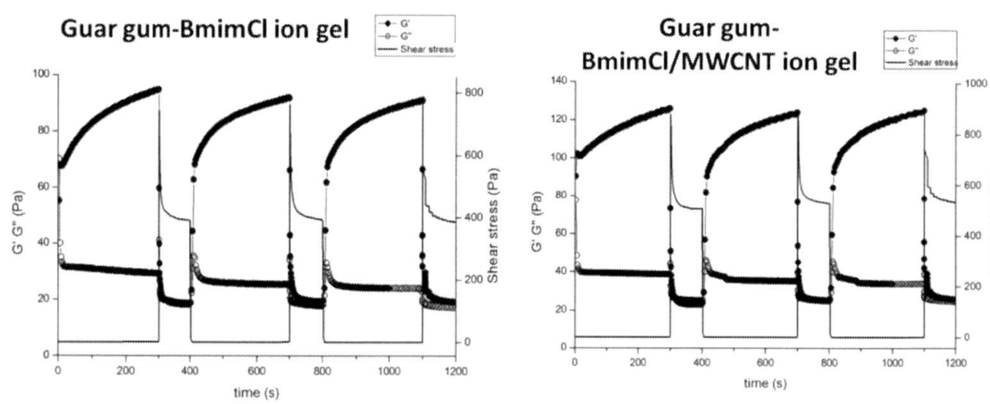

Figure 12. Relaxation studies of guar gum-BmimCl ion-gel (*left*) and its nanocomposite gel with MWCNT (*right*) healed in the presence of polar aprotic solvents [18]. Reproduced with permission from RSC.

Tamarind gum (TG), extracted from endosperm of *Tamarindus indica* seeds [21], is composed of β-(1,4)-D-glucan backbone substituted with side chains of α-(1,4)-D-xylopyranose and [β-D-galactopyranosyl-(1,2)-α-D-xylopyranosyl] (1,6)-linked to glucose residues (Figure 1); glucose, xylose, and galactose units are present in the monomeric units ratio of 2.8:2.25:1.0 [22]. This gum is widely used as a thickening, stabilizing, emulsifying, and gelling agent in food as well as pharmaceutical industry [23]. TG was used to prepare ion gels using synthetic ionic liquids (1-butyl-3-methylimidazolium chloride and 1-butyl-3-methylimidazolium bromide) as well as bio-based ionic liquids (Bio-ILs) (choline acrylate, choline caproate, and choline caprylate) via the heating-cooling process.

Rheological properties of the ion-gels (10 wt% of TG prepared in all the above-listed ILs) were thoroughly evaluated and compared with the properties of the corresponding hydrogel. Shear viscosity of TG ion-gel prepared the ion liquids decreased in the following series of ILs: choline caprylate, BmimBr, choline acrylate, BmimCl, and choline caproate (Figure 13); TG hydrogel showed significantly lower viscosity.

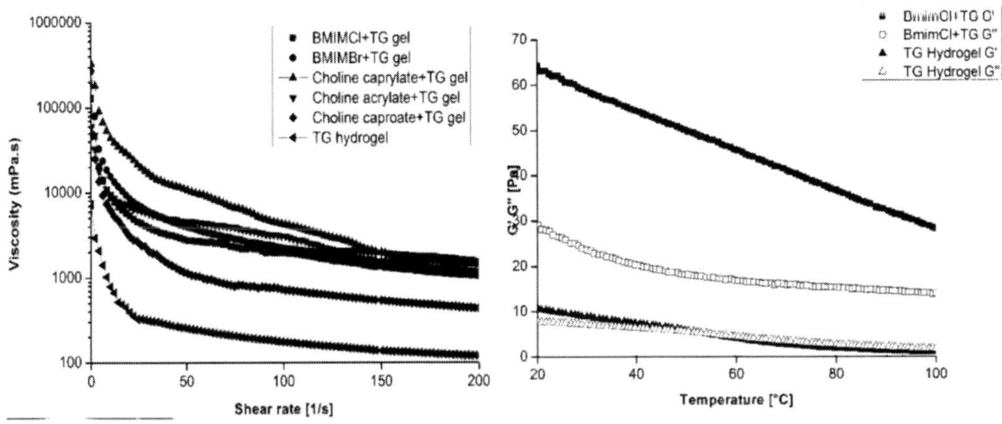

Figure 13. Dynamic shear viscosity and temperature dependence of viscoelastic properties for Tamarind-based ion-gels and the hydrogel prepared in presence of 15% ethanol.

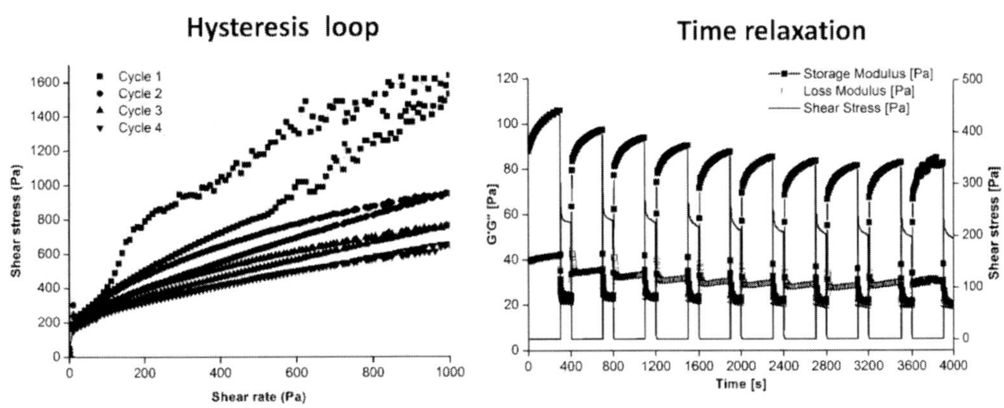

Figure 14. Self-healing ability of tamarind gum–BmimCl ion-gel demonstrated by hysteresis loop (*left*) and time relaxation (*right*) studies [19]. Reproduced with permission from Elsevier.

Ionotropic gelation involving IL–IL interactions along with multiple TG–IL interactions and TG intermolecular hydrogen bonding in TG contributed to formation of a stable associative system revealing long interactions lifetime. Again, the ILs had a dual functionality behaving both as a solvent and a TG chains junction promoter [17, 18]. The effect was not present in case of the hydrogel, and that is why the latter revealed lower shear viscosity. In order to investigate the temperature dependence of the loss and storage modulus of the ion-gels, the representative TG gel prepared in BmimCl evaluated in comparison with the TG hydrogel. The shear modules of the hydrogel showed up a cross-over at 50°C, whereas the modules of the ion-gel were much higher and did not cross-over up to 100°C thus indicating solid-like behavior of the ion-gel over the extended temperature range (Figure 13).

Evolution of the both modules with time further confirmed the solid-like behavior of the ion-gel. Frequency-dependent viscoelastic properties of the ion-gels were typical of viscoelastic material with predominance of the storage modulus over the whole studied frequency range. Dynamic rheology was applied to quantify the self-healing ability of the gel networks after the fracture [24]. In general, the observed properties were similar to those discussed above in the case of guar gum gels. In detail, the ion-gel prepared in BmimCl showed a pronounced thixotropic loop, the thixotropic nature being sustained in the course of three consecutive cycles (Figure 14). Further the loop area decreased with the cycle number due to expulsion of ionic liquid from the gel matrix, making the gels less viscoelastic. It should be noted that the hydrogel did not exhibit any thixotropic behavior.

Summary and Future Prospects

Due to the unique combination of properties, ion-gels have been used in diverse applications, and their application prospects are rapidly expanding. The polysaccharide ion-gels have the advantage of peculiar viscoelastic properties, non-toxicity, and biodegradability. The presence of ionic liquids also increases thermal stability of the gels making them useful for high-temperature applications. Deeper understanding of rheological behavior of the gels is required to spread the gels application over new fields such as biomedical, healthcare, etc. Although rheology of polysaccharide hydrogels have been thoroughly investigated, studies of polysaccharide ion-gels have been scarce. Evidently, much work in this direction is to be done aiming at industrial applications of the new type of polysaccharide ion-gels.

Acknowledgments

KP thanks council of scientific and industrial research (CSIR), New Delhi for the CSIR-Young scientist Awardee project (CSIR-YSP/2011-12) and overall financial support. JB is grateful to AcSIR for PhD registration and Analytical discipline and centralized instrumentation facility of the Institute for overall instrumental analyses. Mukesh Sharma, Dibyendu Mondal, and Chandrakant Mukesh are acknowledged for performing few of the experiments. Prof. Jun-ichi Kadokawa is acknowledged for providing photographs of xanthan and chitin-cellulose hybrid gel.

References

[1] Stryer, L. 1995, Biochemistry, *W. H. Freeman & Company*, New York.

[2] Lapasin, R. and Pricl, S. 1995, *In:* Rheology of Industrial Polysaccharides, *Chapman & Hall*, New York, pp.134–161.

[3] Saha, D. and Bhattacharya S. 2010, *J. Food. Sci. Technol.*, 47, 587–597.

[4] Lodge, T.P. 2008, *Science*, 321, 50–51.

[5] Imaizumi, S., Kokubo, H., and Watanabe M. 2012, *Macromolecules*, *45*, 401.

[6] Swatloski, R.P., Spear, S.K., Holbrey, J.D., and Rogers, R.D. 2002, *J. Am. Chem. Soc.*, 124, 4974-4975.

[7] Liebert, T. and Heinz, T. 2008, *BioResources*, 3, 576–601.

[8] Nakajima, H. and Ohno, H. 2005, *Polymer*, 46, 11499–11504.

[9] Gericke, M., Schlufter, K., Liebert, T., Heinze, T., and Budtova, T., 2009, *Biomacromolecules*, 10, 1188–1194.

[10] Kuang, Q.-L., Zhao, J.-C., Niu, Y.-H., Zhang J., and Wang, Z.-G. 2008, *J. Phys. Chem. B*, 112, 10234–10240.

[11] Prasad, K., Murakami, M., Kaneko, Y., Takada, A., Nakamura, Y., and Kadokawa, J. 2009, *Int. J Biol. Macromol.*, 45, 221–225.

[12] Berezin, Y.A., Chugunov, V.A., and Hutter, K. 2001, *J. Non-Newtonian Fluid Mech.*, 101, 139.

[13] Kadokawa, J., Murakami, M., and Kaneko, Y. 2008, *Carbohydr. Res.*, 343, 769–772.

[14] Prasad, K., Kaneko, Y., and Kadokawa, J. 2009, *Macromol. Biosci.*, 9, 376–382.

[15] Izawa, H. and Kadokawa, J. 2010, *J. Mater. Chem.*, 20, 5235–5241.

[16] Takegawa, A., Murakami, M., Kaneko, Y., and Kadokawa, J. 2010, *Carbohydr. Polym.*, 79, 85–90.

[17] Sharma, M., Mondal, D., Mukesh, C., and Prasad K. 2013, *Carbohydr. Polym.*, 98, 1025–1030.

[18] Sharma, M., Mondal, D., Mukesh, C., and Prasad K. 2013, *RSC Advances*, *3*, 16509–16515.

[19] Sharma, M., Mondal, D., Mukesh, C., and Prasad K. 2014, *Carbohydr. Polym.*, 102, 467–471.

[20] Trivedi, T.J., Srivastava, D.N., Rogers, R.D., and Kumar, A. 2012, *Green Chem.*, 14, 2831–2839.

[21] Glicksman, M. 1986, *In:* Food Hydrocolloids, Glicksman, M. (Ed.), *CRC Press*, Florida, vol. 3, pp.191–202.

[22] Gidley, M.J., Lillford, P.J., Rowlands, D.W., Lang, P., Dentini, M., Crescenzi, V., Edwards, M., Fanutti, C., and Reid, J.S.G. 1991, *Carbohydr. Res.*, 214, 299–314.

[23] Zhang, J., Zeng, J., Zhang, S., Wang, Y., and Wang, Y. 2006, *J. Bioact. Compatible Polym.*, 21, 415–432.

[24] Terech, P., Yan, M., Maréchal, M., Royal, G., Galvez, J., and Velu, S.K.P. 2013, *Phys. Chem. Chem. Phys.*, 15, 7338–7344.

In: Rheology
Editor: Evgeny Karpushkin
ISBN: 978-1-63482-223-7

Chapter 8

SHEAR RHEOLOGY OF POLYMER HYDROGELS. PART I: PROPER MEASUREMENT CONDITIONS

E. Karpushkin[*]
Department of Chemistry, Moscow State University, Russia

ABSTRACT

Efficient employment of rheometry results is possible when the measured values are reproducible and independent of the experiment setup (i.e. reflect the properties of the studied material rather than those of the measuring system). This generally requires that the specimen is tested in the fully equilibrated and steady state, the effects of resonance between the sample and the measuring equipment being excluded. In the case of soft materials, another complication to be taken into account is proper sample loading. This implies finding a reasonable tradeoff between conditions of no slippage of the sample and no microstructure distortion. This important factor is often neglected; however, it can significantly misrepresent the results leading to huge errors, sometimes up to an order of magnitude of the measured value. In this contribution, we summarize and discuss the precautions that should be taken in order to run the rheological measurement of soft samples correctly. Examples are provided, showing the deviation of the measured values from the actual ones if the precautions are neglected.

INTRODUCTION

Hydrogels form a group of materials, properties of which place them at the edge between solid and liquid materials. Like solids, they usually hold a defined shape and thus can be applied in construction of scaffolds, supports, functional surfaces, and artificial tissues or substitutes (especially in ophthalmology). On the other hand, being swollen in liquid aqueous medium, hydrogels are soft and permeable to low-molecular species, which opens for them fields of biomaterials, responsive soft machines, chromatography solid phases, sorbents for concentration of solutes or controlled release of drugs, etc.

[*] e-mail: eukarr@gmail.com

According to the recent IUPAC recommendation [1], hydrogel definition is quite wide ("non-fluid colloidal or polymeric network that is expanded throughout its whole volume by water") and covers a number of materials differing in internal morphology. In the reference literature, various expressions are used for description of hydrogels internal structure. For instance, according to visual appearance they may be called transparent (or clear), semi-transparent (or opaque, translucent), and non-transparent (white). As far as the phase structure is considered, hydrogels are termed homogeneous (or single-phase) and heterogeneous (or phase-separated, multi-phase). For hydrogels containing only part of swelling medium solvating the polymer chains, while part of the solvent is expelled to a distinct phase (porous gels) further accurate definitions are used. According to the IUPAC, macroporous polymers are those that have pores in the range of 50 nm to 1 μm [1], while microporous gels are those with even smaller pores. Materials with pore sizes more than 1 μm are sometimes called supermacroporous, but as the border between these types of materials is very diffuse, both names supermacroporous and macroporous may be used as synonyms to describe systems with pores larger than micrometer range. Often, especially in papers related to biomaterials, macroporous gels are called sponges.

In the scientific literature, some of these classifications are mixed, and terms "clear" and "homogeneous", or "porous" and "non-transparent" are used as synonyms, which may be only partially correct depending on the objects in question. Throughout this work, the following simplified classification was used to refer to different types of hydrogels. Those called "homogeneous gels" were fully transparent with no heterogeneities detectable either visually, by light or electron microscopy. Other (non-transparent) gels revealed several types of morphology under scanning electron microscope (SEM). Non-transparent swollen gels that only revealed a single phase in SEM (that is, the heterogeneities were below electron microscopy resolution) were referred to as "intermediate". Gels showing micrometer-range partially connected droplets of swelling medium surrounded by walls of swollen polymer, were assigned to gels of "interlocking, bicontinuous" morphology. The group of non-transparent gels with communicating pores surrounding well-defined fused spherical particles of swollen polymer constituted a category of "fused particles type" or "particulate gels".

For majority of applications, mechanical properties of hydrogel are of extreme importance. If a material serves as a scaffold for tissue replacement or as a chromatography column filler, hydrogel requires certain mechanical integrity. It was shown in studies of cell growth that substrate modulus can efficiently determine the behavior of cells attached to the surface; a recent example is provided in [2]. The general principle of selection of proper mechanical properties for tissue construction is described in [3]. Briefly, *in vivo* stress/strain histories need to be measured for a variety of activities both in normal and for (sub)failure conditions. These data provide mechanical thresholds for tissue repairs/replacements to hold. The set of mechanical properties must be further selected and prioritized, as the mechanical properties of the designs are not expected to completely duplicate the properties of the native tissues. Then, the effects of physical factors on cellular activity must be determined in engineered tissues. Thus, successful design of a tissue presumes accurate and thorough testing of both natural tissue and an artificial tissue candidates.

Nevertheless, in many application-oriented studies, for low-cost and rapid measurements, simple instruments are used. Several types of forces can be applied simultaneously during such tests, for example, cutting implies both shear and compression. Even though results are reproducible, they are not related to fundamental rheological properties of a material (stresses and strains applied are not defined). To compare data from different laboratories, it is essential to follow all conditions (sample size, magnitude of force or deformation, device design, speed of the probe, and others). This contribution compiles the requirements for such proper testing of polymer hydrogels of varied morphology.

Rheometry of Soft Hydrogels: Common Considerations

To the best of our knowledge, there are no documented protocols completely describing shear testing of soft non-flowing solids (swollen polymer gels). ASTM standards are limited to descriptions of solutions and pastes studies (inks [4] or melts [5]). One of the best descriptions of rheological testing of gels is available in [6]. As discussed there, two precautions are critical for correct food gels rheometry: sample steady state and testing in the linear viscoelastic range. At the same time deformation amplitude should be high enough to provide reasonable signal-to-noise ratio.

Further complication not mentioned in that protocol, neither in other available standards, is the test gap size estimation. Figure 1 illustrates what "gap size" means in rotational rheometry tests.

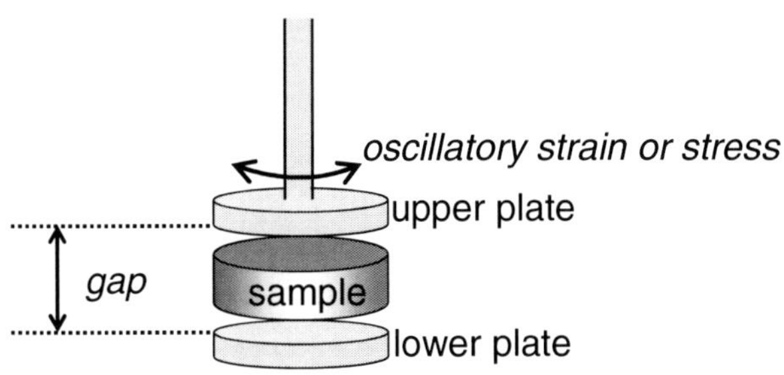

Figure 1. Gap during measurement with plate-plate geometry of rotational rheometer.

If the sample flows, its proper loading into the measuring cell is quite easy. On the other hand, the specimen can be manufactured precisely with required dimensions from a hard material. But for soft non-flowing gel (except the case of preparation directly inside the rheometer cell), the correct measurement gap should be trialed experimentally. Probably, because of the complications of gap estimation and consequent uncertainties in the data, tensile [7–9] and compression stress-strain measurements [7, 8] are currently more popular than shear tests in case of gels. On the other hand, due to construction of measurement instruments, shear testing is easier to be carried out in physiological surrounding (e.g., swollen state, presence of certain substances, normal force loading), and, when properly organized, it provides valuable information. Based on these considerations, the shear testing was selected as a primary technique for hydrogel mechanical properties characterization, and care was taken to achieve relevant and reproducible results.

The rheometry techniques commented on in this chapter are as follows.

Rheological measurements were performed using Bohlin Gemini HR Nano rotational stress-strain rheometer [10]. It offers ultra-low torque measurements (10 nN·m to 200 mN·m) and thus enables the rheological characterization of weakly structured materials such as swollen hydrogels. This top of the range research instrument incorporates a high resolution air bearing of the rotor, that can be driven with very small torques (10 nN·m) providing precise reading of position (stress control mode) or vice versa, the position of the rotor when rotating or oscillating is precisely controlled to cause small strains and corresponding stresses are measured (strain control mode).

All measurements were run with plate-plate geometry on swollen samples immersed in water during the test using a special solvent dish. Samples were conditioned at a pre-set temperature. The presence of water did not influence the measured values (additional friction of rotating part was negligible). The swollen disk diameter was always 20 mm, measuring gap was varied between 1.00 and 3.05 mm depending on the sample dimensions. However, measuring gap did not correspond exactly to the sample thickness as reaching the *no slippage* between the sample and measuring geometry condition demanded certain small compression of the soft gel sample. The value of compression is found qualitatively and precisely from the measuring system response.

The amplitude sweep forced oscillations tests were performed in the strain-control mode in the range of strains between 10^{-5} and 1, the oscillation frequency being 1 Hz. This test served as pre-measurement to find correct value of applied deformation not to depart from linear viscoelasticity region of materials.

The frequency sweep forced oscillations tests were performed in the stress-control mode. Frequency range was typically between 0.05 and 15–80 Hz. The highest frequency was limited by resonance of the sample and rotating part of the rheometer system. The resonance frequency depended on the sample stiffness as already documented in the literature [11]. The stress value in the frequency sweep test was auto-adjusted so that the "target strain" was within the limits of linear viscoelasticity as determined from the amplitude sweep test.

Normal force versus normal deformation measurements were performed by manually changing the gap value and recording the normal force value when it came to equilibrium.

A unique Piezoelectric Rotary Vibrator (PRV) [12] attachment was used to directly measure hydrogels behavior in the high-frequency oscillations range: 10–2000 Hz, in those tests sample thickness could be only around 0.5 mm, and the sample disk diameter was 40 mm. Due to the sample stiffness and thus its thickness limitation, the method was not well suitable for highly porous samples.

Creep tests were performed in the range of shear stresses of 1 to 5000 Pa, the response signal, displacement, was recorded for up to 3600 s. The recovery curve at 0 Pa shear stress was then recorded to prove absence of permanent deformation (either due to flow or sample slippage).

All measurements were performed consecutively at least twice while ensuring complete sample relaxation after loading, changing gap or temperature, and between the measurements.

In this chapter, determination of the correct measurement gap will be commented using two gels based on cross-linked poly(2-hydroxyethyl methacrylate) (polyHEMA) as representative examples. In particular, the gels were prepared either under conditions of no phase separation (to yield homogeneous, nonporous hydrogel) or under conditions of phase separation during polymerization (to yield macroporous gel consisting of fused swollen polymer particles surrounded by the swelling medium in the pores between them). Details of the phase separation features will be covered in the next chapter of this volume.

Noteworthily, the developed approach is quite general and is not limited to the case of polyHEMA hydrogels. For instance, it was applied in further studies of physical gels based on polyacrylonitrile swollen in dimethylsulfoxide–water medium (including those containing carbon nanotubes) as well as water-swollen Nafion membranes. In all the cases, the general procedure to find the measurement gap as described below in this chapter allowed for accurate rheometry results.

For soft solids, gap size determines sample normal deformation and normal force acting on it. This force must be high enough to prevent slippage of the sample and ensure perfect contact with rheometer plates; and sufficiently small not to affect the sample structure. That the sample is not fluid, makes impossible to remove the excess of material after setting a defined gap. The gap issue has been commented in [13] where the gap dependence of shear modulus was discussed for homogeneous dextran methacrylate hydrogels (and, to our knowledge, this is the only paper discussing this issue at all). Acceptable measurement conditions were found at compressions[1] $\varepsilon_{\text{comp}} \approx 20\%$. From our results it follows that it is too much for some samples, especially for porous ones.

FINDING A PROPER MEASUREMENT GAP IN THE CASE OF HOMOGENEOUS GEL

The gel prepared from 40 wt% solution of 2-hydroxyethyl methacrylate in water in the presence of 1 wt% of the crosslinker (diethyleneglycol dimethacrylate) was chosen as a typical representative of strong homogeneous hydrogel.

Thickness measurement with the digital micrometer at several locations (3.0±0.1 mm) was not reliable; mainly due to sample softness. At the gap of 3.1 mm the rheometer showed zero normal force and highly distorted output signal, as the sample was not in contact with the rheometer upper plate. At the gap of 2.9 mm, the normal force was high (7 N or 22 kPa), indicating that the sample might be compressed too much.

Going down stepwise from 3.1 mm, gap_0 was found as a distance at which normal force exceeded the resolution limit of the instrument (0.05 N). At such low normal force, the slippage of the sample (indicated by output signal distortion in the forced oscillation mode) still caused incorrect results. Trials indicated that thrust of ≈ 0.1 N corresponding to the gap of 3.050 mm for the tested sample was enough to prevent sliding (signal harmonic distortion <1%). For comparison, in [13] values of 0.1 to 1 N are reported.

Further we analyzed compression-dependent shear modules by changing the gap in small steps (Figure 2). Starting from $\varepsilon_{\text{comp}}$ of 2%, from the linear part of the normal force plot, Young's modulus was estimated (E=720±20 kPa). Complex shear modulus reached the constant value of G^*=200±10 kPa at $\varepsilon_{\text{comp}}$ of 4–5%.

The reason for non-linear behavior under the incipient normal compression could be either structure change, or imperfect shape of the sample and surface defects. We attributed the behavior of the studied gels to the latter, as the tested deformations were relatively small

[1] Normal deformation that results from compressive stress on a sample was defined as the actual gap size related to the uncompressed state: $\varepsilon_{\text{comp}} = (gap_0 - gap)/gap_0$ where gap_0 is the gap size at the initial contact of the sample with the rheometer plates.

(up to 6%). The shape of the normal force curve was similar to that reported, for example, in [14]. The shear plateau modulus under increasing normal deformation was reported in [13]. In that work the authors also related their observations to the inhomogeneity of the sample surface.

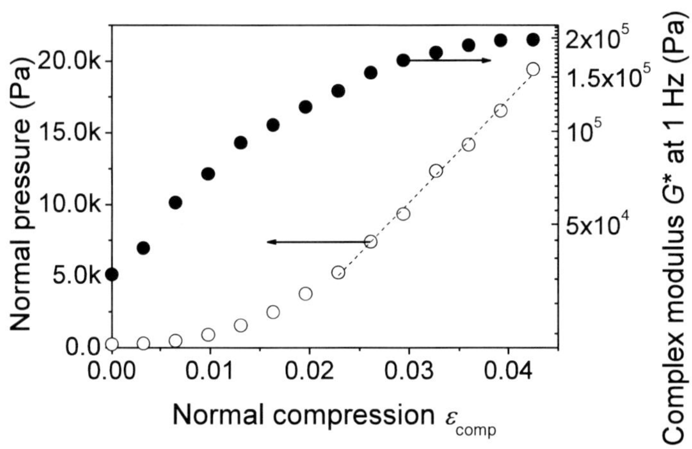

Figure 2. The complex shear modulus variation with increasing compression of the tested homogeneous gel specimen. Zero compression was assigned to a gap at which the normal force was 0.05 N. Young's modulus was calculated from the linear part of the normal pressure curve; shear modulus was estimated as a leveled value of G^* dependence.

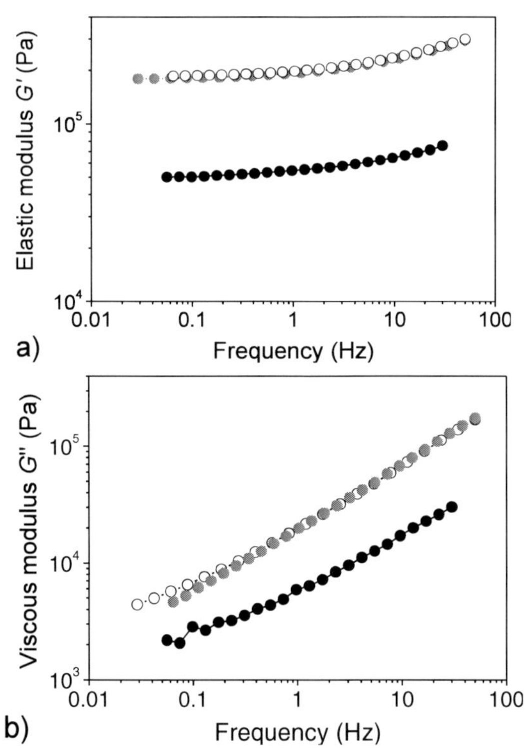

Figure 3. Frequency sweeps of G' (a) and G'' (b) at different normal compressions of the tested homogeneous gel specimen: 0 (black circles), 0.04 (grey circles) and 0.05 (open circles). Zero compression was assigned to a gap at which normal force was 0.05 N.

In the frequency sweep tests in the linear viscoelastic region (LVR), curves at $\varepsilon_{comp} = 0.04$ and $\varepsilon_{comp} = 0.05$) were essentially the same, but differed from that at $\varepsilon_{comp} = 0$ (Figure 3). Thus, at $\varepsilon_{comp} \approx 0.04$ real material properties were measured, not the properties related to experimental setup. That the target ε_{comp} for correct measurement was not too high ensured that the measured values were as close to that of intact sample as possible.

The same ε_{comp} for correct measurement was found from analysis of the shear creep curves in the LVR. Again, at $\varepsilon_{comp} > 0.04$ creep compliance profiles overlapped. After removing the shear stress, creep angle decreased to zero, indicating absence of slippage.

Thus, the gap size should be selected in the "meaningful" region, where the full contact of geometry with the sample was achieved with minimum structure change. For the homogeneous gel under study, the arbitrary choice of the measurement gap size would result in complex shear modulus in the 5–200 kPa range (Figure 2). Correct gap for rheological studies of macroscopically homogeneous hydrogel 40/1 corresponded to the least compressed state at which the shear modulus was independent of ε_{comp}.

In [13] similar conclusion was made but target gap size typically corresponded to 15–20% compression (sandpaper-modified rheometer plates were used, requiring higher normal force for perfect contact with the sample). We observe such "leveling off" at 4–5% of normal compression, and absence of slippage is proved in creep tests. In order to keep the sample structure as intact as possible, using the non-serrated plates that adhere to the tested material (steel for polyHEMA gels) is better than using sandpaper coverage or serrated plates.

FINDING A PROPER MEASUREMENT GAP IN THE CASE OF MACROPOROUS GEL

The gel prepared from 75 wt% solution of 2-hydroxyethyl methacrylate in water in the presence of 1 wt% of the crosslinker (diethyleneglycol dimethacrylate) was chosen as a typical representative of macroporous (particulate) hydrogel. Compression-dependent properties of that gel were studied as an example of hydrogel containing communicating pores filled with water. It was not possible to reach gap-independent mechanical response in this case (compare Figures 2 and 4). Normal force increased nearly proportionally to normal compression only from ε_{comp} of 0.5. Shear modulus did not reach constant value up to ε_{comp} of 0.65. Evidently, normal force induced pushing out of water, thus changing the sample morphology (total pore volume) and affecting its mechanical properties.

To find the proper gap size, the balance should be found between achieving the best sample-plates contact and dewatering of the sample accompanied by pore volume and shape change.

At the amplitude sweeps at ε_{comp} in the range of 0 to 0.65 (Figure 5), the G' steadily increased with compression, and the linear viscoelastic region was wider than at $\varepsilon_{comp} = 0$. Starting from $\varepsilon_{comp} = 0.17$, the LVR range of strains remained constant. We attributed the broadening of LVR to improving the sample-plates contact, while further modulus increase at constant upper limit of LVR was ascribed to sample dewatering. If so, gap size for correct measurement of shear properties corresponded to the lowest normal deformation at which the

LVR reached the constant strain range. However, this method was very approximate and meticulous.

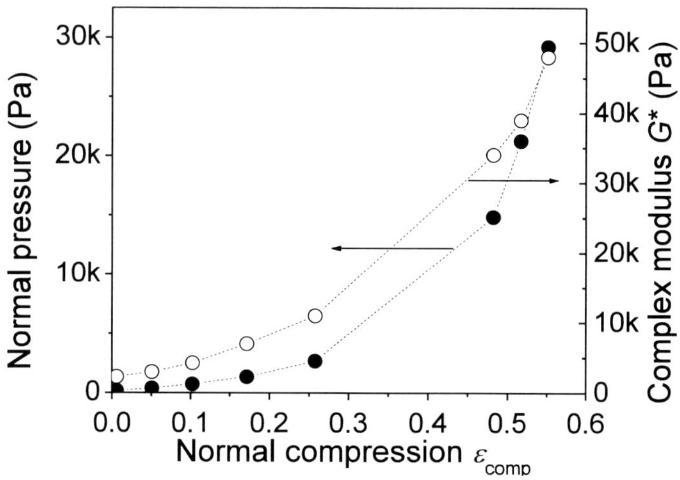

Figure 4. Normal force and complex shear modulus as function of normal compression of the tested macroporous gel specimen. Zero compression was assigned to a gap at which the normal pressure was 0.05 N.

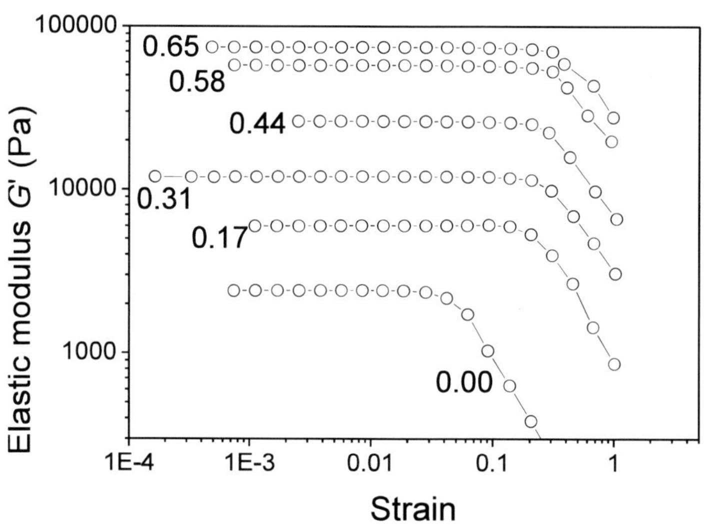

Figure 5. Amplitude sweep oscillatory shear test at different compressive deformations of the tested macroporous gel specimen. Sample compressive deformation is indicated near the curve. Zero compression was assigned to a gap at which normal pressure was 0.05 N.

Analysis of frequency sweeps at different ε_{comp} seemed more convenient. For the particulate sample, the frequency sweep profiles were compression-dependent over the whole range of studied normal deformations. The loss factors at the three selected frequencies are shown in the Figure 6.

At ε_{comp} of 0.00 to 0.03–0.05, the phase angle decreased, due to improving contact of a sample with the rheometer plates and vanishing impact of direct contact with surrounding water. Then phase angle curves showed a plateau (at ε_{comp}=0.03–0.07), when the contact was very good, and the change of hydrogel morphology was negligible. At higher ε_{comp} (0.07 to 0.4), phase angle started to grow; this could be ascribed to pushing out of water. Further phase angle decrease should be ascribed to further morphology changes due to compression of sample with free water in voids practically pushed out.

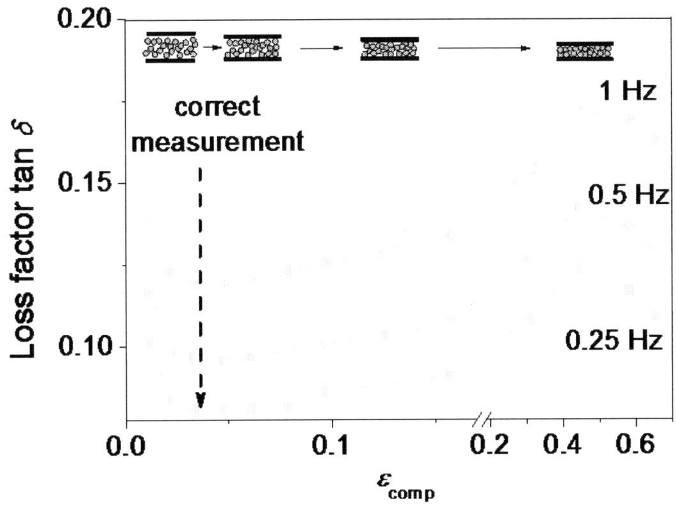

Figure 6 Phase angle profiles at selected frequencies (indicated near the corresponding curves) for the tested macroporous gel specimen at different compressive deformations (0.00 to 0.65). Zero compression was assigned to a gap at which normal pressure was 0.05 N.

Finding a Proper Measurement Gap: Summary

The procedures to determine proper gap are different for the homogeneous and the macroporous hydrogels. In the both cases, the ultimate appeal is to find a balance between improving the contact between the sample and the rheometer plates, and sample dewatering and structure changing.

Correct gap for rheological studies of the macroscopically homogeneous hydrogel corresponds to the least compressed state at which the shear modulus is independent of ε_{comp}.

Correct gap for rheological studies of the particulate sample with communicating pores is found at the least compressed state in the range of ε_{comp} where phase angle is relatively independent of compression.

Noteworthy, for majority of the studied samples target gap corresponded to ε_{comp} of 3–4%, which (taking into account variety of samples morphology and similarity of preparation procedure) supported the assumption that compression dependence of measured rheological properties was due to sample shape and surface roughness.

TIME-TEMPERATURE SUPERPOSITION

Time temperature superposition (or time-frequency superposition, or method of reduced variables) is a procedure to expand the time or frequency range, at which a material properties are studied at a given temperature. This technique involves the use of temperature-dependent shift factors for measured material functions (vertical shift factor, b_T) and for time or frequency (horizontal shift factor, a_T). If time-temperature superposition is obeyed, the use of shift factor will yield a master curve showing viscoelastic behavior over a larger range of time scales than could be studied using a single-temperature measurement. For example, a master curve for the elastic shear modulus G' measured over a range of T is obtained by plotting $b_T G'(T)$ as a function of $a_T\omega$ using logarithmic scales on both axis.

According to [15], the time-temperature superposition is applicable when the various relaxation times related to a given relaxation process have the same temperature dependence which usually holds in terminal and plateau zones, and, to a fair extend, in the transition zone (but the temperature dependence of the shift factors may be different). In practice, the validity of time-temperature superposition may be judged about with several semi-qualitative criteria:

- accuracy of matching of adjacent curves measured at different temperatures, when superimposed;
- validity of the same shift factors for all viscoelastic functions;
- reasonable form (Arrhenius or Williams–Landell–Ferry) of the shift factors temperature dependence.

For the polyHEMA hydrogels, change of swelling (EWC) with temperature and approaching to glass transition at the lower edge of temperatures investigated (5–10 °C) could potentially lead to incorrect time-temperature superposition result.

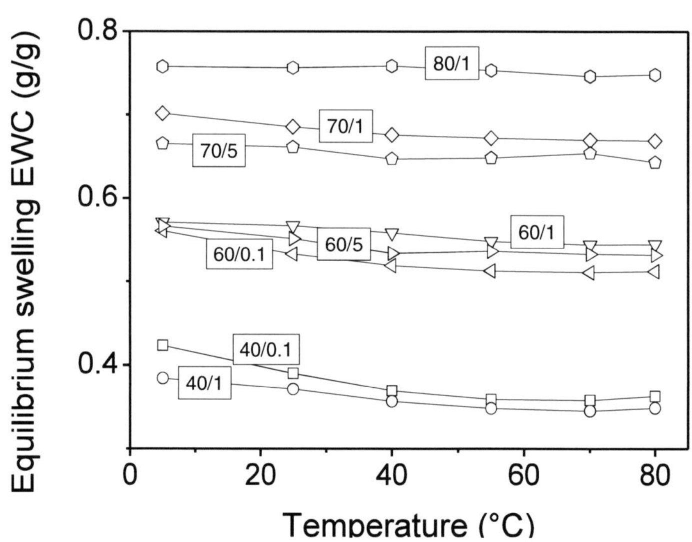

Figure 7. Equilibrium swelling of selected hydrogels in water at different temperatures (see details on the samples coding in the next chapter of this volume).

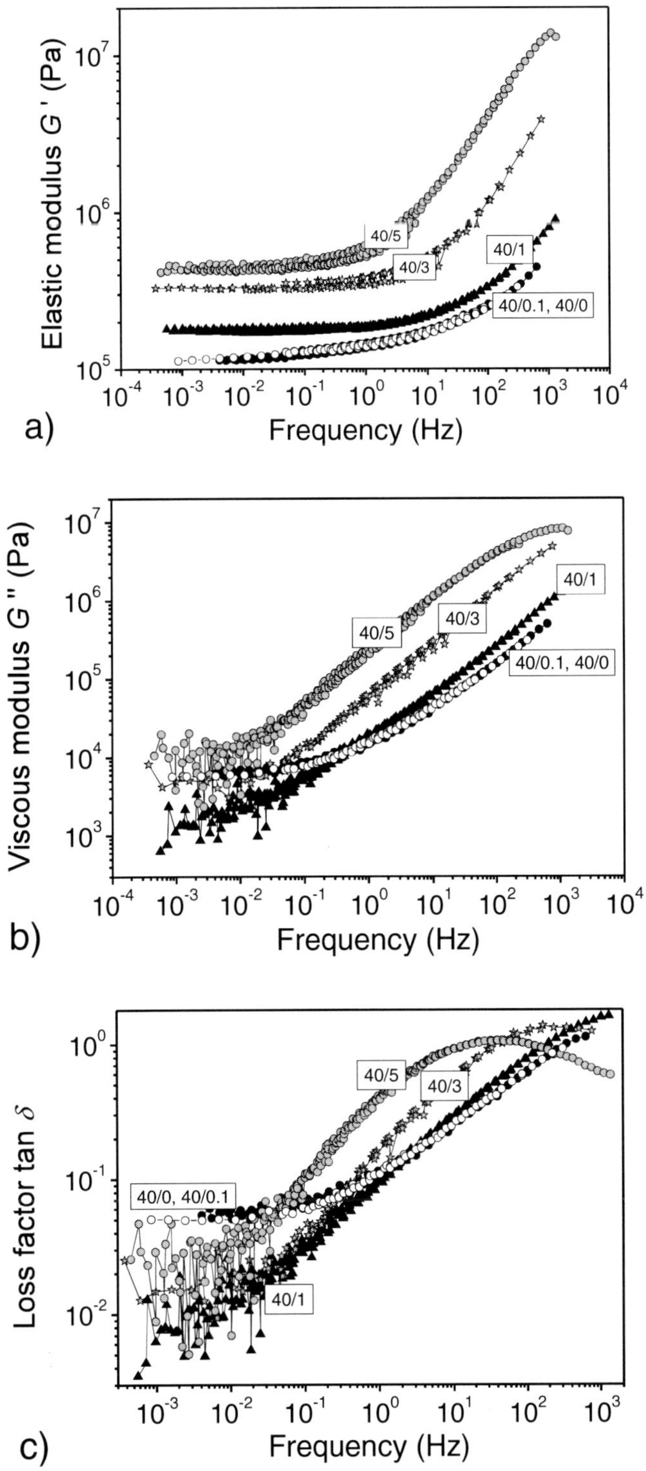

Figure 8. Elastic shear modulus (a), viscous shear modulus (b) and loss factor (c) for hydrogels prepared with 40 wt% of water as diluent and variable concentration of crosslinker, and swollen in water. Curves are result of time-temperature superposition (measurements at 5–70°C were shifted to a reference temperature of 25°C).

According to EWC measured over a range of temperatures (Figure 7), the change in swelling in comparison with the reference temperature (25°C) did not exceed 5% (being the most pronounced for samples prepared at low dilution).

The results of automatic frequency-temperature superposition (only horizontal shift of the experimental curves, as implemented in the rheometer software) compared with manual superposition (arbitrary vertical shifts applied to produce visually best match) showed no significant difference. Therefore, no vertical shift was applied in this work.

The horizontal shift factors calculated for G'' produced also reasonably smooth G' and tan δ master curves (see example in the Figure 8 for series of samples prepared with the initial water concentration of 40 wt% and variable amount of the crosslinker).

The two characteristic samples – 40/0 homogeneous and 80/1 highly porous – were selected for comparison of the superimposed and directly measured loss factors. The behavior of the 40/0 sample was characterized by low tan δ below 1 Hz and well developed increase of tan δ in the high-frequency region towards the glass transition. The 80/1 sample was characterized by larger losses in the low frequency region, with a pronounced increase at high frequency, similar to 40/0, and a weaker increase below 10^{-2} Hz (secondary relaxation).

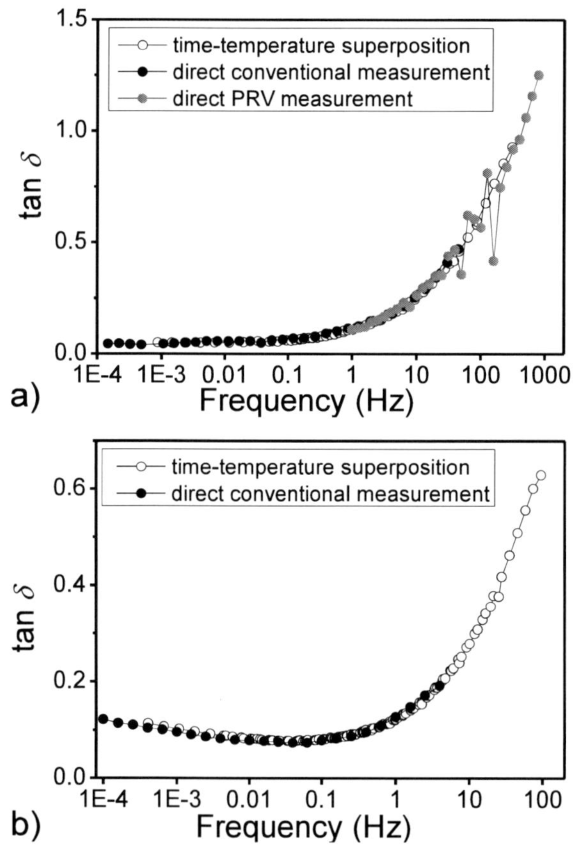

Figure 9. Frequency dependences of loss factor for the gels 40/0 (a) and 80/1 (b) swollen in water to equilibrium. Open symbols are result of time-temperature superposition (measurements at 5–70°C were shifted to a reference temperature of 25°C), filled symbols are result of direct measurements at 25°C in conventional mode (down to 10^{-4} Hz), or in PRV mode (up to 10^{3} Hz, 40/0 sample only). See details of the samples coding in the next chapter of this volume.

In the both cases (Figure 9), the superimposed and directly measured dependences could not be distinguished from one another within the experimental accuracy. Thus, in this special case of swollen polyHEMA network, when the swelling changes were not of large magnitude, the time-temperature superposition could be applied to achieve the enhanced range of shear oscillation frequencies within reasonable testing time.

CONCLUSION

The issue of proper choice of the measurement gap discussed in this chapter for the representative examples of homogeneous and macroporous gels turned to be extremely important for correct measurement of rheological properties of the swollen hydrogels. Noteworthily, this issue has been scarcely commented in the literature so far and, hence, seems to be often neglected or underestimated.

The time-temperature superposition technique, also important for the hydrogels testing, was confirmed to provide reasonable results in the case of swollen samples when their swelling degree was almost independent of the temperature.

Those two important considerations were further used for analysis of the effect of morphology type on rheological properties of polyHEMA hydrogels. This will be discussed in detail in the next chapter of this volume.

ACKNOWLEDGMENTS

This chapter is essentially based on the author's PhD thesis prepared during studies at Charles University (Prague, Czech Republic) and work at the Institute of Macromolecular Chemistry, Academy of Sciences of Czech Republic and reported in [16, 17]. EK cordially thanks his colleagues and especially supervisors (Dr. Miroslava Dušková-Smrčková and prof. D.Sc. Karel Dušek) for continuous support and ideas exchange. Even though they could not directly contribute to the writing of this chapter, their impact cannot be overestimated.

REFERENCES

[1] Jones R.G. 2007, *Pure Appl. Chem.,* 79, 1801–1829.

[2] Saha K., Keung A.J., Irwin E.F., Li Y., Little L., Schaffer D.V., Healy K.E. 2008 *Biophys. J.,* 95, 4426–4438.

[3] Butler D.L., Goldstein S.A., Guilak F. 2000, *T Am. Soc. Mech. Eng,* 122, 570–575.

[4] ASTM Standard D7271, 2006, "Standard Test Method for Viscoelastic Properties of Paste Ink Vehicle Using an Oscillatory Rheometer", ASTM International, West Conshohocken, PA, 2006, DOI: 10.1520/D7271-06, www.astm.org.

[5] ASTM Standard D4440, 2008, "Standard Test Method for Plastics: Dynamic Mechanical Properties Melt Rheology", ASTM International, West Conshohocken, PA, 2008, DOI: 10.1520/D4440-08, www.astm.org.

[6] Wrolstad R.E., editor. Current protocols in food analytical chemistry. New York: Wiley, 2007. Units H3.1-H3.3.

[7] Lou X., van Coppenhagen C. 2001, *Polym. Int.,* 50, 319–325.

[8] Clayton A.B., Chirila T.V., Lou X. 1997, *Polym. Int.,* 44, 201–207.

[9] Chirila T.V., Yu D.Y., Chen Y.C., Crawford G.J. 1995, *J. Biomed. Mater Res,* 29, 1029–1032.

[10] http://www.malvern.com/labeng/products/bohlin/gemini/geminihr/geminihr.htm (accessed 01 October 2012).

[11] Titze I.R., Klemuk S.A., Gray S.D. 2004, *J. Acoust. Soc. Am.,* 115, 392–401.

[12] Kirschenmann L., Pechhold W. 2002, *Rheol. Acta,* 41, 362–368.

[13] Meyvis T.K.L., De Smedt S.C., Demeester J., Hennink W.E. 1999, *J. Rheol.,* 43, 933–950.

[14] Clayton A.B., Chirila T.V., Dalton P.D. 1997, *Polym. Int.,* 42, 45–56.

[15] Ferry J.D. Viscoelastic Properties of Polymers, 3rd edition. New York: John Wiley and Sons, 1980.

[16] Karpushkin E., Dušková-Smrčková M., Remmler T., Lapčíková M., Dušek K. 2012, *Polym. Int.,* 61(2), 328–336.

[17] Karpushkin E., Dušková-Smrčková M., Šlouf M., Dušek K. 2013, *Polymer,* 54(2), 661–672.

In: Rheology
Editor: Evgeny Karpushkin
ISBN: 978-1-63482-223-7

Chapter 9

SHEAR RHEOLOGY OF POLYMER HYDROGELS: PART II: EFFECT OF POLY(2-HYDROXYETHYL METHACRYLATE) HYDROGELS MORPHOLOGY

E. Karpushkin*
Department of Chemistry, Moscow State University, Russia

ABSTRACT

Model series of crosslinked poly(2-hydroxyethyl methacrylate) hydrogels differing in morphology were prepared by polymerization-induced phase separation. Morphology, swelling, and dynamic mechanical properties in broad frequency range of the model gels were tested. The vitrification of gels was found sensitive to microstructure of polymer network, to solvent content, and to polymer-solvent interaction, but not to the macroscopic morphology of the gel. The low-frequency mechanical response was found sensitive to both polymer network properties and hydrogel morphology. Lightly crosslinked hydrogels showed relatively high mechanical losses in the rubbery plateau region, due to physical association of chains. Fused-particles type hydrogels revealed a weak secondary relaxation at low frequency, ascribed to motion of the dangling particles aggregates. Those aggregates, elastically inactive, along with inhomogeneous stress field arisen in porous material, were responsible for underestimation of equilibrium shear modulus of the matrix.

INTRODUCTION

The history of 2-hydroxyethyl methacrylate (HEMA) and its copolymers studies has counted more than half a century so far. The work of Czechoslovak group of Otto Wichterle (started in 1951) has led to recognition of mechanically strong, swellable in water and other physiological media, and optically transparent polyHEMA hydrogels as a valuable biomaterial for soft contact lens production, ophthalmic surgery and dentistry.

* E-mail: eukarr@gmail.com.

During HEMA polymerization in presence of inert diluent not miscible with polyHEMA, phase separation occurs resulting in heterogeneous, non-transparent hydrogels. Due to improved transport of nutrients and metabolism products, and possibility of living cells to accommodate inside the voids filled with the swelling medium, such porous polyHEMA gels are excellent candidates for tissue engineering applications.

Intensive studies on animals have revealed that encapsulation, cell ingrowth, neovascularization and calcification strongly depend on the overall polyHEMA porosity, pores average size, and their connectivity. Moreover, mechanical properties of porous material matrix may affect the cells behavior within a scaffold. Thus, efficient application of hydrogels as biomaterials requires precise control of their morphology and mechanical properties. Realization of this comprises several pieces of knowledge:

- control of final material porosity by preparation procedure;
- influence of preparation conditions on material matrix properties;
- influence of material porosity and matrix properties on the material utilitarian characteristics.

Even though basic principles of preparation–morphology and preparation–mechanical properties relationships have been formulated in the early polyHEMA studies, at that time it was not possible to characterize extremely soft, highly porous hydrogel due to insufficient technology development. The first commercial ESEM instrument to study swollen samples was introduced in 1988, and further important improvements were made even in late 1990s. Controlled-stress rheometers enabling characterization low-modulus materials appeared in 1990s.

In our experience, properties of gels produced by polymerization accompanied by phase separation are sensitive to even small changes in the preparation procedure, including the nature and amount of initiator, or temperature. This is especially so in case of macroporous final products.

Despite numerous studies on polyHEMA hydrogels, the systematic studies on samples of varying morphology are quite limited, and in the best link the preparation conditions with morphology or one of the utilitarian properties. Thus, direct study of relationships between morphology of hydrogels prepared in varied conditions, and their mechanical properties has been lacking till now. This work has been done in order to fill in this gap.

Poly(2-Hydroxyethyl Methacrylate) Hydrogels Overview

Hydrogels form a group of materials, properties of which place them at the edge between solid and liquid materials. Like solids, they usually hold a defined shape and thus can be applied in construction of scaffolds, supports, functional surfaces, and artificial tissues or substitutes (especially in ophthalmology). On the other hand, being swollen in liquid aqueous medium, hydrogels are soft and permeable to low-molecular species, which opens for them fields of biomaterials, responsive soft machines, chromatography solid phases, sorbents for concentration of solutes or controlled release of drugs, etc.

In the referenced literature, various expressions are used for description of hydrogels internal structure. For instance, according to visual appearance they may be called transparent (or clear), semi-transparent (or opaque, translucent), and non-transparent (white). As far as the phase structure is considered, hydrogels are termed homogeneous (or single-phase) and heterogeneous (or phase-separated, multi-phase). For hydrogels containing only part of swelling medium solvating the polymer chains, while part of the solvent is expelled to a distinct phase (porous gels) further accurate definitions are used. According to the IUPAC, macroporous polymers are those that have pores in the range of 50 nm to 1 μm [1], while microporous gels are those with even smaller pores. Materials with pore sizes more than 1 μm are sometimes called supermacroporous, but as the border between these types of materials is very diffuse, both names supermacroporous and macroporous may be used as synonyms to describe systems with pores larger than micrometer range. Often, especially in papers related to biomaterials, macroporous gels are called sponges.

In the scientific literature, some of these classifications are mixed, and terms "clear" and "homogeneous", or "porous" and "non-transparent" are used as synonyms, which may be only partially correct depending on the objects in question. Throughout this work, the following simplified classification was used to refer to different types of hydrogels. Those called "homogeneous gels" were fully transparent with no heterogeneities detectable either visually, by light or electron microscopy. Other (non-transparent) gels revealed several types of morphology under scanning electron microscope (SEM). Non-transparent swollen gels that only revealed a single phase in SEM (that is, the heterogeneities were below electron microscopy resolution) were referred to as "intermediate". Gels showing micrometer-range partially connected droplets of swelling medium surrounded by walls of swollen polymer, were assigned to gels of "interlocking, bicontinuous" morphology. The group of non-transparent gels with communicating pores surrounding well-defined fused spherical particles of swollen polymer constituted a category of "fused particles type" or "particulate gels".

2-Hydroxyethyl methacrylate HEMA is a common monomer for preparation of synthetic hydrogels. This is due to its good availability and low price, simplicity in use, high biocompatibility of polyHEMA, low toxicity of both monomer and polymer, and large variety of prepared hydrogel properties.

HEMA and its polymers were first reported in a patent [2] in 1938. In the next two decades, HEMA was episodically mentioned in the patent literature in connection with novel synthetic pathways or with its potential use as a comonomer for polymeric compositions.

The history of polyHEMA as a biomaterial started in 1950s when Wichterle and Lím [3] synthesized hydrogels based on HEMA copolymers with ethylene dimethacrylate (EDMA) designed for medical applications. After their biocompatibility was proved [4], those hydrogels were widely applied as materials for soft contact lenses. Commercial success of such lenses stimulated further interest in hydrogels, and eventually led to the development of responsive soft materials that could abruptly change size, or even shape upon exposure to an external stimulus [5]. Further successful applications of polyHEMA hydrogels include intraocular lens [6, 7] and dental materials [8].

First polyHEMA hydrogels applied as biomaterials were homogeneous and optically transparent. However, already during early polyHEMA research [2, 3] it was evidenced that at certain conditions porous polyHEMA hydrogels could be prepared. Highly depending on the pores fraction [9], such materials showed great potential in cell cultivating and tissue engineering, as they allowed cells attachment, growth and proliferation.

The *in vivo* evaluation of tissue reaction in the animal models was reported by Czechoslovak researchers [10, 11]: cellular invasion, encapsulation, vascularization, and salts deposition were followed after implantation in rats, rabbits, pigs and mice. The first study to show the influence of polyHEMA hydrogels porosity on their behavior as implanted materials was reported in [9]. At higher porosity, fibrous capsule was narrower and zone of cell ingrowth and vascularization was broader. Also, porosity was the major factor governing cellular invasion and deposition of calcium salts in the implants [12]. Macroporous polyHEMA hydrogels were successfully probed (in animal models) as implants for bone formation [13, 14], in brain surgery [15], and as membranes for artificial pancreas [16, 17].

Due to numerous potential applications of macroporous polyHEMA hydrogels, they are extremely attractive research objects. Even fundamental study of polyHEMA properties not only adds new knowledge about basic theoretical questions of polymer gels behavior, but brings important practical information.

Probably, the first study demonstrating the influence of polyHEMA gels porosity on their behavior as implanted materials was reported in [9]. Hydrogels synthesized with increasing amounts of water (50–90 wt %) crosslinked with 2 wt% of EDMA were implanted in rats and sampled at 30, 60 and 90 days after implantation. The polymers prepared with 50 and 60 wt% water behaved like homogeneous hydrogels (only encapsulation occurred), but those obtained with 70, 80, and 90 wt% water were penetrated by cells and blood vessels. At higher porosity, the fibrous capsule was narrower and the zone of cell ingrowth and vascularization enlarged; no significant calcium salt deposition was noticed. Similar experiments with macroporous HEMA copolymers containing ionic comonomers confirmed these findings [18]. In a more detailed study [12], the phenomenon of calcification was investigated. It was found that variation of porosity, controlled by water amount in polymerizing mixture, was the major factor governing cellular invasion and deposition of calcium salts in the implants.

Detailed studies revealed dependence of specific tissues regeneration on the support porosity [19]. The diameter of cells dictated the minimum suitable pore size. Depending on the application, pore size must be carefully controlled. Experiments revealed optimum pore size of 5 μm for neovascularization, 5–15 μm for fibroblast ingrowth, close to 20 μm for the ingrowth of hepatocytes, 20–125 μm for regeneration of adult mammalian skin, 40–100 μm for osteoid ingrowth, and 100–350 μm for regeneration of bone [20]. Fibrovascular tissues required pores of > 500 μm for rapid vascularization and survival of transplanted cells [21].

Continuity of pores is also important: mass transport and cell migration were inhibited if the pores were not interconnected even if the matrix porosity was high [56]. Cells located more than approximately 200 μm from a blood supply were either metabolically inactive or necrotic due to low oxygenation [22].

A further concern is change in pores structure over time *in vivo*. Even if matrix does not degrade, effective pore size may be reduced by *in vivo* events (invasion of fibrous tissue into the pores or nonspecific adsorption of proteins onto material) [23].

Thus, for majority of macroporous hydrogels applications, porosity characteristics (average size, volume fraction, connectivity, etc.) are important, in addition to other functional characteristics (mechanical properties, presence of functional comonomers, etc.). It is clear that developing well-controlled procedures to produce macroporous materials with desired morphology is desired.

Common methods for preparation of porous gels include: polymerization in presence of diluent which is a non-solvent for formed polymer (phase separation during polymerization), crosslinking polymerization in presence of washable solid substances (salts, sugars, or organic fibers), cryogelation (at temperatures below freeze point of a solvent which serves as a porogen), and polymerization in presence of gas-releasing components.

Each of these methods has advantages and drawbacks. The relative importance of them depends on the particular task, and it is likely impossible to choose the ultimately best method of pores generation.

Using polyHEMA gels as a model, in [24] a comparison of several routes to prepare porous materials was made: morphology, overall porosity, and pores connectivity were checked for prepared samples. Results of the study are summarized in the table below.

Gaseous porogen approach allowed for almost no control of morphology, and the overall porosity was low. Fiber-based gels showed good properties, but procedure was complicated, and it should be developed separately for each new comonomers mixture composition. The methods proved to be most convenient were phase separation polymerization and polymerization in the presence of soluble solid porogen. With the latter method it was easier to fine-tune the pores size and distribution, but overall complexity of this procedure makes polymerization-induced phase separation one of the preferred methods for HEMA-based hydrogels preparation. Basing on the conclusions made in Ref. [24], in this work we used the polymerization-induced phase separation to produce a variety of model polyHEMA hydrogels.

Method	Preparation	Reproducibility	Overall porosity	Adjustability of pores size	Distribution of pores size
Phase separation	very easy	good	20-100%	narrow	broad
Washable porogen	easy	excellent	68-97%	any > 5 μm	narrow
Gas release	usually easy	poor	0-20%	narrow	very broad
Fiber spinning	complicated	good	100%	wide	broad

Phase separation during crosslinking can be induced either by increasing the crosslinker concentration (ν-induced syneresis) or by decreasing the solvating power of the diluent (χ-induced syneresis, [25, 26]). In both cases, the growing polymer network cannot absorb all solvent available in the reaction system (i.e. equilibrium degree of swelling of the network is less than its degree of dilution). Thus, the system collapses to form reaction particles within the separated continuous liquid phase. As the reaction proceeds, new microparticles are continuously generated due to the successive separation of the growing polymer. The agglomeration of the microspheres results in the formation of a heterogeneous gel consisting of continuous phases of a gel and a diluent.

Microsyneresis and macrosyneresis are both usually operative in crosslinking polymerization. Microsyneresis is separation of diluent as droplets inside the gel phase to give dispersion. It is a non-equilibrium process caused by the existence of regions differing in segment and crosslinking density (local incompatibility), and by a slow relaxation of the network. Macrosyneresis includes the decrease of swelling due to the weakening of thermodynamic polymer-diluent interaction. It results in the formation of distinct diluent and gel phases. Microsyneresis usually impacts much on hydrogel structure, since polymerization is faster than gel relaxation, and thermodynamically unstable disperse structure is fixed by crosslinks. A theoretical treatment [26] suggested a greater complexity of the process, in which various factors, such as interaction parameter χ, molar volumes of components, content of diluent in the mixture, and concentration of crosslinking agent all affect phase separation during copolymerization.

Relationship between formation conditions and structure of macroporous networks was studied in [25–30]. The main factors determining the macroporosity of polymer networks were crosslinker concentration, type, amount, and solvating power of diluent, along with temperature and type of initiator. Detailed discussion of those factors falls beyond the scope of this chapter and can be found in the above-cited references.

For majority of applications, mechanical properties of hydrogel are of extreme importance. If a material serves as a scaffold for tissue replacement or as a chromatography column filler, hydrogel requires certain mechanical integrity. It was shown in studies of cell growth that substrate modulus can efficiently determine the behavior of cells attached to the surface; a recent example is provided in [31]. The general principle of proper mechanical properties for tissue construction selection is described in [32]. Briefly, *in vivo* stress/strain histories need to be measured for a variety of activities both in normal and for (sub)failure conditions. These data provide mechanical thresholds for tissue repairs/replacements to hold. The set of mechanical properties must be further selected and prioritized, as the mechanical properties of the designs are not expected to completely duplicate the properties of the native tissues. Then, the effects of physical factors on cellular activity must be determined in engineered tissues. Thus, successful design of a tissue presumes accurate and thorough testing of both natural tissue and an artificial tissue candidates.

Nevertheless, in many application-oriented studies, for low-cost and rapid measurements, simple instruments are used. Several types of forces can be applied simultaneously during such tests, for example, cutting implies both shear and compression. Even though results are reproducible, they are not related to fundamental rheological properties of a material (stresses and strains applied are not defined). To compare data from different laboratories, it is essential to follow all conditions (sample size, magnitude of force or deformation, device design, speed of the probe, and others). Further experimental complications influencing the measured properties were discussed in the previous chapter of this volume. They were taken into consideration to produce the accurate rheometry results discussed hereafter.

Properties of water-swollen crosslinked polyHEMA are controlled by both transient (physical) and permanent (covalent) interactions. Physical hydrophobic interactions viewed as association of stickers (methacrylate moieties) assisted by changes in the iceberg structure of water play an important role in dynamic and static properties of the hydrogels: they are strong enough not to allow chemically uncrosslinked polyHEMA to dissolve in water but only to swell.

Discrimination of permanent and transient networks by examining the storage and loss modules in the low-frequency region was discussed by Kavanagh and Ross-Murphy [33]. The interplay between covalent bonds and attractive physical interactions is not enough clear: the chemical crosslinks can suppress the formation of physical bonds hindering the alignment of chains. Alternatively, increasing number of chemical crosslinks can enhance the weaker, non-specific interactions by increasing the probability of pairing of interacting sites.

Detailed systematic studies of homogeneous polyHEMA hydrogels swelling and equilibrium deformational behavior [34] revealed that it in all cases the actual crosslink density was lower than that calculated that from polymerization feed composition. That was ascribed to two major factors: diffusion and steric limitations decreasing the reactivity of the second vinyl group of crosslinker molecule incorporated into a growing polymer chain, and side reaction of cyclization. Crosslinking efficiency was dependent on the nature of the crosslinker: with 20 wt% of water at preparation, crosslinking efficiency of EDMA was 84%, and for DEGDMA and triethyleneglycol dimethacrylate it was 43% only.

In case of polyHEMA networks swollen in di(ethylene glycol) to various swelling degrees [35, 36], an indication of a low-frequency secondary relaxation was found and ascribed to presence of entanglements. With increasing swelling of the gel, the secondary maximum was weakened and even disappeared.

The level of crosslinking by physical interactions depends on the solvent nature: while water is a relatively poor solvent, many other are better, for instance di(ethylene glycol) used in [35, 36], in which hydrophobic interactions do not exist. Concerning aqueous solutions, ionic salting-in additives [37] or urea [38] minimize the attractive interactions between hydrophobic moieties of polyHEMA and only the effect of covalent crosslinks should be seen. On contrary, additives decreasing the degree of swelling should strengthen attraction between segments.

The effect of crosslink density and dilution at preparation on mechanical properties of homogeneous polyHEMA hydrogels was studied in detail for optimization of contact lenses manufacture. These early findings on polyHEMA gels are summarized in a review [39].

For the porous polyHEMA gels, the situation is more complicated: the topological structure of the matrix gel is not known, and often a working hypothesis is used that it is the same as that formed at dilution at which the gel is homogeneous. This may be questionable especially for macroporous gels and needs verification or rejection. Moreover, for porous polyHEMA hydrogels, their properties depend both on the interactions in the matrix, and the pores volume and geometry.

Majority of studies of porous gels mechanical properties were limited to simple indication of equilibrium modulus decrease as a sign of phase separation. For example, in [40] hydrogels based on poly(N,N-diethylacrylamide) were prepared at various temperatures around critical mixing temperature. With developing porous structure, elastic modulus measured in compression decreased to 1–5 kPa from 50 kPa typical of homogeneous samples. Similar observation was made in [41] where series of copolymer hydrogels based on N-isopropylacrylamide and poly(ethylene glycol) monomethyl ether acrylate were prepared with porogen poly(ethylene glycol) of various M_W. As pores size increased from ten to hundreds of micrometers, G' decreased from 94 to 7 kPa.

In a more detailed study [42] a series of polyHEMA gels and HEMA/N,N-dimethylacrylamide (DMAA) copolymers were prepared via cryogelation and conventional polymerization. Porosity was found to be 94–97% for cryogel, and all the pores were interconnected. As monomer concentration increased, pores in cryogels became smaller, but overall porosity increased. Morphology of conventional hydrogels was not directly studied, but from the preparation conditions a porous structure might be presumed.

For cryogels, the effect of initial dilution on mechanical modulus was less pronounced than for conventional hydrogels. For example, at HEMA/DMAA = 4:1, in the range of initial monomers concentration of 6–20 wt%, the equilibrium modulus measured in compression increased from 4–7 to 25 kPa for cryogels, and from 5 to 300 kPa for conventional gels. Also, conventional gels were found to be more fragile (compression at break 25–55%, versus 86–98% for cryogels).

To conclude, major part of studies on mechanical properties of swollen hydrogels and effects of preparation conditions reported so far were devoted to homogeneous materials. At the same time, systematic studies of mechanical properties of macroporous materials are quite limited, due to experimental complications. At most, the measurements include ultimate tensile properties and equilibrium modulus determination. Even for such measurements, due to non-linearity of compression stress-strain curves, relatively low modulus, and inhomogeneity of the materials, it is often impossible to get information on mechanical properties of porous materials in well-defined conditions. This complication is expected to be overcome in this work by measuring mechanical properties in shear mode taking advantage of sensitive instrument, Bohlin HR Nano rheometer, designed for measurement of low torques. The hydrogel sample could be fully immersed in the swelling medium during measurement, making possible long measurements even at elevated temperature.

EXPERIMENTAL TECHNIQUES

2-Hydroxyethyl methacrylate HEMA (Rohm) and di(ethylene glycol) dimethacrylate DEGDMA (Aldrich) were used as received. HEMA contained 0.07% DEGDMA (bifunctional crosslinker) and 0.44% of di(ethylene glycol) monomethacrylate (according to results of gas chromatography).

Samples Preparation and Coding

HEMA and DEGDMA were copolymerized in the presence of various amounts of diluent (water or aqueous salt) using ammonium persulfate (APS) as initiator and N,N,N',N'-tetramethylethylenediamine (TEMED) as promoter. APS was first mixed with the diluent, HEMA and DEGDMA were added, and the solution was stirred for 5 min (the solutions were always clear at this point, proving full miscibility of components). Then, the sample was bubbled with nitrogen for 5 min, TEMED was added, and the mixture was quickly poured between two glass plates separated with 3 mm silicone rubber sealing. The sample lost its fluidity after 5 to 15 minutes indicating the crosslinked structure formation. The reaction was carried out overnight at room temperature to ensure complete conversion.

After the polymerization the sample was washed several times with an excess of distilled water (at least 5 washes, 24 hours each), and, if necessary, re-swollen in a different medium via gradual solvent exchange. Then, a disk (20 mm in diameter) was cut to be used in the rheology tests. The rest of the sample was used for swelling, morphology and calorimetry studies.

The parameters varied within tested series were: diluent concentration (40 to 80 wt %); DEGDMA amount with respect to HEMA (0.1 to 5.0 mol%); diluent type (water, aqueous salt); and swelling medium (water, DMSO, aqueous salt). APS to monomers ratio was 0.75 wt% for all samples.

The initial water content was expressed either in wt%, (IWC)

$$\mathrm{IWC} = \frac{m_\mathrm{w}}{m_\mathrm{w} + m_\mathrm{p}} \times 100\%$$

or as volume fraction of polymerizable material during polymerization

$$\phi_2^0 = \frac{m_\mathrm{p}/\rho_\mathrm{p}}{m_\mathrm{w}/\rho_\mathrm{w} + m_\mathrm{p}/\rho_\mathrm{p}}$$

where m_p is the mass of the polymer or polymerizable components and m_w is the mass of water. The density of dry polymer was used in calculations instead of monomer density to account for contraction of the monomer during polymerization.

The crosslinker concentration CC, was expressed in molar percent of crosslinker in its mixture with HEMA

$$\mathrm{CC} = \frac{moles_{\mathrm{crosslinker}}}{moles_{\mathrm{crosslinker}} + moles_{\mathrm{HEMA}}}$$

The samples prepared with water as diluent and then swollen in water are coded as follows: *40/1.0,* where the first number (40) is the initial water concentration (IWC) in wt%, and the second (1.0) stands for the crosslinker-to-monomers ratio (CC) in mol%.

For samples prepared with aqueous salt as diluent, the type and concentration of salt is indicated in brackets before the sample code: *(NaCl-0.1M)40/1.0.* For samples re-swollen in solvent other than water before the measurement, the type of swelling medium is indicated in brackets after the sample code: *40/1.0(DMSO).*

Rheometry testing conditions were discussed in the previous chapter of this volume.

Morphology Studies

Morphology of the hydrogels was studied both in the swollen state (light microscopy LM or environmental scanning electron microscopy, ESEM) and in the frozen state (cryo low-vacuum scanning electron microscopy, cryoLVSEM).

SEM was performed with Quanta 200 FEG microscope (FEI, Czech Republic), equipped with a Peltier cooling stage and secondary electrons detector working in low-vacuum conditions (LFD detector). For cryoLVSEM morphology studies, a small piece of wet sample was flash-frozen in a liquid nitrogen, fixed with ice onto the pre-cooled stage (–10°C) of the SEM microscope, the top frozen layer was cut off and the fresh, and frozen, cut surface was observed at cryo low-vacuum conditions (T = –10°C; p = 100 Pa; LFD detector). For ESEM, a hydrogel sample was trimmed with a blade, transferred to a table conditioned at 2°C, and scanned at this temperature. The water vapor pressure in the chamber was adjusted so that a thin water layer at sample surface existed.

LM studies of the samples in their natural state were performed with Leica DM 6000 M (Leica, Germany) microscope. The microscope was equipped with large working distance objectives with extended depth of focus and possibility of bright field and dark field imaging in both transmitted and reflected light. The samples were submerged in water during the whole sample preparation procedure: The bulk piece of hydrogel was cut with a sharp blade to obtain as thin wedge shaped specimen as possible. The slice with a droplet of water was placed between the support and cover glasses and observed in the light microscope.

Equilibrium Water Content

The equilibrium weight content of water, EWC, was determined gravimetrically by weighing the samples in their equilibrium swollen state (m_{sw}) and after drying to constant weight under vacuum at 100°C (m_d). The equilibrium swelling degree was expressed either as weight fraction of water in the swollen sample after extraction of sol fraction:

$$\mathrm{EWC} = \frac{m_{\mathrm{sw}} - m_{\mathrm{d}}}{m_{\mathrm{sw}}}$$

or as the equilibrium volume content of water, EVC,

$$\mathrm{EVC} = 1 - \phi_2 = \frac{(m_{\mathrm{sw}} - m_{\mathrm{d}})/\rho_{\mathrm{w}}}{m_{\mathrm{d}}/\rho_{\mathrm{p}} + (m_{\mathrm{sw}} - m_{\mathrm{d}})/\rho_{\mathrm{w}}}$$

where ρ_w and ρ_p are, respectively, the partial specific gravities of water and the polymer in the swollen sample.

The equilibrium swelling of networks in medium other than water was expressed similarly, as mass of swelling medium in the swollen sample per 1 g of swollen gel.

Accuracy

Gels Preparation

Monomers, initiators and diluent were dosed using the laboratory balance, accuracy of weighing was ± 0.5% of the calculated feed. Thus, the sample composition 40/1 means that the gel was prepared with 40.0 ± 0.2 wt% of water and 1.00 ± 0.01 mol% of the crosslinker, with respect to the monomers. For all the experiments reported in this work, the same batches of HEMA and DEGDMA were used.

Accuracy of experimentally determined values was verified by reproducing of measurement using an independently prepared piece of gel of the same composition.

Equilibrium Swelling Degree

Content of the swelling medium in the homogeneous, intermediate and interlocking gels was reproduced with the relative accuracy of ± 3%. For particulate gels, which were extremely soft, and part of the swelling medium could be inevitably pushed out when blotting the surface, the estimated relative accuracy was within ± 7% of the averaged value. The validity of swelling data was also confirmed by differential thermal analysis of selected water-swollen samples.

Shear Modules

The major source of errors in determining the elastic shear modulus of homogeneous, intermediate and interlocking samples was incorrect placement of the specimen (off-center of the rotating plate). From the replicated measurement the accuracy of measured G' was found to be ± 5% of the averaged value. For particulate samples, finding the correct measurement gap was more complicated, this resulted in determined G' accuracy within ± 10%.

Accuracy of the viscous shear modulus measurement depended largely on G''/G' ratio. With $G''/G' > 0.05$ typically the accuracy of G'' measurement was close to that of G'.

Loss factor $\tan\delta = G''/G'$ was typically determined with accuracy of ± 3% of the averaged value. As elastic and viscous shear modules were similarly affected by the off-center position of the specimen, and this source of errors was partially eliminated in loss factor calculation. However, low values of $\tan\delta$ could not be determined with sufficient accuracy. According to the rheometer specification, loss factor in the range of $0.01 < \tan\delta < 0.05$ could be subjected to significant error, values below 0.01 should be treated as zero.

General Remarks on the Samples Preparation

The miscibility of the reacting systems that is represented by the tertiary mixture monomer/crosslinker/diluent (HEMA/DEGDMA/water) was examined prior their use in model gelling mixtures. If the starting mixture would be immiscible or only partially miscible and thus unstable, some poorly defined and irreproducible effects could alter the final gel morphology. Such effects would be caused by the reactants mixing time, components addition order, intensity and type of stirring, presence or formation of surface active agents. In our study, the intention was to obtain various morphologies by defined and reproducible way by the reaction-induced phase separation, therefore only homogeneous and fully miscible starting mixtures for gel preparation were chosen.

The range of miscibility of HEMA/DEGDMA/water mixture was determined by visual assessment of mixtures opacity. To do so, a mixture of HEMA and water was titrated with DEGDMA till turbidity lasting for at least 10 minutes was detected. In Figure 1, to the left of the experimental curve, there is an area of compositions at which the HEMA/DEGDMA/water system is fully miscible. All gels studied in this work were prepared from such initially homogeneous mixtures.

When the diluent other than water was used (aqueous solution of NaCl or $Mg(ClO_4)_2$), the situation was further complicated. For such mixtures, miscibility of 4-component mixtures (HEMA/DEGDMA/water/additive) was checked to ensure that polymerization started in the initially homogeneous solution.

During crosslinking polymerization, the initially low-viscous mixtures lose fluidity at certain reaction time, corresponding to gel point, i.e. or critical gel conversion. Depending on the mixture composition, the phase separation could also occur at some point, corresponding to critical phase separation conversion. All samples were in the rubbery state during polymerization (non-glassy).

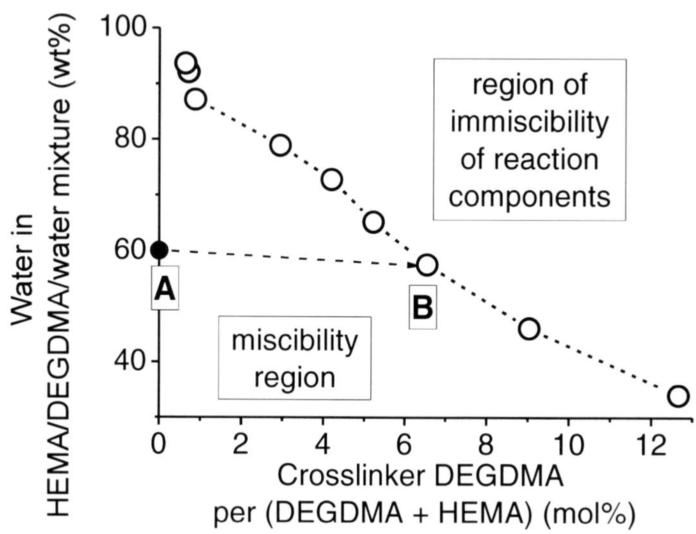

Figure 1. Miscibility of HEMA-DEGDMA-water mixtures (without initiator) at 25°C. Point A shows a starting mixture of HEMA and water. As the addition of DEGDMA decreased water concentration, the actual water concentration at the cloud point (point B) was recalculated.

Gelation times (determined as time at which the sample loses fluidity during polymerization at room temperature) and phase separation times (appearance of visually noticeable turbidity, cloud point) for gels prepared with water as diluent are collected in Figure 2.

Although the time of fluidity loss (Figure 2a) is only an approximation of gelation time, it revealed some important trends. It sharply decreased with adding more crosslinker in the range of 0–2 mol% of DEGDMA, whereas in the range 2–5 mol% of DEGDMA its effect on the gel point time is minor. At initial water content IWC=40–60 wt%, amount of water had only negligible effect on the gel point time, further increase of water concentration at preparation noticeably delayed the gelation.

The cloud point time (Figure 2b) decreased with dilution at IWC = 40–60 wt% of water at preparation, and at constant dilution more crosslinker DEGDMA added accelerated phase separation. With 70–80 wt% of water at preparation, phase separation occurred very fast, and crosslinker effect was negligible.

Comparison of the gel point time and phase separation time for mixtures of the same compositions showed that at relatively low dilution (40 wt% of water at preparation) and crosslinker concentration (< 5 mol% of DEGDMA) phase separation did not occur till polymerization was over (Figure 3a). Increase of initial dilution (to IWC = 50 wt% of water at preparation) and/or crosslinker concentration led to phase separation observed after or close to the gel point, more diluent and/or crosslinker led to even earlier phase separation, before formation of the bulk gel. These interplays between gel and phase separation times led to diverse morphological effects.

The gels were prepared using chain polymerization initiated by redox system enabling initiation with a reasonable rate at room temperature. To estimate reaction time needed to reach complete monomers conversion in all experiments, the reaction rate was measured for the system that showed the slowest gel point time (Figure 2), a mixture containing 80 wt% of water and 20 wt% of monomers, 0.1 mol% of them being crosslinker DEGDMA, i.e. 80/0.1

sample. Conversion of C=C bonds of the monomers was followed with IR ATR technique. From comparison of the initial and final spectra (not shown for the sake of clarity) it followed that polymerization induced a shift of the monomers C=O bands to lower wavenumber, and to complete disappearance of the C=C bands (both C=O and C=C bands of HEMA and DEGDMA were not distinguishable). For the 80/0.1 mixture, the C=C bonds conversion reached nearly 100% within 40 minutes. As the preparation procedure used in this work involved overnight polymerization, it was assumed that full conversion of monomers was reached for all the samples studied. This was directly confirmed by IR ATR studies of several representative final gels: in none of them C=C band was found.

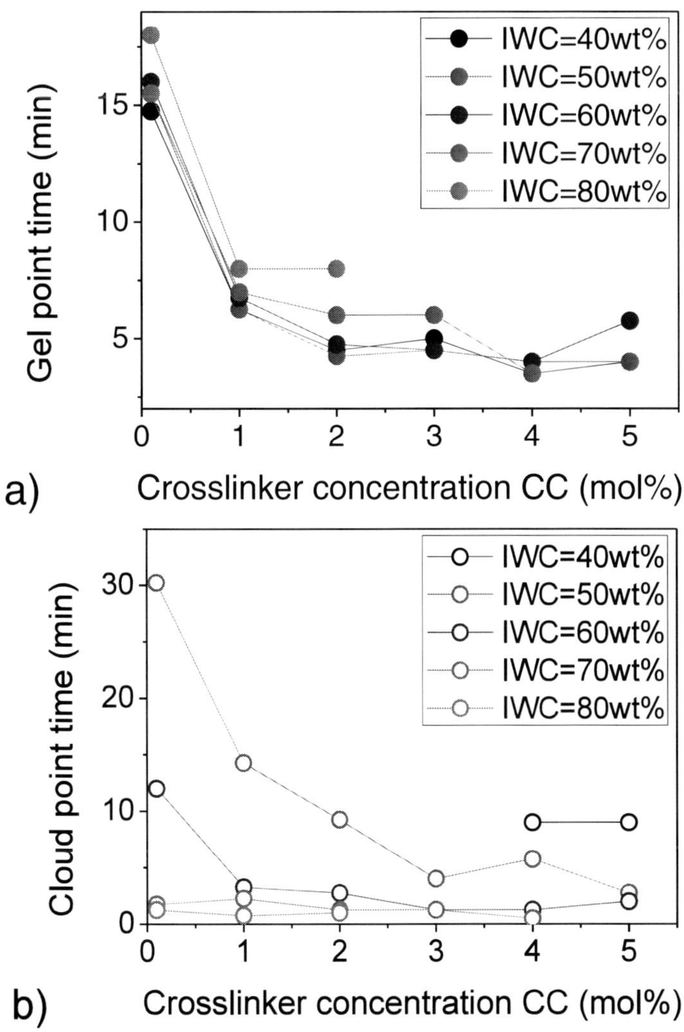

Figure 2. Gelation time (a) and phase separation time (b) determined during polymerization of HEMA/DEGDMA/water mixtures.

The completeness of reaction and extent of network development in these systems can be seen from the sol fraction. Results of solids extraction with water revealed (Figure 3) that only at highest dilution part of solids from the final gel could be extracted (7–8 wt%) whereas for other compositions no more than 3 wt% of the solids were extracted. Because sol content is a function of conversion and reaches zero at full conversion, it was concluded that with acceptable accuracy the composition of polymer corresponded to composition of monomers mixture, and the attained conversion was reasonably high. This also meant that approximately all crosslinker was incorporated into the gel.

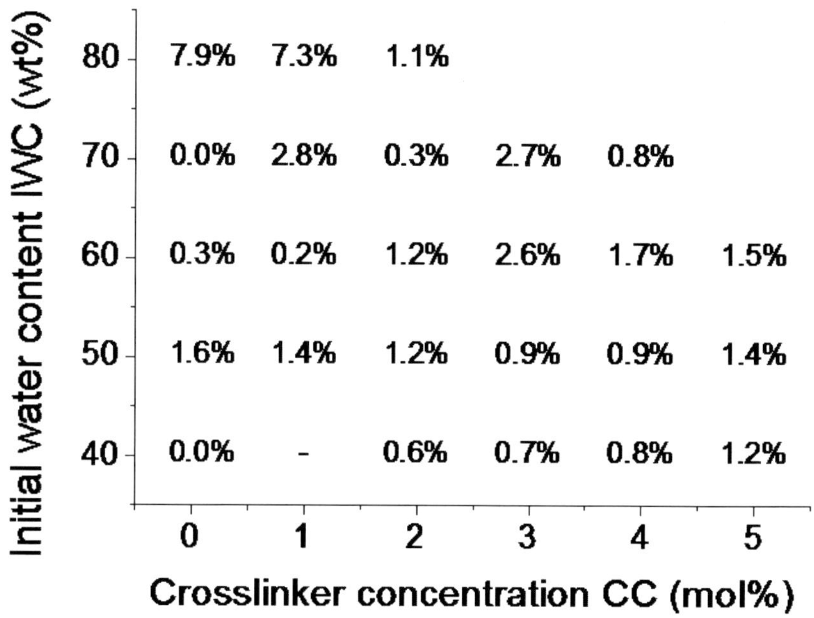

Figure 3. Sol fraction in the hydrogels prepared from HEMA/DEGDMA/water mixtures, determined by extraction with water.

It can be summarized that the preparation procedure used in this work allowed for complete conversion of the monomers during polymerization. That the initial mixtures were always clear, suggested that the phase separation started only due to polymerization, and possible pre-polymerization effects were eliminated.

Hydrogels Morphology

Diluent and Crosslinker Concentration Effects

Typical morphology types of polyHEMA hydrogels observed in this work are as follows (Figure 4). Those hereafter called *"homogeneous gels"* were fully transparent with no heterogeneities detectable either visually, by light or electron microscopy. If the gel was not optically clear, it contained heterogeneities on the order of visible light wavelength (~ 100 nm). With electron microscope, non-transparent gels revealed several types of morphology. Non-transparent swollen gels that only revealed a single phase in SEM or ESEM (that is, the heterogeneities were below electron microscopy resolution) are referred to as *"intermediate"*. Of the gels with heterogeneities visualized by ESEM or SEM, those showing relatively large (micrometer-range) partially connected droplets of swelling medium surrounded by walls of swollen polymer, were assigned to gels of *"interlocking, bicontinuous"* morphology. The group of non-transparent gels with communicating pores surrounding well-defined fused spherical particles of swollen polymer constituted a category of *"fused particles type"* or *"particulate gels"*.

From the map of gel morphologies of hydrogels prepared in water (Figure 4), morphology of the final material changed from homogeneous through intermediate and bicontinuous to particulate, as either initial water concentration or crosslinker concentration were increased.

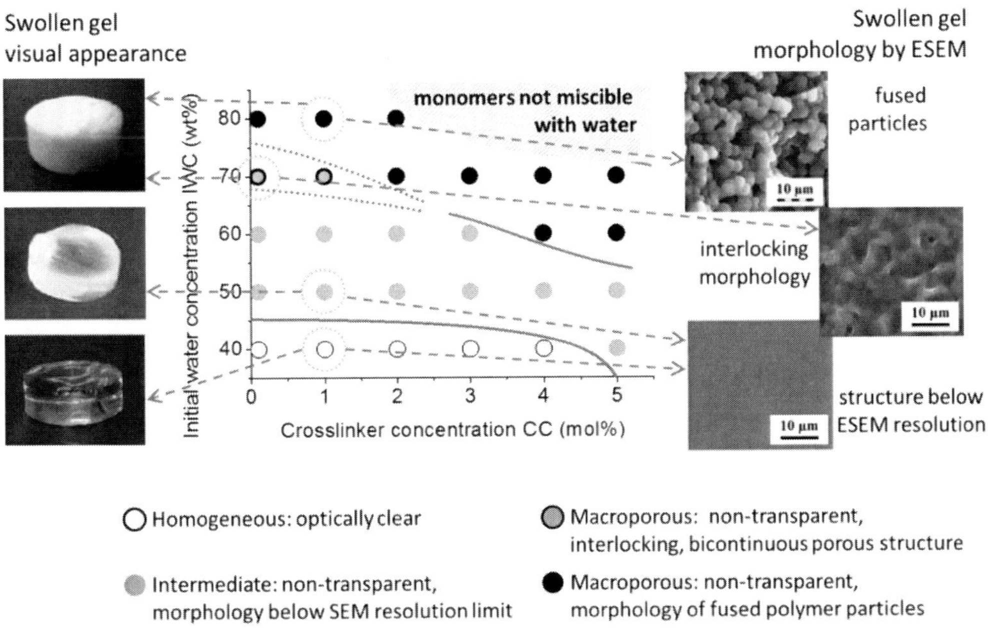

Figure 4. Appearance and morphology of gels prepared with water as diluent, and swollen in water to equilibrium. Each point corresponds to a hydrogel prepared at certain initial dilution and crosslinker concentration.

Increasing crosslinker concentration generally led to smaller particles forming the hydro gel (Figure 5).

Solvating Power of Diluent at Gel Preparation: Effect on Morphology

Thermodynamic quality of the diluent affects significantly the morphology of produced gel samples through miscibility changes. The effect of swelling medium solvating power was studied in detail in [34], where authors analyzed swelling of polyHEMA homogeneous gels in aqueous solutions of various inorganic salts. Ionic environment can cause „salting-in“ or „salting-out“ effects depending on the ions type. Salting-out solutions decreased the swelling medium solvating power, thus reducing swelling of gels; vice versa, salting-in compounds increased swelling of gels due to improved solvating power of the medium. Similar approach was used in [30] to increase porosity of hydrogels prepared for use as cell supports, by addition of a salting-out additive.

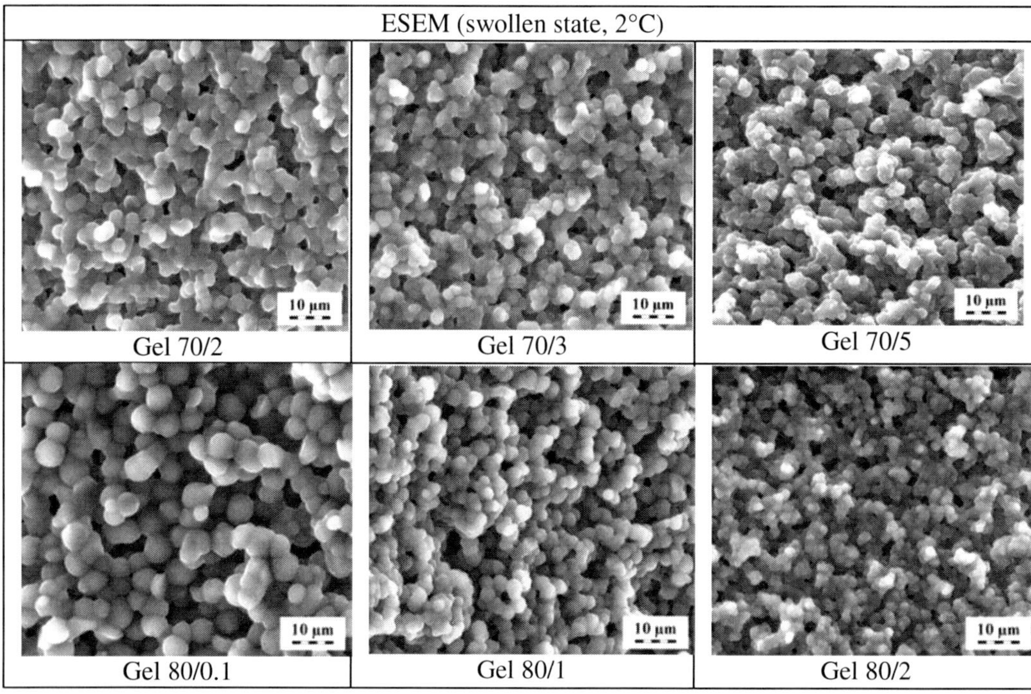

Figure 5. ESEM pictures showing morphology of hydrogels prepared with 70 wt% and 80 wt% of water as diluent and various crosslinker concentrations, observed at equilibrium swollen state in water.

In this work, two ionic additives were used: salting-out sodium chloride and salting-in magnesium perchlorate. Increasing the solvating power of diluent by introducing $Mg(ClO_4)_2$ suppressed the phase separation, and homogeneous hydrogels could be formed at compositions corresponding to phase separation region in pure water (Figure 6). Vice versa, introducing NaCl to polymerizing mixture enhanced the phase separation, and heterogeneous hydrogels were formed at compositions leading to homogeneous gels in pure water (Figure 7).

Fine increase of NaCl concentration in the polymerizing mixtures containing 60 wt% of diluent and 1 mol% of the crosslinker allowed observation of transitional situations, visualizing structural outcome of stepwise enhancing phase separation (Figure 8). With water as diluent, no heterogeneities are observed in SEM picture of a visually non-transparent 60/1 sample (Figure 8a), thus, 60/1 is a typical *"intermediate"* sample. With 0.2 M of NaCl in the diluent at preparation, phase-separated diluent droplets appeared visible in SEM (Figure 8b), and as NaCl concentration in the diluent increased to 0.4–0.425 M, the morphology type changed to *bicontinuous* (Figure 8c–d). With even more NaCl in the diluent (0.45–0.5 M, Figure 8e–g), the material morphology was still bicontinuous but the swollen polymer walls became thicker and porous, and they coexisted with spherical polymer particles fused to their edges. With 0.525 M (or more) NaCl in the diluent (Figure 8h–i), the material morphology changed to *particulate*.

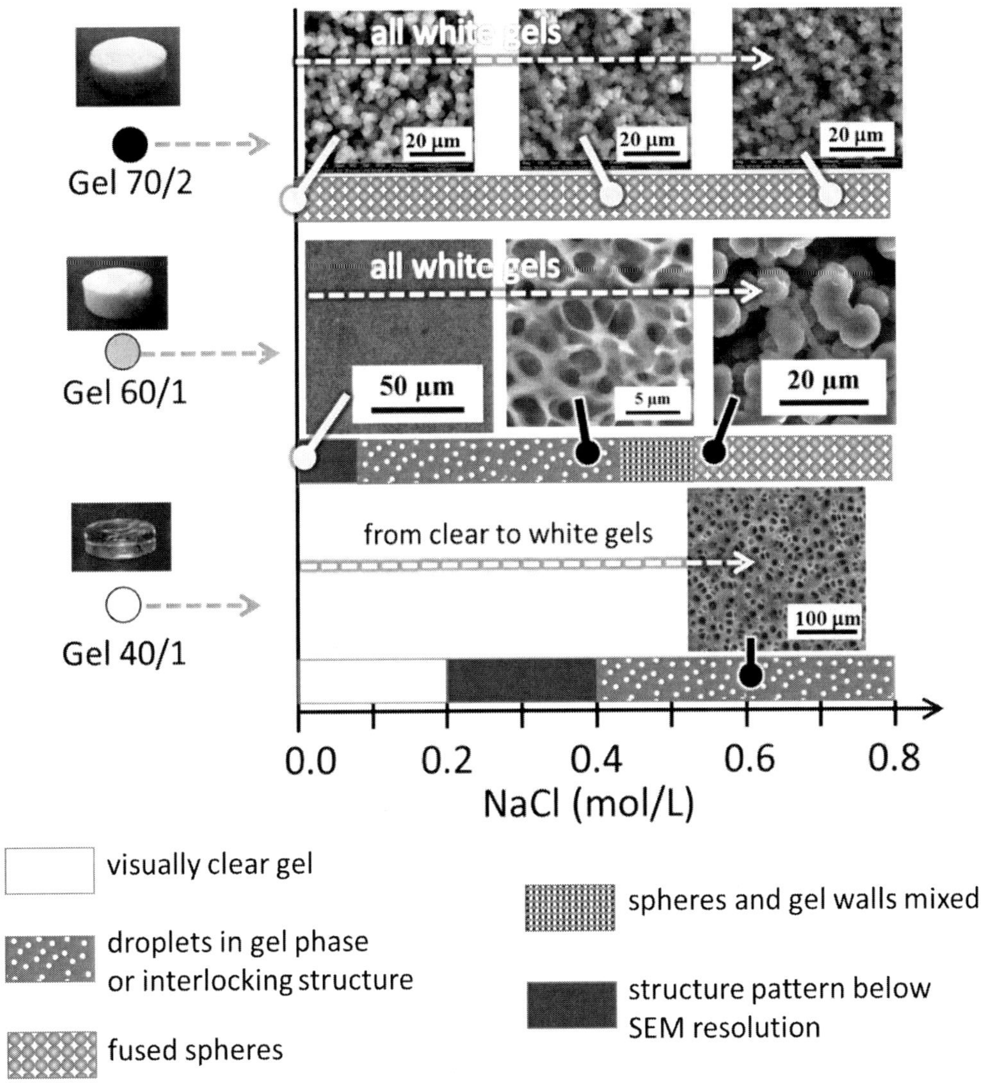

Figure 6. Scheme of gel morphologies (SEM) prepared in diluent of decreasing solvating power, and reswollen in water. The schematic is self-explanatory.

DYNAMIC MECHANICAL RESPONSE OF THE GELS: REPRESENTATIVE EXAMPLES

A typical example of the dynamic mechanical spectra for a homogeneous covalently crosslinked polyHEMA gel swollen in water to equilibrium is presented in Figure 9.

For the 40/1 hydrogel, a traditional material in soft contact lenses applications, at frequency below 1 Hz, elastic shear modulus G' was practically independent of frequency, and was 1–2 orders of magnitude higher than the viscous shear modulus G''. In this range of frequencies, the 40/1 hydrogel is in its rubbery state. The loss factor was close to zero and frequency-independent, typically for chemically crosslinked polymer network. The G' value at such conditions is close to the equilibrium one, and may be used for calculation of the elastically active network chains concentration (crosslink density). Also, elastic modulus at the elasticity plateau (180 kPa for the 40/1 hydrogel) characterizes the overall gel strength.

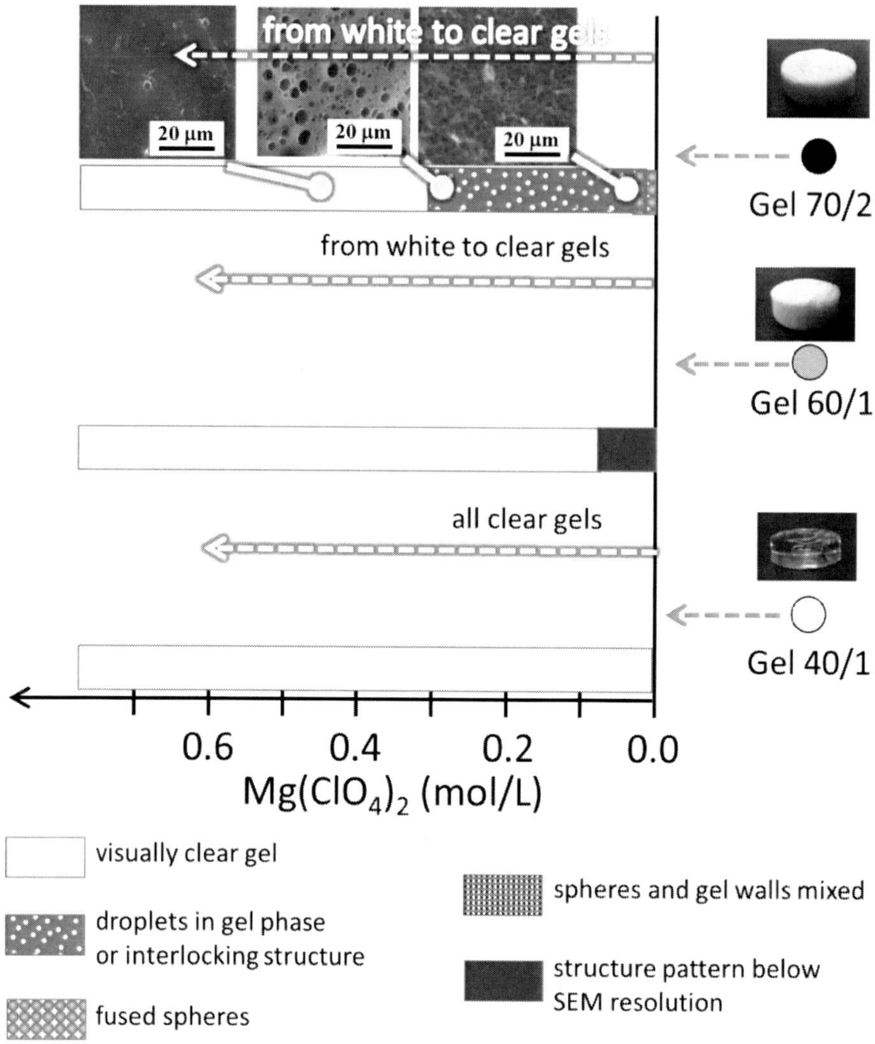

Figure 7. Scheme of gel morphologies (SEM) prepared in diluent of increasing solvating power, and reswollen in water. The schematic is self-explanatory.

G´ started growing with frequency above 1 Hz, marking the onset of the main transition region (also referred to as α-relaxation or mechanical vitrification). *G´´* grew even faster, and at certain frequency (200 Hz) a crossover of elastic and viscous modules occurred. Increased losses (also documented by the loss factor $\tan\delta = G''/G'$) are typical of samples entering the transition region.

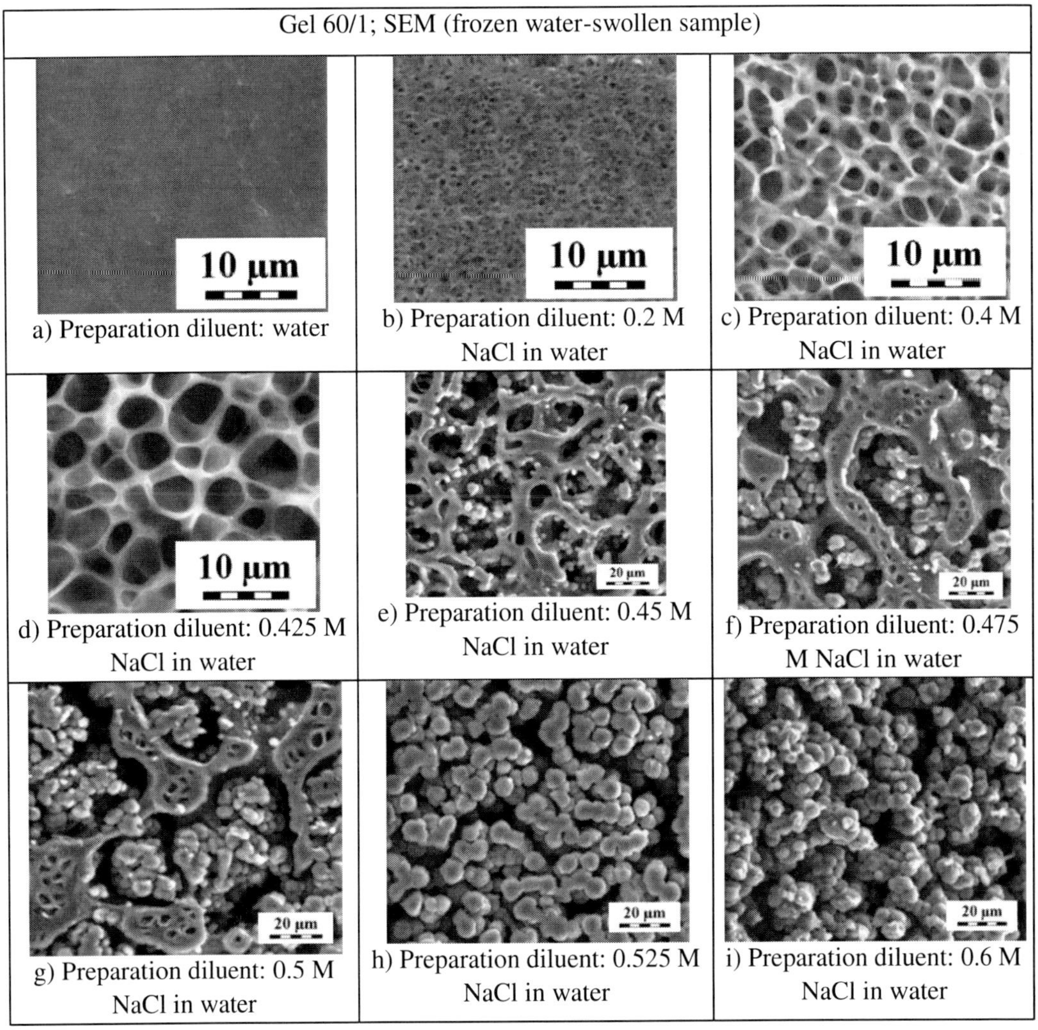

Figure 8. Morphology by SEM of 60/1 hydrogel samples prepared with aqueous NaCl of varied concentration as diluent, and reswollen in water.

Depending on the sample preparation conditions, and thus on its morphology, and/or observation conditions, the dynamic mechanical spectra significantly changed. These effects, discussed in detail in the subsequent sections, are documented in Figures 10–13.

Increase of the crosslinker concentration (Figure 10) led to formation of stronger hydrogel, due to more of elastically active network chains per sample volume, this was reflected in the increase of shear modules of the 40/5 hydrogel (430 kPa at 0.1 Hz) as compared to the 40/1 (180 kPa at 0.1 Hz). Moreover, as network chains of the 40/5 were shorter and more rigid, this sample showed onset of the glassy region at lower frequency (the crossover frequency was reduced to 10 Hz for the 40/5 sample as compared with 200 Hz for the 40/1).

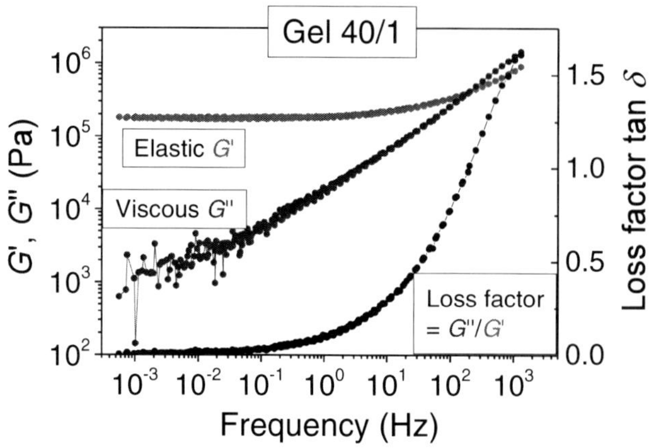

Figure 9. Elastic and viscous shear modules, and loss factor tanδ for water-swollen homogeneous gel 40/1. Equilibrium swelling of the sample was 0.371.

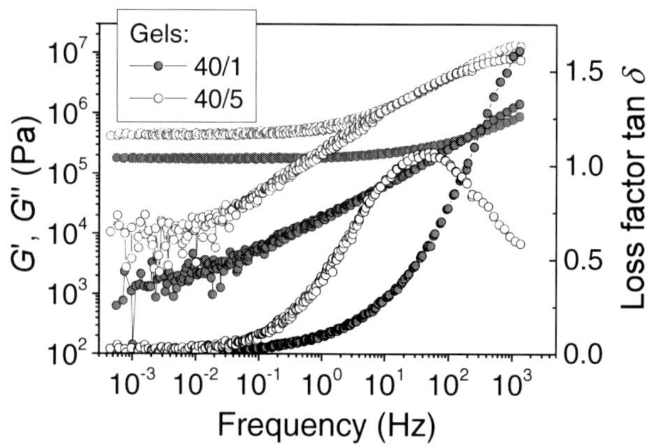

Figure 10. Elastic and viscous shear modules, and loss factor tanδ for water-swollen homogeneous gel 40/1 and water-swollen intermediate gel 40/5. Equilibrium swelling was: 0.371 (40/1) and 0.338 (40/5).

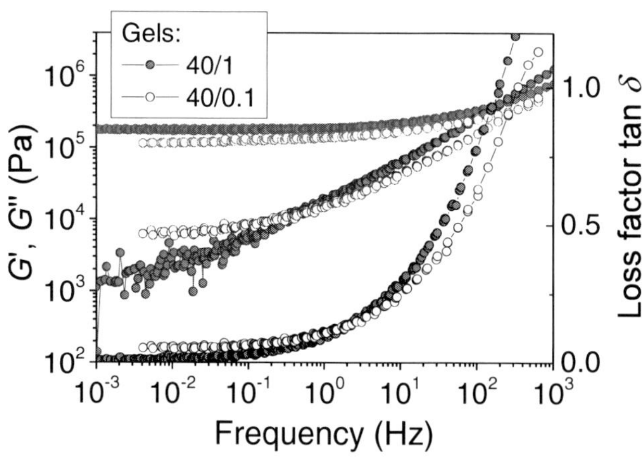

Figure 11. Elastic and viscous shear modules, and loss factor tanδ for water-swollen homogeneous gels 40/1 and 40/0.1. Equilibrium swelling was: 0.371 (40/1) and 0.390 (40/0.1).

Decrease of the crosslinker concentration (viscoelastic responses of samples 40/1 and 40/0.1 are compared in Figure 11) led to slight decrease of the elastic modulus due to less of elastically active chains per sample volume (115 kPa *vs* 180 kPa). Viscous modulus of the 40/0.1 hydrogel as a function of frequency showed a complex trend: with decreasing crosslinker content, G'' increased at lower frequency range (below 0.1 Hz) compared to the 40/1 sample, indicating higher mechanical losses, nearly frequency-independent. However, above 1 Hz viscous modulus of the 40/0.1 sample was lower than that of the 40/1 sample. Overall, this led to shift of vitrification to higher frequency (crossover of G' and G'' for the 40/0.1 sample occurred at 300 Hz).

Morphology of fused spheres was reflected in an abrupt decrease of shear modules (1 kPa *vs* 180 kPa in the plateau region, Figure 12). The main transition onset was observed in roughly the same frequency range. The loss factor tanδ of the particulate 80/1 gel was higher than for the homogeneous 40/1 below ~1 Hz. The onset of secondary relaxation process was observed at very low frequency (~10^{-3} Hz) for the gel constructed of swollen polymer particles, fused together.

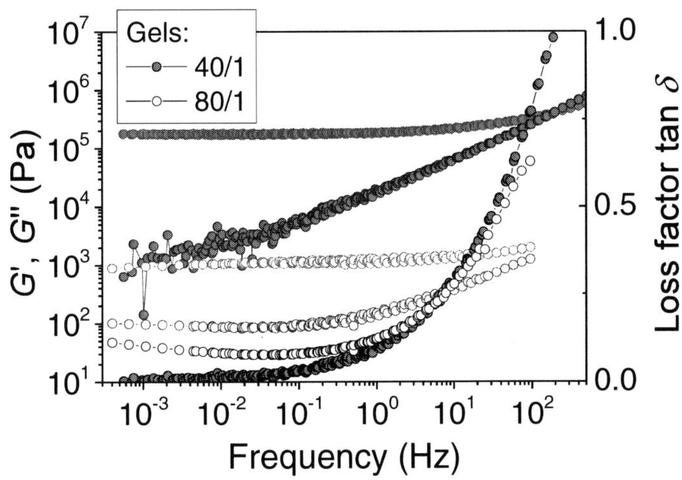

Figure 12. Elastic and viscous shear modules, and loss factor tanδ for water-swollen homogeneous gel 40/1 and water-swollen particulate gel 80/1. Equilibrium swelling was: 0.371 (40/1) and 0.724 (80/1).

Figure 13 shows the effect of swelling medium on mechanical spectra of the homogeneous 40/1 gel. In the medium of greater solvating power than water (DMSO), the elastic modulus was lower due to less polymer material per unit volume of swollen gel. The loss factor was much lower for the DMSO-swollen sample, the onset of α-relaxation was barely seen. Vice versa, swelling in less solvating medium (aqueous 3 M NaCl) led to shift of the α-relaxation to much lower frequency (crossover frequency for the 40/1 gel swollen in 3 M NaCl was ~ 10^{-3} Hz vs 200 Hz for the water-swollen 40/1 gel). Therefore, the elastic modulus was significantly increased over the whole range of frequency, which is typical of a material in glassy state.

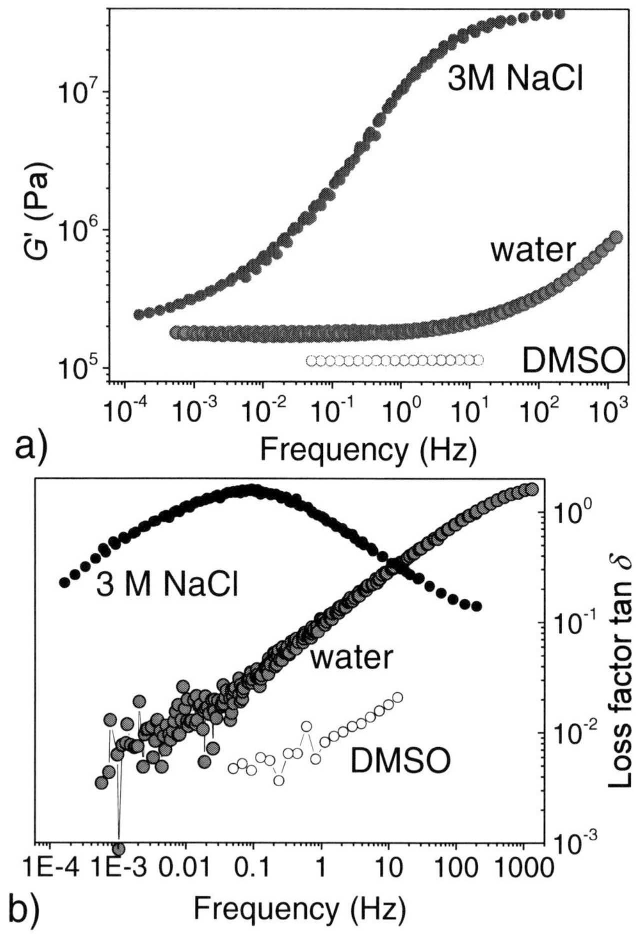

Figure 13. (a) Elastic shear modulus and (b) loss factor tanδ for homogeneous gel 40/1 swollen in water, DMSO or 3M aqueous NaCl.

Equilibrium Elastic Properties

Gels Prepared with Water As Diluent and Swollen in Water

Change of morphology from homogeneous or intermediate to particulate type was accompanied by an abrupt drop of the elastic shear modulus in the region of elasticity plateau.

Equilibrium elastic shear modules for all studied hydrogels polymerized in water and swollen in water to equilibrium are collected in Figure 14.

- The materials showed strikingly different equilibrium shear modulus, in the range between 0.8 and 430 kPa.
- At constant crosslinker concentration, equilibrium shear elastic modulus always decreased with initial dilution.
- Change of morphology to fused particles type was reflected in decrease of equilibrium elastic shear modulus, significantly steeper than that caused by the

change of dilution for homogeneous and intermediate gels. The initial dilution corresponding to abrupt decrease of elastic shear modulus was shifted to lower water concentration with more crosslinker in the gel.

- Outside the border between particulate and intermediate hydrogels, at constant initial dilution, equilibrium shear modulus increased with concentration of crosslinker.
- However, when the change of morphology was caused by increasing crosslinker concentration, corresponding decrease of *G´* might be masked by the crosslinker effect (compare modules in pairs 70/1, 70/2 and 60/3, 60/4).

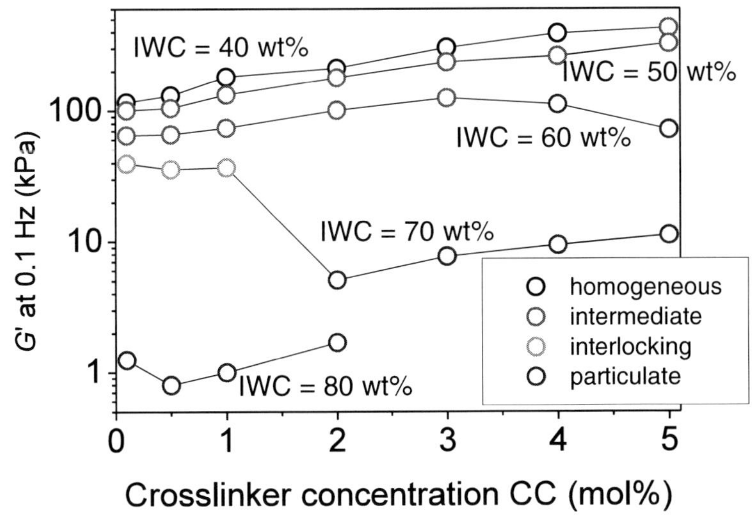

Figure 14. Storage shear modulus at 0.1 Hz as a function of crosslinker concentration for hydrogels prepared with water as diluent and swollen in water. Numbers at curves show initial water content (wt%) for corresponding series. Black: homogeneous, red: intermediate, green: interlocking, blue: particulate gels.

Effect of Polymerization Diluent on Equilibrium Elastic Properties of Hydrogels

In order to study the effect of the solvating power of diluent at preparation on the structure and elastic properties of final hydrogels, ionic compounds with salting-in ($Mg(ClO_4)_2$) or salting-out (NaCl) effect were used. PolyHEMA gels were polymerized in aqueous solutions of one of these salts, and the effect of the salt concentration on the elastic shear modulus was studied after reswelling of prepared samples in water. Results are reported in Figure 15 along with the equilibrium swelling of corresponding hydrogels.

For compositions 60/1 and 70/2, addition of salting-in compound led to increase of the equilibrium elastic shear modulus, with hydrogels morphology changing to homogeneous (Figure 15a). Constant value of *G´* was not reached for these compositions even at higher concentration of $Mg(ClO_4)_2$, 0.5 M.No effect of $Mg(ClO_4)_2$ addition was observed in case of 40/1 series, the final hydrogels were always optically clear within this series.

From Figure 15b, homogeneous hydrogels prepared with the salting-in compound showed similar equilibrium swelling in water (0.37–0.42 g/g) independently of the mixture composition. Still, the equilibrium modulus was different (between 70 kPa for the 70/2 gel at 0.4 M $Mg(ClO_4)_2$ and 180 kPa for 40/1 gel). Thus, introduction of a salting-in additive is an easy way to produce homogeneous hydrogels softer than could be manufactured with pure water as diluent at any initial dilution/crosslinker content combination.

As expected, the effect of salting-out NaCl was opposite (Figure 15c, d). With decreasing solvating power of the diluent the phase separation was enhanced, and at some point the morphology of final hydrogel was changed to that of particulate type, which was reflected in abrupt drop of the elastic modulus.

For particulate hydrogels, further increase of NaCl concentration in polymerizing mixture led to increase of the elastic shear modulus at practically constant equilibrium swelling. This increase of *G´* was weaker than that due to suppressing phase separation. This effect could be explained by appearance of smaller particles fraction, leading to better connectivity of the swollen polymer phase and thus, higher elastic modulus.

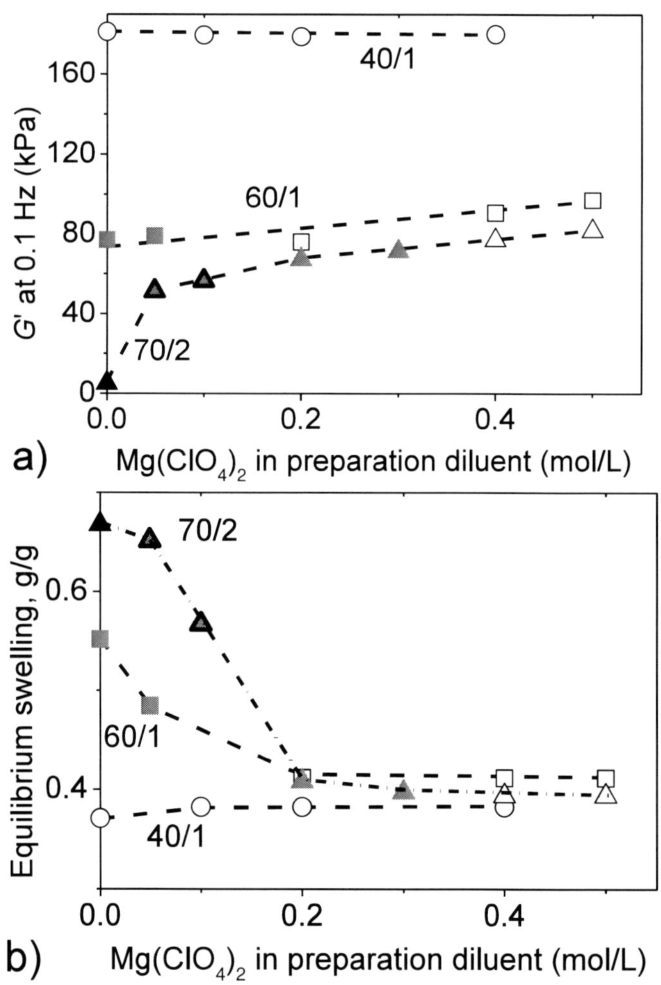

Figure 15. (Continued).

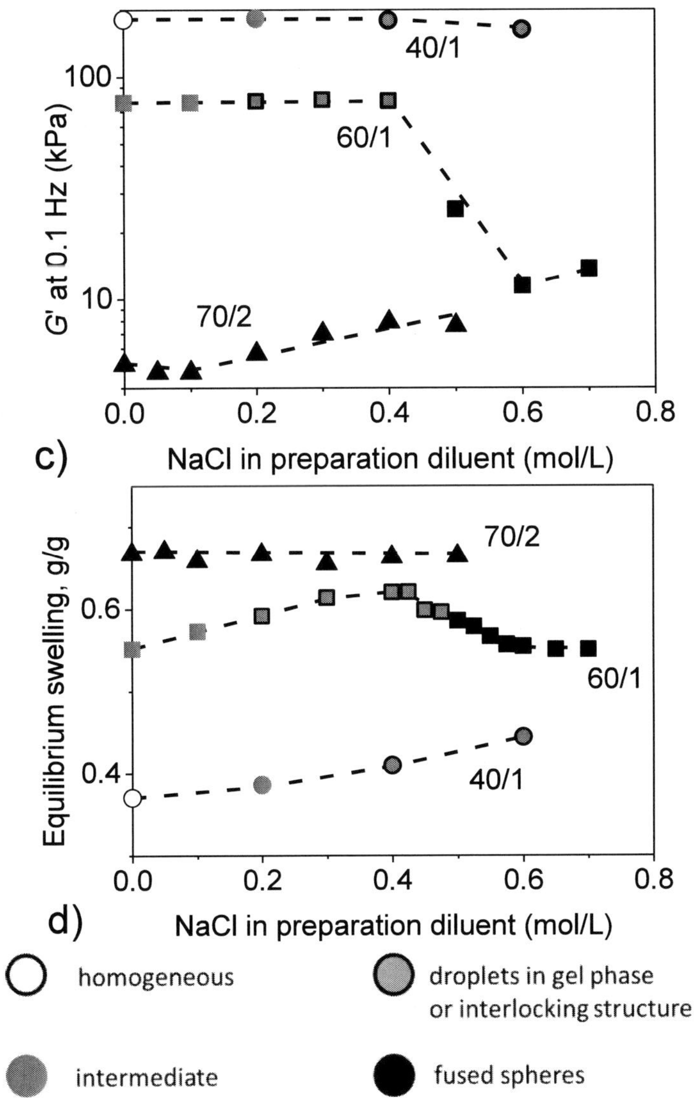

Figure 15. (a, c) Elastic shear modulus at 0.1 Hz and (b) equilibrium swelling in water for hydrogels 40/1, 60/1, 70/2 prepared with aqueous (a, b) $Mg(ClO_4)_2$ or (c, d) NaCl of varied concentration as diluent, and reswollen in water. Samples composition is indicated by code near the curve. Symbol shape shows the hydrogel morphology.

Homogeneous Gels: Elastically Active Network Chains Concentration

Concentration of elastically active network chains (EANC) can be calculated from the equilibrium elastic shear modulus G_e and equilibrium swelling data, according to Equation 1:

$$G_e = RT\nu_e(\phi_2^0)^{2/3}(\phi_2)^{1/3} \quad (1)$$

where ν_e is the EANC concentration, in moles per dry unit volume of the network ϕ_2 is volume fraction of polymer in the gel swollen to equilibrium and ϕ_2^0 is volume fraction of polymerizable substances at preparation, R and T are the gas constant and absolute temperature, respectively.

In this section, results of such calculations for series of gels prepared with 40 wt% of water as diluent (40/X series) will be discussed. All these hydrogels are homogeneous, except for the 40/5 sample, which is intermediate. For majority of 40/X gels, the storage modulus is constant at frequency below 0.1 Hz, and can be considered equilibrium shear modulus, G_e. However, for gels with the lowest crosslinker concentration 40/0 and 40/0.1 constant G' value was not reached even at 10^{-3} Hz. It was assumed that at frequency 10^{-3} Hz $G' \approx G_e$ with reasonable accuracy, and the concentration of elastically active network chains was determined taking this value for calculations. Since for homogeneous gels $\phi_2^0 = \phi_2$, Equation 2 holds

$$G_e = RT\nu_e\phi_2^0 \tag{2}$$

Similar calculations were performed for the same 40/X gels reswollen in DMSO, swelling medium of greater solvating power than water. For polyHEMA gels swollen in DMSO, already at 0.1 Hz G' reached its equilibrium value.

The ideal concentration of elastically active network chains ν_{chem} was calculated from the composition of polymerization mixture assuming that each consumed double bond of the crosslinker contributed to one EANC. The impact of the crosslinker DEGDMA present in commercial HEMA (0.07 wt% by gas chromatography) was also considered.

Results for water- and DMSO-swollen gels are summarized in Figure 16.

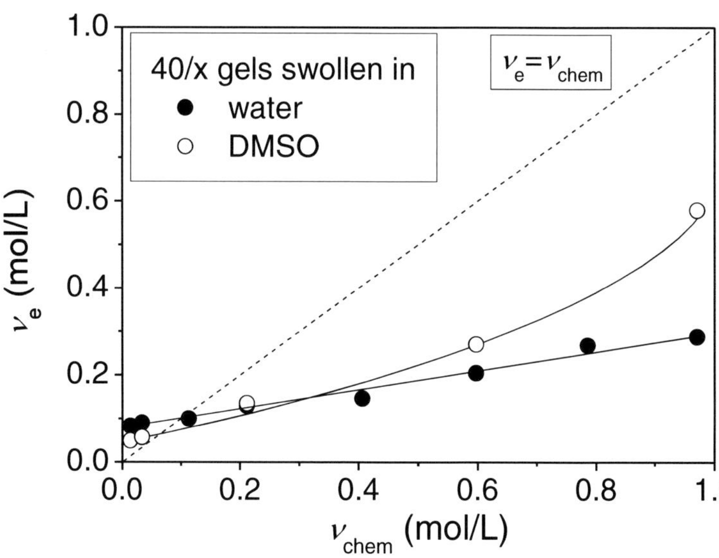

Figure 16. Concentration of elastically active network chains determined experimentally and ideal concentration of elastically active network chains.

The dependence of ν_e on ν_{chem} for water-swollen samples was close to linear with slope of 0.22 (due to incomplete utilization of crosslinker in forming EANCs). That slow relaxations caused likely by regrouping of hydrophobic clusters are still operative in case of water-swollen gels, is documented by a relatively high value of tan δ for gels with lowest crosslinker content.

Dependences for both water-swollen and DMSO-swollen 40/X series showed a positive intercept for $\nu_{chem}\rightarrow 0$ which could be due to still ongoing relaxations of hydrophobic clusters and/or permanent constraints (trapped entanglements). In DMSO, these hydrophobic clusters cannot exist, tan δ falls down to a very small value; thus, the positive intercept in case of DMSO-swollen samples is due to trapped entanglements.

Phase-separated Gels: Elastically Active Network Chains Concentration

Unlike the homogeneous hydrogels series 40/X, the porous samples contained voids filled with swelling medium. When considering the elastic response of the matrix, it should be related to the amount of polymeric material. Assuming that swelling of the matrix equaled to swelling of homogeneous hydrogel with the same crosslink density, a trivial correction for the modulus was applied: $G_e = G'\phi_2(40)/\phi_2(XX)$ with $\phi_2(40)$, volume fraction of polymer in the reference homogeneous sample and $\phi_2(XX)$, volume fraction of the porous gel counting also the solvent in pores. That corrected value of the modulus was used for calculating the concentration of elastically active network chains using Equation (1) with $\phi_2^0 = \phi_2 = \phi_2(40)$.

Results of calculations are presented in Figure 17. With increasing dilution (IWC = 40–60 wt%) EANC concentration and the slope of linear part of the curves steadily decreased, likely due to increasing impact of side reactions of cyclization. Developing of particulate hydrogel morphology caused a striking decrease of the EANC concentration calculated from experimental data.

However, the values of ν_e for particulate gels should be called "fictive" because the trivial correction for the pores fraction was not sufficient for calculating of the concentration of EANCs of the matrix in porous gels. Inhomogeneous stresses arising from varying thickness of the stressed macroporous material and participation of deformation modes other than shear counted among the other factors. Inhomogeneous stresses arisen from the fact that the force applied to the outer surface concentrated at a small fraction of the cross-section of the porous material. This situation is visualized by Figure 18. The necks connecting the fused spheres are the effective load bearing parts of the macroporous material. The stress concentration is strongly enhanced by dangling structures, which means more material per the same cross-section of the neck. The stress concentration results in larger deformation and, therefore, in lower equilibrium modulus for the matrix compared to the true modulus of the matrix.

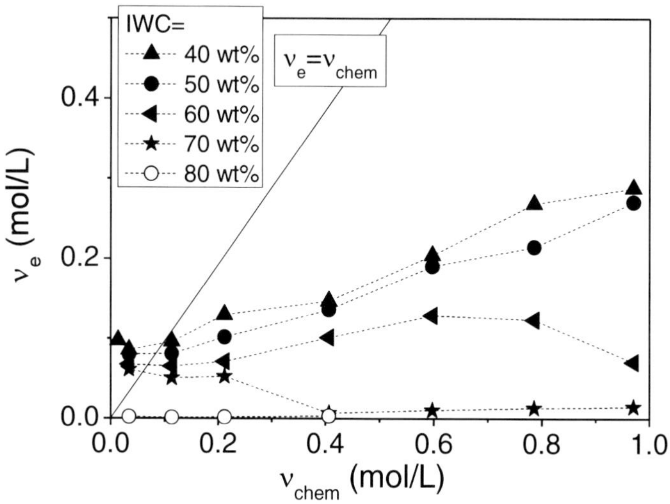

Figure 17. Concentration of EANC determined experimentally for water-swollen gels and ideal concentration of EANC. Solid line shows equality of experimentally determined and calculated EANC concentrations.

DYNAMIC MECHANICAL PROPERTIES

Examples presented in Figures 10–13 demonstrate that analysis of viscoelastic properties (particularly, loss factor tanδ behavior) in a range of frequencies adds to the information gained from the equilibrium properties.

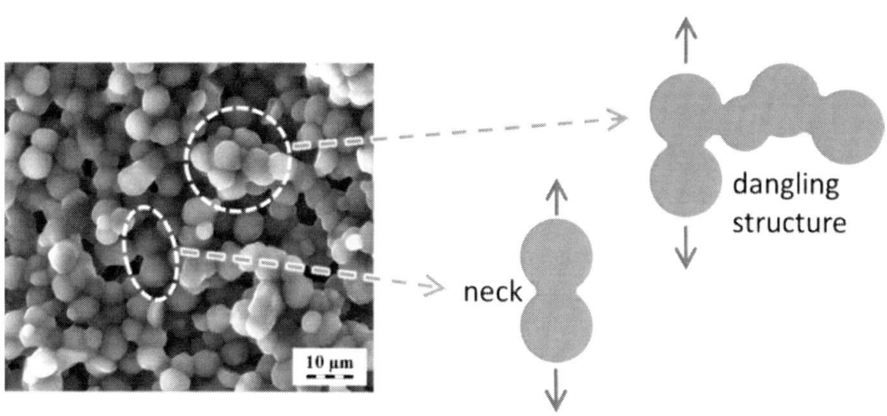

Figure 18. SEM image of fused-sphere like macroporous gel (sample 80/0.1). Sketch visualizes necks connecting particles as loci of higher stress causing larger deformation, and large dangling structures.

Results of dynamic mechanical studies of gels prepared with water as diluent and swollen in water are collected in Figure 19 (hydrogels prepared at a fixed crosslinker concentration and varied initial dilution) and Figure 20 (hydrogels prepared at a fixed initial dilution and varied crosslinker concentration) in form of loss factor frequency dependences. Series of samples with different initial dilution or crosslinker content showed similar trends.

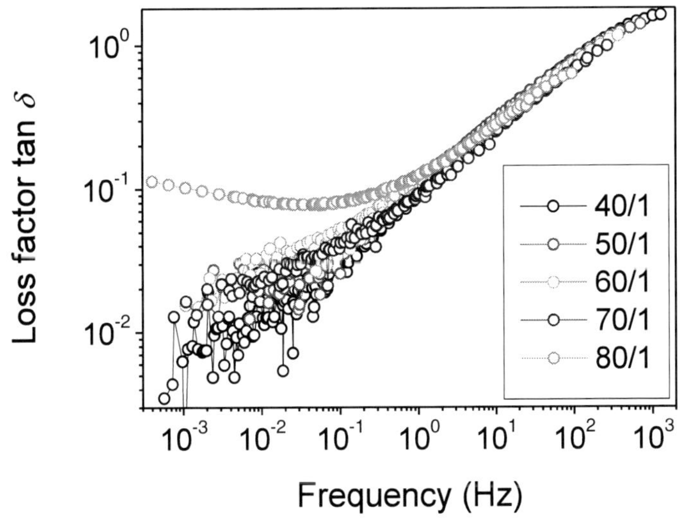

Figure 19. Frequency dependences of loss factor for hydrogels prepared with 1 mol% of crosslinker and variable initial concentration of water as diluent, and swollen in water. Corresponding sample codes are shown near the curves.

Within a series prepared with 1 mol% of crosslinker and variable initial water concentration, the loss factor showed the onset of α-relaxation around 5 Hz for all the samples. Above 1 Hz the difference between the loss factors was negligible. Oppositely, increasing crosslinker concentration at constant initial dilution resulted in the shift of α-relaxation to lower frequencies. Thus, IWC had minor effect on the high-frequency part of the loss factor profiles, corresponding to the α-relaxation process, whereas crosslinker content was one of major factors governing the α-relaxation onset frequency. The described trends are visualized also in Figure 21a showing loss factors at 10 Hz.

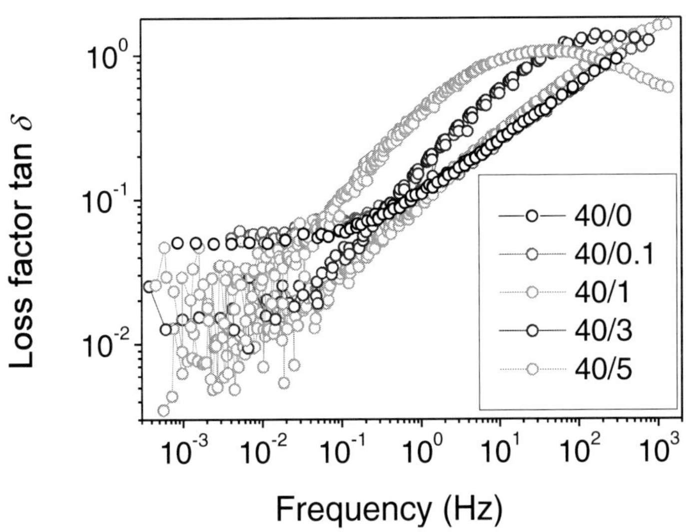

Figure 20. Frequency dependences of loss factor for hydrogels prepared with 40 wt% of water as diluent and variable concentration of crosslinker, and swollen in water. Corresponding sample codes are shown near the curves.

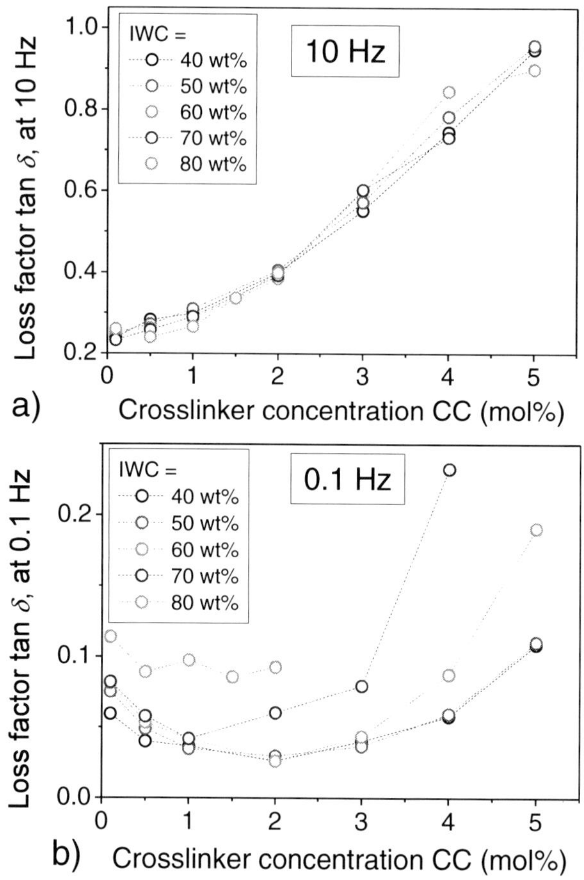

Figure 21. Loss factor measured at 10 Hz (a) and at 0.1 Hz (b) as function of crosslinker concentration CC and initial water content IWC for hydrogels prepared with water as diluent and swollen in water.

At lower frequencies, the effect of changing initial dilution IWC was more significant (loss factors at 0.1 Hz are shown in Figure 21b). The loss factor dependence on crosslinker concentration showed a minimum at 1–2 mol% of DEGDMA. Loss factors for homogeneous, intermediate and interlocking hydrogels formed a single master curve, while loss factors for particulate samples were higher. The structural origin of these effects will be discussed in this section.

The trends revealed in mechanical behavior studies in the forced oscillation mode were in general confirmed by shear creep studies.

Main Transition (Vitrification) of Swollen Hydrogels

The increase of loss factor with frequency above 0.1–1 Hz means that a swollen gel approaches the glass transition, possibly facilitated by hydrophobic aggregates or by reorganization of water molecules [43, 44].

In the α-relaxation region, the loss factor was mainly governed by polymer chains segmental mobility: experimentally, the onset of main transition was shifted to lower frequency with increasing crosslinker concentration.

That the change of main relaxation frequency with crosslinker content was indeed due to shift of the swollen network glass transition region, was confirmed by dynamic differential scanning calorimetry measurements (detailed discussion and the plots are omitted for the sake of brevity).

The effect of swelling medium on the main transition process is documented in Figure 13 for a homogeneous 40/1 gel. Trends for gels prepared at other initial dilutions and/or crosslinker concentrations (including particulate gels) were similar. In all cases, swelling of samples in medium of less solvating power (aqueous NaCl) shifted the main relaxation to lower frequencies (less than 10^{-2}–10^{-3} Hz), while medium of higher solvating power (DMSO) caused onset of frequency relaxation barely seen even at 1–10 Hz.

Thus, the main transition process of swollen polyHEMA gels was found to be very sensitive to concentration of the crosslinker used for network preparation (and associated change of chain segmental mobility), and to the solvating power of the swelling medium (in poor solvents the α-relaxation is facilitated by hydrophobic aggregation). Interestingly, the initial dilution showed negligible effect on the α-relaxation. Basing on this information, conclusion was made on the morphology of non-transparent hydrogels below the SEM resolution limit (intermediate gels).

Morphology of Heterogeneous Hydrogels with Structure below SEM Resolution

Intermediate hydrogels, non-transparent, heterogeneous samples with structure below electron microscopy resolution, were produced in a vast area of polymerizing mixture compositions. In general, those gels can either incorporate very small solvent-filled droplets or microphases differing in the polymer and/or crosslinker concentration. It was demonstrated [45] that non-porous hydrogels with inhomogeneous distribution of polymer material and crosslinker (semi-interpenetrating networks) possessed higher swelling degree and lower elastic modulus than hydrogels with same average concentration of crosslinker units, but evenly distributed. Thus, both morphology types described above would explain swelling and equilibrium elastic behavior of intermediate polyHEMA hydrogels.

The dynamic mechanical behavior in the main transition frequency range supported the morphology of solvent droplets within homogeneous polymer phase. Indeed, the main relaxation process occurred in the same frequency range in case of particulate (80/1), interlocking (70/1), intermediate (50/1, 60/1) and homogeneous (40/1) water-swollen samples at the same level of the crosslinker at preparation (Figure 19). That was an indication that the matrix in phase-separated hydrogels had the same swelling properties in water as did homogeneous hydrogel of corresponding concentration of crosslinker. The equality of the degrees of swelling in water of the matrix in porous gels and respective homogeneous gels might imply that also the network structure was the same. Such conclusion was corroborated by the closeness of α-relaxation maxima for gels of various morphologies reswollen in 3 M NaCl solution (Figure 22).

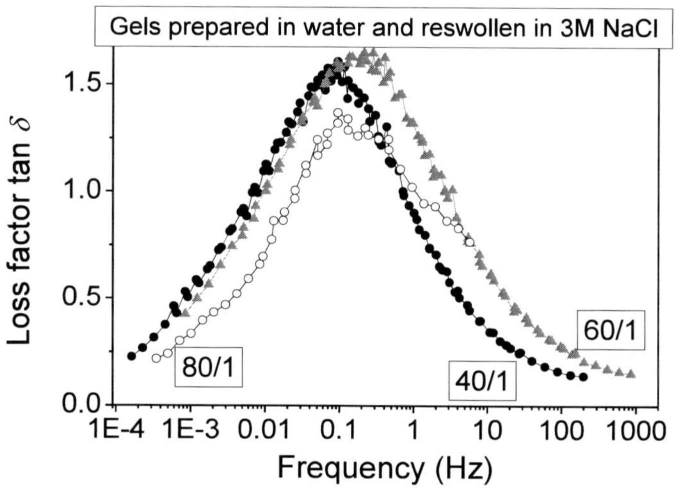

Figure 22. Loss factor tan δ for hydrogels prepared at 1 mol% crosslinker and variable concentration of water, reswollen in 3M NaCl to equilibrium. Sample compositions are shown near the curves.

That swollen polyHEMA matrix in an intermediate gel is the same as homogeneous gel with the same crosslinker concentration, implies that intermediate gels are most likely constituted of the continuous swollen polymer matrix incorporating small droplets of swelling medium. This conclusion was supported by differential scanning calorimetry of selected hydrogel samples (detailed discussion and the plots are omitted for the sake of brevity).

Low-frequency Viscoelasticity Reveals Dangling Chains in PolyHEMA Network and Particulate Morphology of Hydrogels

The studied hydrogels are divided into three categories according to their low-frequency loss factor behavior. In case of non-particulate gels prepared with moderate to high (1–5 mol%) concentration of the crosslinker, tanδ dropped to very low values (below 0.01, equipment resolution limit) at 10^{-2}–10^{-3} Hz. The loss factor of water-swollen hydrogels prepared with concentration of the crosslinker below 1 mol% was significantly higher (0.04–0.08) at frequencies < 10^{-2} Hz, and was frequency-independent. Finally, particulate samples studied showed an onset of a broad and weak secondary relaxation maximum corresponding to ~10^{-4}–10^{-5} Hz (at 25°C). These three types of behavior are illustrated in Figure 23.

The loss factor in a good solvent might be affected by removal (dissolution) of some clusters or associates that are present in poor solvent (water) due to physical interactions, but relaxations due to entanglements and dangling chains, if any, should not be influenced. After reswelling of hydrogels in DMSO, better solvent than water, the loss factors were at least an order of magnitude lower than for corresponding water-swollen gels (representative example is shown in Figure 13). As DMSO is known to disrupt hydrogen bonds present in water, hydrogen bonds between segments of readily available dangling chains are likely responsible for high losses at low frequencies for water-swollen lightly crosslinked gels.

The low-frequency tanδ behavior documented in Figure 23 for the 80/1 sample was characteristic of all studied particulate hydrogels. Some examples are presented in Figure 24. Apparently, the intensity of secondary relaxation increased with higher initial dilution of the polymerized mixture. At the same time, crosslink density had less impact on this secondary relaxation process. Same conclusion were made from analysis of low-frequency behavior of loss factors of hydrogels prepared with 1 mol% of crosslinker and 60 wt% of diluent of varied solvating power - pure water or aqueous NaCl solution.

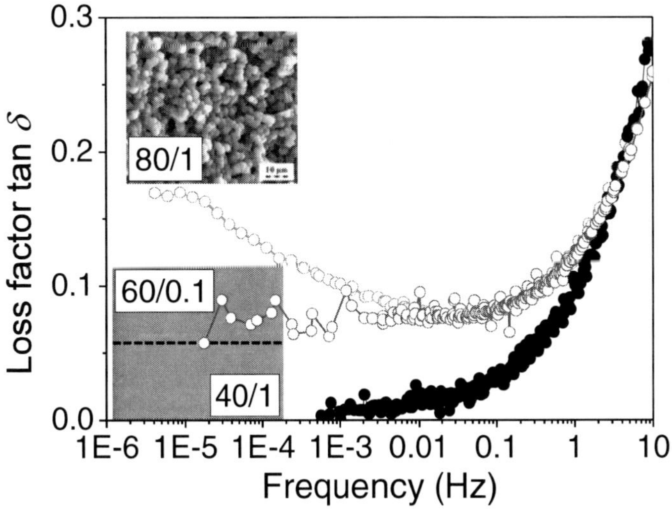

Figure 23. Low-frequency loss factors for representative examples of particulate, lightly crosslinked and other hydrogels, swollen in water to equilibrium.

Dangling chains were reported to be responsible for the low frequency relaxation in polyHEMA gels limitedly swollen with ethylene glycol [35], and that relaxation disappeared with increased swelling (in fact, shifted to higher frequencies and overlapped with the onset of the main relaxation). Similar approach was tested in this work: the samples prepared with water as diluent were swollen in much better solvent, DMSO, and tested in shear at low frequency. Indeed, the secondary relaxation maximum for sample 80/1 disappeared with increasing equilibrium swelling.

Although the consequence of swelling increase was the same in [35] and in this work, the mechanism behind seems to be different. In this work, the low-frequency relaxation maximum was observed only for particulate water-swollen gels, which contained fused polymer gel particles, surrounded by water. Disappearance of relaxation maximum after reswelling in DMSO might be connected with morphology changes at micrometer scale rather than with that at the level of chain segments. In the light microscopy pictures of water- and DMSO-swollen samples 80/1, with increase of overall swelling the particles became larger, and voids shrank.

That the low-frequency relaxation maximum was only observed for particulate samples, suggests that long-range coordinated movements of substructures (dangling aggregates of collated polymer particles, Figure 18) are responsible of it, and void (solvent-filled) space facilitates such movements. After swelling in better solvent, the movement of such aggregates is constrained due to either less void space or due to increase of the moving part weight (which should lead to relaxation at much lower frequency, not reached experimentally).

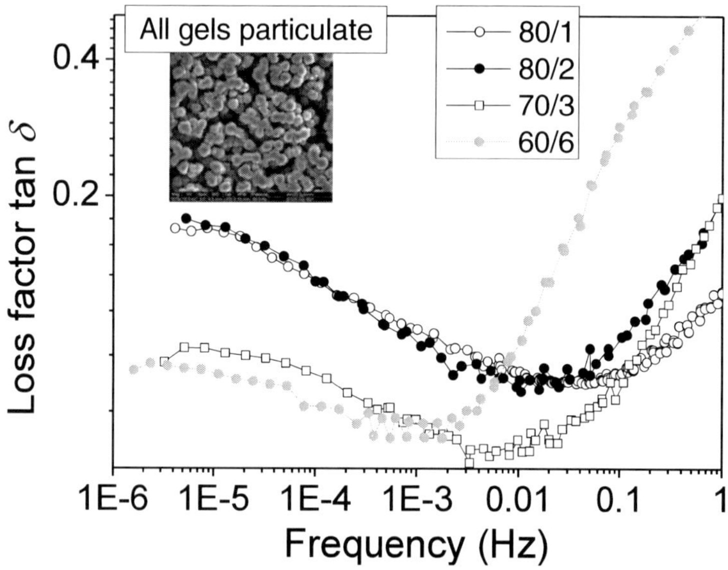

Figure 24. Low frequency behavior of loss factor for fused-particles type hydrogels swollen in water to equilibrium.

CONCLUSION

PolyHEMA hydrogels, including porous ones, have been intensively studied, and many publications about their morphology and properties are available. However, information about relationship of porous polyHEMA gels mechanical properties has been limited; this work adds some more knowledge on this.

Homogeneous and porous polyHEMA gels were prepared by crosslinking polymerization of HEMA in the presence of water at various dilutions and crosslinker contents. Large changes in morphology were achieved when water as diluent was replaced by salt solutions ($Mg(ClO_4)_2$ and NaCl) which improved or deteriorated, respectively, solvation power of the diluent. Varying of salt concentration made possible to fine tune the morphology and to control the passage from isolated droplets morphology to bicontinuous interlocking and fused-spheres like hydrogel structures.

Swelling and mechanical behavior of hydrogels were investigated in relation to their morphology and matrix structure. The physical associates existing in water-swollen gels due to hydrophobic interactions limit the solubility of polyHEMA and swelling of polyHEMA gels. However, these associates were mobile enough to enable reaching mechanical equilibrium within the available frequency range (down to 10^{-3} Hz). This was valid except of gels of the lowest crosslink density and fused-spheres type hydrogels. Reswelling of the gels in a good solvent dissolved the associates and removed the slow relaxations.

From the elastic modulus in the rubbery region, concentration of the elastically active network chains was calculated. For homogeneous gels, its dependence on the crosslinker-to-HEMA ratio revealed that hydrophobic associates added to the crosslink density, acting as physical crosslinks. The impact of permanent constraints (trapped entanglements) on the equilibrium elastic modulus was also noticeable.

spindle are waiting for the starting signal, straining in a state of dynamic tension from a tug-of-war between the sisters' kinetochores pulling toward opposite poles.[3,5,13,147] The starter signal they are waiting for is being suppressed by a limitedly diffusing signal emitted by the kinetochores of those pairs that have not yet attached to the spindle and been put under tension.[3,123a,142,148] In the simplest possible terms, the assembly of a set of so-called Bub and Mad proteins on unattached, unstrained kinetochores includes the Bub1(Mps1; 3F3/2 epitope) protein kinase that recruits and phosphorylates Mad2 protein, which then binds to local APC·p55CDC(Cdc20) complexes and prevents them from starting anaphase.[148] When microtubules are tethered to the kinetochore by the CENP-E and MCAK KRP motors, the combination of microtubule attachment and tension shuts off the protein kinases and kicks Mad2 and its APC-like accomplices off the kinetochores into the cytoplasm.[148] When all of the chromosomes have been attached and their kinetochore signaling silenced, the now unsuppressed spindle-associated APC·p55CDC (Cdc20) can start anaphase when activated by the Anaphase-GO! signal.

The p55CDC(Cdc20) protein that has been accumulating since late in the S phase combines with APC to make the functional APC·p55CDC(Cdc20) complexes that will start anaphase when activated by a protein kinase, Plk (Polo-like kinase), after the expulsion of Mad2.[124–126,148] Until now, the cohesins that are tying the sister chromatids together have been untouched because the separin (Esp1) scissors for cutting these tethers have been locked up by securin (Pds1/cut2).[123a] The Anaphase-GO! signal must unlock the cohesin cutters. It must also get rid of the mitosis-starting Cdk1·cyclin B that up to this point has been driving the condensation of the chromosomes, otherwise the untethered sister chromatids will separate all right, but they will not decondense to form new daughter nuclei.[5,13,14] But before anything can happen, the APC·p55CDC(Cdc20)-restraining PKA activity also must drop to enable APC·p55CDC(Cdc20) to be phosphorylated and activated by Plk, which has been stimulated by Cdk1·cyclin B.[124,125] However, it seems that although the cyclic AMP/PKA surge inactivated APC·p55CDC(Cdc20), it also activated the proteasome that will be ready to shred the polyubiquitinated components when they are produced by the Plk-activated APC·p55CDC(Cdc20).[75,149] In the next chapter, we will learn that the calmodulin-dependent protein kinase II (CaMK II) is the prime mover of Cdk1·cyclin B destruction.[150]

One of APC·p55CDC(Cdc20)'s targets for polyubiquitination and consignment to the proteasome is the securin that has been restraining separin, the cohesin cutter.[123a,145,146] However, cutting cohesin tethers is not enough for chromosome separation — there are still the tangles holding the sister chromosomes together at their centromeres that must be removed by topoisomerase II.[5,14,123a,151,152] Cyclin B is targeted for polyubiquitination by APC·p55CDC(Cdc20) because of its *N*-terminal D (for destruction) box.[2,5,13,14] The Cdk1 liberated from Cdk1·cyclin B by the destruction of its partner cyclin B is inactivated (it becomes unable to bind cyclin A or B) by having its Thr^{161} (which was phosphorylated by CAK) dephosphorylated.[2,13,150] Once the sister chromosomes are untethered and the CDK has been destroyed, the "Pac Man" motors of the chromosomes' kinetochores are switched on and the kinetochores start chewing their way up the spindle fibers to the spindle poles, each dragging its cargo of genes and other things behind it.

The APC activation and the exit from mitosis is ultimately triggered by Cdk1·cyclin B, which is thus the agent of its own destruction.[5,150] But how does the cell stop Cdk1·cyclin B from doing this too early and abort the proper packaging of chromosomes for assembly on the metaphase spindle? Of course, one of the ways it does this is to cause a burst of APC·p55CDC(Cdc20)-inhibiting PKA activity.[145] Also, this may be one of the jobs of its prophase-peaking co-worker and immediate predecessor, Cdk1·cyclin A, before the cyclin A crash just before metaphase (Fig. 1.2). Cdk1·cyclin A does not trigger its own destruction and it prevents Cdk1·cyclin B from triggering the APC-cyclin-destroying mechanism through activated Plk.[150]

During chromosome replication along the assembly lines of the replication factories, the pre-replication complexes were dispersed when phosphorylated by the Cdk2·cyclin A kinases in the synthesomes. While the ORC complexes immediately re-attached to the duplicated origins, the MCM and the unstable hyperphosphorylated Cdc6 proteins could not reattach and were then discarded. No replacements could get into the nucleus. This prevented reinitiation and multiple rounds of DNA replication.[106,107] When the chromosomes have been ratcheted up the chromosomal microtubule bundles to the opposite poles (anaphase A) by their microtubule-depolymerizing Pac-Man-like MCAK kinase/motors, and the spindle poles have finished being pushed apart by the sliding of the interzonal spindle fibers along one another driven by MKLP1/CHO1 KRP motors lodged in the spindle midbody (anaphase B), the daughter cells enter telophase when they decondense their chromosomes and make new nuclei.[13,14,145a,148] It is around this time that they can re-license their chromosomes for replication. The nuclear membrane has not yet been assembled, the Cdk1·cyclin B kinases have long since gone, and unphosphorylated Cdc6 protein level has peaked.[106–110] Cdc6 can now associate with the potentially functional origins as they re-emerge in the decondensing chromosomes. The new nucleus now has the RLF-B1 replication licensor, which has been in the mother cell's cytoplasm but could not get through the intact nuclear envelope. The new cell will be able to replicate its chromosomes *if* it is signaled to start a cycle.

By anaphase the cell has wrapped a wonderful device around its middle just underneath the plasma membrane — a contractile belt of actomyosin fibrils, which has been kept relaxed until the right moment by having certain residues of the light chains of the myosin motor heads phosphorylated by Cdk1·cyclin B.[13,152a] However, this stops when the kinase is destroyed by the APC-exit mechanism and the inhibitory phosphates it planted on key residues of the myosin motors are removed by a surge of protein phosphatase activity. Then, a small G-protein, Rho, is activated during anaphase by being converted by a guanine nucleotide-exchanger from GDP·Rho to GTP·Rho and starts the myosin motors by enhancing the phosphorylation of another set of myosin light chain residues by calmodulin-dependent protein kinase II by inactivating myosin light chain phosphatase.[13,14,150] As the myosin motors pull on the actin filaments, they tighten the belt which guides vesicles for new membrane production into the cleavage furrow the belt produces as it pinches the large, elongating cell into two, usually normal size, daughters.[152a] The fate of the daughter cells will depend on the genes they are expressing (i.e., their transcriptome), the signals they receive from their neighbors, and the hormones and other factors flowing into the extracellular matrix from the blood (Fig. 1.1). If the daughters are normal, their Rb-family pocket proteins were dephosphorylated and reactivated during mitosis, which will prevent the initiation of a G_1 buildup until growth factor signaling resumes. If the replication-related genes are now locked shut, the cells may differentiate, work for a while, and then kill themselves. If their chromosomes are not licensed for replication in the replication factories, they will be unable to proliferate regardless of whether or not their proliferogenes are locked shut. Could this be why mature neurons can function for decades without proliferating despite often intense signaling by neurotrophic and other factors that would start a G_1 buildup in other cells?

The neurons in the brains of the victims of Alzheimer's disease, choking with neurofibrillary tangles and nesting among clumps of β-amyloid protein, the perilously poised neurons in the penumbra of a stroked brain, and other crippled neurons of other neurodegenerative disorders bear witness to the terrible things that happen when proliferatively disabled cells respond to damage signals by trying to proliferate to replace lost cells and repair their damaged tissue.[153–164] While a tiny few neurons might succeed,[165] most try and *die*! Their abortive attempt to start a cycle results in the disastrously piecemeal expression of only some of the CDKs or their components such as cyclin D1, cyclin E, and cyclin B1, Cdk1, Cdk4, Cdk5 as well as replication components such as PCNA and the nuclear Ki-67 protein. The cell might even try to go directly into mitosis by expressing the mitotic CDK, Cdk1·cyclin B1.[163] One consequence of this is the appearance of the Cdk1·cyclin B1-phosphorylated tau protein of the neurofibrillary tangles of Alzheimer neurons, the phospho-epitopes of which are the same in mitotic cells.[163] However, since the stimulated neurons cannot replicate their chromosomes, the appearance of CDKs such as the prophase-triggering Cdk1·cyclin B1 would lethally trigger premature prophase.

A common reason for this attempt to force neurons to suicidally start building up to DNA replication by the piecemeal expression of CDKs and their cyclin components is likely to be the CDK-induced inactivation of the Rb pocket protein and the inappropriate expression E2F1, which, as we learned earlier, is the one member of the E2F family that can induce a $p19^{ARF}$-driven stabilization and accumulation of p53(TP53). Without the checks, balances, and protector proteins of a complete or balanced G_1 buildup, the accumulating p53(TP53) stimulates the expression of the $p21^{Cip1/WAF1}$ CKI and other apoptogenic genes.[47–50] Indeed, the accumulation of p53(TP53) could explain the observed expression of $p21^{Cip1/WAF1}$ in Alzheimer brains.[155]

References

1. Boulikas T. Phosphorylation of transcription factors and control of the cell cycle. Crit Rev Eukaryotic Gene Expression 1995; 5: 1–77.
2. Bowen ID, Bowen SM, Jones AH. Mitosis and Apoptosis. London, Chapman & Hall, 1998.
3. Elledge SJ. Cell cycle checkpoints: preventing an identity crisis. Science 1996; 274: 1664–1672.
4. Hunter T. Regulation of the Eukaryotic Cell Cycle. Chichester, John Wiley and Sons, 1992: 1–289.
5. King RW, Deshaies RJ, Peters J-M, et al. How proteolysis drives the cell cycle. Science 1996; 274: 1652–1658.
6. Murray A, Hunt T. The Cell Cycle. New York, WH Freeman, 1993: 1–256.
7. Nasmyth K. Viewpoint: putting the cell cycle in order. Science 1996; 274: 1643–1645.

8. Pestell RG, Albanese C, Reutens AT, et al. The cyclins and cyclin-dependent protein kinase inhibitors in hormonal regulation of proliferation and differentiation. Endocrine Revs 1999; 20: 501–534.
9. Pines J. Cyclins and cyclin-dependent kinases: a biochemical view. Biochem J 1995; 308: 697–711.
10. Pines J. Cyclins, CDKs and cancer. Sem Cancer Biol 1995; 6: 63–72.
11. Sherr CJ. Cancer cell cycles. Science 1996; 274: 1672–1677.
12. Stillman B. Cell cycle control of DNA replication. Science 1996; 274: 1659–1664.
13. Whitfield JF. Calcium in Cell Cycles and Cancer. Boca Raton, CRC Press, 1995: 1–232.
14. Whitfield JF. Calcium: Cell Cycle Driver, Differentiator and Killer. Austin / New York, Landes Bioscience / Chapman & Hall, 1997.
15. Whitfield JF, Bird RJ, Chakravarthy BR, et al. Calcium — cell cycle regulator, differentiator, killer, chemopreventor, and maybe, tumor promoter. J Cell Biochem 1995; Suppl 22: 74–91.
16. Beijersbergen RL, Bernards R. Cell cycle regulation by the retinoblastoma family of growth inhibitory proteins. Biochim Biophys Acta 1996; 1287: 103–120.
17. Cobrinik D. Regulatory interactions among E2Fs and cell cycle control proteins. Curr Topics Microbiol Immunol 1996; 208: 31–61.
17*a*. Humbert PO, Verona R, Trimarchi JM, et al. E2F3 is critical for normal cellular proliferation. Genes Dev 2000; 14: 690–703.
18. Kaelin WG. Functions of the retinoblastoma protein. BioEssays 1999; 21: 950–958.
19. Sardet C, LeCam L, Fabbrizio E, et al. E2Fs and the retinoblastoma protein family. In: Yaniv M, Ghysdael J, editors. Oncogenes as Transcription Regulators, Vol 2: Cell Cycle Regulators and Chromosomal Translocation. Basel, Birkhäuser Verlag, 1997: 1–62.
20. Sigler P, Giordano A. Big brothers are watching: the retinoblastoma family and growth control. Prog Mol Subcell Biol 1998; 20: 25–42.
21. Sidle A, Palaty C, Dirks P, et al. Activity of the retinoblastoma family proteins, pRB, p107, and p130, during cellular proliferation and differentiation. Crit Rev Biochem Mol Biol 1996; 31: 237–271.
22. Slansky JE, Farnam PJ. Introduction to the E2F family. Curr Topics Microbiol Immunol 1996; 208: 1–30.
22*a*. Greasley PJ, Bonnard C, Amati B. Myc induces the nucleolin and BN51 genes: possible implications in ribosome biogenesis. Nucleic Acids Res 2000; 28: 446–453.
22*b*. Osthus R, Shim H, Kim S, et al. Deregulation of glucose transporter 1 and glycolytic gene expression by c-Myc. J Biol Chem 2000; 275: 21797–21800.
23. Kaldis P. The cdk-activating kinase (CAK): from yeast to mammals. Cell Mol Life Sci (CMLS) 1999; 55: 284–296.
24. Holt PR, Philipova R, Moss S, et al. Calcium/calmodulin-dependent phosphorylation and activation of human Cdc25-C at the G_2/M phase transition in HeLa cells. J Biol Chem 1999; 274: 7958–7968.

25. Saha P, Eichbaum Q, Silberman ED, et al. p21[Cip1] and Cdc25A: competition between an inhibitor and an activator of cyclin-dependent kinases. Mol Cell Biol 1997; 17: 4338–4345.

25*a*. Bouchard C, Thieke K, Maier A, et al. Direct induction of cyclin D2 by Myc contributes to cell cycle progression and sequestration of p27. EMBO J 1999; 18: 5321–5333.

25*b*. Hermeking H, Rago C, Shuhmacher M, et al. Identification of CDK4 as a target of c-MYC. Proc Natl Acad Sci USA 2000; 97: 2229–2234.

26. Su TT, Follette PJ, O'Farrell PH. Qualifying for the license to replicate. Cell 1995; 81: 828.

27. Depoortere F, Dumont JE, Roger PP. Paradoxical accumulation of the cyclin-dependent kinase inhibitor p27kip1 during the cAMP-dependent mitogenic stimulation of thyroid epithelial cells. J Cell Sci 1996; 109: 1759–1764.

28. Smith E, Frenkel B, MacLachlan TK, et al. Post-proliferative cyclin E-associated kinase activity in differentiated osteoblasts: inhibition by proliferating osteoblasts and osteosarcoma cells. J Cell Biochem 1997; 66: 141–152.

29. Whitfield JF, Morley P, Willick GE. The Parathyroid Hormone: An Unexpected Bone Builder for Treating Osteoporosis. Austin, RG Landes Company, 1998.

30. Peters J-M. SCF and APC the Yin and Yang of cell cycle regulated proteolysis. Curr Opin Cell Biol 1998; 10: 759–768.

31. March KL, Wilensky RL, Roeske RW, et al. Effects of thiol protease inhibitors on cell cycle and proliferation of vascular smooth muscle cells in culture. Circ Res 1993; 72: 413–423.

32. Mellgren RL. Evidence for participation of a calpain-like cysteine protease in cell cycle progression through late G1 phase. Biochem Biophys Res Commun 1997; 236: 555–558.

33. Yang J, Kornbluth S. All aboard the cyclin train: subcellular trafficking of cyclins and their CDK partners. Trends Cell Biol 1999; 9: 207–210.

34. Jaumont M, Estanyol JM, Serratosa J, et al. Activation of cdk4 and cdk2 during rat liver regeneration is associated with intranuclear rearrangements of cyclin–cdk complexes. Hepatology 1999; 29: 385–395.

35. Depoortere F, Van Keymeulen A, Lukas J, et al. A requirement for cyclin D3-cyclin-dependent kinase (cdk)-4 assembly in the cyclic adenosine monophosphate-dependent proliferation of thyrocytes. J Cell Biol 1998; 140: 1427–1439.

35*a*. Aguda BD, Tang Y. The kinetic origins of the restriction point in the mammalian cell cycle. Cell Prolif. 1999; 32: 321–335.

35*b*. Winston JT, Chu C, Harper JW. Culprits in the degradation of cyclin E apprehended. Genes Devel 1999; 13: 2751–2757.

36. Dou Q-P, Pardee AB. Transcriptional activation of thymidine kinase, a marker for cell cycle control. Progr Nucl Acid Res 1996; 53: 197–217.

37. Ford HL, Pardee AB. The S phase: beginning, middle, and end: a perspective. J Cell Biochem 1998; Suppl 30/31: 1–7.

37*a*. Davie JR, Spencer VA. Signal transduction pathways and the modification of chromatin structure. Progr Nucleic Acid Res Mol Biol 2001; 65: 299–340.

38. Slansky JE, Farnham PJ. Transcriptional regulation of the dihydrofolate reductase gene. BioEssays 1996; 18: 55–62.
39. Saville MK, Watson RJ. B-Myb: a key regulator of the cell cycle. Adv Cancer Res 1998; 72: 109–140.
40. Hinchcliffe EH, Li C, Thompson EA, et al. Requirement of Cdk2-cyclin E activity for repeated centrosome reproduction in *Xenopus* egg extracts. Science 1999; 283: 851–854.
41. Lacey KR, Jackson PK, Stearns T. Cyclin-dependent kinase control of centrosome duplication. Proc Natl Acad Sci USA 1999; 96: 2817–2822.
42. Marshall W. Centrosome birth control. Trends Cell Biol 1999; 9: 214.
43. Santella L, Kyozuka K, De Riso L, et al. Calcium, protease action and the regulation of the cell cycle. Cell Calcium 1998; 23: 123–130.
44. Bartsch O, Horstmann S, Toprak K, et al. Identification of cyclin A/Cdk2 phosphorylation sites in B-Myb. Eur J Biochem 1999; 260: 384–391.
45. Dedhar S. Integrins and signal transduction. Curr Opin Hematol 1999; 6: 37–43.
46. Diehl JA, Zindy F, Sherr CJ. Inhibition of cyclin D1 phosphorylation on threonine-288 prevents rapid degradation via the ubiquitin–proteasome pathway. Genes Dev 1997; 11: 957–972.
46*a*. Howe AK, Juliano RL. Regulation of anchorage-dependent signal transduction by protein kinase A and p21-activated kinase. Nature Cell Biol 2000; 2: 593–600.
47. Adams PD, Kaelin WG Jr. The cellular effects of E2F overexpression. Curr Top Microbiol Immunol 1996; 208: 79–93.
48. Juven-Gershon T, Oren M. Mdm2: the ups and downs. Mol Med 1999; 5: 71–83.
48*a*. Sherr CJ. Tumor surveillance via the ARF-p53 pathway. Genes Dev 1998; 12: 2984–2991.
49. Kowalik TF, DeGregori J, Leone G, et al. E2F1-specific induction of apoptosis and p53 accumulation, which is blocked by Mdm2. Cell Growth Differ 1998; 9: 113–118.
50. Pan H, Yin C, Dyson NJ, et al. Key roles for E2F1 in signaling p53-dependent apoptosis and in cell division within developing tumors. Mol Cell 1998; 2: 283–292.
51. Wasylyk C, Salvi R, Argentini M, et al. p53 mediated death of cells overexpressing MDM2 by an inhibitor of MDM2 interaction with p53. Oncogene 1999; 18: 1921–1934.
52. Freedman DA, Wu L, Levine AJ. Functions of the MDM2 oncoprotein. Cell Mol Life Sci (CMLS) 1999; 55: 96–107.
53. Shtutman M, Zhurinsky J, Simcha I, et al. The cyclin D1 gene is a target of the β-catenin/LEF-1 pathway. Proc Natl Acad Sci USA 1999; 96: 5522–5527.
54. Orford K, Orford C, Byers SW. Exogenous expression of β-catenin regulates contact inhibition, anchorage-independent growth, anoikis, and radiation-induced cell cycle arrest. J Cell Biol 1999; 146: 855–868.
55. Frisch SM, Vuori K, Ruoslahti E, et al. Control of adhesion-dependent cell survival by focal adhesion kinase. J Cell Biol 1996; 134: 793–799.
56. Bianco C, Tortora G, Baldassare GT, et al. 8-Chloro-cyclic AMP inhibits autocrine and angiogenic growth factor production in human colorectal and breast cancer. Clin Cancer Res 1997; 3: 439–448.

57. Ciardiello F, Damiano V, Bianco R, et al. Antitumor activity of combined blockade of epidermal growth factor receptor and protein kinase A. J Natl Cancer Inst 1996; 88: 1770–1776.
58. Ciardello F, Tortora G. Interactions between the epidermal growth factor receptor and type I protein kinase A: biological significance and therapeutic implications. Clin Cancer Res 1998; 4: 821–828.
59. Tortora G, Damiano V, Bianco C, et al. The RIα subunit of protein kinase A (PKA) binds to Grb2 and allows PKA interactions with the activated EGF-receptor. Oncogene 1997; 14: 923–928.
60. Tortora G, Pepe S, Bianco AR, et al. The RIα subunit of protein kinase A controls serum dependency and entry into cell cycle of human mammary epithelial cells. Oncogene 1994; 9: 3233–3240.
61. Tortora G, Pepe S, Cirafici AM, et al. Thyroid-stimulating hormone-regulated growth and cell cycle distribution of thyroid cells involve type I isozyme of cyclic AMP-dependent protein kinase. Cell Growth Diff 1993; 4: 359–365.
62. Sheffield LG. Oligonucleotide antisense to catalytic subunit of cAMP-dependent protein kinase inhibit mouse mammary epithelial cell DNA synthesis. Exp Cell Res 1991; 192: 307–310.
63. Gadbois DM, Grissman HA, Tobey RA, et al. Multiple kinase arrest points in the G_1 phase of non-transformed mammalian cells are absent in transformed cells. Proc Natl Acad Sci USA 1992; 89: 8626–8630.
64. Villone G, De Amicis F, Veneziani BM, et al. Sustained versus transient cyclic AMP intracellular levels: effect on thyrotropin-dependent growth of thyroid cells. Cell Growth Differ 1997; 8: 1181–1188.
65. Boynton AL, Whitfield JF. The role of cyclic AMP in cell proliferation: a critical assessment of the evidence. Adv Cyclic Nucleotide Res 1983; 15: 193–294.
66. Roger PP, Reuse S, Maenhaut C, et al. Multiple facets of the modulation of growth by cAMP. Vitamins and Hormones 1995; 51: 59–191.
67. Desdouets C, Matesic G, Molina C, et al. Cell cycle regulation of cyclin A gene expression by the cyclic AMP-responsive transcription factors CREB and CREM. Mol Cell Biol 1995; 15: 3301–3309.
68. Foulkes NS, Sassone-Corsi P. Transcription factors coupled to the cAMP-signalling pathway. Biochim Biophys Acta 1996; 1288: F101–F121.
69. Saeki K, You A, Takaku F. Cell-cycle-regulated phosphorylation of cAMP response element-binding protein: identification of novel phosphorylation sites. Biochem J 1999; 338: 49–54.
70. Shaywitz AJ, Greenberg ME. CREB: a stimulus-induced transcription factor activated by a diverse array of extracellular signals. Annu Rev Biochem 1999; 68: 821–861.
71. Vairo GS, Argyriou AM, Bordun G, et al. Inhibition of the signalling pathways for macrophage proliferation by cyclic AMP. J Biol Chem 1990; 265: 2692–2701.
72. Hegde AN, Inokuchi K, Pei W, et al. Ubiquitin C-terminal hydrolase is an immediate–early gene essential for long-term facilitation in Aplysia. Cell 1997; 89: 115–126.

73. Sikorska M, Brewer LM, Youdale T, et al. Evidence that mammalian ribonucleotide reductase is a nuclear membrane-associated glycoprotein. Biochem Cell Biol 1990; 68: 880–888.
74. Whitfield JF, Sikorska M, Youdale T, et al. Ribonucleotide reductase new twists to an old tale. Adv Enzyme Reg 1989; 28: 113–123.
75. Murray EJ, Bently GV, Grisanti MS, et al. The ubiquitin–proteasome system and cellular proliferation and regulation in osteoblastic cells. Exp Cell Res 1998; 242: 460–469.
76. Chinery R, Borckman JA, Dransfield DT, et al. Antooxidant-induced nuclear translocation of CCAAT/enhancer-binding protein beta. A critical role for protein kinase-A-mediated phosphorylation of Ser299. J Biol Chem 1997; 272: 30356–30361.
77. Gauthier-Rouvière C, Vandromme M, Lautredou N, et al. The serum-response factor nuclear localization signal: general implications for cyclic AMP-dependent protein kinase activity in control of nuclear translocation. Mol Cell Biol 1995; 15: 433–444.
78. Mishra K, Parnaik VK. Essential role of protein phosphorylation in nuclear transport. Exp Cell Res 1995; 216: 124–134.
79. Moffett RB, Webb TE. Regulated transport of messenger ribonucleic acid from isolated liver nuclei by nucleic acid binding proteins. Biochemistry 1981; 20: 3253–3262.
80. Shirakawa F, Mizel SB. In vitro activation and nuclear translocation of NF-kappa B catalyzed by cyclic AMP-dependent protein kinase and protein kinase C. Mol Cell Biol 1989; 9: 2424–2430.
81. Vandromme M, Carnac G, Gauthier-Rouviére C, et al. Nuclear import of the myogenic factor MyoD requires cAMP-dependent protein kinase activity but not the direct phosphorylation of MyoD. J Cell Sci 1994; 107: 613–620.
82. Xiao CY, Hubner S, Elliot RM, et al. A consensus cAMP-dependent protein kinase (PK-A) site in place of the CcN motif casein kinase II site simian virus 40 large T-antigen confers PK-A-mediated regulation of nuclear import. J Biol Chem 1996; 271: 6451–6457.
83. Baldin V, Roman AM, Bosc-Bierne I, et al. Translocation of bFGF to the nucleus is G1 phase cell cycle specific in bovine aortic endothelial cells. EMBO J 1990; 9: 1511–1517.
84. Bustamante JO. Nuclear ion channels in cardiac myocytes. Pflügers Arch 1992; 421: 473–485.
85. Panté N, Aebi U. Molecular dissection of the nuclear pore complex. Crit Rev Biochem Mol Biol 1996; 31: 153–199.
86. Harada H, Becknell B, Wilm M, et al. Phosphorylation and inactivation of BAD by mitochondria-anchored protein kinase A. Mol Cell 1999; 3: 413–422.
87. Kirkwood T. Time of Our Lives. New York, Oxford University Press, 1999: 91–92.
88. Bates S, Vousden KH. Mechanism of p53-mediated apoptosis. Cell Mol Life Sci (CMLS) 1999; 55: 28–37.
89. Chen F, Peterson SR, Story MD, et al. Disruption of DNA-PK in Ku80 mutant xrs-6 and the implications in DNA double-strand break repair. Mutat Res 1996; 362: 9–19.

90. Hartley KO, Geell D, Smith GC, et al. DNA-dependent protein kinase catalytic subunit: a relative of phosphatidylinositol 3-kinase and the ataxia telangiectasia gene product. Cell 1995; 82: 849–856.
91. Hunter T. When is a lipid kinase not a lipid kinase? When it is a protein kinase. Cell 1995; 83: 1–4.
92. Janus F, Albrechtensen N, Dornreiter I, et al. The dual role model for p53 in maintaining genomic integrity. Cell Mol Life Sci (CMLS) 1999; 55: 12–27.
93. Oren M, Prives C. p53: upstream, downstream and off stream. Review of the 8th p53 workshop (Dundee, July 5–9, 1996). Biochim Biophys Acta 1996; 1288: R13–R19.
94. Oren M, Rotter V. p53 — the first twenty years. Cell Mol Life Sci (CMLS) 1999; 55: 9–11.
95. Poltoratsky VP, Shi X, York JD, et al. Human DNA-activated protein kinase (DNA-PK) is homologous to phosphatidylinositol kinases. J Immunol 1995; 155: 4529–4533.
96. Weaver DT. What to do at an end: DNA double-strand-break repair. Trends Genet 1995; 11: 388–392.
97. Rensberger B. Life Itself. New York, Oxford University Press, 1996: 238.
98. Rathmell WK, Kaufmann WK, et al. DNA-dependent protein kinase is not required for accumulation of p53 or cell cycle arrest after DNA damage. Cancer Res 1997; 57: 68–74.
99. Sang N, Baldi A, Giordano A. The roles of tumor suppressors pRB and p53 in cell proliferation and cancer. Mol Cell Diff 1995; 3: 1–29.
99*a*. Carr AM. Piecing together the p53 puzzle. Science 2000; 287: 1765–1766.
100. MacLachlan TK, Sang N, Giordano A, et al. Cyclins, cyclin-dependent kinases and cdk inhibitors: implications in cell cycle control and cancer. Crit Revs Eukaryotic Gene Expression 1995; 5: 127–156.
100*a*. Tanaka H, Arakawa H, Yamaguchi T, et al. A ribonucleotide reductase gene involved in a p53-dependent cell-cycle checkpoint for DNA damage. Nature 2000; 404: 42–49.
101. Bellamy COC, Malcomson RDG, Harrison DJ. Cell death in health and disease: the biology and regulation of apoptosis. Sem Cancer Biol 1995; 6: 3–16.
102. Canman CE, Kastan MB. Induction of apoptosis by tumor suppressor genes and oncogenes. Sem Cancer Biol 1995; 6: 17–25.
102*a*. Oda K, Arakawa H, Tanaka T, et al. p53*AIP1*, a potential mediator of p53-dependent apoptosis, and its regulation by Ser-46-phosphorylated p53. Cell 2000; 102: 849–862.
103. Jung Y, Miura M, Yuan J. Suppression of interleukin-1 beta-converting enzyme-mediated cell death by insulin-like growth factor. J Biol Chem 1996; 271: 5112–5117.
104. Ruben R, Baserga R. Biology of disease: insulin-like growth factor-I receptor. Lab Invest 1995; 73: 311–331.
105. Sell C, Baserga R, Rubin R. Insulin-like growth factor I (IGF-I) and the IGF-I receptor prevent etoposide-induced apoptosis. Cancer Res 1995; 55: 303–306.

106. DePamphilis, ML. Initiation of DNA replication in eukaryotic chromosomes. J Cell Biochem 1998; Suppl 30/31: 8–17.
106*a*. Davey MJ, O'Donnell M. Mechanisms of DNA replication. Curr Opin Chem Biol 2000; 4: 581–586.
107. Tada S, Chong JPJ, Mahbubani HM, et al. The RLF-B component of the replication licensing system is distinct from Cdc6 and functions after Cdc6 binds to chromatin. Curr Biol 1999; 9: 211–214.
108. Tye BK. MCM proteins in DNA replication. Annu Rev Biochem 1999; 68: 649–686.
108*a*. Reddy GPV. Regulation of DNA replication and S phase. In: Stein GS, Baserga R, Giordano A, Denhardt DT, eds. The Molecular Basis of Cell Cycle and Growth Control. New York, Wiley–Liss, 1999: 80–184.
109. Johnson LH, Masai H, Sugino A. First the CDKs now the DDKs. Trends Cell Biol 1999; 9: 249–252.
110. Malkas LH. DNA replication machinery of the mammalian cell. J Cell Biochem 1998; Suppl 30/31: 18–29.
111. Brush GS, Anderson CW, Kelly TJ. The DNA-activated protein kinase is required for the phosphorylation of replication protein A during simian virus 40 DNA replication. Proc Natl Acad Sci USA 1994; 91: 12520–12524.
112. Pan ZQ, Amin AA, Gibbs E, et al. Phosphorylation of the p34 subunit of human single-stranded-DNA-binding protein in cyclin A-activated G1 extracts is catalyzed by cdk – cyclin A complex and DNA-dependent protein kinase. Proc Natl Acad Sci USA 1994; 91: 8343–8347.
113. Graeme C, Smith M, Jackson SP. The DNA-dependent protein kinase. Genes Dev 1999; 13: 916–934.
114. Cook PR. The organization of replication and transcription. Science 1999; 284: 1790–1795.
115. Hickey RJ, Malkas LH. Mammalian cell DNA replication. Crit Revs Eukaryotic Gene Express 1997; 7: 125–127.
116. Kelman Z, Hurwitz J. Protein–PCNA interactions: a DNA-scanning mechanism? Trends Biochem Sci 1998; 236–238.
117. Kornberg A, Baker TA. DNA Replication, 2E. New York, WH Freeman and Company, 1991.
117*a*. Urquidi V, Tarin D, Goodison S. Role of telomerase in cell senescence and oncogenesis. Ann Rev Med 2000; 51: 65–79.
118. Kirkwood T. Time of Our Lives. New York, Oxford University Press, 1999: 110–113.
119. Baylin SB, Herman JG, Graff JR, et al. Alterations in DNA metabolism: a fundamental aspect of neoplasia. Adv Cancer Res 1998; 72: 141–196.
120. Breivik J, Gaudernack G. Carcinogenesis and natural selection: a new perspective to the genetics and epigenetics of colorectal cancer. Adv Cancer Res 1999; 76: 185–212.
121. Chuang LS-H, Ian H-I, Koh T-W, et al. Human DNA-(cytosine-5) methyltransferase – PCNA complex as a target for $p21^{WAF1}$. Science 1997; 277: 1996–2000.

122. Jones PA. The DNA methylation paradox. Trends Genet 1999; 15: 34–37.
123. Feil R, Khosla S. Genomic imprinting in mammals. Trends Genet 1999; 15: 431–435.
123*a*. Clarke DJ, Giménez-Abián JF. Checkpoints controlling mitosis. BioEssays 2000; 22: 351–363.
124. Kotani S, Tugendreich S, Fujii M, et al. PKA and MPF-activated polo-like kinase regulates anaphase promoting activity and mitosis progression. Mol Cell 1998; 1: 371–380.
125. Morgan DO. Regulation of the APC and the exit from mitosis. Nature Cell Biol 1999; 1: E47–E53.
126. Page AM, Hieter P. The anaphase-promoting complex: new subunits and regulators. Annu Rev Biochem 1999; 68: 583–609.
127. Bischoff FR, Maier G, Tilz G, et al. A 47-kDa human nuclear protein recognized by antikinetochore autoimmune sera is homologous with the protein encoded by RCC1, a gene implicated in onset of chromosome condensation. Proc Natl Acad Sci USA 1990; 87: 8617–8621.
128. Bischoff FR, Ponstingl H. Catalysis of guanine nucleotide exchange on Ran by the mitotic regulator RCC1. Nature 1991; 354: 80–82.
129. Clarke PR, Klebe C, Wittinghofer A, et al. Regulation of Cdc2/cyclin B activation by Ran, a Ras-related GTPase. J Cell Sci 1995; 108 (Pt. 3): 1217–1225.
130. Dasso M, Nishitani H, Kornbluth S, et al. RCC1, a regulator of mitosis, is essential for DNA replication. Mol Cell Biol 1992; 12: 3337–3345.
131. Dasso M, Smythe C, Milarski K, et al. DNA replication and progression through the cell cycle. Ciba Found Symp 1992; 170: 161–180.
132. Klebe C, Bischoff FR, Ponstingl H, et al. Interaction of the nuclear GTP-binding protein Ran with its regulatory proteins RCC1 and RanGAP1. Biochemistry 1995; 34: 639–647.
133. Melchior F, Guan T, Yokoyama N, et al. GTP hydrolysis by Ran occurs at the nuclear pore complex in an early step of protein import. J Cell Biol 1995; 131: 571–581.
134. Miyabashira J, Sekiguchi T, Nishimoto T. Mammalian cells have two functional RCC1 proteins produced by alternative splicing. J Cell Sci 1994; 107 (Pt. 8): 2203–2208.
135. Moroianu J. Molecular mechanisms of nuclear protein transport. Crit Revs Eukaryotic Gene Expression 1997; 7: 61–72.
136. Nishijima H, Seki T, Nishitani H, et al. Premature chromatin condensation caused by loss of RCC1. Progr Cell Cycle Res 2000; 4: 145–156.
136*a*. Nishimoto T. Upstream and downstream of ran GTPase. Biol Chem 2000; 381: 397–405.
137. Ohba T, Seki T, Azuma Y, et al. Premature chromatin condensation induced by loss of RCC1 is inhibited by GTP- and GTPγS-Ran, but not GDP-Ran. J Biol Chem 1996; 271; 14665–14667.
138. Seino H, Nishitani H, Seki T, et al. RCC1 is a nuclear protein required for coupling activation of cdc2 kinase with DNA synthesis and for start of the cell cycle. Cold Spring Harb Symp Quant Biol 1991; 56: 367–375.

139. Seki T, Yamashita K, Nishitani H, et al. Chromosome condensation caused by loss of RCC1 function requires the cdc25C protein that is located in the cytoplasm. Mol Biol Cell 1992; 3: 1373–1388.
139*a*. Tachibana T, Hieda M, Miyamoto Y, et al. Recycling of important alpha from the nucleus is suppressed by loss of RCC1 function in living mammalian cells. Cell Struct Funct 2000; 25: 115–123.
140. Tachibana T, Imamoto N, Seino H, et al. Loss of RCC1 leads to suppression of nuclear protein import in living cells. J Biol Chem 1994; 269: 24542–24545.
141. Uchida S, Sekiguchi T, Nishitani H, et al. Premature chromosome condensation is induced by a point mutation in the hamster RCC1 gene. Mol Cell Biol 1990; 10: 577–584.
142. Nishimoto T, Kai R, Sekiguchi T, et al. Molecular cloning of human genes that complement the temperature-sensitive cell-cycle mutants of BHK-21 cells. In: Beach D, Basilico C, Newport J, editors. Cell Cycle Control in Eukaryotes. Cold Spring Harbor, Cold Spring Harbor Laboratory, 1988: 169–179.
143. Yasuda H, Matsumoto Y, Mita S, et al. A mouse temperature-sensitive mutant defective in H1 histone phosphorylation is defective in deoxyribonucleic acid synthesis and chromosome condensation. Biochemistry 1981; 20: 4414–4419.
144. Patel R, Holt M, Philipova R, et al. Calcium/calmodulin-dependent phosphorylation and activation of human Cdc25-c at the G2/M phase transition in HeLa cells. J Biol Chem 1999; 274: 7958–7968.
145. Hirano T. SMC-mediated chromosomal mechanics: a conserved scheme from bacteria to vertebrates. Genes Devel 1999; 13: 11–19.
145*a*. Endow S. Spindle and chromosome motility. The Kinesin Home Page. 1998: http://www.proweb.org/kinesin/.
146. Biggins S, Murray AW. Sister chromatid cohesion in mitosis. Curr Opin Cell Biol 1998; 10: 769–775.
147. McIntosh JR. Structural and mechanical control of mitotic progression. Cold Spring Harb Symp Quant Biol 1991; 56: 613–619.
148. Maney T, Ginkel L, Hunter AW, et al. The kinetochore of higher eucaryotes: a molecular view. Int Rev Cytol 2000; 194: 67–131.
149. Grieco D, Porcellini A, Avvedimento EV, et al. Requirement for cAMP-PKA pathway activation by M phase-promoting factor in the transition from mitosis to interphase. Science 1996; 271: 1718–1723.
150. Lorca T, Abrieu A, Means A, et al. Ca^{2+} is involved through type II calmodulin-dependent protein kinase in cyclin degradation and exit from metaphase. Biochim Biophys Acta 1994; 1223: 325–332.
151. Larsen A, Skadanowski A, Bojanowski K. The roles of DNA topoisomerase II during the cell cycle. Progr Cell Cycle Res 1996; 2: 229–239.
152. Warburton PE, Earnshaw WC. Untangling the role of DNA topoisomerase II in mitotic chromosome structure and function. BioEssays 1997; 19: 97–99.
152*a*. Hales KG, Bi E, Wu J-Q, et al. Cytokinesis: an emerging unified theory for eukaryotes? Curr Opin Cell Biol 1999; 11: 717–725.
153. Busser J, Geldmacher DS, Herrup K. Ectopic cell cycle proteins predict the sites of neuronal cell death in Alzheimer's disease brain. J Neurosci 1998; 18: 2801–2807.

154. Herrup K, Busser JC. The induction of multiple cell cycle events precedes target-related neuronal death. Development 1995; 121: 2385–2395.
155. McShea A, Wahl AF, Smith MA. Re-entry into the cell cycle: a mechanism for neurodegeneration in Alzheimer disease. Med Hypoth 1999; 52: 525–527.
156. Nagi Z, Esiri MM. Neuronal cyclin expression in the hippocampus in temporal lobe epilepsy. Exp Neurol 1998; 150: 240–247.
157. Nagy Z, Esiri MM, Cato A-M, et al. Cell cycle markers in the hippocampus in Alzheimer's disease. Acta Neuropath 1997; 94: 6–15.
158. Nagy Z, Esiri MM, Smith AD. The cell division cycle and the pathophysiology of Alzheimer's disease. Neuroscience 1998; 87: 731–739.
159. Park DS, Morris EJ, Padmanabhan J, et al. Cyclin-dependent kinases participate in death of neurons evoked by DNA-damaging agents. J Cell Biol 1998; 143: 457–467.
160. Raina AK, Monteiro MJ, McShea A, et al. The role of cell cycle-mediated events in Alzheimer's disease. Int J Exp Path 1999; 80: 71–76.
161. Ross ME. Cell division and the nervous system: regulating the cycle from neural differentiation to death. Trends Neurosci 1996; 19: 62–68.
162. Smith TW, Lippa CF. Ki-67 immunoreactivity in Alzheimer's disease and other neurodegenerative disorders. J Neuropath Exp Neurol 1995; 54: 297–303.
163. Vincent I, Jicha G, Rosado M, et al. Aberrant expression of mitotic cdc2/cyclin B1 kinase in degenerating neurons of Alzheimer's disease brain. J Neurosci 1997; 17: 3588–3598.
164. Zhu X, Raina AK, Smith AK. Cell cycle events in neurons. Proliferation or death? Am J Pathol 1999; 155: 327–329.
165. Gould E, Reeves AJ, Graziano MSA, et al. Neurogenesis in the neocortex of adult primates. Science 1999; 286: 548–552.

2

Calcium — A Cell Cycle Driver

The cycle-driving Ca^{2+} signals: a quick preview

"Calcium signals are used in a large proportion of regulatory pathways and are thus considered ubiquitous in cell biology, yet they are also selective or specific in the control exerted among myriad functions within a common cytoplasm. How this complex messaging is achieved through a single ion, calcium, remains unsolved."[1]

Now that we have finished our quick tour of the cell cycle, we will turn around and go back to see where, and maybe how, Ca^{2+} promotes the key cell cycle transitions. Essential though Ca^{2+} and its sensors and effectors, most especially CaM (calmodulin), are for driving cells through their cycle, they have *never* been at the forefront of cell cycle research![1a,2] Indeed, they are rarely, if ever, mentioned in reviews of the CDKs and events such as the initiation of DNA replication and prophase where in fact they play major, if not principal, triggering roles. Thus, the best we can do here is weave what little information we have into a tantalizing tapestry of "educated guesses" of how the ion and its effectors trigger DNA replication, mitosis, and drive cell division (cytokinesis).

The roles of Ca^{2+} transients and oscillations and the corresponding cresting and crashing waves of CaM are just as important as the cresting and crashing waves of CDKs in the $G_0 \rightarrow G_1$, $G_1 \rightarrow S$, $G_2 \rightarrow$prophase, and metaphase→anaphase transitions and the pinching of the cell into its two daughters. So far, we only know that what crudely seem to be identical Ca^{2+} surges or oscillations drive the major parts of the cell cycle. However, shifting patterns of spatially restricted Ca^{2+} sparks, spikes, and waves (Ca^{2+}'s diffusion radius is only about 0.1 μm) will light up different cellular regions and targets as the cell progresses through the different stages of the cycle.[3–5] Indeed, mapping the four-dimensional topography of Ca^{2+} signals during the cell cycle is the first prerequisite for understanding the ion's grasp on the cell cycle controls. Only now are we on the verge of being able to do this.[1–2]

Ca^{2+} and starting the cycle

We start with a snoozing cell as we did in the last chapter. It may be a cultured cell that has lapsed into reversible, G_0 quiescence because of a lack of serum growth factors and will start a cycle upon the appearance of the appropriate growth factor(s), or it may be a hepatocyte in a normal adult liver working hard to maintain essential body-wide functions while starting a cycle. Or it may be a small, minimally functioning small lymphocyte with a thin rim of cytoplasm wrapped around a small nucleus with highly condensed, relatively inactive masses of chromatin, making and then wasting ribosomal RNA. We must note that quiescent, G_0 cultured cells, unlike the differentiated cells, are ready to respond to a mitogen by keeping a pool of untranslated messages from cycle-starter genes. This is why the so-called "delayed–early" genes in regenerating liver are the "immediate–early" genes in cultured liver cells of established lines.[6]

The quiescent cell has triggered its cycle-shutdown program. It has planted in its membrane the Gas-1 protein that somehow contributes to the suppression of proliferation.[7–9] It also has installed in its nuclear membrane the 56-kDa phosphoprotein statin complexed with a 45-kDa serine/threonine protein kinase.[10–16] It has upregulated the expression of $p27^{Kip1}$, which would block any low-level surge of Cdk4(or 6)·cyclin D activity.

The silencing of growth factor receptors has affected all aspects of RNA and protein metabolism as well as the mitochondrial energy supply. One key to how a cell switches in and out of the cycling mode is the low priority for access to the translation machinery of the messages from genes encoding growth-regulating transactivators and other regulators. The low position of these transcripts on the translation totem pole is due to their having extensive secondary structure in "capped," 5′-upstream untranslated regions, which have to be "melted" by growth-factor-activated translation initiation factors belonging to the eIF-4 family before they can bind to 43-S pre-initiation complexes and be translated.[17–35] The shutdown mechanism stops the processing of the proliferation-driving gene transcripts, demobilizes the "weak" messages for growth-promoting peptides, reduces the total mRNA content, decreases ribosomal RNA synthesis, increases ribosomal RNA degradation, increases protein degradation by lysosomes, reduces the flow into the nucleus of short-lived non-histone proteins such as transcription factors, and lowers mitochondrial activity.[36] Only the "strong" messages (with their low 5′-secondary structure) for basic housekeeping proteins, proteins for specific tissue functions, and quiescence-maintaining proteins are translated while the number of polyribosomes drops dramatically as the cell shunts much of its "weak" mRNA into a non-polysomal pool.[36] The drop in the translatability of the "weak" transcripts is due mainly to two things. When growth-factor-activated receptor protein kinases fall silent, components of the eIF-4 factor, which limit the translation rate, are dephosphorylated and inactivated.[17–35] Thus, there is a lack of phosphoSer^{209}-eIF-4E in the eIF-4 complexes to bind to the 5′-m^7GpppX caps of the growth-regulated transcripts and to let the 50-kDa eIF-4A helicase component of eIF-4 complexes melt the transcripts' excessively structured 5′-untranslated regions to enable them to bind to eIF-2·GTP·Met-tRNAf·40-S ribosome pre-initiation complexes.[17–35] The silencing of the growth factor receptors, and consequently the Rsk Ser/Thr protein kinases, results in the dephosphorylation and inactivation of the 40-S ribosome S6 protein components that are needed to form the eIF-2·GTP·Met-tRNAf·40-S pre-initiation complexes.[24]

A signal to a quiescent, cultured cell from a growth factor such as PDGF (platelet-derived growth factor), to a hepatocyte in the adult rat liver from an endogenous growth factor such as the hepatocyte growth factor released in response to partial hepatectomy, or to a small T-lymphocyte from an macrophage-proffered antigen or a plant lectin, starts the cells growing to a state of competence in which they can respond to a second signal from receptors for an endogenous autocrine (self-produced, self-stimulating) or an exogenous paracrine (neighboring cells-stimulating) factor to start the final run-up to DNA replication. This first "wake-up" signal triggers a veritable explosion of events. The signal-aroused cell will need much energy, which, as we learned in the previous chapter, it gets from glucose, the uptake of which is enhanced by c-Myc. The energy is supplied in the form of ATP by mitochondria lying against the Ca^{2+}-loaded endoplasmic reticulum.[36a] This snuggling of mitochondria up to the endoplasmic reticulum is a clever way to ensure that signal-triggered activities will be adequately fueled. The signals from the receptors target the endoplasmic reticulum and cause it to dump a large amount of Ca^{2+} onto the adjacent mitochondria, which increases its ATP production when it picks up this Ca^{2+}.[36a,37] The surging Ca^{2+} also activates the cytoskeleton, a striking effect of which is cell surface ruffling, and the nuclear actomyosin motor is switched on to reposition the chromatin to receive and respond to incoming transcription factors and protein kinase and phosphatase modulators of nuclear transcription factors already attached to their target genes' promoters.[38,39] As we learned in the last chapter, at the heart of the cascade are the c-Myc·Max complexes that stimulate the expression of the genes for BN51, eIF-4E, nucleolin, ODC (ornithine decarboxylase) and through their products the mobilization and (or) translation of the transcripts for cycle-starters such as CAK, Cdk4, cyclin D2, c-Fos, c-Fra, ODC, and those of c-Myc itself.[36] To make these and all of the other proteins needed for growth, the cell stops degrading ribosomal RNA and lets it accumulate. Lysosomal degradation stops and a large number of housekeeping proteins and cycle drivers, such as Cdk4 and cyclin Ds to make the cycle-starting CDK, Cdk·cyclin D and CAK to activate it; transcription factors such as c-Fos, c-Myc, NFAT (nuclear factor of activated T-lymphocytes), NF-κB, Stat3; and non-histone chromatin-decondensing proteins are rapidly shipped to the nucleus to reconfigure the chromatin, stimulate the first wave of proliferogene expression, and start the cycle-driving parade of CDKs.[6,36] Putrescine-ODC subunits of RNA polymerase I, addressed specifically to the nucleolus and along with nucleolin, are shipped into the nucleus to stimulate ribosomal RNA production.[36]

What gives this wake-up call and how does it do it? Among the most common cellular alarm clocks are the growth factor "velcroceptors." Ligand binding to the extracellular parts of the monomeric velcroceptors (e.g., EGF, PDGF), heterodimeric velcroceptors (e.g., IGF-I, insulin), or multi-component or modular velcroceptors (e.g., the T-cell receptor, or the IL-6 receptor with its associate JAK protein tyrosine kinase) causes the receptor molecules to activate their cytoplasmic PYK (protein tyrosine kinase) domains, and cross-phosphorylate each other's tyrosine residues.[40,41] Of course, the monomeric receptors and modular receptors must dimerize and (or) collaborate with PYKs to get their tyrosine residues phosphorylated, while the heterodimeric receptors are already configured for cross-phosphorylation. This *velcroizes* the receptor, which can then hook onto a variety of co-signalers by shoving its phosphorylated tyrosines into the co-signalers' SH2 (Sarc Homologous 2) pockets where the tyrosines' phosphates are grabbed and held by the terminal guanido groups of an arginine residue.[42–47] Velcroization is a wonderful way of luring a team of signal enzymes and adaptors to a place where they can be coordinately phosphorylated by the receptor dimers' protein tyrosine kinase domains, and interact with each other and other membrane-associated molecules to switch on several differently targeted signal mechanisms. In other words, velcroization enables a single receptor to coordinately activate a specific panel of signal mediators, the velcrocluster, and thus simultaneously turn on several different signal mechanisms.[41–47] The kinds of SH2-bearing co-signalers in a velcrocluster or velcroceptor component depends on the amino acids flanking the velcroceptor's phosphotyrosine residues and the co-signaler's SH2 pockets.[41–47] For example, the phospho-Tyr^{739} residue of the β-PDGF receptor, with its associated phospho-Tyr–Asn–Ala–Phe–Tyr sequence, grabs Ras-GAP (GTPase-activating protein); the receptor's phospho-Tyr^{1021}, with its associated phospho-Tyr–Ile–ile–Phe–Tyr sequence, hooks phospholipase-Cγ_1; phospho-Tyr^{708} and phospho-Tyr^{719}, with their associated phospho-Tyr–Met–Asp–Met–Ser and phospho-Tyr–Val–Phe–Met–Leu sequences, grab the 85-kDa regulatory subunit of PtdIns-3K; and the phospho-Tyr^{579} residue, with its associated phospho-Tyr–Cys–Val–Asp sequence, hooks Src-family protein tyrosine kinases.[36] Thus, differently sited phospho-velcroceptors, and the different adaptors, such as IRS1 and Shc, that they phosphorylate, recruit different sets of signalers, which in turn trigger different signal cascades in different parts of the cell.[41,48,49]

To awaken a quiescent cell, we may induce it to express the first set of proliferogenes by activating appropriate mitogenic velcroceptors on its surface. The signals from such receptors will activate the transcription factors that will turn these genes on. In this type of activation, things start when Grb2 adaptors shove the receptor's appropriately identified phospho-Tyr residues into their SH2 pockets, which causes the adaptor to expose its SH3 motif.[41,50] It then binds and thereby activates mSos, which, in turn, activates membrane-bound Ras molecules by stimulating them to replace their attached GDP with GTP. The now activated GTP·Ras interacts with and activates the 74-kDa Raf-1 Ser/Thr protein kinase.[36,41,48–51] The activated Raf-1 kinase then phosphorylates and activates an enzyme known as MEK, which is a dual-function Ser/Thr/Tyr-protein kinase.[41,52] MEK then activates the Ser/Thr MAP kinase (MAPK) by phosphorylating the enzyme's Thr^{183} and Tyr^{185} residues.[52,53] The activated MAP kinase can enter the nucleus to phosphorylate and thus activate a variety of transcription factors such as the multipurpose c-Jun and the exclusively proliferation-related c-Myc whose targets are key proliferogenes such as the Cdk4·cyclin D-activating CAK.[54–59] Another key target of the MAP kinases are the S6 Ser/Thr protein kinases, which are activated and phosphorylate five serine residues of the S6 subunits of 40-S ribosomes.[36] These five residues are on the "platform" of the 40-S ribosome, which is involved in codon:anticodon tRNA binding. This phosphorylation triggers a configurational change, which increases the 40-S subunit's message-binding ability.[24] Yet, another result of this cascade is the phosphorylation and activation of eIF-4E's Ser^{209}, which enables the protein to form complexes with the eIF-4A helicase that mobilize growth-driving messages by melting the extensive secondary structure of their capped 5′-untranslated regions and binding them to the primed, S6-phosphorylated, 43-S pre-initiation complexes.[32–36] All of this results in a wave of protein synthesis, particularly the synthesis of the key proteins that will eventually start the parade of cycle-driving cyclins and cyclin-dependent protein kinases.

However, I have not mentioned the hero of this chapter and book, Ca^{2+}. Indeed, velcroceptor-started cycles (there are other starters as we will see further on) start with a Ca^{2+} spike or oscillations (Fig. 2.1). The Ca^{2+} story starts with the auto(Tyr) phosphorylated velcroized receptor, grabbing phospholipase-Cγ by its SH2 pockets. The activated phospholipase stimulates the breakdown of PtdIns(4,5)P_2 (phosphatidylinositol-4,5-bisphosphate) into Ins(1,4,5)P_3 (inositol-1,4,5-trisphosphate) and a first wave of short-lived diacylglycerols.[36]

Fig. 2.1. The Ca^{2+} signals that have been found to trigger the major stages of the cell cycle.[36,71,196] A Ca^{2+} surge is involved in starting the G_1 buildup. In some cells, this Ca^{2+} comes mainly from internal stores, while in other cells an essential part of the Ca^{2+}comes through membrane channels. Therefore, external Ca^{2+} deprivation prevents cells with inadequate internal Ca^{2+} stores from starting the G_1 buildup, while it stops other cells starting out with adequate stores only when they near end of the buildup where Ca^{2+} oscillations are involved in triggering chromosome replication. Then come the Ca^{2+} surges that trigger prophase and later anaphase. This figure is actually just Fig. 1.1 with both the checkpoints (● ●) and Ca^{2+} signaling points indicated.

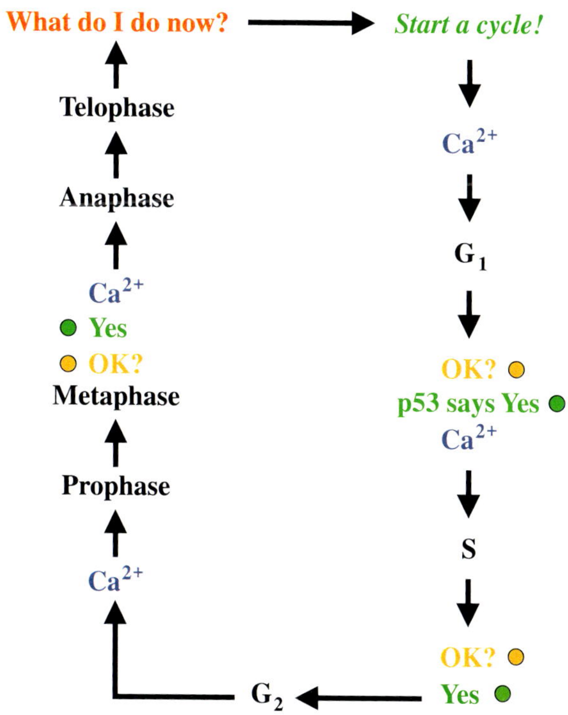

First, let us look at the diacylglycerols. They stimulate nine (α, βI, βII, γ, δ, η, ϵ, θ, μ) of a family of 12 Ser/Thr protein kinases, the PKCs (protein kinase-Cs).[60,61] At low concentrations, the diacylglycerols stimulate any inactive PKCs that the cell might be harboring in its plasma membrane, but at higher concentrations they can drive more PKCs from the cytosol onto activation sites in the plasma membrane with the help of an Ins(1,4,5)P_3 -induced release of Ca^{2+} from internal stores.[62] The membrane-associated PKCs phosphorylate and inhibit some receptors such as the EGF/TGF-α receptors and enzymes such as the phospholipase-Cγ that generated them and then phosphorylate and modulate the activities of a host of other enzymes, signal mechanisms, and gene transcription factors around which a huge literature has collected and will not be discussed until the next chapters.[36,62] The first wave of diacylglycerols crashes within a few seconds as the membrane-associated PKCs phosphorylate and silence phospholipase-Cγ.[36] However, while they have shut down phospholipase-Cγ, they have stimulated another phospholipase, phospholipase-D, which breaks phosphatidyl choline down into phosphatidic acid and choline.[36] The phosphatidic acid is then dephosphorylated to diacylglycerol, which re-activates PKCs, thus setting up a positive-feedback stimulation that produces a second, but now sustained, surge of membrane-associated PKCs activity.[36]

The transient surge of Ins(1,4,5)P_3 that accompanies the brief burst of phospholipase-Cγ activity does a very important thing. It binds to 260-kDa Ins(1,4,5)P_3 receptors/channels on Ca^{2+}- storage vesicles, which may bud from, or remain as specialized regions of, the endoplasmic reticulum.[63–68] The emptying of the Ca^{2+} stores triggers, via a message from the Ins(1,4,5)P_3 receptors' heads to the cell membrane, the opening of CRAC (Ca^{2+}-release-activated) or SOC (stores-operated) channels.[63,65,69] The upshot of this is the transient burst of intracellular Ca^{2+} oscillations that marks the start of the G1 buildup in a wide variety of cells.[36,63,70,71] The Ins(1,4,5)P_3 generated at the velcroceptor sites in the cell membrane spreads to the underlying nucleus and triggers the release of Ca^{2+} from the nuclear envelope stores into the nucleus through Ins(1,4,5)P_3 receptor/channels.[72–76] The surging Ca^{2+} forms complexes with a variety of cytoskeletal and other proteins, perhaps the most important of which are the complexes with CaM (calmodulin), Ca^{2+}·CaM. The Ca^{2+} surge also facilitates the movement of intermediately-sized (around 10 kDa) molecules into the nucleus by removing a central plug or gate in the nuclear membrane pores.[77,78]

The surging Ca^{2+} and with it a wave of Ca^{2+}·CaM complexes activate a formidable array of enzymes such as calcineurin (a protein Ser/Thr phosphatase), CaMKII (calmodulin-dependent Ser/Thr protein kinase), CaMKIV, and MLCK (myosin light-chain kinase). The nuclear Ca^{2+} content shoots upwards and the resulting Ca^{2+}·CaM complexes do two important things. They activate MLCK and thus stimulate the nuclear actomyosin machinery to move the chromatin into a responsive configuration.[38,39] And they stimulate the expression in the quiescent cell of at least one key proliferogene, c-*myc*, by binding to and displacing the transcriptionally inactive Max/Max homodimers from the gene's promoter, which enables their replacement by transcriptionally active Myc/Max heterodimers.[58,79] Some cells, such as corticotrophs, lymphocytes, neurons (which can respond to proliferogenic signals by growing processes, but generally cannot replicate their possibly unlicensed chromosomes [see previous chapter]), and reproductive cells, have CaMKIV, which, unlike most CaMKII isoforms (although δ-CaMKII does have a nuclear localization motif), can enter the nucleus where, among other things, it can mimic cyclic AMP by activating an "immediate–early" cyclic AMP-responsive gene such as c-*fos* by phosphorylating and activating the CREB (cyclic AMP-responsive element [CRE]-binding protein) protein sitting on the CaRE (Ca^{2+}-responsive element with a consensus CRE motif) element of its promoter in response to a nuclear Ca^{2+} surge.[80–82] In another example, the Ca^{2+} surge triggered by the modular T-cell velcroceptor activates the calcineurin phosphatase, which associates with and dephosphorylates the cytoplasmic NFAT1 (Nuclear Factor of Activated T-cells), which then translocates to the nucleus where it attaches to five sites in the promoter of the IL(interleukin)-2 gene, at four of which it joins with the AP-1 (c-Fos/c-Jun) transactivation complex.[83–89] The result of this is the production of IL-2, which triggers the second stage of the growing T-cell's G_1 buildup.[36]

We learned in the previous chapter that Myc·Max transactivates the genes for two very important things needed to turn on Cdk4(or 6)·cyclin D, the G_1-starting CDK. First are the components of Cdk7·cyclin H, CAK, which besides being involved in activating

Cdk4(or 6)·cyclin D, is part of the RNA polymerase II transcriber's pre-initiation complex when associated with the MAT1(p36) protein.[57] Second is the early G_1-specific, CDKs-activating STY phosphatase Cdc25A, which like its G_2-specific isoform, Cdc25C, might be activated by being phosphorylated by Ca^{2+}·CaM-stimulated CaMKII.[90] Indeed, specifically blocking CaMKII activity with KN-93 greatly reduces the level of cyclin D1 and blocks the G_1 buildup in NIH3T3 mouse cells.[70–91]

Starting a cycle can be lethal for a normal cell if there is only an incomplete, thus "unbalanced," G_1 buildup.[92–99] Thus, for example, hyperexpressing the c-*myc* proliferation-starter gene in normal fibroblasts without the support of protector signals from serum factors indeed leads to the overexpression of E2F1, which stimulates the accumulation of the apoptogenic p53 and also the expression of cycle-driving components such as cyclin A and Cdk1·cyclin B1, which are apoptogenic when expressed out of context and without protectors such as Bcl-2 and the anti-p53 Mdm2.[54–56,92,93,96,97] The signal must trigger *in the proper order* the first wave of cyclins, the cyclin Ds, and the assembly of active Cdk4·cyclin D protein kinases that start the G_1 buildup (Fig. 1.2). However, it must also supply the protectors and downregulate the promiscuous CDK inhibitor $p21^{Cip1/WAF1}$ that is stimulated to increase by the mitogenic signal.[92–99] To do this it must receive converging streams of signals from serum factors and a subset of β1 integrin receptors that are activated by binding to substrate proteins such as fibronectin and vitronectin.[92–99] As we learned in the previous chapter and will continue to learn in the next chapters, the integrin signals provide the MAPK, PtdIns-3K, and ILK activities that stimulate the expression of the Bcl-2 protector, suppress the apoptogenic BAD protein, suppress the apoptogenic caspase proteases, and snatch β-catenin from the jaws of the proteasome shredder so that it can get into the nucleus to stimulate its several proliferogenic target genes. Therefore, detaching various normal activated cells from culture vessel or their appropriate basal lamina triggers the apoptogenic mechanism, a response that we learned in the previous chapter is called *anoikis* (from the Greek word for homelessness).[98] This suicide is extremely important for a multicellular society where a proliferatively competent, unattached cell and its progeny would wreak social havoc, but in the next chapter we will see how anoikis has been co-opted for the *diffpoptosis* (differentiation-apoptosis) of keratinocytes. Ca^{2+} is involved in the final mechanism that ultimately kills all apoptosing cells by inducing the collapse of the potential across the inner mitochondrial membrane ($\Delta\psi_m$), which opens Bcl-2-blockable permeability transition pores (megachannels) through which a 50-kDa caspase (formerly known as ICE[Interleukin-1β-like Converting Enzyme]-like cysteine protease) moves from the mitochondrion into the cytoplasm and triggers the apoptogenic cascade of protease activities.[98,100–107] How can this happen? Why must the cell activate a parallel survival mechanism to protect itself from the mitogenic signal? The answer may lie in the Ca^{2+} oscillations and oxidizing radicals triggered by the burst of velcroceptor signaling.[36,108,109] This causes the mitochondria to accumulate Ca^{2+},[108] which can convert the membrane's ADP/ATP transporter into a large channel (i.e., the channel has a very large conductivity for cations of 300–600 pSiemens)[110–112], which, if not controlled by Bcl-2 and (or) other factors enhanced by the survival arm of the cycle-driving mechanism, will trigger the irreversible common caspase-mediated execution phase of apoptosis. (Incidentally, megachannel formation in the inner mitochondrial membrane is enhanced by inorganic phosphate,[113] which explains Whitfield's 36-year-old observation that radiation-induced apoptosis in thymic lymphocytes is potently primed by inorganic phosphate,[36,114] and Meleti et al.'s much more recent observation that inorganic phosphate induces apoptosis in cultured human osteoblasts.[114a])

However, there is a different cycle-starting pathway, the specialized cyclic AMP/PKA (cyclic AMP-dependent protein kinase) pathway, which, like the velcroceptors' MAPKs/PKCs default pathway, converges at the surge of Cdk4·cyclin D activity needed to liberate E2F from the grip of the Rb-family pocket proteins and starts the waves of gene expressions and CDKs, leading to chromosome replication.[36,115–118] The cyclic AMP/PKA pathway is used by many types of cell, including many, if not most, differentiated epithelial secretory cells such as adrenocortical cells, parotid gland acinar cells, pancreatic cells, pituitary somatotrophs, and thyrocytes.[36,115–118] Such cells have a special mitogenic mechanism that allows them to replace lost functioning cells or to respond to an increased functional demand by triggering a transient burst of proliferation, which is followed by the prompt withdrawal of the new cells into the functional pool.[36,115–118]

The cyclic AMP/PKA mechanism has been most intensely studied in dog thyroid cells stimulated to proliferate by TSH (thyroid-stimulating hormone).[116,118] Adding an optimally effective dose of TSH to a primary culture of dog thyrocytes triggers a 48-h surge of cyclic AMP and proliferation-specific PKA activity (TSH's receptor is a slowly downregulated member of the serpentine or 7-TMD [transmembrane domain] family of G-protein-activating receptors) with none of the protein tyrosine kinase activation, phospholipase-Cγ activation, PtdIns(4,5)P_2 breakdown, surge of PKCs activities on the plasma membrane or in the nucleus, or Ca^{2+} transient that would be triggered in the same cells by EGF or hepatocyte growth factor velcroceptors.[116–120] Like the signals from velcroceptors, the cyclic AMP/PKA signal from TSH receptors downregulate the expression of a statin-related gene and stimulate the expression of the c-*myc* gene, but unlike this gene's prolonged elevation in response to signals from mitogenic velcroceptors, the c-*myc* response to the mitogenic cyclic AMP signal peaks at 1 h only to crash by 3 h.[116,118,121] Cyclic AMP/PKA initially lifts a transcriptional elongation block, which is followed 30 min later by the stabilization of the mRNA transcripts.[121] In a second stage, and in stark contrast to the sustained c-*myc* expression in velcroceptor-stimulated cells, the c-*myc* transcripts are destabilized and transcription is again shut off by the appearance of a protein that limits the mitogenic response to allow the resumption of function.[116,118,121] The mitogenic, Ca^{2+}-mobilizing velcroceptor signals strongly stimulate both c-*jun* and *jun D* gene expression, while the cyclic AMP/PKA signal inhibits c-*jun* and stimulates *jun D* expression.[116] Since cyclic AMP/PKA actually inhibits Raf-1; there is *no* Ras/Raf 1-mediated MAP kinase cascade in the cyclic AMP/PKA-stimulated thyrocytes as there is in velcroceptor-stimulated thyrocytes; there is no phosphorylation/activation of c-Jun proteins by activated phospho-MAP kinases surging into the nuclei; and because of this plus the failure to increase c-Jun production, there is none of the high, differentiation-suppressing AP1 activity that characterizes velcroceptor-stimulated cells.[116]

The cyclic AMP/PKA signal starts the cells growing and expressing and accumulating the RIα PKA regulatory subunits to make the PKAI isozymes needed to drive the G_1 buildup and sets the cell on a path that must include the Cdk4/6·cyclin D-triggered cresting and crashing waves of CDKs' activities.[116,122–125] However, cyclic AMP/PKA do not stimulate, and may even reduce, the expression of the cyclin Ds![126] But there is a very big problem with this! How do they stimulate proliferation when, as we learned in the previous chapter, activated Cdk4/6·cyclin Ds must accumulate above a critical level in the nucleus to inactivate the pocket proteins and release E2F to stimulate the cycle-starting genes? The answer is delightfully simple. Cyclic AMP/PKA enhances the assembly of Cdk4·cyclin D3 by increasing cyclin D3's affinity for Cdk4 and stimulates the transport of the Cdk4·cyclin D3s into the nucleus.[126]

These cyclic AMP/PKA-stimulated cells, just like cells stimulated by a mitogenic velcroceptor, now briefly need second signal*s* from a velcroceptor to stimulate growth and the accumulation of these and other components such as Cdk2·cyclin E and Cdk2·cyclin A to their critical threshold levels and the Ca^{2+} surge needed to drive the cells into the S phase.[116,122–125] In velcroceptor-activated thyrocytes, this second velcroceptor signal is from insulin or IGF-I receptors, but in the cyclic AMP/PKA-activated thyrocytes the second signal is from insulin velcroceptors that have been accumulating during the G_1 buildup.[127]

But how do cyclic AMP and the PKAs start the drive to DNA replication and mitosis? Cultured dog thyroid cells, Schwann cells, and Swiss albino 3T3 mouse cells have both the cyclic AMP/PKA proliferogenic response mechanism and the PtdIns(4,5)P_2/Ras/Raf-1/MEK/MAP default proliferogenic mechanism. Other cells such as NIH3T3 cells have only the default mechanism: they cannot be stimulated to proliferate by cyclic AMP/PKA.[128] In the cyclic AMP-responsive cells, the cyclic nucleotide-triggered mechanism bypasses the head of the default mechanism to start the G_1 buildup by activating Ras, which in turn stimulates its effectors RalGDS and PtdIns-3K (phosphoinositide-D3 kinase), which phosphorylates and stimulates $p70^{s6k}$ after $p70^{s6k}$ has relocated to the cell membrane.[128–133] The activated $pp70^{s6k}$ protein kinase then enters the nucleus and activates the expression of key cycle-starting genes by phosphorylating CREM (cyclic AMP-response element modulator) transactivators on cyclic AMP-response elements in their promoters.[134] This protein kinase has the added feature of enhancing protein synthesis by phosphorylating the target for which it was named, the ribosomal S6 subunit discussed above.

So far we have assumed that the TSH-stimulated cyclic AMP surge in a thyrocyte starts the buildup to chromosome replication by only activating PKAs. Nevertheless, it now looks as if things are not so "simple." While a cyclic AMP surge stimulates thyrocyte proliferation, microinjecting active PKA catalytic subunits does not.[134*a*] It seems likely that cyclic AMP operates in part by binding directly to a protein, CNrasGEF (Cyclic Nucleotide ras Guanine nucleotide Exchange Factor), that activates Ras by stimulating the exchange of the GDP in inactive GDP·Ras for GTP to produce active GTP·Ras complexes (Fig. 2.2).[134*b*] The activated GTP·Ras complexes then operate through RalGDS and PtdIns-3K to activate $p70^{s6k}$, which bypasses the Raf-1/MAPK default mechanism (Fig. 2.2) and collaborates with equally essential PKA-stimulated processes to start the buildup to chromosome replication. These recent observations suggest that expressing CNrasGEF is why some cells but not others can be stimulated to proliferate by things that cause cyclic AMP surges.

During the last quarter century, it has become clear that Ca^{2+} released from internal stores with or without Ca^{2+} flowing in through opened CRAC (SOC) channels is necessary to start the cycles of a variety of cells with velcroceptor-activating mitogens[36,70] but,

Fig. 2.2. Why can cyclic AMP stimulate some cells such as thyrocytes to start building up to chromosome replication when it (via the PKAs it activates) interferes with growth-factor velcroceptors (V) signals from activating the MEK/MAP kinase mechanism that starts the buildup in many other types of cell? A possible answer resides in the ability of some cells to express CNrasGEF, which directly binds cyclic AMP and then activates Ras by stimulating the exchange of GDP for GTP as shown in the drawing. The activated GTP·Ras then uses the alternative or bypass mechanism shown here to switch on the required final common mechanism involving the CDKs, the stimulation of replication-related genes, and the various actions of Ca^{2+} and Ca^{2+}·CaM.

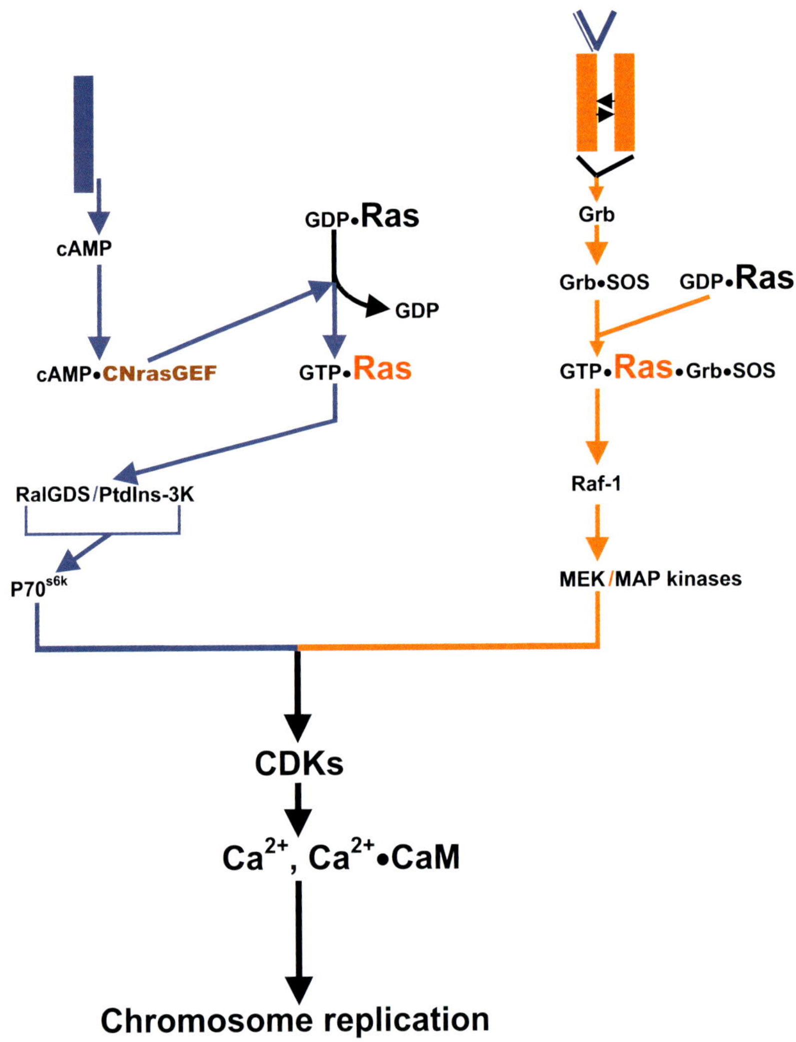

as we have just seen, apparently *not* in epithelial cells that have been additionally equipped with an activated cyclic AMP/PKA mitogenic mechanism. The velcroceptor/PLC/Ca^{2+}/PKC mechanism is the default mechanism that suppresses differentiation and promotes sustained cycling, while the cyclic AMP/PKA mechanism is designed for limited cycling and maintenance of differentiated functions. However, as discussed in the previous chapter, there is a seemingly universal later G1 cyclic AMP surge in normal cells and some cancer cells which has been proven to be needed to complete the buildup to DNA replication.[36,115] But it should be kept in mind that there is no distinct later G_1 surge in TSH -stimulated thyrocytes because of the prolonged elevation of the cyclic AMP level by the slowly deactivating TSH receptor, although there is a small, second cyclic AMP surge in isoproterenol-activated parotid gland acinar cells which is necessary because of the rapid deactivation of the adenylyl cyclase-activating, cyclic AMP-elevating β-adrenergic receptors.[36] This later G_1 surge is controlled by external Ca^{2+} in the hepatocytes of primary neonatal rat liver cultures. These cells, just like hepatocytes activated by partial hepatectomy in the adult rat liver, generate both an early and the later G_1 cyclic AMP surge when stimulated by EGF in serum-free, synthetic medium containing 1.0 mM Ca^{2+}, but while they still make the same amounts of cyclic AMP when stimulated by EGF in medium containing only 0.02 mM Ca^{2+}, they dump the cyclic nucleotide into the medium and thus cannot generate the much-needed later internal G_1 surge.[135] Thus, external Ca^{2+} and therefore maybe signals from the hepatocyte's surface Ca^{2+} sensor/receptor, which we will formally meet further on, restrain the activity of the transporter-mediated mechanism that drives cyclic AMP efflux.[135a] However, while such a severe external Ca^{2+} deprivation is attainable in cell cultures, it is not attainable in a rat; the blood Ca^{2+} concentration does not fall below 0.5 mM after removal of the parathyroid glands. In the rat, the relatively small (50%) reduction of the plasma Ca^{2+} concentration and the resulting severe drop in plasma $1\alpha,25(OH)_2$ vitamin D_3 concentration that develop by 3 days after parathyroid removal do *not* prevent the later G_1 cyclic AMP surge in hepatocytes after their proliferative activation by partial hepatectomy, but they prevent the surge of the PKAs' common catalytic subunit without affecting the surge of their RI subunits and thus contribute to the failure of the hepatocytes to initiate DNA replication.[36] As we learned in the previous chapter, the responses to this failure are likely to be failures to stimulate cyclin A expression, inadequate delivery of cyclin A and other key components into the nucleus, and consequently an inability to initiate DNA replication.

Ca^{2+} and the $G_1 \rightarrow S$ transition

As the cell, be it a hepatocyte in the regenerating rat liver, a T51B rat liver cell, a phytohemagglutinin-activated T-lymphocyte, a NRK kidney-derived rat cell, or a thyrocyte, advances to the Cdk2·cyclin E- and the Cdk2·cyclin A-driven stages of the G_1 buildup, it starts expressing one or more of its three non-allelic calmodulin genes with identical (80–85%) coding regions but different non-coding regions that produce an identical product.[135b] The cell accumulates this key sensor/signaler in preparation for the Ca^{2+}-dependent triggering of DNA replication.[36,136–143] As the normal cell nears the end of the G_1 buildup, it becomes extremely sensitive to external Ca^{2+} deprivation. For example, hepatocytes activated by partial hepatectomy in a rat 24 h after it has had its plasma Ca^{2+} concentration halved or nearly halved by thyroparathyroidectomy cannot accumulate ribonucleotide reductase subunits and thymidylate synthase and thus cannot make the deoxyribonucleoside triphosphates needed to start replicating DNA, but injecting $CaCl_2$ (or PTH — hepatocytes express PTH/PTHrP receptors[144]) intraperitoneally to produce a transient pulse of plasma Ca^{2+} between 12 and 15 h after the partial hepatectomy enables the cells to start replicating DNA on time between 18 and 20 h; but a later Ca^{2+} injection does not work: a window of opportunity has been closed.[36,145]

For the last two decades, it seemed that normal hepatocytes in the liver of a rat could sense and respond like parathyroid gland chief cells to changes in the extracellular Ca^{2+} concentration. Now we know that hepatocytes do indeed have the same low-affinity CaRs (Ca^{2+}or divalent cation receptors) (high-affinity receptors would be locked in the active state by even the lowest possible blood Ca^{2+} concentrations) as the parathyroid cells and, as we shall see in the next two chapters, keratinocytes and colon cells.[146–147] Evidently, cells in various tissues of the body need to be able to sense the external Ca^{2+} concentration. Thus, it seems at least worth considering the possibility that an increasing volume of signals from CaRs accumulating on the plasma membrane during the G_1 buildup might stimulate PLC-γ1 to produce an Ins(1,4,5)P_3-induced release of Ca^{2+} from internal stores and diacylglycerol-stimulated membrane-associated PKCs activity, which are needed at a critical point in the G_1 buildup. Consequently, halving the extracellular Ca^{2+} concentration by removing the parathyroid glands would stop the buildup by preventing the CaR signaling from reaching a critical level.

Ca^{2+} injection (or a *single* injection of PTH) by itself does not enable the hepatocytes in prolonged (e.g., for 72 or 96 h) hypocalcemic TPTX (thyroparathyroidectomized) rats to complete the construction of replication factories and start replicating DNA because of the severe shortage of 1α,$(OH)_2$ vitamin D_3 that develops after TPTX, and the consequent failure to accumulate enough PKA catalytic subunits and CaRs if 1α,$(OH)_2$ vitamin D_3 stimulates the expression of the CaR gene in hepatocytes as it does in keratinocytes.[36,146–149] However, when both the blood Ca^{2+} and 1α,$(OH)_2$ vitamin D_3 concentrations are kept from falling after TPTX by feeding the rats a low-phosphorus diet, hepatocytes can replicate their DNA at the normal time even when they are stimulated by partial hepatectomy at 72 h after TPTX.[150]

Lowering the external Ca^{2+} concentration to a sub-optimal level also stalls BALB/c 3T3 mouse cells, C3H10T1/2 mouse fibroblasts, primary rat thigh muscle cells, T51B rat liver cells, and human WI38 fibroblasts late in the G_1 buildup.[36,70] This Ca^{2+}-sensitive period is extremely short in Swiss albino 3T3 cells.[151] Ca^{2+} deprivation does not stop these cells from starting a cycle, probably because they, like some other cells driven by the velcroceptor-activated default pathway, release enough Ca^{2+} from their internal stores without needing to depend on Ca^{2+} flowing in from stores-operated CRAC channels.[70,152] (But it does stop cells from starting the G_1 buildup if they do not have large enough internal stores.[70,152]) Then, 5-min exposures to Ca^{2+} deprivation prevent the initiation of DNA replication only when carried out within a couple of hours before the expected onset of replication, but is totally ineffective *before* or *after*.[151] In other words, some critical thing requiring external Ca^{2+} appears briefly, does its job, and then disappears and with it the need for external Ca^{2+}. However, if the external Ca^{2+} is removed during this critical time, the cells cannot initiate DNA replication, and re-adding Ca^{2+} does not enable them to complete the buildup and initiate DNA replication; an unstable component(s) has been lost and the cell resets itself into a G_0 state.[151] BALB/3T3 cells and T51B rat liver cells in confluent cultures, like the cells in the regenerating rat liver in the hypocalcemic, 24-h post-thyroparathyroidectomized rat, also cannot stay long in a late G_1 block imposed by an external Ca^{2+} shortage, and they rapidly become unable to immediately restart and finish the buildup if the external Ca^{2+} level is raised; they must reset themselves and be stimulated to start over again.[36] Yet, another indicator of a Ca^{2+}-dependent late G_1 mechanism is a subpopulation of lymphoblasts in the rat thymus that are blocked in late G_1 poised to start replicating DNA in less than an hour after the addition of Ca^{2+} or an agent that can trigger a Ca^{2+} surge (as well as surges of cyclic AMP or cyclic AMP-elevators).[36]

After accumulating in the cytosol, CaM is shipped into the nucleus, and the nuclear calmodulin content, for example, in regenerating liver cells, as much as triples.[72] This CaM may stimulate the transport of Cdk4·cyclin D into the nucleus where it can associate with the chromosomes and hyperphosphorylate and disable the gene-suppressing pocket proteins we met in the previous chapter.[153–155] CaM does not simply diffuse through the nuclear pores, although it is small enough to do so.[39] Instead, it binds to p21[Cip1/WAF1]· CDk4·cyclin D complexes, which would expose p21's nuclear localization sequence and use it to trigger the shipment of the CaM·p21[Cip1/WAF1]·CDk4·cyclin D complex into the nucleus.[153]

The nuclear envelope, a specialized extension of the ER (endoplasmic reticulum), has the ER's Ins(1,4,5)P_3 receptor/channels, ryanodine receptors (opened by cyclic AMP-ribose), and Ca^{2+} pumps.[36a,39] Something now triggers the late-G_1 Ca^{2+} signaling (Fig. 2.1) that in C127 mouse mammary cells starts in the perinuclear region probably upon the emptying of the Ins(1,4,5)P_3-responsive stores in the nuclear membrane.[1,72–75,141] Hepatocytes in the regenerating liver prepare for this stage by radically shifting the expression of their IP_3Rs (Ins(1,4,5)P_3 receptor/channels) from the initially predominant IP_3R2s located on apical stores to IP_3R1s concentrated on perinuclear stores.[155a] This shift does two things. First, it ensures that Ca^{2+} boluses are aimed directly at nuclear targets. Second, it enables the extra-store Ca^{2+}concentration to oscillate because Ins(1,4,5)P_3-activated/opened IP_3R1s close to enable the stores to refill and thus lower the extra-store Ca^{2+} concentration when the extra-store concentration exceeds a certain level, and then while still activated by Ins(1,4,5)P_3 they open again to empty the refilled stores when the extra-store Ca^{2+} concentration falls below the critical level.[155a] Activated IP_3R2s do not close when the extra-store Ca^{2+} concentration rises as the stores empty: their activation results simply in a sustained increase in the Ca^{2+}concentration.[155a]

Soon the nuclear IP_3R1s are activated and the intracellular Ca^{2+} concentration starts oscillating — the cell starts singing its initiating Ca^{2+} song.[71] This signaling depends on an adequate pre-loading of the Ca^{2+} stores by the ER's Ca^{2+} pump (the so-called SERCA pump[36a]). (In regenerating hepatocytes, the SERCA 1 isoform does this job while the cells for some reason downregulate the SERCA 3 isoform.[155a]) Thus, switching off the ER pump's gene delays the onset of DNA replication, which arouses the slumbering $G_1 \rightarrow S$ gatekeeper p53(TP53), which increases the expression of the p21[Cip1/WAF1] cyclin-dependent protein kinase inhibitor.[156] Moreover, emptying the Ins(1,4,5)P_3-sensitive Ca^{2+} stores (by inhibiting the stores'-refilling ATPase pumps with thapsigargin) of diploid human fibroblasts just before the time when DNA replication should begin prevents the expression of the cyclin A needed to produce the Cdk2·cyclin A that drives DNA replication in the factories.[157]

What drives this final stage of the G_1 buildup? From the results of experiments on BALB/c 3T3 cells and vascular smooth muscle cells, it seems that things start with the expression and installation in the cell membrane of Ca^{2+}-permeable channels[158] and the late-G_1 expression of c- or B-Myb.[159] The c-Myc induced at the start of the buildup by the mitogenic velcroceptor (e.g., the PDGF receptor) stimulates the expression of the IGF-I velcroceptor gene and the subsequent buildup of the receptors to a threshold level at which, when they are activated by exogenous or autocrine IGF-I, their signals are loud enough to open the newly installed channels in the cell membrane and enable Ca^{2+} to flow into the cell.[158–161] The IGF-I-induced Ca^{2+} influx and the resulting internal Ca^{2+} surge is reinforced by a Myb-induced reduction of Ca^{2+} efflux.[159] However, there is much more than this to the signal from a velcorceptor such as the IGF-I receptor. The Ras-mediated signal from the IGF-I receptors may also selectively target the nuclear envelope and trigger the nuclear PtdIns(4,5)P_2 breakdown cascade, as it seems to do in regenerating rat liver cells and Swiss albino 3T3 cells,[162,163] thereby triggering the release of Ca^{2+} from perinuclear stores and the start of the Ca^{2+} oscillations that are needed to start DNA replication. The activated IGF-I receptors also have specific domains that are not involved in the cycle-driving signals, but prevent apoptogenic consequences of the Ca^{2+} oscillations by maintaining the level of the megachannel-blocking Bcl-2 protector protein and preventing a murderous, apoptogenic caspase (ICE-like protease) cascade.[160,164,165] The signals from adhesion-activated β1-integrins and ILK (the integrin-linked protein kinase we will meet in the next chapter) and their targets collaborate with the IGF-I receptors to suppress a caspade (caspase cascade).[99]

We have left the Cdk4·cyclin D part of the buildup and are now in the Cdk2·cyclin E and Cdk·cyclin A stages of the buildup. While the exact relations between the late Cdk2·cyclins, Ca^{2+}/PKCs surges, the expression of the replication genes, and the construction of replication factories are far from known, we do have the vague outlines of them. We know that the release of Ca^{2+} from Ins(1,4,5)P_3-responsive stores and adhesion-induced signals from the β1-integrin-associated kinases are needed for the expression of cyclin A.[98,99,166–170] We also know that CaM is needed to start replication.

The cell has been loading its nucleus with CaM and the IGF-I/Myb-triggered Ca^{2+} oscillations around the nucleus spawn Ca^{2+}·CaM complexes that stimulate the ATPase pumps in the nuclear envelope to pump Ca^{2+} into the nucleus.[72,74] This Ca^{2+} surge also might directly stimulate calpain protease isoforms, which could shred the CKIs that have been restraining the CDKs.[171] Indeed, inhibiting calpain activity stops the G_1 buildup in cells such as vascular smooth muscle cells.[1,171] The ion also collaborates with cyclic AMP/PKA[171,172] to provide the open nuclear pores needed at this critical nuclear loading time by binding to the EF-handed, Ca^{2+}-binding gp210 pore protein.[39] When it surges into the nucleus, it produces nuclear Ca·CaM complexes and causes CaM to move to the nuclear periphery.[72]

The surging nuclear Ca^{2+}·CaM complexes somehow trigger a very late step in the initiation sequence. Thus, DNA replication can be prevented in COS-7 cells by selectively blocking nuclear Ca^{2+}·CaM action with an engineered peptide consisting of multiple tandem repeats of the Ca^{2+}·CaM-binding domain of myosin light chain kinase and multiple repeats of a nuclear localization sequence to load it into the nucleus.[173] The Ca^{2+}·CaM blocker, W-7, a naphthalene sulfonamide, blocks Chinese hamster ovary cells in so late a stage of the G_1 buildup that they start making DNA immediately after removal of the blocker.[174] Exposing serum-activated T51B rat liver cells in confluent cultures to another naphthalene sulfonamide, W-13, at 10 h after serum addition, which is just before the expected onset of DNA replication, prevents initiation, but the cells will almost immediately start replicating their DNA if the inhibitor was to be removed 10 h later.[175] In other words, the liver cells can wait for 10 h with their huge replication factories ready to start production when cued by Ca^{2+}·CaM. What are the cells waiting for? Perhaps it was for Ca^{2+}·CaM to unblock the transcription of the PCNA gene.[176,177] Whatever it is, it is not the same as the earlier external Ca^{2+}/CaR-dependent mechanism, which is too unstable to wait around for 10 h, which forces a buildup reset.

Ca^{2+}·CaM is also needed for DNA replication, and it has many targets in the nucleus that are involved in replication.[39,70] One of the targets is CaMKII, which would activate Cdk2·cyclin E and (or) Cdk2·cyclin A by phosphorylating and stimulating the Cdc25A CDK activator.[70,178] However, the CaMKII blocker, KN 93, prevents HeLa human cervical adenocarcinoma cells from initiating DNA replication seemingly *after* the surge of Cdk2·cyclin A activities has begun, which would indicate that Ca^{2+}·CaM-activated CaMKII stimulates something critical in or for the DNA factories after the CDKs' activation by Cdc25A (Fig. 1.2).[70,141] Another target is a 68-kDa Ca^{2+}·CaM-binding protein that surges along with CaM into the nucleus and binds to nuclear matrix-associated replication complexes.[179–182] The binding of Ca^{2+}·CaM·p68 to the replication factories on the matrix is associated with the stimulation of replication.[181,182] And CaM does seem to regulate both DNA polymerases α and δ.[183] Moreover, expressing a selectively CaM-binding inhibitor protein during the S-phase stops proliferation and reduces DNA replication, which indicates a continuing need for Ca^{2+}·CaM or CaM apoprotein during replication.[184] Yet, other Ca^{2+}·CaM targets might be the nuclear contractile machinery, consisting of actin, α-spectrin that binds actin to the inner leaflet of the nuclear envelope, caldesmon, α-fodrin, myosin light chain kinase, and myosin, which could be involved in orienting and configuring the replication foci and the Ca^{2+}·CaM-binding proteins that attach CaM to the nuclear matrix upon the onset of replication.[38,39,72] In particular, the surging Ca^{2+}·CaM would bind to caldesmon and release actin from its complex with caldesmon, which would allow actin to associate with myosin to form a contractile machine for nuclear and chromatin reconfiguring.

However, there is another Ca^{2+}-binding protein associated with the DNA-replication complexes that probably responds to the oscillating Ca^{2+}. This is the heavy chain of annexin II (also known as calpactin I or lipocortin II). The gene encoding annexin II is growth-regulated, and its product associates with four other components to make the primer-recognizing RF-C (replication factor-C) complex that, as we learned in the previous chapter, stimulates the assembly of the homotrimeric PCNA into a ring that clamps itself around the DNA strands and keeps DNA Pol δ progressively replicating the passing template strand.[36,185,186] Reducing the expression of annexin II in HeLa, 293, and 293T cells with antisense oligonucleotides reduces ongoing DNA replication, and immunodepleting annexin II in *Xenopus* egg extracts prevents DNA replication.[186,186a]

While we have been focusing on CaM flowing into the nucleus, we have missed something important happening on the cell surface. While CaM accumulates during the G_1 buildup in the cytoplasm of cells such as hepatocytes in the regenerating rat liver, K562 human leukemic lymphocytes, and T51B rat liver cells, its level *drops* (!) just before the onset of DNA replication.[187–189] Where does it go? Is it destroyed? Apparently not! The cell seems to dump it outside after the internal wave has crested despite the molecule not being addressed for secretion.[188] However, there are indications that it stimulates something on the cell's surface that is required to start replication. Thus, either adding Ca^{2+} or CaM can stimulate Ca^{2+}-deprived T51B rat liver cells to start making DNA, and this DNA-synthetic response to Ca^{2+} can be prevented by anti-calmodulin antibody, which is virtually certain not to get into the unpermeabilized cell let alone reach the nucleus.[187,190,191] Proof of this for at least one type of cell is the ability of anti-calmodulin antibody or the CaM inhibitor W7 bound to agarose beads (to stop it from getting into the cell) to prevent the initiation of DNA replication and stop the proliferation of K562 leukemia cells.[188]

This external CaM action cannot be the same as the CaR-mediated action because the latter is blocked in the presence of reduced external Ca^{2+} concentrations (e.g., 0.5 mM in the regenerating liver of the thyroparathyroidectomized rat) that would not be low enough to affect the formation of active Ca^{2+}·CaM complexes from any dumped CaM. Indeed, it appears that the progression to the CaM step depends on the external Ca^{2+}/CaR step. Thus, there is no wave of CaM or initiation of DNA replication when serum-deprived T51B rat liver cells are stimulated to start the G_1 buildup by fresh, low-Ca^{2+} (0.01 mM) serum, but there is a CaM wave which crests and crashes between 12 and 16 h and the initiation of DNA replication between 14 and 16 h if Ca^{2+} is added 4 h after serum.[187] Thus, it seems that the external Ca^{2+}/CaR-triggered step leads to the accumulation of CaM, some of which goes to the nucleus and the rest of which is dumped out of the cell and onto some surface receptor which drives an essential part of the buildup.

As a general rule, established largely by the results of experiments carried out at the National Research Council over the last 30 years, tumor cells greatly reduce or even lose their requirement for *external* Ca^{2+} to build up to DNA replication.[36] However, since CaM and other Ca^{2+}-binding proteins are parts of the DNA-replicating factories, tumor cells simply could not be expected to be able to do without them. And, of course, they can't, as indicated by the malignant HeLa cell's need for CaMKII to start replicating DNA.[141] They are also needed by fibroblasts transformed by the constitutive hyperactivity of the avian sarcoma virus's 60-kDa v-Src protein tyrosine kinase. This wild hyperactive enzyme eliminates the need for adhesion-triggered integrin signals to restrain apoptogenesis and express CDKs such as Cdk2·cyclin A needed to trigger replication, and they can replicate DNA and proliferate normally and indefinitely in medium containing only 0.02 mM Ca^{2+}, which would silence a putative CaR and thus prevent normal cells from even starting replication.[36,188] However, they still need Ca^{2+}·CaM, because although they can proliferate rapidly and indefinitely with very little external Ca^{2+} they become *extremely vulnerable* to the Ca^{2+}·CaM inhibitors, R24571and W7, just before (e.g., 2 h) DNA replication begins.[36,188,192]

The Ca^{2+} oscillations stop just after the onset of replication — the cell stops singing its "let's-replicate-DNA" song. When Ca^{2+}, Ca^{2+}·CaM, CaMKII, and the other Ca^{2+}-binding parts of the replication machines have done their jobs, phosphoCdk2·cyclin A accumulates in replication factories where it phosphorylates and activates both the initiator helicase that starts the firing of the first battery of replicons by preparing the DNA for being pulled along the assembly line by prying the DNA strands apart at the replication origins and the RP-A complex that then holds the single template strands apart for the polymerases. At the same time, the kinase has also de-licensed the new chromatids to prevent multiple rounds of replication by phosphorylating, and thus expelling the origin-bound, licensors, Cd6, and the MCM proteins. When the chromosomes are all replicated and there is no red light and wailing siren from an alerted G_2 checkpoint mechanism, the cell shreds the G_1/S-specific Cdk2 and starts expressing and using the mitotic Cdk1 which has been suppressed by Cdk2·cyclin A while the chromosomes were being replicated. It also starts expressing the last of the cyclins, the mitotic cyclin Bs.

Ca^{2+} in mitosis

The Cdk1·cyclin-driven breakdown of the nuclear membrane and all that goes with it are triggered by the strictly localized (punctate) release of Ca^{2+} from thousands of endoplasmic reticular vesicles crowding around the nucleus (Fig. 2.3).[1,1a,15,193–197] In other words, the trigger is a shower of Ca^{2+} "puffs" and "sparks," which would last for only a second at the most if they were away from the nucleus, but at this point they can last as long as 10 s because they are perinuclear and invade the nucleus.[197a] Preventing the Ca^{2+} surges prevents prophase, while the Ca^{2+}-mobilizing caffeine or microinjected $Ins(1,4,5)P_3$ triggers premature prophase.[15,193–197] The phospholipase-C activity needed to generate $Ins(1,4,5)P_3$ does appear in the nucleus during the G_2 buildup.[197b] However, $Ins(1,4,5)P_3$ may not be the real physiological Ca^{2+} liberator because it is freely diffusible, which the real liberator is not, and the PLC that would be needed to generate it is not in the membranes of the Ca^{2+}-storage/regulating vesicles in the prophase mitotic apparatus.[1] Instead, it could be leukotriene B4, which is confined to the vesicles along with the PLA_2 and the other enzymes needed to make it.[1]

The mediation through Ca^{2+}·CaM of the responses to the barrage of Ca^{2+} from the thousands of perinuclear vesicles is suggested by the results of experiments with *ts*BN2 hamster kidney cells, which express a thermolabile mutant of the nuclear/chromosome-associated RCC1 protein,[197c] the disappearance of which at a non-permissive temperature such as 39.5°C triggers the premature initiation of prophase.[198] The disappearance of RCC1 is accompanied by a large CaM surge, the importance of which for the initiation of premature prophase is established by the ability of a Ca^{2+}·CaM blocker, W-7, to prevent this prophase response.[198] Exactly what Ca^{2+}·CaM ultimately does appears to be unknown. However, it does it by activating CaMKII.[193,194] When this remarkable multi-subunit enzyme with its approximately 10 α- and β- or β^1-subunits is activated by Ca^{2+}·CaM, it autophosphorylates itself and becomes prolongedly active without further prompting by Ca^{2+}.[199] Thus, the shower of Ca^{2+} transients triggers another, more

Fig. 2.3. How a surge of Ca^{2+} into the nucleus from the nuclear envelope might trigger prophase. The surge produces Ca^{2+}·CaM complexes, which activate CaMKII, which in turn activates Cdc25BorC, the G_2-specific dual-function phosphatase that removes inhibitory phosphate from Cdk1 and thus activates the mitosis-specific CDKs. The CDKs phosphorylate the chromosomes' H1 histones, which causes the chromosomes to reconfigure themselves to admit the chromosome-condensing condensins, which have also been phosphorylated and "activated" by the CDKs. Meanwhile, the Ca^{2+} spike has also stimulated the calpain protease, which starts breaking down the nuclear envelope (NEB).[1,39]

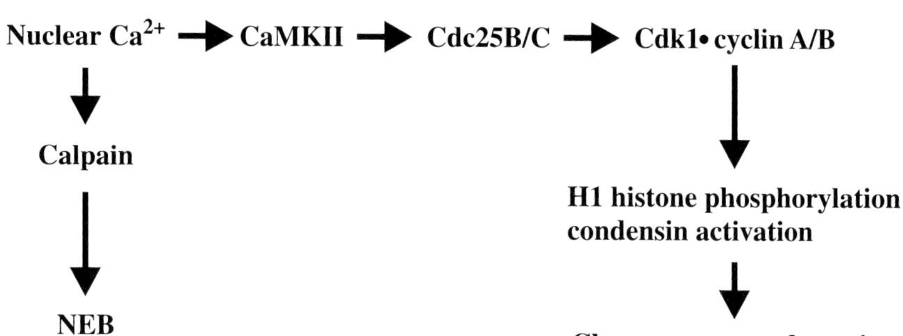

sustained, wave of CaMKII activity. But what does it do? It turns on the late-G_1-specific Cdc25C, a STY (serine/threonine/tyrosine) dual function phosphatase, which activates the Cdk1·cyclins A/B that in turn activate the chromosome-condensing condensins and phosphorylate H1 and H3 histones to reconfigure the chromosomes to receive the phospho-condensins that will condense them (Fig. 2.3).[90]

The Ca^{2+} barrage may trigger nuclear envelope breakdown at least in part by directly stimulating the protease calpain (Fig. 2.3).[1a,39] In fact, inhibiting calpain blocks the $G_2 \rightarrow M$ transition at least in some cells.[200] This is supported by some evidence that, as mitosis approaches, calpain II relocates from the plasma membrane to the nucleus and its entourage of Ca^{2+}-loaded vesicles.[201] There, the calpain may participate in breaking down the nuclear envelope under the barrage of Ca^{2+}.

Another key contributor to the nuclear envelope breakdown is the Ca^{2+}-dependent PKC-βII, which is involved in the disassembly of the lamin network that lines the inner surface of the nucleus and provides "hitching posts" on the nuclear envelope for chromosomes. During the G_2 buildup, PKC-βII is translocated to the nucleus and plugged into the nuclear membrane alongside its target lamin B.[201a] Then, the membrane-tethered kinase is selectively activated when a product of nuclear membrane lipid breakdown, a specific type of nuclear phosphatidylglycerol, binds to its C-terminal region.[201a] The activated PKC-βII then triggers the disassembly of the lamin network and the release of chromosomes from the nuclear envelope by phosphorylating lamin B.[201b]

When the chromosomes have *all* assembled (congressed) at the mitotic spindle's mid-zone and the last of the cycle's checkpoint mechanisms (Fig. 2.1) that monitors and responds to signals from unattached kinetochores has been switched off, and the tension on the straining sister chromatids from the spindle fibers has reached a critical threshold, the Anaphase-GO! signal is given.[202] This is another Ca^{2+} spike or spikes (Fig. 2.1).[39,195] It seems likely that Cdk1·cylin B induces the spike(s) and the Ca^{2+} spills out of storage vesicles associated with the spindle.[1,1a,36] Once again, a Ca^{2+} spike activates CaMKII, which, in turn, activates the spindle-restricted APC (anaphase-promoting complex) that contains a ubiquitin-conjugating E2 enzyme specific for mitotic cyclins and the securin that has been holding back the cohesin-cutting separin.[203–205] One result of this is the polyubiquitination of the mitotic cyclins which are marked for this and consequently the proteasome shredding by a conserved N-terminal motif known as the "destruction box."[36,205] The loss of the cyclin exposes the now naked Cdk1 catalytic subunit to further inactivation by having its Thr*161* residue dephosphorylated by type 2A protein phosphatase, which prevents the subunit from grabbing another cyclin. This cyclin destruction and the disabling of Cdk1 allow the chromosomes to decondense at telophase, but they are not needed for chromatid separation.[205] This requires the APC-driven destruction of the cohesins that have been tying the sister chromatids together[203,204], as well as the removal of residual DNA tangles by Ca^{2+}-stimulated, centromere-associated topoisomerase II.[206,207] The subsequent poleward ratcheting of the separated sister chromosomes along the spindle fibers by the "PacMan"-like action of their kinetochore engines is also driven at least in part by Ca^{2+}.[36]

Another important consequence of the CaMKII-triggered cyclin destruction and elimination of CDK activity is the enabling of the licensing of the new chromosomes for replication by the establishment of pre-replication complexes on the new replication origins by replacing the proteins that were expelled from the origin and eliminated as a consequence of being phosphorylated by Cdk2·cyclin A during the passage of the DNA through the replication factory assembly line.[208] The new licensing proteins have accumulated in the cytoplasm, but they could not get at the chromosomes until the breakdown of the nuclear membrane, but even if they had somehow got into the nucleus before prophase they could not have attached themselves to the target origins because they would have been phosphorylated by the Cdk1·cyclins that took over from Cdk2·cyclin A.[157] However, as soon as a nuclear envelope is established in the daughter cells further access of cytoplasmic licensing factors is cut off.

Ca^{2+} may also be involved in forming the new nuclear envelope during telophase. The fusion of pre-envelope vesicles (e.g., from the disassembled nuclear envelope in a mitosing *Xenopus* egg) into a new nuclear envelope requires the flux of Ca^{2+} through the vesicles' Ins(1,4,5)P_3 receptor/channels.[39,209]

While all of this has been going on, the cell has been putting a belt of cortical actomyosin filaments around its middle. This belt has been kept from tightening prematurely by having its myosin motors turned off by the phosphorylation of certain residues in their light chains by Cdk1·cyclin B.[36] The motors are activated by the destruction of Cdk1·cyclin B and then started by the stimulation of their MLC (myosin light chain kinase) by the Ca^{2+}·CaM complexes generated by a final prolonged Ca^{2+} surge.[36,195,205] The belt tightens and pinches the elongating cell into two daughters, each with licensed chromosomes awaiting instructions from their surroundings to cycle or differentiate, or from the spontaneous goading by inner voices to start cycling if they are malignant or are on their way to malignancy.

And now we must hurry on to skin keratinocytes and their Ca^{2+} diffpoptosis.

References

1. Silver RB. Imaging structured space–time patterns of Ca^{2+} signals: essential information for decisions in cell division. FASEB J 1999; 13: S209–S215.

1*a*. Santella L. The roles of calcium in the cell cycle: facts and hypotheses. Biochem Biophys Res Commun 1998; 244: 317–324.

2. Whitfield JF. Calcium: Cell Cycle Driver, Differentiator and Killer. Austin / New York, Landes Bioscience / Chapman & Hall, 1997.
3. Allbritton NL, Meyer T. Localized calcium spikes and propagating calcium waves. Cell Calcium 1993; 14: 691–697.
4. Allbritton NL, Meyer T, Stryer L. Range of messenger action of calcium ions and inositol 1,4,5-trisphosphate. Science 1992; 258: 1812–1815.
5. Berridge MJ, editor. Microdomains and elemental events in calcium signalling. Cell Calcium 1996; 20: 95–226.
6. Taub R. Transcriptional control of liver regeneration. FASEB J 1996; 10: 413–427.
7. Del Sal G, Ruaro EM, et al. The growth arrest-specific gene, gas1, is involved in growth suppression. Cell 1992; 70: 595–607.
8. Del Sal G, Ruaro EM, et al. Gas1-induced growth suppression requires a transactivation-independent p53 function. Mol Cell Biol 1995; 15: 7152–7160.
9. Schneider C, King RM, Philipson L. Genes specifically expressed at growth arrest of mammalian cells. Cell 1988; 54: 789–793.

10. Ansari B, Dover R, Gillmore CP, et al. Expression of the nuclear membrane protein statin in cycling cells. J Pathol 1993; 169: 391–396.
11. Bissonnette R, Lee MJ, Wang E. The differentiation process of intestinal epithelial cells is associated with the appearance of statin, a non-proliferation-specific nuclear protein. J Cell Sci 1990; 95 (Pt. 2): 247–254.
12. Ching G, Wang E. Characterization of two populations of statin and the relationship of their syntheses to the state of cell proliferation. J Cell Biol 1990; 110: 255–261.
13. Lee MJ, Sandig M, Wang E. Statin, a protein specifically present in nonproliferating cells, is a phosphoprotein and forms a complex with a 45-kilodalton serine/threonine kinase. J Biol Chem 1992; 267: 21773–21781.
14. Pellicciari C, Mangiarotti, R, Bottone MG, et al. Identification of resting cells by dual-parameter flow cytometry of statin expression and DNA content. Cytometry 1995; 21: 329–337.
15. Sandig M, Bissonnette R, Liu CH, et al. Characterization of 57-kDa statin as a true marker for growth arrest in tissue by its disappearance from regenerating liver. J Cell Physiol 1994; 158: 277–284.
16. Wang E, Pandey S. Down-regulation of statin, a nonproliferation-specific nuclear protein, and up-regulation of c-myc after initiation of programmed cell death in mouse fibroblasts. J Cell Physiol 1995; 163: 155–163.
17. Altman M, Trachsel H. Regulation of translation initiation and modulation of cellular physiology. Trends Biochem Sci 1993; 18: 429–432.
18. Frederickson RM, Sonenberg N. Signal transduction and regulation of translation initiation. Semin Cell Biol 1992; 3: 107–115.
19. Gingras A-C, Raught, Sonenberg N. eIF-4 initiation factors: effectors of mRNA recruitment to ribosomes and regulators of translation. Annu Rev Biochem 1999; 68: 913–963.
20. Joshi B, Cai AL, Keiper BD, et al. Phosphorylation of eukaryotic protein synthesis initiation factor 4E at Ser-209. J Biol Chem 1995; 270: 14597–14603.
21. Koromilas AE, Lazaris-Karatzas A, Sonenberg N. mRNAs containing extensive secondary structure in their 5′ non-coding region translate efficiently in cells overexpressing initiation factor eIF-4E. EMBO J 1992; 11: 4153–4158.
22. Mader S, Sonenberg N. Cap binding complexes and cellular growth control. Biochimie 1995; 77: 40–44.
23. Minich WB, Balasta ML, Goss DJ, et al. Chromatographic resolution of the in vivo phosphorylated and nonphosphorylated eukaryotic translation initiation factor eIF-4E: increased cap affinity of the phosphorylated form. Proc Natl Acad Sci USA 1994; 91: 7668–7672.
24. Morley SJ, Thomas G. Intracellular messengers and the control of protein synthesis. In: Taylor CW, editor. Intracellular Messengers. Oxford, Pergamon Press, 1993: 447–483.
25. Polunovsky VA, Rosenwald IB, Tan AT, et al. Translational control of programmed cell death: eukaryotic translation initiation factor 4E blocks apoptosis in growth-factor-restricted fibroblasts with physiologically expressed or deregulated Myc. Mol Cell Biol 1996; 16: 6573–6581.

26. Rhoads RE. Regulation of eukaryotic protein synthesis by initiation factors. J Biol Chem 1993; 268: 3017–3020.
27. Rhoads RE, Joshi B, Minich WB. Participation of initiation factors in the recruitment of mRNA to ribosomes. Biochimie 1994; 76: 831–838.
28. Rhoads RE, Joshi-Barve S, Rinker-Schaeffer C. Mechanism of action and regulation of protein synthesis initiation factor 4E: effects on mRNA discrimination, cellular growth rate, and oncogenesis. Prog Nucleic Acid Res Mol Biol 1993; 46: 183–219.
29. Rosenwald IB. Deregulation of protein synthesis as a mechanism of neoplastic transformation. BioEssays 1996; 18: 243–250.
30. Rosenwald IB, Kaspar R, Rousseau D, et al. Eukaryotic translation initiation factor 4E regulates expression of cyclin D1 at transcriptional and post-transcriptional levels. J Biol Chem 1995; 270: 21176–21180.
31. Rousseau D, Gingras AC, Pause A, et al. The eIF4E-binding proteins 1 and 2 are negative regulators of growth. Oncogene 1996; 13: 2415–2420.
32. Rousseau D, Kaspar R, Rosenwald I, et al. Translation initiation of ornithine decarboxylase and nucleocytoplasmic transport of cyclin D1 mRNA are increased in cells overexpressing eukaryotic initiation factor 4E. Proc Natl Acad Sci USA 1996; 93: 1065–1070.
33. Smith MR, Jaramillo M, Liu Y-L, et al. Translation initiation factors induce DNA synthesis and transform NIH3T3 cells. New Biol 1990; 2: 648–654.
34. Smith MR, Jaramillo M, Tuazon P, et al. Modulation of the mitogenic activity of eukaryotic translation initiation factor-4E by protein kinase. New Biol 1991; 3: 601–607.
35. Sonenberg N. Translation factors as effectors of cell growth and tumorigenesis. Curr Opin Cell Biol 1993; 5: 955–960.
36. Whitfield JF. Calcium in Cell Cycles and Cancer. Boca Raton, CRC Press, 1995: 9–58.
36*a*. Brini M, Carafoli E. Calcium signaling: a historical account, recent developments and future perspectives. Cell Mol Life Sci 2000; 57: 354–370.
37. Williams RJP. Calcium: the developing role of its chemistry in biological evolution. In: Carafoli E, Klee C, editors. Calcium as a Cellular Regulator. New York, Oxford University Press, 1999: 3–27.
38. De Boni U. The interphase nucleus as a dynamic structure. Int Rev Cytol 1994; 150: 149–171.
39. Santella L, Bolshover S. Calcium in the nucleus. In: Carafoli E, Klee C, editors. Calcium as a Cellular Regulator. New York, Oxford University Press, 1999: 487–511.
40. Fante WJ, Johnson DE, Williams LT. Signalling by tyrosine kinases. Annu Rev Biochem 1993; 62: 453–486.
41. Malarkey K, Belham CM, Paul A, et al. The regulation of tyrosine kinase signalling pathways by growth factor and G-protein-coupled receptors. Biochem J 1995; 309: 361–375.
42. Birge RB, Hanafusa H. Closing in on SH2 specificity. Science 1993; 262: 1522–1524.

43. Chook YM, Gish GD, Kay CM, et al. The Grb2–mSos1 complex binds phosphopeptides with higher affinity than Grb2. J Biol Chem 1996; 30472–30478.
44. Mayer BJ. Why two heads are better. Structure 1995; 3: 977–980.
45. Mayer BJ, Baltimore D. Signalling through SH2 and SH3 domains. Trends Cell Biol 1993; 3: 8–13.
46. Pawson T. New impressions of Src and Hck. Nature 1997; 385: 582–585.
47. Pawson T, Nash P. Protein–protein interactions define specificity in signal transduction. Genes Devel 2000; 14: 1132–1145.
48. Myers MG, White MF. Insulin signal transduction and the IRS proteins. Annu Rev Pharmacol Toxicol 1996; 36: 615–618.
49. Rubin R, Baserga R. Insulin-like growth factor-I receptor. Lab Invest 1995; 73: 311–331.
50. Buday, L. Membrane-targeting of signaling molecules by SH2/SH3 domain-containing adaptor proteins. Biochim Biophys Acta 1999; 1422: 187–204.
51. Magnuson NS, Beck T, Vahidi H, et al. The Raf-1 serine/threonine protein kinase. Semin Cancer Biol 1994; 5: 247–253.
52. Guan K-L. The mitogen activated protein kinase signal transduction pathway: from the cell surface to the nucleus. Cellular Signalling 1994; 6: 581–589.
53. Kortenjann M, Shaw PE. The growing family of MAP kinases: regulation and specificity. Crit Rev Oncogenesis 1995; 6: 99–115.
54. Barrett J, Lewis BC, Hoang AT. Cyclin A links c-Myc to adhesion-independent proliferation. J Biol Chem 1995; 270: 15923–15925.
55. Henriksson M, Lüscher B. Proteins of the Myc network: essential regulators of cell growth and differentiation. Adv Cancer Res 1996; 68: 109–182.
56. Hoang AT, Cohen KJ, Barrett JF. Participation of cyclin A in Myc-induced apoptosis. Proc Natl Acad Sci USA 1994; 91: 6875–6879.
57. Kaldis P. The cdk-activating kinase (CAK): from yeast to mammals. Cell Mol Life Sci (CMLS) 1999; 55: 284–296.
58. Latchman DS. Eukaryotic Transcription Factors. 3rd ed. London, Academic Press, 1998: 1–374.
59. Ryan K, Birnie GD. *Myc* oncogenes: the enigmatic family. Biochem J 1996; 314: 713–721.
60. Liu J-P. Protein kinase C and its substrates. Mol Cell Endocrinol 1996; 116: 1–29.
61. Martelli AM, Sang N, Borgatti P, et al. Multiple biological responses activated by nuclear protein kinase C. J Cell Biochem 1999; 74: 499–521.
62. Chakravarthy BR, Whitfield JF, Durkin JP. Inactive protein kinase Cs: a possible target for receptor signalling. Biochem J 1994; 304: 809–816.
63. Berridge MJ, Lipp P, Bootman MD. The calcium entry pas de deux. Science 2000; 287: 1604–1605.
64. Furuichi T, Michikawa T, Mikoshiba K. Intracellular calcium channels. In: Carafoli E, Klee C, editors. Calcium as a Cellular Regulator. New York, Oxford University Press, 1999: 200–248.
65. Putney JW, Ribiero CM. Signaling pathways between the plasma membrane and endoplasmic reticulum calcium stores. Cell Mol Life Sci 2000; 57: 1272–1286.

66. Rossier MF, Bird GSJ, Putney JW. Subcellular distribution of the calcium-storing inositol-1,4,5-trisphosphate-sensitive organelle in rat liver. Biochem J 1991; 274: 643–650.
67. Rossier MF, Putney JW. The identity of the calcium-storing inositol 1,4,5-trisphosphate-sensitive organelle in non-muscle cells: calciosome, endoplasmic reticulum or both? Trends Neurosci 1991; 14: 310–314.
68. Taylor CW, Richardson A. Structure and function of inositol trisphosphate receptors. In: Taylor CW, editor. Intracellular Messengers. Oxford, Pergamon Press, 1993: 199–254.
69. Gill DL, Waldron RT, Rys-Sikora KE. Calcium pools, calcium entry, and cell growth. Biosci Rep 1996; 16: 139–157.
70. Means A, Kahl CR, Crenshaw DG, et al. Traversing the cell cycle: the calcium/calmodulin connection. In: Carafoli E, Klee C, editors. Calcium as a Cellular Regulator. New York, Oxford University Press 1999: 512–528.
71. Lu KP, Means AR. Regulation of the cell cycle by calcium and calmodulin. Endocrine Rev 1993; 14: 40–58.
72. Bachs O, Agell N, Carafoli E. Calcium and calmodulin function in the cell nucleus. Biochim Biophys Acta 1992; 1113: 259–270.
73. Hennager DJ, Welsh MJ, DeLisle S. Changes in either cytosolic or nucleoplasmic inositol 1,4,5-trisphosphate levels can control nuclear Ca^{2+} concentration. J Biol Chem 1995; 270: 4959–4962.
74. Nicotera P, McConkey D, Jones DP et al. ATP stimulates Ca^{2+} uptake and increases the free Ca^{2+} concentration in isolated liver nuclei. Proc Natl Acad Sci USA 1989; 86: 453–457.
75. Stehno-Bittel L, Luckhoff A, Clapham DE. Calcium release from the nucleus by $InsP_3$ receptor channels. Neuron 1995; 14: 163–167.
76. Stehno-Bittel L, Perez-Terzic C, Luckhoff A, et al. Nuclear ion channels and regulation of the nuclear pore. Soc Gen Physiol Ser 1996; 51: 195–207.
77. Perez-Terzic C, Pyle J, Jacomi M, et al. Conformational states of the nuclear pore complex induced by depletion of nuclear Ca^{2+} stores. Science 1996; 273: 1875–1877.
78. Stehno-Bittel L, Perez-Terzic C, Clapham DE. Diffusion across the nuclear envelope inhibited by depletion of the nuclear Ca^{2+} store. Science 1995; 270: 1835–1838.
79. Corneliussen B, Hol M, Waltersson Y, et al. Calcium/calmodulin inhibition of basic-helix–loop–helix transcription factor domains. Nature 1994; 368: 760–764.
80. Antoine M, Gaiddon C, Loeffler JP. Ca^{2+}/calmodulin kinase types II and IV regulate c-fos transcription in the At20 corticotroph cell line. Mol Cell Endocrinol 1996; 120: 1–8.
81. Foulkes NS, Sassone-Corsi P. Transcription factors coupled to the cAMP-signalling pathway. Biochim Biophys Acta 1996; 1288: F101–F121.
82. Matthews RP, Guthrie CR, Wailes LM, et al. Calcium/calmodulin-dependent protein kinase types II and IV differentially regulate CREB-dependent gene expression. Mol Cell Biol 1994; 14: 6107–6116.

83. Genot E, Cleverley S, Henning S, et al. Multiple p21 ras effector pathways regulate nuclear factor of activated T cells. EMBO J 1996; 15: 3923–3933.
84. Luo C, Burgeon E, Carew JA, et al. Recombinant NFAT1 (NFATp) is regulated by calcineurin in T cells and mediates transcription of several cytokine genes. Mol Cell Biol 1996; 16: 3955–3966.
85. Luo C, Carew JA, Kim J, et al. T-cell receptor stimulation elicits an early phase of activation and a later phase of deactivation of the transcription factor NFAT1. Mol Cell Biol 1996; 16: 3945–3954.
86. Loh C, Shaw KT, Carew J, et al. Calcineurin binds the transcription factor NFAT1 and reversibly regulates its activity. J Biol Chem 1996; 271: 10884–10891.
87. Luo C, Shaw KT, Raghavan A, et al. Interaction of calcineurin with a domain of the transcription factor NFAT1 that controls nuclear import. Proc Natl Acad Sci USA 1996; 93: 8907–8912.
88. Rooney JW, Sun YL, Glimcher LH, et al. Novel NFAT sites that mediate activation of the interleukin-2 promoter in response to T-cell receptor stimulation. Mol Cell Biol 1995; 15: 6299–6310.
89. Woodrow M, Ceipstone NA, Cantrell D. $p21^{ras}$ and calcineurin synergize to regulate the nuclear factor of activated T-cells. J Exp Med 1993; 178: 1517–1522.
90. Holt PR, Philipova R, Moss S, et al. Calcium/calmodulin-dependent phosphorylation and activation of human Cdc25-C at the end of G_2/M phase transition in HeLa cells. J Biol Chem 1999; 274: 7958–7968.
91. Morris TA, De Lorenzo RJ, Tombes RM. CaMK-II inhibition reduces cyclin D1 levels and enhances the association of $p27^{Kip1}$ with Cdk2 to cause G1 arrest in NIH 3T3 cells. Exp Cell Res 1998; 240: 218–227.
92. Bellamy COC, Malcomson RDG, Harrison DJ, et al. Cell death in health and disease: the biology and regulation of apoptosis. Semin Cancer Biol 1995; 6: 3–16.
93. Bowen ID, Bowen SM, Jones AH. Mitosis and Apoptosis. London, Chapman & Hall,1998.
94. Canman CE, Kastan MB. Induction of apoptosis by tumor suppressor genes and oncogenes. Semin Cancer Biol 1995; 6: 17–25.
95. Duttaroy A, Qian J-F, Smith JS, et al. Up-regulated $P21^{CIP1}$ expression is part of the regulation quantitatively controlling serum deprivation-induced apoptosis. J Cell Biochem 1997; 64: 434–446.
96. Kowalik TF, DeGregori J, Leone G, et al. E2F1-specific induction of apoptosis and p53 accumulation which is blocked by Mdm2. Cell Growth Differ 1998; 9: 113–118.
97. Kowalik TF, DeGregori J, Schwarz JK. E2F1 overexpression in quiescent fibroblasts leads to induction of cellular DNA synthesis and apoptosis. J Virology 1995; 69: 2491–2500.
98. Meredith JE, Schwartz MA. Integrins, adhesion and apoptosis. Trends Cell Biol 1997; 7: 146–150.
99. Wary KK, Mainiero F, Isakoff SJ, et al. The adaptor protein Shc couples a class of integrins to the control of cell cycle progression. Cell 1996; 87: 733–743.
100. Castedo M, Hirsch T, Susin SA, et al. Sequential acquisition of mitochondrial and plasma membrane alterations during early lymphocyte apoptosis. J Immunol 1996; 157: 512–521.

101. Kroemer G, Petit P, Zamzami N, et al. The biochemistry of programmed cell death. FASEB J 1995; 9: 1277–1287.
102. Marchetti P, Castedo M, Susin SA, et al. Mitochondrial permeability transition is a central coordinating event of apoptosis. J Exp Med 1996; 184: 1155–1160.
103. Marchetti P, Hirsch T, Zamzami N, et al. Mitochondrial permeability transition triggers lymphocyte apoptosis. J Immunol 1996; 157: 4830–4836.
104. Martins LM, Earnshaw WC. Apoptosis: alive and kicking in 1997. Trends Cell Biol 1997; 7: 111–114.
105. Petit PX, Susin SA, Zamzami N, et al. Mitochondria and programmed cell death: back to the future. FEBS Lett 1996; 396: 7–13.
106. Rustenbeck I, Munster W, Lenzen S. Relation between accumulation of phospholipase A_2 reaction products and Ca^{2+} release in isolated liver mitochondria. Biochim Biophys Acta 1996; 1304: 12138.
107. Susin SA, Zamzami N, Castedo M, et al. Bcl-2 inhibits the mitochondrial release of an apoptogenic protease. J Exp Med 1996; 184: 1331–1341.
108. Hoek JB, Farber JL, Thomas AP, et al. Calcium ion-dependent signalling and mitochondrial dysfunction: mitochondrial calcium uptake during hormonal stimulation in intact liver cells and its implications for the mitochondrial permeability transition. Biochim Biophys Acta 1995; 1271: 93–102.
109. Suzuki YJ, Forman HJ, Sevanian A. Oxidants as stimulators of signal transduction. Free Rad Biol Med 1997; 22: 269–285.
110. Brustovetsky N, Klingenberg M. Mitochondrial ADP/ATP carrier can be reversibly converted into a large channel by Ca^{2+}. Biochemistry 1996; 35: 8483–8488.
111. Evtodienko Yu V, Teplova V, Khawaja J, et al. The Ca^{2+}-induced permeability transition pore is involved in Ca^{2+}-induced mitochondrial oscillations. A study on permeabilized Ehrlich ascites tumour cells. Cell Calcium 1994; 15: 143–152.
111. Massari S. Kinetic analysis of the mitochondrial permeability transition. J Biol Chem 1996; 271: 31942–31948.
113. Kowaltkowski AJ, Castilho RF, Grijalba MT. Effect of inorganic phosphate concentration on the nature of inner mitochondrial membrane alterations mediated by Ca^{2+} ions. A proposed model for phosphate-stimulated lipid oxidation. J Biol Chem 1996; 271: 2929–2934.
114. Whitfield JF, Kellerer S, Brohée H, et al. The Feasibility of a New Procedure for Biological Dosimetry. Euratom Report EUR 2505.e. Brussels, Presses Academiques Européennes, 1965.
114*a*. Meleti Z, Shapiro IM, Adams CS. Inorganic phosphate induces apoptosis of osteoblast- like cells in culture. Bone 2000; 27: 359–366.
115. Boynton AL, Whitfield JF. The role of cyclic AMP in cell proliferation: a critical assessment of the evidence. Adv Cyclic Nucleotide Res 1993; 15: 193–294.
116. Dumont JE, Lamy F, Roger P, et al. Physiological and pathological regulation of thyroid cell proliferation and differentiation by thyrotropin and other factors. Physiol Revs 1992; 72: 667–697.
117. Lukas J, Bartkova J, Bartek J. Convergence of mitogenic signalling cascades from diverse classes of receptors at the cyclin D-cyclin-dependent kinase-pRb-controlled G1 checkpoint. Mol Cell Biol 1996; 16: 6917–6925.

118. Roger PP, Reuse S, Maenhaut C, et al. Multiple facets of the modulation of growth by cAMP. Vitamins and Hormones 1995; 51: 59–191.
119. Lamy F, Wilkin F, Baptist M, et al. Phosphorylation of mitogen-activated protein kinases is involved in the epidermal growth factor and phorbol ester, but not in the thyrotropin/cAMP, thyroid mitogenic pathway. J Biol Chem 1993; 268: 8398–8401.
120. Raspé E, Reuse S, Roger PP, et al. Lack of correlation between the activation of the Ca^{2+}-phosphatidylinositol cascade and the regulation of DNA synthesis in the dog thyrocyte. Exp Cell Res 1992; 198: 17–26.
121. Pirson I, Coulonval K, Lamy F, et al. c-Myc expression is controlled by the mitogenic cAMP-cascade in thyrocytes. J Cell Physiol 1996; 59–70.
122. Baptist M, Lamy F, Gannon J, et al. Expression and subcellular localization of CDK2 and cdc 2 kinases and their common partner cyclin A in thyroid epithelial cells: comparison of cyclic AMP-dependent and -independent cell cycles. J Cell Physiol 1996; 166: 256–273.
123. Taton M, Dumont JE. Dissociation of the stimuli for cell hypertrophy and cell division in the dog thyrocyte: insulin promotes protein accumulation while TSH triggers DNA synthesis. Exp Cell Res 1995; 221: 530–533.
124. Tortora G, Pepe S, Cirafici AM, et al. Thyroid stimulating hormone-regulated growth and cell cycle distributions of thyroid cells involve type I isozyme of cyclic AMP-dependent protein kinase. Cell Growth Differ 1993; 4: 359–365.
125. Yamamoto K, Hirai A, Ban T. Thyrotropin induces G1 cyclin expression and accelerates G1 phase after insulin-like growth factor I stimulation in FRTL-5 cells. Endocrinology 1996; 137: 2036–2042.
126. Depoortere F, Van Keymeulen A, Lukas T. A requirement for cyclin D3-cyclin-dependent kinase (cdk)-4 assembly in the cyclic adenosine monophosphatekinase-dependent proliferation of thyrocytes. J Cell Biol 1998; 140: 1427–1439.
127. Burikhanov R, Coulonval K, Pirson I, et al. Thyrotropin via cyclic AMP induces insulin receptor expression and insulin co-stimulation of growth and amplifies insulin and insulin-like growth factor signaling pathways in dog thyroid epithelial cells. J Biol Chem 1996; 271: 29400–29406.
128. Cass LA, Meikoth JL. Differential effects of cyclic adenosine 3′,5′-monophosphate on p70 ribosomal S6 kinase. Endocrinology 1998; 139: 1991–1998.
129. Cass LA, Summers SA, Prendergast GV, et al. Protein kinase A-dependent and -independent signaling pathways contribute to cyclic AMP-stimulated proliferation. Mol Cell Biol 1999; 19: 5882–5891.
130. Vetter JR, Linnemann T, Wohlgemuth S, et al. Structural and biochemical analysis of Ras-effector signaling via RalGDS. FEBS Lett 1999; 451: 175–180.
131. Withers DJ. Signalling pathways involved in the mitogenic effects of cAMP. Clin Sci (Colch) 1997; 92: 448–451.
132. Withers DJ, Bloom SR, Rozengurt E. Dissociation of cAMP-stimulated mitogenesis from activation of the mitogen-activated protein kinase cascade in Swiss 3T3 cells. J Biol Chem 1995; 270: 21411–21419.
133. Wolthuis RM, Bos JL. Ras caught in another affair: the exchange factors for Ral. Curr Opin Genet Dev 1999; 9: 112–117.

134. deGroot RP, Ballou LM, Sassone-Corsi P. Positive regulation of the cAMP-responsive activator CREM by the p70 S6 kinase: an alternative route to mitogen-induced gene expression. Cell 1994; 79: 81–91.
134*a*. Tsyganova OM, Kupperman E, Wen W, et al. Cyclic AMP activates Ras. Oncogene 2000; 19: 3609–3615.
134*b*. Pham N, Cheglakov I, Koch CA, et al. The guanine nucleotide exchange factor CNrasGEF activates ras in response to cAMP and cGMP. Curr Biol 2000; 10: 555–558.
135. Armato U, Ribecco M, Guerriero C, et al. Extracellular calcium modulates pre-replicative cyclic AMP surges in EGF-stimulated primary neonatal rat hepatocytes. J Cell Physiol 1994; 161: 55–62.
135*a*. Orlov SN, Maksimiva NV. Efflux of cyclic adenosine monophosphate from cells: mechanisms and physiological implications. Biochemistry (Moscow) 1999; 64: 164–173.
135*b*. Toutenhoofd SL, Strehler EE. The calmodulin multigene family as a unique case of genetic redundancy: multiple levels of regulation to provide spatial and temporal control of calmodulin pools? Cell Calcium 2000; 28: 83–96.
136. Bosch M, Lopez-Girona A, Bachs O, et al. Protein kinase C regulates calmodulin expression in NRK cells activated to proliferate from quiescence. Cell Calcium 1994; 16: 446–454.
137. Colomer J, Agell N, Engel P, et al. Calmodulin expression during proliferative activation of human T lymphocytes. Cell Calcium 1993; 14: 609–618.
138. Colomer J, Lopez-Girona A, Agell N, et al. Calmodulin regulates the expression of cdks, cyclins and replicative enzymes during proliferative activation of human T lymphocytes. Biochem Biophys Res Commun 1994; 200: 306–312.
139. Means AR, Rasmussen CD. Calcium, calmodulin and cell proliferation. Cell Calcium 1988; 9: 313–319.
140. Rasmussen CD, Means AR. Calmodulin is required for cell-cycle progression during G1 and mitosis. EMBO J 1989; 8: 73–82.
141. Rasmussen G, Rasmussen C. Calmodulin-dependent protein kinase II is required for G_1/S progression in HeLa cells. Biochem Cell Biol 1995; 73: 201–207.
142. Takuwa N, Zhou W, Takuwa Y. Calcium, calmodulin and cell cycle progression. Cellular Signalling 1995; 7: 93–104.
143. Wang J, Moreira KM, Campos B, et al. Targeted neutralization of calmodulin in the nucleus blocks DNA synthesis and cell cycle progression. Biochim Biophys Acta 1996; 1313; 223–228.
144. Watson PH, Kisiel MA, Natale BV, et al. PTH/PTHrP receptor protein and mRNA in kidney and non-traditional PTH target tissues. Bone 1998; 23: S567.
145. Rixon RH, Whitfield JF. The control of liver regeneration by parathyroid hormone and calcium. J Cell Physiol 1976; 87: 147–156.
146. Brown EM, Quinn SM, Vassilev PM. The plasma membrane calcium sensor. In: Carafoli E, Klee C, editors. Calcium as a Cellular Regulator. New York, Oxford University Press, 1999: 295–310.
146*a*. Riccardi D. Cell surface, Ca^{2+}(cation)-sensing receptor(s): one or many? Cell Calcium 1999; 26: 77–83.

147. Canaff L, Petit J-L, Kisiel M, et al. The calcium-sensing receptor is expressed in rat hepatocytes: coupling to intracellular calcium mobilization and stimulation of bile flow. J Biol Chem; online pre-publication manuscript, Nov. 8, 2000.
148. Bikle DD. Vitamin D. A calciotropic hormone regulating calcium-induced keratinocyte differentiation. J Am Acad Dermatol 1997; 37: S42–S52.
149. Ratnam AV, Bikle DD, Cho JK. 1,25 dihydroxyvitamin D_3 enhances the calcium response of keratinocytes. J Cell Physiol 1999; 178: 188–196.
150. Rixon RH, Isaacs RJ, Whitfield JF. Control of DNA polymerase-α activity in regenerating rat liver by calcium and $1\alpha,25(OH)_2D_3$. J Cell Physiol 1989; 139: 354–360.
151. Damluji R, Riley PA. Time dependency of a crucial effect of EGTA on DNA synthesis in cultures of Swiss albino 3T3 cells. Exp Cell Biol 1979; 47: 446–453.
152. Whitfield JF, MacManus JP, Boynton AL, et al. Futures of calcium, calcium-binding proteins, cyclic AMP and protein kinases in the quest for an understanding of cell proliferation and cancer. In: Corradino RA, editor. Functional Regulation at the Cellular and Molecular Levels. New York, Elsevier North Holland, 1982: 61–87.
153. Jaumont M, Estanyol JM, Serratosa J, et al. Activation of cdk4 and cdk2 during rat liver regeneration is associated with intranuclear rearrangements of cyclin–cdk complexes. Hepatology 1999; 29: 385–395.
154. Taulés M, Rius E, Talaya D, et al. Calmodulin is essential for cyclin-dependent kinase 4 (cdk 4) activity and nuclear accumulation of cyclin D1 – cdk 4 during G_1. J Biol Chem 1998; 273: 33279–33286.
155. Taulés M, Rodríguez-Vilarrupla A, Rius E, et al. Calmodulin binds to $p21^{Cip1}$ and is involved in the regulation of its nuclear localization. J Biol Chem 1999; 274: 24445–24448.
155*a*. Magnino F, St-Pierre M, Lüthi M, et al. Expression of intracellular calcium channels and pumps after partial hepatectomy in rat. Mol Cell Biol Res Commun 2000; 3: 374–379.
156. Cheng G, Liu B-F, Yu Y. The exit from G_0 into the cell cycle requires and is controlled by sarco(endo)plasmic reticulum Ca^{2+} pump. Arch Biochem Biophys 1996; 329: 65–72.
157. Takuwa N, Zhou W, Kumada M, et al. Involvement of intact inositol-1,4,5-trisphosphate-sensitive Ca^{2+} stores in cell cycle progression at the G1/S boundary in serum-stimulated human fibroblasts. FEBS Lett 1995; 360: 173–176.
158. Kanzaki M, Shibata H, Mogami H, et al. Expression of calcium-permeable cation channel CD20 accelerates progression through the G_1 phase in Balb/c 3T3 cells. J Biol Chem 1995; 270; 13099–13104.
159. Simons M, Ariyoshi H, Salzman EW, et al. c-*myb* affects intracellular calcium handling in vascular smooth muscle cells. Am J Physiol 1995; 268: C856–C868.
160. Rubin R, Baserga R. Insulin-like growth factor-I receptor. Lab Invest 1995; 73: 311–331.
161. Sugimoto T, Kanatani M, Kano J, et al. IGF-I mediates the stimulatory effect of high calcium concentration on osteoblastic cell proliferation. Am J Physiol 1994; 266: E709–E716.

162. Banfic H, Zizak M, Divecha N, et al. Nuclear diacylglycerol is increased during cell proliferation in vivo. Biochem J 1993; 290: 633–636.
163. Divecha N, Banfic H, Irvine RF. The polyphosphoinositide cycle exists in the nuclei of Swiss 3T3 cells under the control of a receptor (for IGF-1) in the plasma membrane, and stimulation of the increased nuclear diacylglycerol apparently induces translocation of protein kinase C to the nucleus. EMBO J 1991; 10: 3207–3214.
164. O'Connor R, Kauffmann-Zeh A, Lin Y, et al. Identification of domains of the insulin-like growth factor I receptor that are required for protection from apoptosis. Mol Cell Biol 1997; 17: 427–435.
165. Singleton JR, Dixit VM, Feldman EL. Type I insulin-like growth factor receptor activation regulates apoptotic proteins. J Biol Chem 1996; 271: 31791–31794.
166. Guan J-L, Chen H-C. Signal transduction in cell-matrix interactions. Int Rev Cytol 1996; 168: 81–121.
167. McNamee HM, Ingber DE, Schwartz MA. Adhesion to fibronectin stimulates inositol lipid synthesis and enhances PDGF-induced inositol lipid breakdown. J Cell Biol 1993: 121: 673–678.
168. Guadagno TM, Ohtsuka M, Roberts JM, et al. A link between cyclin A expression and adhesion-dependent cell cycle progression. Science 1993; 262: 1572–1575.
169. Symington BE. Fibronectin receptor modulates cyclin-dependent kinase activity. J Biol Chem 1992; 267: 25744–25747.
170. Symington BE. Growth signalling through the α_5 β_1 fibrinogen receptor. Biochem Biophys Res Commun 1995; 208: 126–134.
171. March KL, Wilensky RL, Roeske RN, et al. Effects of thiol protease inhibitors on cell cycle and proliferation of vascular smooth muscle cells in culture. Circ Res 1993; 72: 413–423.
172. Bustamante JO. Nuclear ion channels in cardiac myocytes. Pflügers Arch 1992; 421: 473–485.
173. Wang J, Moreira KM, Campos B, et al. Targeted neutralization of calmodulin in the nucleus blocks DNA synthesis and cell cycle progression. Biochim Biophys Acta 1996; 1313: 223–228.
174. Hidaka H, Sasaki Y, Tanaka T, et al. *N*-(6-Amino-hexyl)-5-chloro-1-naphthalene sulphonamide, a calmodulin antagonist inhibits cell proliferation. Proc Natl Acad Sci USA 1981; 78: 4354–4357.
175. Whitfield JF, Boynton AL, Rixon RH, et al. The control of cell proliferation by calcium, Ca^{2+}-calmodulin, and cyclic AMP. In: Boynton AL, Leffert HL, editors. Control of Animal Cell Proliferation, Vol I. Orlando, Academic Press, 1985: 331–365.
176. Lopéz-Girona A, Bachs O, Agell N. Calmodulin is involved in the induction of DNA polymerases α and δ activities in normal rat kidney cells activated to proliferate. Biochem Biophys Res Commun 1995; 217: 566–574.
177. Lopéz-Girona A, Bosch M, Bachs O, et al. Addition of calmodulin antagonists to NRK cells during G1 inhibits proliferating cell nuclear antigen expression. Cell Calcium 1995; 18: 30–40.
178. Jinno SK, Suto A, Nagata M, et al. Cdc25A is a novel phosphatase functioning early in the cell cycle. EMBO J 1994; 13: 1549–1556.

179. Cao QP, McGrath CA, Baril EF, et al. The 68-kDa calmodulin-binding protein is tightly associated with the multiprotein DNA polymerase α-primase complex in HeLa cells.
180. Reddy GP. Cell cycle: regulatory events in G1→S transition of mammalian cells. J Cell Biochem 1994; 54: 379–386.
181. Reddy GP, Reed WC, Deacon DH, et al. Growth factor modulated calmodulin-binding protein stimulates nuclear DNA synthesis in hemopoietic progenitor cells. Biochemistry 1994; 33: 6605–6610.
182. Subramanyam C, Hona SC, Reed WC, et al. Nuclear localizarion of 68 kDa calmodulin-binding protein is associated with the onset of DNA replication. J Cell Physiol 1990; 144: 423–428.
183. Lopez-Girona A, Bachs O, Agell N. Calmodulin is involved in the induction of DNA polymerases α and δ activities in normal rat kidney cells activated to proliferate. Biochem Biophys Res Commun 1995; 217: 566–574.
184. King KL, Moreira KM, Babcock GF, et al. Temporal inhibition of calmodulin in the nucleus. Biochim Biophys Acta 1998; 1448: 245–253.
185. Chiang Y, Davis RG, Vishwanatha JK. Altered expression of annexin II in human B-cell lymphoma cell lines. Biochim Biophys Acta 1996; 1313: 295–301.
186. Keutzer JC, Hirschhorn RR. The growth-regulated gene 1B6 is identified as the heavy chain of calpactin I. Exp Cell Res 1990; 188: 153–159.
186*a*. Chiang Y, Rizzino A, Sibenaller ZA, et al. Specific down-regulation of annexin II expression in human cells interferes with cell proliferation. Mol Cell Biochem 1999; 199: 139–147.
187. Boynton AL, Kleine LP, Durkin JP, et al. Mediation by calcicalmodulin and cyclic AMP of tumor promoter-induced DNA synthesis in calcium-deprived rat liver cells. In: Boynton AL, McKeehan WL, Whitfield JF, editors. Ions, Cell Proliferation, and Cancer. New York, Academic Press, 1982: 417–431.
188. Crocker G, Dawson RA, Barton CH, et al. An extracellular role for calmodulin-like activity in cell proliferation. Biochem J 1988; 253: 877–884.
189. MacManus JP, Braceland BM, Rixon RH, et al. An increase in calmodulin during growth of normal and cancerous liver in vivo. FEBS Lett 1981; 133: 99–102.
190. Boynton AL, Whitfield JF, Mac Manus JP. Calmodulin stimulates DNA synthesis by rat liver cells. Biochem Biophys Res Commun 1980; 95: 745–749.
191. Jones A, Boynton AL, MacManus JP, et al. Ca-calmodulin mediates the DNA-synthetic response of calcium-deprived liver cells to the tumor promoter TPA. Exp Cell Res 1982; 138: 87–93.
192. Durkin JP, Whitfield JF, MacManus JP. The role of calmodulin in the proliferation of transformed and phenotypically normal *ts*-ASV-infected rat cells. J Cell Physio 1983; 115: 313–319.
193. Baitinger C, Alderton J, Poenie M, et al. Multifunctional Ca^{2+}/calmodulin-dependent protein kinase is necessary for nuclear envelope breakdown. J Cell Biol 1990; 111: 1763–1773.
194. Lindsay HD, Whitaker MJ, Ford CC. Calcium requirements during mitotic cdc2 kinase activation and cyclin degradation in *Xenopus* egg extracts. J Cell Sci 1995; 108: 3557–3568.

195. Silver RB. Calcium, BOBS, QEDS, micodomains and a cellular decision: control of mitotic cell division in sand dollar blastomeres. Cell Calcium 1996; 20: 161–179.
196. Steinhardt RA, Alderton J. Intracellular free calcium rise triggers nuclear envelope breakdown in sea urchin embryos. Nature 1988; 332: 364–366.
197. Twigg J, Patel R, Whitaker M. Translational control of $InsP_3$-induced chromatin condensation during the early cell cycles of sea urchin embryos. Nature 1988; 332: 366–369.
197*a*. Bootman MD, Thomas D, Tovey SC, et al. Nuclear calcium signaling. Cell Mol Life Sci 2000; 57: 371–378.
197*b*. Sun B, Murray NR, Fields AP. A role for nuclear phosphatidylinositol-specific phospholipase C in the G2/M phase transition. J Biol Chem 1997; 272: 26313–26317.
197*c*. Heald R, Weis K. Spindles get the Ran around. Trends Cell Biol 2000; 10: 1–4.
198. Nishimoto T, Ajiro K, Hirata M, et al. The induction of chromosome condensation in *ts*BN2, a temperature-sensitive mutant BHK21, is inhibited by the calmodulin antagonist, W-7. Exp Cell Res 1985; 156: 351–358.
199. Meyer T, Hanson PI, Stryer L, et al. Calmodulin trapping by calcium-calmodulin-dependent protein kinase. Science 1992; 256: 1199–1202.
200. March KL, Wilensky RL, Roeske RW, et al. Effects of thiol protease inhibitors on cell cycle and proliferation of vascular smooth muscle cells in culture. Circ Res 1993; 72: 413–423.
201. Schollmeyer JE. Calpain II involvement in mitosis. Science 1988; 240: 911–913.
201*a*. Fields AP, Thompson LJ. The regulation of mitotic nuclear envelope breakdown: a role for multiple lamin kinases. Prog Cell Cycle Res 1995; 1: 271–286.
201*b*. Murray NR, Fields AP. Phosphatidylglycerol is a physiologic activator of nuclear protein kinase C. J Biol Chem 1998; 273: 11514–11520.
202. Elledge SJ. Cell cycle checkpoints: preventing an identity crisis. Science 1996; 274: 1664–1672.
203. Clarke DJ, Giménez-Abián JF. Checkpoints controlling mitosis. BioEssays 2000; 22: 351–363.
204. King RW, Deshaies RJ, Peters J-M, et al. How proteolysis drives the cell cycle. Science 1996; 274: 1652–1659.
205. Lorca T, Abrieu A, Means A, et al. Ca^{2+} is involved through type II calmodulin-dependent protein kinase in cyclin degradation and exit from metaphase. Biochim Biophys Acta 1994; 1223: 325–332.
206. Osheroff N, Zechiedrich EL. Calcium-promoted DNA cleavage by eukaryotic topoisomerase II: trapping the covalent enzyme-DNA complex in an active form. Biochemistry 1987; 26: 4303–4309.
207. Warburton PE, Earnshaw WC. Untangling the role of DNA topoisomerse II in mitotic chromosome structure and function. BioEssays 1997; 19: 97–99.
208. Stillman B. Cell cycle control of DNA replication. Science 1996; 274: 1659–1664.
209. Sullivan KMC, Busa WB, Wilson KL. Calcium mobilization is required for nuclear vesicle fusion in vitro: implications for membrane traffic and IP_3 receptor function. Cell 1993; 73: 1411–1422.

3
Calcium and Keratinocyte Diffpoptosis

Stem cells and cell stacks

The epidermis is maintained by the proliferation of basement membrane-attached cells in its basal layer. This cell production is balanced by *diffpoptosis*, a differentiation program of a tandemly changing transcriptome, the products of which are ultimately brought together by an apoptosis-like process only after the basal cells have detached from the basal lamina, stopped cycling, and stepwisedly accumulated the products of key Ca^{2+}-activated genes as well as Ca^{2+}-activated enzymes.[1] The keratinocytes are stacked layers of differentiating cells topped by horny-shelled corneocytes.[2–10] In the basal layer, there are clusters or patches of clonogenic stem cells lying on the basement membrane above the dermal papillae.[2–10] The clustered stem cells are tightly linked to each other by homodimers of the transmembrane Delta1 protein and the clusters are surrounded by a network of "transit amplifier" cells.[2–10]

The stem cell has a large nucleus and a small amount of cytoplasm. It is the only indefinitely self-renewable keratinocyte. Normally, it cycles only intermittently in its special niche (established by the combined efforts of many things among which are an adjacent epidermal Langerhans cell, the underlying dermal cells, and basal lamina components) to replenish the supply of transit amplifying cells while replacing itself.[3,9,10] Because of its small cytoplasm and initially limited load of the things needed for the G_1 buildup, it, like the similarly built small lymphocyte,[11,12] takes a long time (35-40 h in culture) to start the accumulation of the pocket protein-inactivating, cycle-starting Cdk4/6·cyclin Ds we met in Chap. 1, and to gear up for DNA replication and mitosis in the first cycle after receiving a "GO!" signal.[9] The stem cell has another peculiarity which is characteristic of tumor cells: it is stimulated to proliferate by the PKCs-stimulating, tumor-promoting TPA (12-*O*-tetradecanoyl phorbol-13 acetate) rather than differentiate as would its cycling progeny.[13,14] The stem cell makes types 5 and 14 keratins and has many proliferation-driving, $\alpha_2\beta_1$and $\alpha_3\beta_1$ integrins and strictly basal, keratin-linked hemidesmosomal $\alpha_6\beta_4$ integrins in focal adhesion complexes on its underside with which it binds itself tightly to the kalinin and laminin proteins in its basement membrane niche.[4–6,15–18] The larger, faster proliferating cells in the bottom layer of the cell stack are the morphologically serrated transit amplifying precursor cells, which express the ECaBP (Epidermal Ca^{2+}-Binding Protein), and they are attached less strongly to the basement membrane because they have many fewer β_1 integrins on their surfaces than the ancestral stem cell.[4,6,15–20] This difference between the surface densities of $\beta1$ integrins and adhesion strengths can be used to locate stem ("bright") cells and transit amplifying ("dull") cells at the epidermal–dermal junction with fluorescently tagged anti-β1 subunit antibodies. The much stronger, integrin-mediated adhesion of stem cells to substrates such as collagen can be exploited to separate them from the transit amplifying cells in very small pieces of residual normal skin and use them to repair burns and other injuries.[5,6,10,11]

The stem cell has a considerable, if not unlimited, proliferative potential that depends at least in part on its expression of telomerase (a reverse RNA transcriptase that protects the chromosomes' ends from progressive shortening at each round of replication by adding to their G-rich, 3′-ends (Fig. 3.1).[21–23] Therefore, it can produce large colonies in culture.[5] An unlimited proliferative potential and the many rounds of its replication that occur over many years demands that the stem cell genome also be protected by the vigilance of the products of the mismatch repair genes that Kinzler and Vogelstein have called the genome's "caretakers."[24] However, the more loosely attached, although bigger and faster cycling, transit amplifying and differentiation-committed cells have only a limited proliferative potential and consequently can form only much smaller colonies in culture.[5] They seem to have some kind of cycle counter which starts out with five on it. After five cycles, the number drops to 0 in the counters of the 32 progeny (even in those of an isolated *cultured* transit amplifier cell), which are programmed to switch off their cell cycle genes, dismantle their cell cycle engines, stop cycling, start stratifying, and set out on the diffpoptotic road leading to the maintenance of the permeability barrier and dead, keratin-packed corneocytes. But these "terminal" cells are not irreversibly "terminal" because they can be "immortalized" (i.e., reacquire a stem cell's proliferative potential) by expressing the adenoviral E1 protein, which can switch off or reset the cycle counter.[25]

As we learned in Chap. 1, normal cells such as the basal cells need the signals from β_1 integrin receptors clustered in focal adhesion complexes of several signaling enzymes and the tyrosine kinase receptors for soluble growth factors such as TGF-α to express c-Myc, make Cdk4/6·cyclin Ds to inactivate pocket proteins and thus activate E2F target genes, which results in surges of cycle-driving Cdk2·cyclin E and Cdk2·cyclin A.[5,11,25*a*] Also as we learned in Chap. 1, the cytoplasmic domains of the β1 subunits of these adhesion-activated integrins are attached to the 59-kDa serine/threonine protein kinase, ILK (Integrin-Linked Kinase) (Fig. 3.1).[26–28] ILK is activated when it contacts PtdIns(3,4,5)P_3 made in the cell membrane from PtdIns(4,5)P_2 by PtdIns-3Ks (type I phosphoinositide-D3 kinase). PtdIns-3K is stimulated when GTP·Ha-Ras binds to its 110-kDa catalytic component or when the phosphotyrosines of integrin-clustered/activated FAKs (Focal Adhesion complex protein tyrosine Kinases) are inserted into the NSH2 and CSH2 pockets of the enzyme's 85-kDa regulatory component.[25*a*,26,28–31]

Signals from the membrane-associated ILKs stimulate the translocation of β-catenin into the nucleus to suppress the expression of the gene for the β-catenin-binding E-cadherin and to stimulate the expression of the genes for cell cycle drivers such as cyclin D1 and c-Myc and, through c-Myc, the cyclin D2 gene and the gene for the cyclin Ds' partner, Cdk4, in the cycle-starting CDKs, Cdk4·cyclin D1 or D2.[31*a*–36] The PtdIns(3,4,5)P_3-activated ILK and PtdIns(3,4,5)P_3-activated PDK(phosphoinositide-dependent protein kinase)-1 kinase give a "double whammy" to the cell by phosphorylating and activating PKBs (protein kinases B) attached to membrane PtdIns(3,4,5)P_3 by their PH (pleckstrin homology) domains, which also stimulate the

Fig. 3.1. The skin and its Ca^{2+} gradient (the values are from human skin and expressed as mmol/kg dry weight). Basal keratinocytes (BC) need a low external Ca^{2+} concentration and signals from β_1 integrins attached to the basal lamina to proliferate. These signals stimulate ILK kinase and prevent the destruction of β-catenin, which stimulates cell cycle genes such as c-*myc* and suppresses E-cadherin expression. The stem cells and their basal transit amplifying progeny avoid apoptosis by making Bcl-2 or Bcl-X_L proteins and suppressing the Bax killer protein. They also express telomerase (Telo), which prevents their chromosomes' telomeres from shortening with each round of replication.

Reducing β1 integrin expression causes the cell to lift off the basal lamina and generate a Ca^{2+} surge that stops proliferation and starts the cell making the things needed to become a corneocyte. The emerging spinous cell (SC) stops expressing Bcl-2, Bcl-X_L, and Telo and starts expressing Bax. It also starts making full-length PTHrP and chops it into fragments. The PTHrPs stimulate the dermal fibroblasts to express and secrete IGF-I and KGF. IGF-I promotes the basal cell survival and proliferation. KGF also promotes basal cell proliferation by inducing the cells to express TGF-α and prevents the spinous cells with their high density of KGF receptors from prematurely initiating apoptosis.

The final stages of differentiation in the granular cells (GC) are driven by signals from the CaRs (Ca^{2+} receptors) activated by a sudden rise in the external Ca^{2+} concentration when the cell nears the head of the transepidermal Ca^{2+} gradient. At the top of the gradient, the CaR signaling in the transitional cells (TC) triggers the apoptosis-like mechanism that converts the cells into the corneocytes of the horny layer (or stratum corneum [C]).

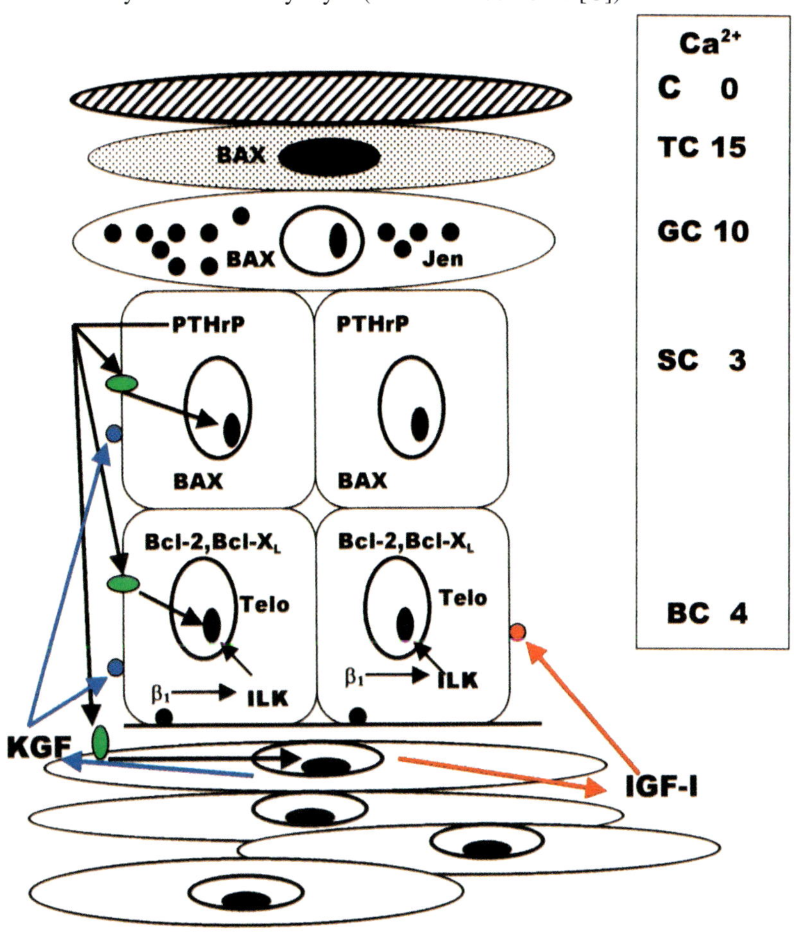

translocation of β-catenin into the nucleus as well as fend off apoptosis (by stimulating Bcl-2 production and inactivating the pro-apoptosis BAD protein) and increase the activity of the E2F transcription factor, which, as we learned in the previous chapter, targets and switches on the genes for key DNA-replicating enzymes.[26,28,37,38]

β-Catenin does two very important, yet very different, things, depending on where the cell is in the epidermal stack. First, it couples the cytoplasmic C-terminal tails of the Ca^{2+}-binding/activated E-cadherin with α-catenin, vinculin, α-actinin, and F-actin to make intracellular cables that tie the suprabasal spinous cells together.[39–41] However, this locking of β-catenin into cables must be stopped by ILK and PDK-1 if basal cells are to proliferate. What is most important for a *cycling* basal cell is β-catenin's role as a *transcription factor* for cycle driver genes. Suppressing E-cadherin expression and cable-making with ILK increases the amount of free β-catenin that can be shipped into the basal cell's nucleus (in a Ran-independent manner)[42] to join an "architectural" transcription factor such as one of the LEFs (lymphocyte enhancer-binding factors) / TCFs, which is bound to specific DNA motifs and has sharply bent the promoters/enhancers of the target genes to bring together widely separated DNA motifs to make a platform for assembling the components of transcription-initiating or inhibiting complexes.[34,39,43] Without β-catenin, LEF/TCF associates with CBP (CREB-binding protein) and the Groucho and CtBP (C-terminal TATA box-binding) co-repressor proteins in a complex that, for example, shuts off the c-*myc* gene.[44,44a] However, when β-catenin surges into the nucleus it replaces Groucho to make a β-catenin·LEF/TCF complex on the promoter DNA, which then assembles an active transcription–initiation complex with RNA polymerase II.[40,43–44a] In anticipation of things to come in the next chapter, colon cells climbing up the crypt wall have been found to make another protein besides E-cadherin that arranges for destruction of free β-catenin rather than just locking it up. This protein is the tumor-suppressing APC (adenomatous polyposis coli).[32,39,40,44–46] It binds to and reconfigures a protein known as *axin*.[44a,46,47] The re-configured axin acts as a "marriage broker" which brings β-catenin and GSK(glycogen synthase kinase)-3β to the same "table."[44a,46,47] GSK-3β then phosphorylates β-catenin, a 'kiss of death' which marks it for ubiquitination and dumping into the proteasome shredder we met in the previous chapter.[44,44a,48] But both ILK itself and the PKB it activates phosphorylate and thus inhibit GSK-3β.[26,38] This saves β-catenin from the shredder and enables it to get into the nucleus to make cycle-driving β-catenin·LEF-1 complexes.[27,44,44a,49]

The β_1 integrins also link to the Shc adaptor protein via caveolin in the caveolar cell membrane pouches to provide a docking site which has been velcroized with phosphorylated tyrosine residues to collect a cluster of SH2/SH3-containing components, such as Grb·mSos complexes, that contribute to the sustained adhesion-dependent signaling that keeps the cell cycling and prevents it from killing itself prematurely with the apoptogenic cascade of caspase proteases discussed in the previous chapter.[25a,49a,50] These signals stimulate the cells: to load their membranes with the key phospholipase-Cγ1 signaling substrate PtdIns(4,5)P_2 needed for signal generation by the growth factor velcroceptors; to make critical stage-specific CDKs; to stimulate the proliferogenic, gene-activating, Ras/Raf-1-triggered, MAP kinase cascade alongside growth factors; to reconfigure the cytoskeleton to enable the cell to respond to growth factors.[51–55] Perhaps surprisingly, the integrins also bind and activate the receptor protein tyrosine kinases for soluble growth factors such as EGF/TGF-α and PDGF.[56] In fact, integrin-induced clustering can activate

PDGF receptors even without their ligand, PDGF, and integrin-induced co-clustering can also activate EGF/TGF-α receptors and enhance the EGF-dependent activation of the MAPK cascade.[56] Of course, these β1 integrin-induced clusterings of the soluble growth factor kinases amplify PtdIns-3K activity and the activity of the all-important β-catenin-sparing, PtdIns(3,4,5)P_3-activated ILK and PKB.

However, the proliferating basal cells must protect themselves from a premature activation of the apoptogenic mechanism by the proliferation-driving c-Myc·Max complexes and signal byproducts, such as reactive oxygen intermediates if there should be any unschedeuled interruption of the stream of survival signals from the integrins and their clustermates. They do this by making anti-apoptosis protector proteins in response to signals from their β_1 integrins. The stem cells make the anti-apoptosis Bcl-2 protector that we met in the two previous chapters, while their progeny may supplement it or replace it with the related Bcl-X_L protector protein.[57,58] They are also smart enough not to express the dangerous, pro-apoptosis Bax protein.[57,58] Indeed, Bcl-2 may be a stem cell marker.[59]

Adhesion-induced clustering of the cytoplasmic domains of integrin β_1 subunits in focal adhesion complexes produces integrin·talin complexes, which in turn form complexes with the pivotal 125-kDa FAK protein tyrosine kinase.[51–55,60] FAK components of the integrin:talin-induced oligomers then velcroize each other by transphosphorylating their Tyr^{397} residues located in their [-Tyr-Ala-Glu-Ileu-] motifs, which are specific docking sites for c-Src or other members of the Src family such as c-Fyn.[52] An approaching c-Src protein tyrosine kinase then shoves a FAK phospho-Tyr^{397} into its SH2 pocket. The now docked and locked Src then phosphorylates, and fully activates, the FAK by phosphorylating Tyr residues in other motifs, such as the [-Tyr-Met-X-Met-] motif that specifically binds the NSH2 and CSH2 pockets of the 85-kDa regulatory subunit of PtdIns-3K, the resulting activity of which is needed to prevent apoptosis and promote proliferation by producing PtdIns(3,4,5)P_3 to stimulate ILK, PDK-1, PKB, PKCs δ, ϵ, η, and ς or to mediate the stimulation of the MAP protein kinase cascade by growth factors and the C-terminal [-Tyr^{925}-Glu-Asn-Val-] motif that specifically binds the SH2 pocket of the Grb-2 adaptor.[25a,26,52,61] Grb2 in turn binds with its SH3 group to the mSos Ras GDP/GTP exchanger protein in the cell membrane. This triggers the replacement of GDP with GTP to produce active membrane-bound GTP·Ras complexes. These Ras complexes can then directly stimulate the neighboring PtdIns-3Ks and trigger the G_1 buildup-driving Raf-1/MEK/MAP kinase cascade. Other signaling enzymes that are directly or indirectly turned on include the following: PLA_2, with a resulting arachidonate surge and production of eicosanoids; PLC-γ1, which breaks PtdIns(4,5)P_2 down into Ca^{2+}-mobilizing Ins(1,4,5)P_3 and PKCs-stimulating diacylglycerols; and the Rho-mediated stimulation of PtdIns(4)P-5 kinase that loads the membrane with PtdIns(4,5)P_2 with which the activated PI-3K makes the ILK and PKB-stimulating PtdIns(3,4,5)P_3.[28,37,51–54,56] A very important upshot of this adhesion-driven cascade is the expression of several genes, one of which is the gene for cyclin A (but not the genes for cyclin D, cyclin E, Cdk1, or Cdk2), which, as we have seen in the previous chapters, is a component of the Cdk2·cyclin A protein kinase that is generated specifically at the end of the G_1 buildup and surges into the nucleus to work throughout the S phase in the replication factories activating replication origins.[55,56,60,62–64] The adhesion signals may also suppress the expression of the growth-arrest-specific gene, *gas*1, and its membrane-associated cycle-suppressing protein product.[65,66] As we shall soon see, switching off the β1 integrin signals streaming into the basal cell triggers the diffpoptosis program.

It must be assumed that the basal keratinocyte's β1-integrin-primed cycle is driven in the same way by the same machinery, cycle clocks, waves of CDKs and Ca^{2+} transients that drive the cycles of other cells we met in the previous chapters. It also includes the continuous expression and actions of the Id(inhibitor of differentiation)-1, -2, -3 proteins that suppress the activation of the diffpoptosis program by blocking the action of bHLH (basic Helix–Loop–Helix) transcription factors.[66a] Signals from the 7-TMD receptors for β-adrenergic agonists and the EGF/TGF-α velcroceptors, a pack of paracrine factors (extracellular matrix-bound bFGF, hepatocyte growth factor/scatter factor, IGF-I, IGF-II, KGF [keratinocyte growth factor, a.k.a. FGF-7]) secreted by dermal fibroblasts, and probably the most important of all, the keratinocytes' autocrine/paracrine TGF-α, drive keratinocyte proliferation by stimulating the parade of stage-specific CDKs and suppressing the expression of the diffpoptosis drivers. However, the effectiveness of these agents depends on the membrane's stores of substrates for receptor protein kinases such as PtdIns(4,5)P_2 and the expression of, among many things, the cyclin A gene by adhesion-activated β1 integrins.[1,51,54,63,64,67,68] The very important TGF-α is active both as the tip of a large transmembrane precursor sticking out of the cell surface, which stimulates proliferation by binding to EGF/TGF-α velcroceptors on adjacent cells, a "*juxtacrine*" activation mechanism, and as the soluble small "paracrine" TGF-α peptide, which when snipped off the precursor diffuses toward the velcroceptors on neighboring cells.[69,70] Keratinocytes also need a small amount of external Ca^{2+}, such as 0.05 mM,[1] to cycle, but as we shall soon see, cycling stops and different parts of the diffpoptosis program start when the external Ca^{2+} concentration rises above a certain level.

The Ca^{2+} gradient

A single, major, if not *the* major, diffpoptosis driver is the striking, steep epidermal Ca^{2+} gradient where the Ca^{2+} level at the granular/horny layer interface is at least 4–5 times higher than in the basal and spinous layers (Fig. 3.1).[71–73] In the human epidermis, the total calcium contents of the basal and spinous zones are about 3–4 mmol/kg dry weight, but the content jumps to about 10 mmol/kg dry weight in the granular layer and finally to 15 mmol/kg dry weight in the lower horny layer (Fig. 3.1).[71] This gradient is an ancient (probably established around 350×10^6 years ago along with the parathyroid glands as two of the sine qua nons for the invasion of dry land) and elegant way of ensuring that the parts of the diffpoptosis program are triggered at the right times and places in the epidermal stack and that the vital permeability barrier is set up and maintained. But how is the gradient loaded at the top when the only source of Ca^{2+} in a dry land-dwelling animal, the blood, is at the *foot* of the gradient?

The gradient develops relatively late in mouse (day 19) and rat (day 20) fetal development as the fetus nears the time when it will need the double protection of the permeability barrier and the horny layer to face the air and insults.[65] It is established along with the permeability barrier in a stratified horny layer.[72] A clue to how it might be done lies in the observations of Menon et al.[74,75] who used an oxalate-pyroantimonate probe to precipitate and localize Ca^{2+}. Ca^{2+} was found in the cytosol, mitochondria, and the nuclei of some basal and spinous cells but *not* in the intercellular spaces in either layer. The heightened total Ca^{2+} level in the granular layer observed by Grundin et al.[71] points to this being the place where *extracellular* Ca^{2+} starts driving diffpoptosis. Indeed, in the experiments of Menon et al.[74,75] *inter*cellular Ca^{2+} first became detectable in the lower granular layer (Ca^{2+} had disappeared from the nuclei and was restricted to the mitochondria [maybe an indication of the start of the apoptogenic megachannel formation in the mitochondrial inner membranes] and to the limiting membranes and discs of the lipid-rich lamellar bodies[76]) and then rose dramatically in the mid-granular layer. This was followed by a surge of intracellular Ca^{2+} in upper granular cells. By contrast, most horny-coated corneocytes had no demonstrable Ca^{2+} — they had pumped out the Ca^{2+} they had accumulated in the upper granular layer before they destroyed their plasma membranes and their Ca^{2+} pumps in the final apoptotic frenzy. This distribution with the highest Ca^{2+} levels in the granular and lower horny layers and low levels in the upper horny layer has also been seen by Vicanova et al.[77] in human epidermis cultured in a serum-free medium with vitamins and lipids, which was optimal for rebuilding the permeability barrier.

Although the rapidly cycling basal keratinocytes are exposed to a very low *external* Ca^{2+} concentration, they have substantial amounts of cytosolic and even nuclear Ca^{2+} (which probably reflects the transport of Ca^{2+} into the nucleus, which we learned in the previous chapter occurs at the end of the G_1 buildup) with which to drive their cycling. They may use their ECaBP to get this Ca^{2+} by sucking up and sequestering Ca^{2+} coming up from the blood and seeping down from the granular and lower horny layers and thus produce the low *external* Ca^{2+} concentration in the basal layer needed for optimal integrin activity (integrin receptors do need a small amount of Ca^{2+} to assume their preactivation configurations, but it is the high-Ca^{2+}-inhibitable action of Mg^{2+} and (or) Mn^{2+} that is then needed for the receptors' active configurations[78]). At the same time, there is a large enough internal Ca^{2+} store to produce the Ca^{2+} spikes and oscillations that trigger the major events of the cell cycle as well as enough external Ca^{2+} to enable the cell to progress through the brief, possibly Ca^{2+}·CaR-dependent, critical later stage of the G_1 buildup.[1,79] The accumulated Ca^{2+} is ultimately expelled from upper granular cells' internal stores by signals from the CaRs (Ca^{2+}-responsive polyvalent cation receptors)[80–84] to trigger the cornifying apoptosis, and all of it is then pumped out of the developing corneocytes by Ca^{2+}·CaM-activated membrane pumps before the cell membrane is destroyed and replaced by the horny envelope. This prevents a potentially steady and massive leakage of Ca^{2+} from the body in shedding corneocytes. Thus, all of the Ca^{2+} collected by the cells during their journey from the bottom to the top of the stack is dumped out to trickle down into the intercellular spaces of the granular layer and activate the Ca^{2+}-receptors on the upcoming granular cells as it goes (Fig. 3.2). This recycling Ca^{2+} is, therefore, a *downwardly diffusing diffpoptosis driver* (Fig. 3.2).

What triggers diffpoptosis?

It appears that the process starts with the progressive loss of surface β1 integrins, the reduction of ILK and PKB activities that keep the transit amplifiers cycling as they migrate or are pushed along the basement membrane away from the stem cell niche.[6,15,16] Indeed, when the volume of integrin/ILK and PKB signaling drops below a critical level, Ca^{2+} spills out of internal stores, which probably triggers the influx of external Ca^{2+} through SOCs (Stores-Operated non-specific Na^+/Ca^{2+} Channels[83]). The resulting

Fig. 3.2. How could the epidermis make the top-down Ca^{2+} gradient when the Ca^{2+} must come from dermal blood vessels (BV) at the bottom of the gradient? One possibility is this "*load-up-then-dump-out*" mechanism in which the cell accumulates and locks up Ca^{2+} to maintain a low intercellular concentration while moving from the basal layer through the spinous (SPIN) layer into the lower granular (GRAN) layer. Midway into the granular layer, the cell's CaRs are activated (as indicated by the sparks from a representative CaR) by a dramatically increased intercellular Ca^{2+} concentration. This signaling produces intracellular Ca^{2+} surges that drive the conversion of the cell into a keratin-packed corpse. But before it dies, the cell uses its Ca^{2+}-activated pumps to dump the accumulated Ca^{2+} into the intercellular space to maintain the head of the gradient.

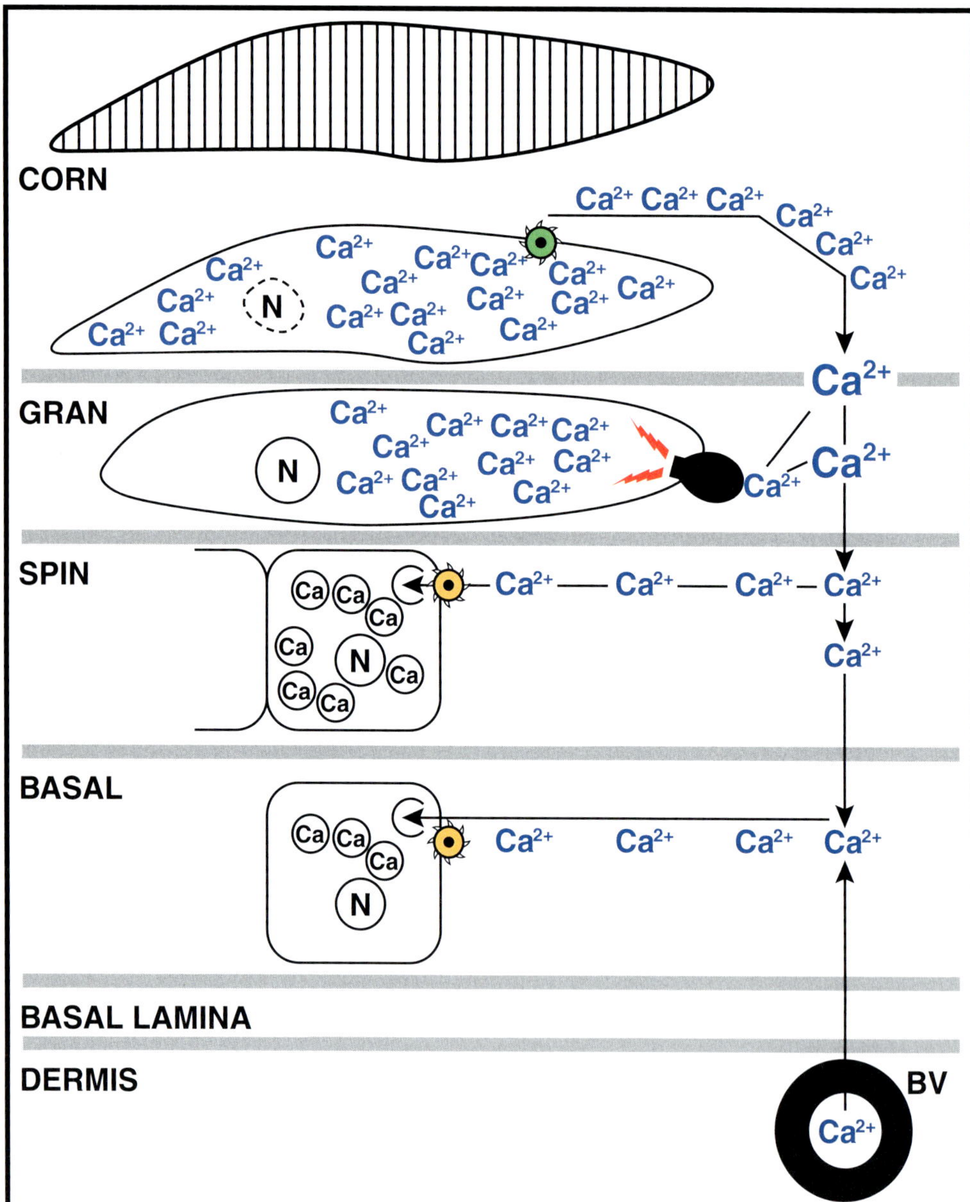

internal Ca^{2+} surge (it must be understood that the external Ca^{2+} concentration even in the basal layer is much higher than the nanomolar free Ca^{2+} level inside the cell) shuts down the production of the Id-1, -2, and –3 diffpoptosis suppressors and stimulates the expression of the gene coding for the spinous type 1 keratin, which, in turn, starts the production of the type 10 keratin when it reaches a critical level. This scenario is suggested by the observation of Li et al.[66a,84] that suspending primary BALB/c mouse keratinocytes *in a low-Ca^{2+} (0.5 mM) medium* triggered internal Ca^{2+} surges resulting from the emptying of Ca^{2+} stores plus a flow of Ca^{2+} into the cell. These suspension-triggered Ca^{2+} surges stopped proliferation and triggered the expression of the keratin 1 gene (which has a Ca^{2+}-responsive enhancer), the keratin 10 gene, the loricrin gene, the filaggrin gene, and the accumulation of their products despite an overall *reduction* of protein synthesis.[85] Ca^{2+} shock (from raising the external Ca^{2+} concentration from 0.05 to 1.4 mM) in mouse keratinocytes suppresses TGF-β1 expression and stimulates TGF-β2 expression and secretion,[86] but detachment stimulates TGF-β1 expression and secretion by human keratinocytes.[87] Thus, a detachment-triggered Ca^{2+} surge stimulates the production and secretion of autocrine/paracrine cycle-stopping TGF-β1 or TGF-β2 and other diffpoptosis-related proteins. The detachment-triggered signal might also activate the *gas*1 gene and the accumulation at the cell membrane of its cycle-suppressing protein product as occurs in colon cells.[66] However, the downregulation of ILK expression and the silencing of the proliferogenic, apoptosis-preventing FAK, ILK, and PKB kinase signaling should be enough, and is indeed enough, to stop cycling by preventing the appearance of cyclin A and the DNA replication-triggering Cdk-2·cyclin A protein kinase complexes at the end of the G_1 buildup.[88] Thus, for example, detaching murine keratinocytes from their substrate drops the cyclin A level, which as would be expected from the role of Cdk2·cyclin A in driving DNA replication stops the cells in S-phase.[88] The resulting detachment-triggered Ca^{2+} surge might also reduce the anti-apoptotic Ras activity induced by the proliferogenic signals from the integrin-receptor clusters.[89]

Ca^{2+} and diffpoptosis in the unreal world of the culture dish

Where and how Ca^{2+} drives diffpoptosis in the real world of the skin are hard to see in the large body of information from experiments on cultured primary keratinocytes or keratinocytes of established lines in the unreal world of the culture dish. The differentiation program in cultured primary human foreskin keratinocytes can be triggered either by shocking the cells with a large increase in the external Ca^{2+} concentration or by inducing Ca^{2+} signals by cell–cell contact or detachment from the culture dish. Thus, raising the Ca^{2+} concentration in the medium removes a major barrier to diffpoptosis by suppressing the expression of the anti-apoptosis, Bcl-X_L protein, and stimulating the production of the pro-apoptosis Bax protein.[58] Cells grown in a low-Ca^{2+} (0.07 mM) medium start making the horny envelope constituent involucrin (from the Latin involucrum, envelope) and envelope-making transglutaminases to get ready for making a cornified envelope only one week *after* they start touching each other and stop proliferating when their culture becomes confluent.[90] By contrast, cells cultured in a shock medium containing 1.2 or 2.4 mM Ca^{2+} make involucrin and transglutaminases *before* confluence. This enables them to start making, or trying to make, horny envelopes without prior contact. In other words, the Ca^{2+} shock signals were enough to switch on the appropriate genes to start and then drive differentiation.[90]

In most experiments on cultured cells, the Ca^{2+} concentration in the medium is simply raised from a proliferation-supporting 0.05 to 1.0 or 1.2 mM or higher. This Ca^{2+} shock triggers an internal Ca^{2+} surge and a highly compressed (or "zipped" to use a computer term) set of events, ranging from proliferative shutdown to cornified envelope formation.[1,84,85,91,92] While a Ca^{2+} shock can trigger diffpoptosis in cultured cells, the preparations for cornification and barrier maintenance in the real world of the skin would have to be started at or near the basal→spinous transition by internal Ca^{2+} surges triggered by a drop in adhesion-induced integrin-orchestrated signaling in the presence of a low external Ca^{2+} concentration simply because the Ca^{2+} concentration in the gradient does not start rising until the cell is halfway through the granular layer (Figs. 3.1 and 3.2).[71,74,75,77]

There is evidence that the different stages of diffpoptosis can be selectively triggered or suppressed by different levels of Ca^{2+} shock (Fig. 3.3). Thus, the shock given by raising the Ca^{2+} concentration in the medium from 0.05 to 0.12 mM causes mouse keratinocytes to switch off their cell cycle genes and the genes coding for the small, cycling-compatible keratins 5 and 14 and to switch on the genes for the larger keratins 1 and 10 and other events such as surface display of the Ca^{2+}-configured/activated E-cadherins (at the expense of the β1- integrins) and associated desmosome assembly responsible for the cell–cell-adhesion and desmosome-attached heterodimeric keratin 1/10 tonofibrillar bundles that are the morphological hallmarks of spinous cells (Fig. 3.3).[1,15,92,93] The stronger shock from raising the Ca^{2+} concentration in the medium to 0.3 or 1.2 mM, or triggering an internal Ca^{2+} surge by emptying internal Ca^{2+} storage vesicles by exposing

Fig. 3.3. Parts of the diffpoptosis program in cultured keratinoctyes can be selectively stimulated or stopped by raising the Ca^{2+} concentration in the culture medium by certain amounts. While these in vitro findings cannot be directly extended to the skin, they may be models for, and thus mimic, the responses to a rising intensity of Ca^{2+} signaling in vivo, first by detachment from the basal lamina, then by autocrine and paracrine factors produced in the spinous layer, and finally by Ca^{2+} receptors (CaRs) in the granular and lower corneum layers, which trigger the stepwise initiation of events and inhibition of preceding events leading from basal cell to corneocyte.

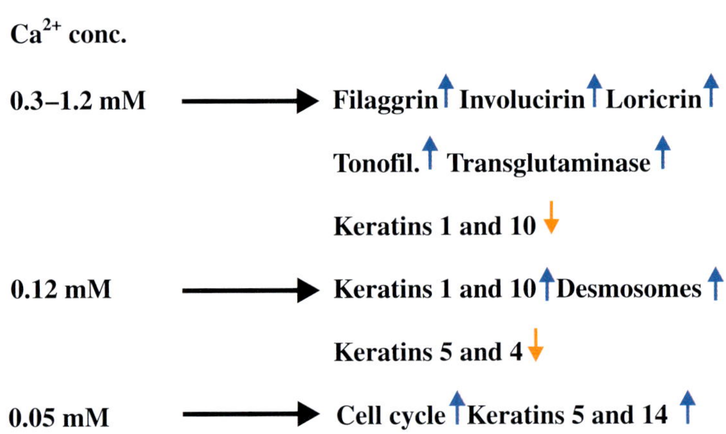

cells to an inhibitor of the storage vesicles' inwardly directed Ca^{2+} pump, thapsigargin, shuts down the keratins 1 and 10 genes and stimulates the expression of the genes of the granular cell/corneocyte parts of the diffpoptosis program that code for filaggrin, involucrin, loricrin, and the keratinocyte-specific TG_K (the keratinocyte-specific transglutaminase) (Fig. 3.3).[85] Raising the Ca^{2+} concentration from 0.05 to 1.0 mM causes the cells to bypass earlier stages and directly become partially cornified corpses. The need for a Ca^{2+} gradient explains why keratinocytes stratify and differentiate better in floating, air-exposed cultures (liquid medium/air interface similar to skin) than in submerged cultures.[1,10,90] The floating cultures can establish the physiological, or near-physiological, Ca^{2+} gradient and permeability barrier. However, in submerged cultures the Ca^{2+} concentration would equalize with the concentration in the medium. From these observations on isolated cells, it seems that epidermal diffpoptosis can be driven stepwise by superficially similar (although undoubtedly both temporally and spatially very different) Ca^{2+} signals from different receptors to different targets appearing at different stages of the program.

Ca^{2+} and diffpoptosis in the real world of the skin

Are these observations on cultured keratinocytes relevant to keratinocyte differentiation in the real world of the skin? All they tell us is that a Ca^{2+} shock can stop keratinocytes from proliferating and can differentiate to a varying extent, depending on the strength of the shock. However, there are some striking differences between the ultimate products of differentiation in cultures and skin, one of which, for example, is an inability of Ca^{2+}-shocked cultured keratinocytes to make the rigid envelopes of the mature horny cells in the skin.[94] So let us try to decompress or "unzip" the responses of the shocked cells and put them together with observations on keratinocytes in the skin to try to see how a newborn transit amplifying keratinocyte can produce a pile of water-repellant, lipid-coated horny cadavers. From the outset, we must keep in mind that during this diffpoptotic journey specific mRNA transcripts appear when the cell enters a layer only to vanish when the cell enters the next layer/stage, but their many protein products stay on and accumulate to collaborate with the next batch of proteins to ultimately produce corneocytes and maintain an effective permeability barrier. In other words, pulses of stage-specific mRNAs drive the buildup of the many things needed to make a corneocyte.

The diffpoptosis program starts on the border of a stem cell cluster where cells have lowered their expression of the transmembrane Delta1 protein that binds the stem cells together in the body of the cluster.[10,94a] The signal to become a "transit amplifying" cell is given to the border cell when its transmembrane Notch receptors bind to Delta1 proteins on neighboring cells of the "transit amplifying" network.[10,94a] As a result of the Delta1–Notch signal, the cells loses its "immortality": its cycle counter (registering 5) has been switched on. Its cycling accelerates, but it has lost the ability to produce large clones, and its descendants will stop cycling when their cycle counters register 0. Armed against apoptosis by the PKB-promoted expression of the anti-apoptosis Bcl-2 and $Bcl\text{-}X_L$ proteins and the inhibition by PKB of the pro-apoptotic ICE-like proteases[37,58], it is still tightly bound to the basal lamina by its β1 integrins which densely decorate its surface.[6] The decision to start a new cycle is governed by the number on its cycle counter, the volume of flow of agents through gap-junctions (intercellular channels made of hexamers of connexin-43) connecting its interior to those of overlying spinous and adjacent basal cells[95–102] and by various growth factors such as IGF-I and KGF secreted by underlying PTHrP-stimulated dermal fibroblasts. KGF also causes the cell to make the autocrine proliferogen TGF-α.[103] (However, we shall learn further on that it is the spinous cell and *not* the cycling basal cell that has the greatest density of KGF receptors on its surface.) Transit amplifying cells pick up other factors from the underlying dermal cells in micropinocytic vesicles formed between the hemidesmosomes of their serrated basal surfaces.[104]

If the decision is "GO!," the cell will continue the cycle countdown by starting the G_1 buildup to DNA replication and mitosis. It will of necessity express the Id-1, -2, and -3 diffpoptosis-suppressors and the master replication-related transcription factors such as c-Myc with its vast array (as many as 198) of target genes such as those for CAK, cyclin D1, and E2F1, which results in the production and activation of Cdk4·cyclin D1 that starts the G_1 buildup by inactivating the Rb-family "pocket proteins," which releases E2F1 from E2F·pocket protein complexes to start the cresting and crashing waves of CDKs that drive the expression of replication-related genes.[35,66*a*,104*a*–106] Of course, it will generate the Ca^{2+} spikes that start the G_1 buildup and trigger DNA replication, mitotic prophase, and finally the metaphase→anaphase transition. Besides their ILK-, PKB-, and β-catenin-mediated upregulation of c-*myc* and other cycle-driving genes, the signals from the β1 integrins stimulate the cell to express the Bcl-2 and Bcl-X_L proteins and suppress the expression of pro-apoptosis Bax protein to prevent it from prematurely killing itself through the actions of reactive oxygen intermediates and mitochondrial Ca^{2+} cycling generated by growth factor receptor activities and the apoptogenic arm of c-Myc action.[1,35,37,49,55,107–111] Near the end of the G_1 buildup, the cell starts its replication-initiating Ca^{2+} oscillations.[1,79] The integrin and IGF-I signals stimulate the expression of the cyclin A gene.[55,62–64] Cyclin A now downregulates and replaces cyclin E to form the Cdk2·cyclin A protein kinases that will terminate the Cdk2·cyclin E/E2F·DP-1-driven expression of the genes for the DNA-replicating enzymes and at the same time activate DNA-replication origins and start and then maintain the sequential firing of the replicon batteries during the DNA-synthetic S phase.[112,113] After replicating its chromosomes, the cell will get rid of its Cdk2·cyclin A protein kinases and make the Cdk1·cyclin A and Cdk1·cyclin B protein kinases that, together with Ca^{2+} transients, bursts of phosphatase activity, and help from the proteasome shredder, will drive the cells into and through mitosis.

The first step: from basal cell to spinous cell

Depending on the balance of the "GO!" and "NO GO!" signals, the new cells will start new cycles and move, or be pushed, further away from the ancestral cluster on the basal lamina. As they do so, they express the ECaBP and start to reduce their complement of β1 integrins, which progressively weakens their attachment to the basal lamina and reduces the activities of PtdIns-3K, ILK, and PKB.[4–6,20,21,26,37] Along with this, the cells stop making the transmembrane Delta1 protein and thus stop the signaling from the Delta1·Notch complexes that triggered the transition from stem cells to transient amplifying cells.[94*a*] This fading of signaling from the adhesion complexes and Delta1·Notch creates a subpopulation of basal cells that have begun changing their transcriptome for next stage of the program as indicated by the appearance of keratins K1 and K10.[4] Important factors in the maintenance of their proliferative activity may be their expression of TGF-β1, which specifically stimulates the expression of the basal cell-specific keratins 5 and 14,[114] a low-level production of the inhibitor, TGF-β2, and its retention in the cytoplasm.[115,116]

When the signaling from β1-integrins drops below a critical level, probably before the cell has actually lifted from the basement membrane, the tyrosine kinase receptors / FAK / Ras-driven PtdIns-3K activity drops, PtdIns(3,4,5)P_3-dependent ILK and PKB signaling stops, and there is an internal Ca^{2+} surge resulting from the discharge of internal stores and a coupled influx of Ca^{2+} through SOC/CRAC cation channels in the cell membrane.[30,31,84] *This is likely what silences the Id-1, -2, -3 protein diffpoptosis suppressors, starts the diffpoptosis program, and sets the cell on the road to the permeability barrier, the head of the Ca^{2+} gradient, and ultimately a horny death.*

Before the transcriptome can shift to start the diffpoptosis program, the cell cycle-driving Cdk·cyclins are silenced by members of the Cip/Kip family of CKIs (cyclin-dependent kinase inhibitors). As we learned in the previous chapters, these CDK inhibitors stop the cell cycling in several different ways: by inhibiting the Cdk2·cylin E protein kinase that otherwise would stimulate the expression of the replication enzymes; by inhibiting the Cdk2·cyclin A protein kinase that would activate batteries of origin-attached DNA replication complexes throughout the S-phase; and by inhibiting PCNA, the accessory DNA strand-clamp for the replication factories' DNA polymerase-δ. The first of these CKIs to go into action is the multifunctional $p21^{Cip1/WAF1}$. Things start with the p53(TP53)-*independent* (as in intestine[117] p53(TP53) which normally stimulates p21 expression actually *crashes*), p300 protein-mediated, activation of the $p21^{Cip1/WAF1}$ gene's Ca^{2+}-responsive promoter.[118–121] p21's job is to knock the Cdk·cyclins out of action, which prevents the inactivation (by hyperphosphorylation) of the pRb cycle suppressor, and stops proliferation. As we shall also see in the intestine,[117] p21 is not involved in starting diffpoptosis, but it must be removed of as soon as it has done its job, otherwise it would block the expression of at least one of the major diffpoptosis-related genes, involucrin, possibly through the interaction of p21·CDK complexes with the gene's promoter.[119] Therefore, after the CDKs have been silenced, and cycling has stopped, p21 is dropped into the ubiquitin-proteasome shredder to allow diffpoptosis to start.[119] Then, p57 and another multifunctional member of the Cip/Kip family of CKIs, $p27^{Kip1}$, surge upwards.[120–122] But $p27^{Kip1}$'s job is very different from that of p21. Despite being a CKI, it is not involved in cycle-stopping (p21 has already done this), but instead it is needed to start the diffpoptosis program.[122]

Detachment-triggered Ca^{2+} signals also stimulate the expression and secretion of TGF-β1 in human keratinocytes,[87] but probably suppress TGF-β1 expression and secretion and stimulate the synthesis and secretion of TGF-β2 in mouse skin.[86,114] A surge of TGF-β1 or β2 could contribute further to the proliferative shutdown by stimulating the later expression of the Cdk 4/6-specific inhibitor, $p15^{INK\,4b}$, which displaces another inhibitor, $p27^{Kip1}$ from Cdk4 and Cdk6 that can then bind to Cdk2 and block the G_1 buildup.[118,123,124] We might expect that the detachment signal would stop c-*myc* gene expression, but surprisingly it seems that it does not.[125–127] Indeed, a continued expression of c-*myc* in a cell expressing its death machinery without other G_1 drivers to provide survival signals could start prematurely the apoptotic arm of diffpoptosis.[109] Moreover, detachment alone is dangerous and would trigger apoptosis in any normal keratinocyte unprotected by Bcl-2, Bcl-X_L, or an activated mutant Ha-Ras protein.[30]

Daughters of the penultimate descendants of a transit amplifying cell will probably not load their DNA replication origins with ORC·Cdc45·MCMs pre-initiation complexes that have been licensed for replication by RLF-B: they have lost their license for replication. Also, as we learned in Chap. 1, the rapidly cycling transit amplifying cells would have DNA-synthetic complexes known as synthesomes (the operational parts of the huge nuclear matrix-bound chromosome replication factories), which they retain even if they should temporarily stop cycling.[128,129] This complex contains CaM, DNA polymerases α and δ, primase, replication factor C (with its Ca^{2+}-binding 36-kDa annexin II subunit), replication protein A, helicases, poly(ADPribose) polymerase, PCNA (proliferating cell nuclear antigen), DNA ligase, and topoisomerases I and II.[128,129] However, if a keratinocyte is like other terminally differentiating cells, one of the first things it does when it starts diffpoptosis is to stop making parts for the replication factories, such as DNA polymerase-α, replication factor C, and PCNA, and get rid of existing factory parts.[129] Along with this is a shutdown of the expression of telomerase that protected the chromosomal ends from end-erosion by replication over the generations from the ancestral stem cell (Fig. 3.1).[21–23]

Although basal keratinocytes have functional death machinery, their high levels of Bcl-2 and Bcl-X_L (induced by ILK-activated PKBs and the PKB's inhibition of pro-apoptotic ICE-like proteases) keep it switched off (Figs. 3.1 and 3.2).[37,56] Progressive loosening of attachment, the fading of adhesion-induced signals, the loss of the Ha-Ras/PtdIns-3K suppression of apoptosis, and the resulting Ca^{2+} surge could probably lift the Bcl-2/Bcl-X_L-imposed barrier to the mobilization of the protease components of the death machinery, such as the mitochondrial-derived caspase (ICE [pro-interleukin 1β converting enzyme] cysteine protease[s]), as does the detachment of mammary epithelial cells.[56,130] Detachment also downregulates Bcl-2 expression and upregulates Bax protein expression so that the death machinery can eventually be switched on by the loudly signaling CaRs (Ca^{2+}-activated divalent cation receptors) when the cell reaches the head of the Ca^{2+} gradient in the upper granular layer.[58,131–135] Another possible consequence of the loss of β1-adhesion complex signaling and the lack of CDK activity is the accumulation of the hypophosphorylated Rb protein, which has been found to trigger apoptosis in some cells.[56] This suicidal detachment-triggered apoptosis is known as *anoikis* (from the Greek word for homelessness, hence *homelessness-induced programmed cell death*), which we first met in Chap. 1.[136]

Another reason for anoikis might be the detachment of DCCs, receptors belonging to the immunoglobulin superfamily that were first found to be deleted in the later stages of colon carcinogenesis and subsequently found to be expressed in navigating neurons and the basal layers of many epithelia including skin.[137] Neurons use their DCC receptors to bind to the netrin landmarks that guide their growth cones to their targets.[137] Disconnecting DCC receptors from their binding sites in the extracellular matrix triggers aniokis in normal, but not in neoplastic, cells.[138]

Clearly, detachment from the basal lamina and a changing transcriptome usher in a perilous period for the seemingly defenseless new spinous cell. However, the cell defends itself from anoikis by greatly increasing its complement of KGF receptors and making PTHrP (see below) to stimulate the underlying dermal fibroblasts to make KGF.[139] It is this KGF from the fibroblasts that takes over from the departing Bcl-2, Bcl-X_L, and the Ha-Ras/PtdIns(3,4,5)P_3/PDK-1-stimulated PKB to suppress the premature onset of such terminal diffpoptotic events as the activation of membrane-associated transglutaminase, horny envelope construction, and chromosome chopping during the spinous stage.[32,121]

The detachment-triggered Ca^{2+} signal stimulates the expression and shipment into the nucleus of transcription factors for the first set of differentiation-driving genes. For example, the signal stimulates the expression of the gene for the membrane-anchored, 22.5-kDa, S100-like, Ca^{2+}-binding, CLED (Calcium-Linked Epithelial Differentiation) protein.[139a] In another example, a cycling basal cell has a lot of C/EBPβ (a bZIP-family CCAAT enhancer-binding protein) and some CHOP9C/EBP-homologous protein factors in its cytoplasm but little in its nucleus.[140] However, the detachment-triggered signal causes C/EBPβ and CHOP to move into the nucleus.[140] This signal stimulates through the gene's Ca^{2+}-responsive promoter, the first of the differentiation events, the expression of the spinous keratin-1 gene (located in the cluster of type II keratin genes in human chromosome 12q13[141]).[142,143] When keratin-1 reaches a critical level, it triggers the expression of keratin 10, the gene for which has been primed by the incoming C/EBPβ binding to a specific site in the gene's 5′-regulatory region.[1,140,143a] The Ca^{2+} signal may also enhance C/EBPβ's transcriptional activity by phosphorylating the protein with a Ca^{2+}·CaM-activated protein kinase such as CaMKIV.[144,145] The cell stops making the Ca^{2+}-transporting ECaBP protein, which will not be needed when the cell reaches the upper levels of the Ca^{2+}gradient.[19] At this point, the cell is still expressing β1 integrins but the intracellular transport of newly synthesized β1 integrins to the cell surface for installation in the membrane stops as the new proteins are held at the endoplasmic reticulum by the reticulum's Ca^{2+}-binding protein, calnexin; and β1 receptors already on the surface are endocytosed and delivered to the lysosomes for destruction.[146,147]

At the same time, detachment and the internal Ca^{2+} surge do the key things for starting diffpoptosis. When the cell was attached to the basal lamina, the β-catenin·LEF-1 transcription complexes generated by its β1-integrin-activated ILK suppressed the expression of the E-cadherin gene, which has an inhibitory LEF-1-binding motif in its regulatory region.[40,148] The detachment of the cell and the consequent silencing of ILK and the PKBs results in the dephosphorylation and activation of GSK-3β. GSK-3β then phosphorylates free β-catenin, which sends it to the proteasome shredder via ubiquitination. This destruction of β-catenin shuts down key cycle-driving genes and turns on the E-cadherin gene.[7,26–29,32,36,37,39,43–49] The emerging cadherins now suppress the expression of the β1 integrins and ILK.

The Ca^{2+} signal causes the protein tyrosine kinase Fyn (or in its absence Src) to move to the cell membrane where it promotes the formation of E-cadherin–F-actin cables by phosphorylating the E-cadherins and the β-catenin and γ-catenin (plakoglobin) linkers.[149] Outside the cell, the emerging cadherins, shifting into their optimal adhesive configuration when their five Ca^{2+}-binding motifs (the "cadherin repeats") meet the relatively high (with respect to internal) external Ca^{2+} concentration, cause the lifting cell to bind to the E-cadherins protruding from its neighbors' surfaces.[146,150–152] If keratinocyte E-cadherins behave like those of EMT/6 mammary carcinoma cells, they may add yet another cycle-suppressor by increasing the activity of the protein kinase inhibitor, $p27^{Kip}$, which reduces cyclin D1 production and thus "activates" Rb by preventing its hyperphosphorylation.[153] However, Alani et al.[118] have found that Cdk4 complexes do not change during the differentiation of human keratinocytes in the culture dish. But most important for our Ca^{2+} story is the fact that it is at this point in the diffpoptosis program that *external* Ca^{2+} becomes a major player in the shutting down and dismantling of the cell cycle engine because the Ca^{2+}-driven interlinking of the emerging E-cadherins to those on adjacent cells triggers the locking of β-catenin into intracelluar E-cadherin·α-catenin·β-catenin·vinculin·α-actinin·F-actin cables, which, of course, keeps the β-catenin from getting into the nucleus and driving cell-cycle gene expression.

The cell now becomes a keratin-making machine that produces larger differentiation-compatible keratins such as the spinous-specific types 1 and 10 that will make up more than 85% of the total protein of the ultimate horny cadaver.[93,154] The keratins 1 and 10 combine into 1·10 heterodimers, which form a network by building onto the preexisting network of keratin 5·14 heterodimers left over from the proliferating basal stage.[154–156] The Ca^{2+}-dependent interlinking of cells by the emerging cadherins sends a signal to stimulate the expression of new proteins such as desmoglein, a cadherin that is specifically associated with desmosomes, which are the cellular "pop-rivets" that attach the fledgling spinous cell to its neighbors.[1,40,150,152,157] Each desmosome consists of cell–cell-interlinking extracellular domains of the desmosome-specific cadherins (the 100–120-kDa desmocollins and 130–150-kDa desmoglein) whose cytoplasmic tails in turn bind to a dense cytoplasmic plaque (consisting of a core of 240-kDa desmoplakin I, 215-kDa desmoplakin II, and 82-kDa plakoglobin bordered by the huge 680-kDa desmoyokin) into which the keratin 1·10 heterodimers loop and bind to specific plaque proteins to form what is effectively a resilient supracellular structure that interlinks the cells of the epithelial sheet to provide great flexibility combined with a strong resistance to rupture.[40,157] The cell starts making the full-length, 106-kDa, keratinocyte-specific transglutaminase (TG_K) and the 68-kDa rod-shaped involucrin, both of which it sends to the plasma membrane for use much later to make the horny envelope.[94,158–162]

The promoter of the involucrin gene on chromosome 1q21-22 in humans, like the Ca^{2+}-responsive promoter of the gene for the cycle-stopping $p21^{Cip/WAF1}$, is co-activated by p300.[120] The human involucrin gene has a Ca^{2+}-responsive, 5′-regulatory element between nucleotides –2456 and –1272 and a PKCs-responsive, AP-1 transcription factor-stimulated regulatory element between nucleotides –159 and –1.[163] Between these two promoter regions, there is a –651 to –160 transcription silencing region that interacts with a spinous-specific Fra-1(or -2)·Jun D AP-1 heterodimer as well as the YY1 transcription silencer.[163,163a] It would be reasonable to guess that YY1 attached to the silencer sequence represses the expression of involucrin in proliferating keratinocytes, but the silencer is inactivated when cell cycling stops and the detachment-triggered Ca^{2+}/PKCs/AP-1(Fra-1(or -2)·Jun D) signal stimulates the gene's Ca^{2+}-responsive regulatory region. The cell also starts making the small proline-rich SPRR1 protein (the human homolog of cornifin in rabbits), one of a family of 12 small proteins which, when the time comes, will cross-link the various larger proteins of the cornified envelope.[160,161] The SPRR1 (cornifin) gene, like the involucrin gene, is another member of the EDC (epidermal differentiation complex), a remarkable cluster of genes on the q21-22 region of human chromosome 1 (Fig. 3.4).[164–166]

Fig. 3.4. The epidermal differentiation gene complex (EDC) on human chromosome 1q21-22. The members of this gene cluster can be divided into three groups: genes coding for cornified envelope constituents (IVL, involucrin; LOR, loricrin; SPRR, small proline-rich proteins); genes coding for macrofilament-producing keratin intermediate filament cross-linkers (FLG, profilaggrin; THH, trichohyalin); and genes coding for the Ca^{2+}-binding S100 proteins. Their tight clustering has enabled the profilaggrin and trichohyalin genes to fuse with the neighboring S100 genes and to each acquire two S100-like Ca^{2+}-binding motifs in their regulatory elements and with them a responsiveness to Ca^{2+} signals from the granular cell's divalent cation receptors. This figure from Whitfield.[1]

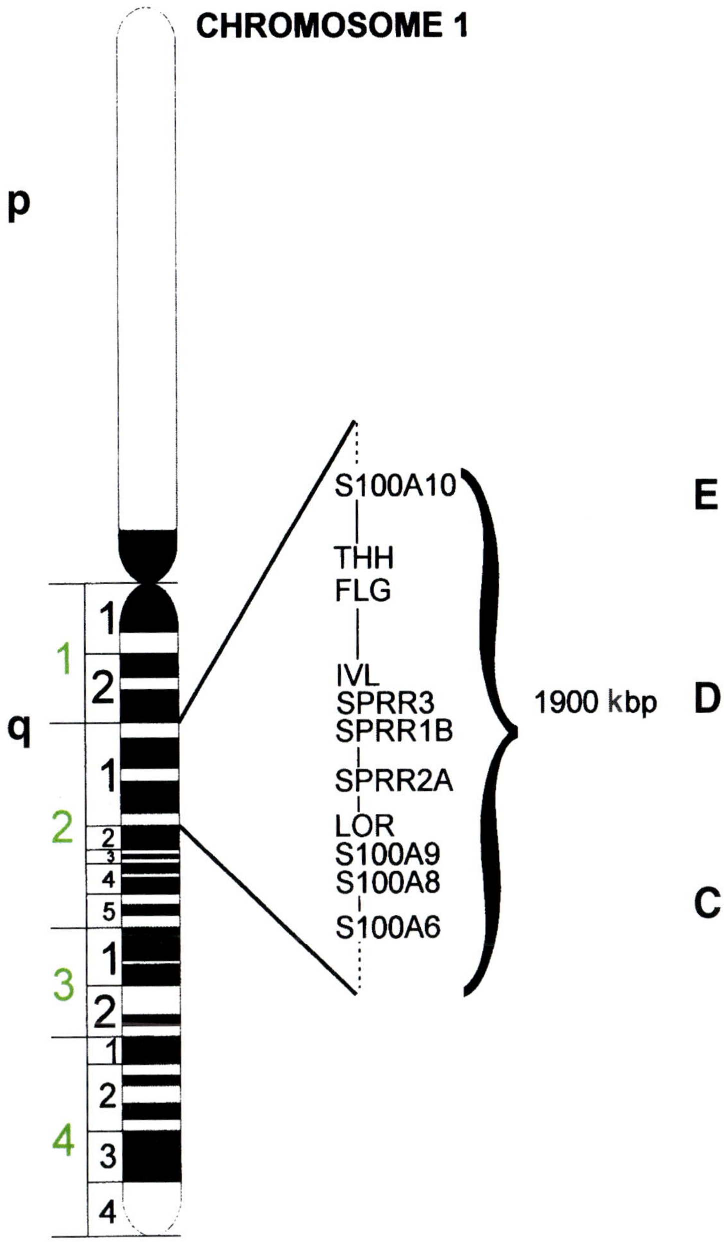

The members of this gene cluster encode involucrin, loricrin, profilaggrin, SPRR1B, SPRR2A, SPRR3, S100A6, S100A8, S100A9, S100A10, and trichohyalin (Fig. 3.4).[164–166] Their products can be divided into three groups: the cornified envelope components involucrin, loricrin, and the SPRR proteins; the macrofilament-making keratin intermediate filament cross-linkers, profilaggrin and trichohylalin; and the S100 Ca^{2+}- binding regulators. This intimate mingling of cornified envelope-making genes with the S100 genes in human and other animals (Fig. 3.4) has enabled two of the key players in keratinocyte diffpoptosis, profilaggrin and trichohyalin (which are next to S100A10 and S100A11), to borrow S100 coding sequences and thus become responsive to Ca^{2+} signals. Both of these proteins have acquired two N-terminal S100 Ca^{2+}-binding motifs, which are followed by tandem repeats characteristic of the cornified envelope proteins.[165–170] Although tightly clustered on chromosome 1, the EDC genes are not coordinately regulated: they are turned on at different points in the diffpoptotic program, and some, such as the SPRR genes, are controlled by a different set of transcription factors and are expressed in different regions of the skin.[164]

The human spinous cell also starts making the protease-inhibitor keratolinin (Gk: *keratos*, horny; *lininos*, inner surface cover) or cystatin A, which will play an important supporting role in the run-up to cornified envelope formation.[160] As the cell nears the top of the spinous layer, it starts accumulating the 300–500-nm ellipsoidal lamellar bodies, each containing a battery of still inactive hydrolases (e.g., acid phosphatase, proteases, lipases, and glycosidases) and packed with a much-folded ("accordioned") lipid membrane the components of which (glucosylceramides, free sterols, phospholipids) will later be released from the cell, processed, and made into the lipid permeability barrier.[72,76] The cell also starts making cholesterol sulfatase separately in order to produce the cholesterol sulfate required to "glue" the cell to its neighbors until desquamation.[76,104,171] At this point, the expression of gap junction proteins in mouse spinous cells switches from connexins 26 and 43 to connexins 31 and 31·1.[95] This restricts junctional permeability and the transjunctional chattering and exchanges between themselves and underlying basal cells by replacing the basal cell-type gap junctions with junctions that still transfer small metabolites such as nucleotides and amino acids as effectively as the basal cell gap junctions, but transfer larger molecules such as neurobiotin, carboxyfluorescein, and Lucifer yellow less effectively than the basal cells' junctions.[95]

At this point, we must address the roles of TGF-β2 and the IGF-II-selective IGFBP-6 (IGF-binding, hence inhibiting, protein-6) in starting the differentiation program. Basal cells must not make or secrete these potent autocrine/paracrine cell cycle stoppers. However, suprabasal cells in the skin are known to start secreting TGF-β2 into intercellular spaces and continue into the upper layers.[116] Experiments on human HaCat keratinocytes indicate that these cells constitutively secrete *latent* TGF-β2 in a low-Ca^{2+} (0.03 mM) medium, but secrete *active* TGF-β2 when the medium contains 1 mM Ca^{2+}.[116] A high external Ca^{2+} concentration plus a high cell density also stimulates the secretion of IGFBP-6.[116] While these factors could account for the inhibition of the proliferation of cultured keratinocytes by high-Ca^{2+}, they probably are not key proliferative inactivators of suprabasal cells in the real world of the skin because detachment from the basement membrane and the cessation of β1-integrin signaling should be more than enough to stop cycling. Moreover, neither the total nor the intercellular Ca^{2+} concentration rises until the cell reaches the granular layer (Figs. 3.1 and 3.2).[71–75]

Besides loading up with heavy keratins, welding itself to its neighbors with desmosomes and E-cadherins, and starting to make the lipids of the permeability barrier, micellar granules, and future cornified envelope components (e.g., SPRR1 and involucrin), the neophyte spinous cell now upregulates the expression, and production, of the potent signaler, PTHrP (Fig. 3.1).[172,173] PTHrP appears to have been involved in making skins from the beginning of vertebrate evolution because it is expressed in the skins of the jawless lamprey, lungfish, and sharks,[174] which suggests that it or its progenitor was already working in the epidermis of the amphioxus-like, proto-vertebrate, *Pikaia*, and the more recently discovered ancient chordates *Haikouella* and *Yunnanozoan* that swam in the warm equatorial Cambrian seas more than 0.5 Ga ago.[174a,175]

Unlike the blood-borne endocrine PTH, with which it shares 8 of its first 13 N-terminal amino acids and a functionally identical receptor-binding/activating configuration in its first 34 amino acids, PTHrP does not circulate in the blood of *normal* adult animals and humans but is locally produced in many tissues where it operates as an autocrine/paracrine factor.[176,177] Depending on the type of producer cell, the PTHrP gene can be transcribed into one of three mRNAs encoding a mature 139-, 141-, or 173-amino acid protein with 5–7 endoprotease target motifs that enable it to be processed, like pre-pro-vasopressin or proopiomelanocortin, into smaller bioactive products with different receptors and different functions.[176,177] Some of these fragments with the intact PTHrP N-terminus will bind in order to then activate the type I PTH/PTHrP receptors on the producer cell's surface, the surfaces of its neighbors, as well as on the surfaces of the more distant, although still local, dermal fibroblasts (Fig. 3.1).[178]

PTHrP is a polyprotein, other fragments of which may activate different receptors on the surfaces of both the producers and other local cells. Stem cells do not express their PTH/PTHrP receptor genes.[179] Expression starts in the early, slowly cycling transit amplifying basal cells.[179] The functioning, differentiation-modulating product of this gene is not the usual type I receptor found on other cells such as dermal fibroblasts, osteoblasts, and proximal kidney tubule cells.[178] In other cells, the conventional type I receptor is coupled to both the G_s protein that activates cyclic AMP synthesis by adenylyl cyclase and to one of the $G_{q/11}$ proteins that activates membrane-associated phospholipase-Cβ, which breaks PtdIns(4,5)P_2 down into diacylglycerols that stimulate membrane-associated PKCs and Ins(1,4,5)P_3, which in its turn triggers the release of Ca^{2+} from the internal stores bearing the Ins(1,4,5)P_3 receptors we met in the previous chapter. But the normal keratinocyte's PTH/PTHrP receptor is the product of a different gene. Keratinocytes do not express their gene for the conventional type 1 receptor (PTHR1), and the receptor they do express activates a phospholipase-Cβ but not adenylyl cyclase despite an ample supply of functional G_s proteins as indicated by the ability of the β-adrenergic agonist isoproterenol to induce the cells to strongly stimulate adenylyl cyclase.[180–184] However, they have not irretrievably lost the ability to make the conventional receptor because during the course of their "immortalization" human RHEK-1 keratinocytes began expressing conventional PTH/PTHrP, adenylyl cyclase-stimulating PTH/PTHrP receptors.[185]

There could be another reason for this unconventional responsiveness of keratinocytes to PTH. HEK (human embryonic kidney)/W cells seem to respond to hPTH-(1-34) like keratinocytes.[186] Their adenylyl cyclase is not inactivated, although as in keratinocytes, it can be strongly stimulated by forskolin or prostaglandin E_2. However, they do respond with a biphasic release of Ca^{2+} from internal stores, which is not triggered by phospholipase-C-generated Ins(1,4,5)P_3. The Ca^{2+} release is triggered by a different, ryanodine-sensitive mechanism. This "peculiar" response is somehow due to the HEK/W cells having a very low (in fact not measurable) PTHR1 density. Raising the receptor density to 350 000/cell to produce HEK/T cells can replace or override this response with the conventional one consisting of a strong stimulation of adenylyl cyclase and a biphasic, Ins(1,4,5)P_3-triggered release of Ca^{2+} from internal stores. Does this mean that the keratinocyte response to PTH might be due, or partly due, to a low conventional receptor density rather than a differently coupling receptor?

Other fragments containing the PTHrP molecule's C-terminal T(Thr)^{107}-R(Arg)^{108}-S(Ser)^{109}-A(Ala)^{110}-W(Tryp)^{111} motif or even just the small TRSAW fragment itself stimulate membrane-associated PKCs activity without affecting adenylyl cyclase.[187,188] Do keratinocytes express different receptors for N-terminal PTH/PTHrP fragments and fragments bearing the TRSAW motif?[188,188a] Or do they express a receptor which is activated by both kinds of fragment? We eagerly await the answer to these questions.

We don't know whether keratinocytes in the skin make C-terminal PTHrP fragments with the (*107–111*) TRSAW motif. They might do so because the spinous cells in a reconstructed organotypic culture of human foreskin appear to chew up the holoprotein and specifically make a fragment that binds an antibody to PTHrP's *34–53* mid-region.[189] The signals from the receptors stimulated by TRSAW-bearing PTHrP fragments are extremely potent *two-way modulators* of keratinocyte proliferation.[188] Thus, when BALB/MK mouse keratinocytes have stopped proliferating because of EGF withdrawal from their medium, one 30-min incubation each day in a medium containing 10 nM TRSAW (and the brief burst of membrane-associated PKCs activity it causes) followed by return to the orginal EGF-free medium will at least *partially replace* EGF and stimulate them to start, and then continue, proliferating.[188] On the other hand, if the cells are already cycling in an EGF-containing medium, *only one* 30-min exposure to 10 nM TRSAW is enough to stop them proliferating for at least 5 days, perhaps by starting the cycle-stopping part of the differentiation program.[188] Primary adult human keratinocytes are even more sensitive to TRSAW. As little as 10^{-14} M TRSAW stops EGF-driven proliferation, and exposure to 10^{-10} M maximally stimulates the EGF-deprived human cells to start proliferating.[188]

PTHrP is a rather special signaler like bFGF and PDGF.[190,191] Besides being just a conventional surface receptor-activating autocrine/paracrine factor operating through a surface receptor, it is an "intracrine" factor that is selectively transported into the nucleus and particularly to the dense fibrillar core of the nucleolus where it affects various nuclear functions such as ribosomal RNA production, which would increase the cell's protein-producing capacity.[165–167] The mature, unprocessed protein has a bipartite nuclear localization (NLS) motif in its *87–107* region (i.e., just upstream from the TRSAW region).[192–194] The NLS of PTHrP binds to the 90-kDa importin-β, and the resulting complex docks on nucleoporins in the nuclear pore complex, which starts the GTP-fueled transport of the PTHrP through the pore and into the nucleoplasm.[195–197] But how does nascent PTHrP get to the nuclear border to be ferried into the nucleus? Surely this must be the wrong way because PTHrP has a signal sequence which should, and in fact does, direct it into the endoplasmic reticulum and from there through the Golgi apparatus where it is addressed, packaged for secretion, and eventually dumped into the extracellular space. When the secreted PTHrP attaches to a receptor, the PTHrP·receptor complex can be endocytosed and then delivered rapidly to the nucleus.[196] However, the protein can also be delivered directly to the nucleus without leaving the cell.[196] This direct delivery might be due to the signal sequence being altered during translation in which case the protein would bypass the endoplasmic reticulum and be shipped into the nucleus.[194,194a]

What triggers PTHrP expression? The short answer at the beginning of 2001 is — *We don't know*. It seems to require the low-Ca^{2+} concentration characteristic of the basal and spinous layers, and its production could be stimulated by one of the factors secreted by the dermal fibroblasts. It could be induced by the detachment-triggered Ca^{2+} signal. This possibility is suggested by the Ca^{2+}·CaM-mediated stimulation of PTHrP gene expression in the K^+-depolarized cerebellar granule cell by a surge of Ca^{2+} through voltage-opened L-type Ca^{2+} channels in the cell membrane beside or directly on top of the nucleus.[198] It could be induced by autocrine TGF-β2, which, as we have seen, is specifically expressed (at least in mouse skin) and secreted suprabasally, because it has been shown that TGF-βs strongly stimulate PTHrP production by human foreskin keratinocytes and cells of a squamous carcinoma-derived cell line.[199,200]

It follows from these observations that the autocrine/paracrine 139-173-amino acid PTHrP polyprotein holomolecule and (or) one or more of its several differently bioactive fragments, which are induced in the emerging spinous cell possibly by the detachment-triggered $Ca^{2+}$and Ca^{2+}·CaM surges, activate the keratinocyte's unconventional PTH/PTHrP receptor(s). The PLC-β-activating signals from the surface receptors would trigger more Ca^{2+} and Ca^{2+}·calmodulin surges and a surge of membrane-associated PKCs' activity, which together could stop proliferation and stimulate the production of involucrin, keratins 1 and 10, and transglutaminase as has been observed in the cells of Ca^{2+}-shocked primary cultures of human keratinocytes and cultures of BALB/MK mouse keratinocytes.[180,181,188,201] The mediation of these responses to Ca^{2+} shock by PTHrP is indicated by the ability of the suppression of PTHrP expression in human HPK1A keratinocytes by antisense RNA to accelerate proliferation and prevent the shock from inducing the production of keratins 1 and 10, involucrin, and transglutaminase and from stimulating the expression of the involucrin gene.[201] At the same time, those endogenous PTHrP molecules or PTHrP molecules from neighboring cells with the (*87–107*) NLS motif would be shipped to the nucleolus on endocytosed receptors to increase ribosomal RNA production to enhance the translation of the transcripts of the genes for differentiation-related proteins. Of course, PTHrP might also be a transcription factor or cofactor.

An important clue to the function of PTHrP is the ability of the phosphorylation of Thr^{85} near the NLS to prevent the protein's import into the nucleus.[201a] Thr^{85} is targeted by CDKs containing Cdks 1 and 2, which results in PTHrP being *excluded* from the nucleus, starting from the Cdk2·cyclin E-driven stage of the G_1 buildup to the Cdk1·cyclins A/B-driven G_2/M.[201a] Thus, PTHrP would not be expected to affect certain nuclear functions in the cycling basal cells, but would move into the nucleus and start affecting gene expression and other activities only with the disappearance of the CDKs in the post-proliferative suprabasal cells.

According to the most recent information,[202] it seems, as first suggested by Whitfield,[1] that PTHrP controls keratinocyte proliferation and differentiation in *mature* mouse skin in the same way it controls endochondral bone formation and drives the growth and branching of mammary gland ducts.[203,204] Prehypertrophic chondrocytes make and secrete a paracrine factor called Indian hedgehog, which stimulates pre-articular perichondrial cells to make and secrete paracrine PTHrP, which feeds back onto the receptor-bearing prehypertrophic cells and by promoting the expression of the anti-apoptogenic Bcl-2 protector protein restrains their further terminal differentiation into the hypertrophic cells of the calcifying and vascular invasion zones that die through the action of the apoptogenic Bax protein.[205] A failure to make PTHrP results in a failure to express the Bcl-2 protector and consequently premature terminal differentiation and Bax protein-promoted apoptosis, which results in reduced resting and proliferative zones and a deformed rib cage, shortened mandible, and fore-shortened limbs.[202,205] In PTHrP-null mice (which avoid a lethal rib cage failure because they carry a procollagen II-PTHrP transgene, which is specifically expressed in chondrocytes but not keratinocytes), there is premature diffpoptosis that produces a thin and hyperkeratotic skin without a drop in the fraction of cycling basal cells.[202] In other words, the transit time from basal cell to corneocyte is accelerated, which depletes the populations of cells in the intermediate stages of differentiation and thus the thickness of the skin. In these animals, K1 keratin expression, which is normally limited strictly to the spinous (suprabasal) layer, is initiated in some basal cells and loricrin is expressed in the upper spinous layer as well as in the granular and horny layers.[202] By contrast in mice in which keratinocyte PTHrP is selectively overexpressed in basal cells because of the activity of a K14-PTHrP transgene (for the proliferation-related keratin 14), the fraction of cycling cells is doubled (presumably not because of the protein's intrakine nuclear functions in these cycling cells), basal cells are enlarged, the basal cell layer is expanded, and differentiation is held back, as indicated by the expression of K14 even in the lower spinous layer and a displacement of K1- and loricrin-expressing cells to higher suprabasal layers.[202] Because of this the PTHrP-overexpressing skin is much thicker with marked acanthosis and papillomatosis.[202] In summary, PTHrP maintains epidermal thickness (and the complement of hair follicles and sebaceous glands) by appropriately balancing proliferation and differentiation by delaying terminal apoptosis just as it does in cartilage.

PTHrP-driven feedback circuits between keratinocytes and dermal cells are the actual balancers of keratinocyte proliferation and differentiation. Dermal fibroblasts express the conventional, adenylyl cyclase-activating type I PTH/PTHrP receptors, and the PTHrP coming down from the spinous cells stimulates the fibroblasts to proliferate and make fibronectin, an important integrin-binding (hence ILK-stimulating) component of the basal lamina.[177,206,207] Human and rat dermal fibroblasts make a soluble paracrine factor(s) that stimulates keratinocytes to produce PTHrP.[208] Therefore, the PTHrP fragments trickling down from the spinous layer stimulate the dermal fibroblasts to make and secrete growth factors such as KGF, which promotes basal cell proliferation by inducing them to make autocrine TGF-α while at the same time keeping the spinous cells from initiating the final, apoptotic stage of diffpoptosis until they have collected enough of the different components needed to become proper corneocytes possibly by promoting the expression of the Bcl-2 protector as it does in the prehypertrophic cartilage cells (Fig. 3.1).[103,139,177,205–209] Adenylyl cyclase-stimulating N-terminal PTHrP fragments also stimulate the dermal fibroblasts to make IGF-I, which the keratinocytes do *not* make but which provides "survival signals" and promotes the G_1 buildup in the basal cells as well as promotes extracellular matrix synthesis (Fig. 3.1).[210] Clearly, a flood of autocrine/paracrine factors produced in response to overexpressed PTHrP would thicken the epidermis by increasing the flow of cells into the spinous layer and then delaying their differentiation and death, while the lack of KGF and IGF-I in a PTHrP-null skin would thin the skin through the accelerated diffpoptosis of the cells flowing normally into the spinous layer.

This PTHrP-mediated interaction between epithelial and mesenchymal stromal cells is not restricted to skin. PTHrP made by mammary ductal epithelial cells targets receptor-bearing stromal cells, which, when stimulated by the PTHrP, drive mammary ductal growth and morphogenesis via the same factors such as KGF and IGF-I that regulate keratinocyte proliferation and diffpoptosis.[204]

Another major driver of the diffpoptosis program is $1\alpha,25(OH)_2$ vitamin D_3. It operates through a nuclear receptor that binds to the so-called VDRE (vitamin D-responsive element) in the promoters of its target genes. However, it also activates surface receptors, which, like the PTHrP receptors, stimulate a PLC-β1-induced Ca^{2+}/PKCs cascade in keratinocytes and other cells.[211–217] In cultured keratinocytes, $1\alpha,25(OH)_2$ vitamin D_3 downregulates the expression of the EGF receptor and c-*myc* and can therefore stop proliferation.[218] But it also enhances the Ca^{2+}-mediated expression of involucrin, transglutaminase, and the formation of the cornified envelope by stimulating the expression of the CaR gene and the genes encoding PLCs such as PLCs-β2, -γ1, and -δ1 but particularly PLC-β1.[218] Indeed, $1\alpha,25(OH)_2$ vitamin D_3's main mission is to keep the evolving keratinocyte highly responsive to the Ca^{2+} gradient.[219] Of course, the steroid would also increase the sensitivity of the cells to the various possible PTHrP fragments, all of which activate PLC-β1.

What does $1\alpha,25(OH)_2$ vitamin D_3 do in the real world of the skin? It appears that there is a paracrine system in the skin in which $1\alpha,25(OH)_2$ vitamin D_3 is made and secreted by less differentiated cells, but it is the more differentiated cells that have the VDR mRNA and VDR.[220] It would seem reasonable to expect that signals from the receptors for the autocrine/paracrine PTHrP produced by the spinous cells could stimulate $1\alpha,25(OH)_2$ vitamin D_3 synthesis as they do in proximal rat kidney cells.[218,221] The $1\alpha,25(OH)_2$ vitamin D_3 could then, for example, directly stimulate the expression of VDR-responsive genes and CaR expression in the overlying granular cells.

From spinous cell to granular cell: CaRs and upward to the head of the Ca^{2+} gradient

When the cell has acquired a granular cell transcriptome and has reached the granular layer it hits a wall of Ca^{2+} — the external Ca^{2+} concentration triples (Fig. 3.1).[71] By now, it has loaded its surface with CaRs to detect and respond to this big jump in the external Ca^{2+} concentration.[80–82,222–230] Functional CaRs are needed to trigger the final stages of diffpoptosis in the granular layer. Without them, there is a reduced expression of loricrin and profilaggrin, abnormal granule morphology, defective lamellar body secretion, and a disorganized horny layer.[223,224]

The human keratinocyte's CaR appears to be the same as the parathyroid gland chief cells' low-affinity CaR, which monitors the blood Ca^{2+} concentration and triggers PTH secretion if the Ca^{2+} concentration should fall below a critical level and reduces PTH secretion if the Ca^{2+} concentration should rise above a critical level.[80,81] The CaR must have a low affinity for Ca^{2+} to be able to detect and respond to changes in the extracellular Ca^{2+} concentration because if it had the high affinity of an ultrasensitive sensor such as intracellular CaM ($K_d \approx 10^{-6}$ M) it would be permanently activated by the millimolar levels of extracellular Ca^{2+}. The CaR is made up of two different peptide structures, the N-terminal one of which has come from the bacteria and the other of which has come from the eukaryotes. The large, extracellular, N-terminal Ca^{2+}-trapping domain is descended from the very ancient family of bacterial periplasmic ion and nutrient detectors that grab ions (like a Venus flytrap grabs flies) and then appropriately steer the cell's swimming engines toward or away from the target, while the part of the molecule with its 7-transmembrane α-helices come from the huge family of thousands of eukaryotic G-protein-coupled serpentine receptors (to which the PTH/PTHrP receptors also belong albeit to a different clan from the CaR's clan).[80,231–233a] The cation-binding ability of the protein's extracellular N-terminal domain resides in several regions rich in negatively charged amino acids.[80,81] The human keratinocyte's CaR, like the parathyroid cell's CaR, is really a polyvalent cation receptor that responds to Ba^{2+} and Sr^{2+} as well as to Ca^{2+}, but in practice in the body it is effectively a selective Ca^{2+} sensor.[223] It triggers the release of Ca^{2+} from internal stores as well as a prolonged flow of Ca^{2+} into the cells by opening non-specific cation channels in the cell membrane known as SOCs (stores-operated channels) or CRACs (Ca^{2+}- release-activated channels).[234] It is probably coupled to phospholipase-Cγ1, which breaks PtdIns(4,5)P_2 down into Ins(1,4,5)P_3 that releases Ca^{2+} from internal stores and diacylglycerols which stimulate PKCs.[225,226] The set point of the BALB/MK keratinocyte's cation receptor for responding to Ca^{2+} appears to be about 0.7 mM.[227] Raising the external Ca^{2+} concentration from 0.05 to 1.8 mM triggers a Ca^{2+} transient in these cells that peaks 3 min later and a burst of membrane-associated PKCs activity (from diacylglycerol-induced activation of PKCs already attached to the cell membrane and a Ca^{2+}-induced translocation of PKCs from the cytosol to the diacylglyerol-generating/activation sites in the membrane) that peaks 7 min *after* Ca^{2+}.[208] This long delay between the Ca^{2+} peak and the peak PKCs activity makes the signal from the CaR rather special. The expected response to such a phospholipase-C-coupled receptor would consist of simultaneous or nearly simultaneous Ins(1,4,5)P_3-induced surges of internal Ca^{2+} and diacylglycerol-stimulated membrane-associated PKCs activity and the phosphorylation by these PKCs of the 85-kDa MARCKS (Myristylated Alanine-Rich C-Kinase Substrate) protein bound electrostatically to patches of PtdIns(4,5)P_2 that it has organized in the membrane near PLC-γ1, but the relative slowness of the PKCs' response to activation of the CaR gives the Ca^{2+}·CaM complexes produced by the surging Ca^{2+} in time to reach and block the MARCKS proteins' phosphorylation site-domains (and probably those of some other signal-mediating proteins) before the activated PKCs can get to them.[227,228] The functional significance of this striking blockage of MARCKS phosphorylation and possibly other major signaling components is unknown, but it is probably a very important characteristic of CaR signaling. The CaR-induced Ca^{2+}·CaM triggers a wide-ranging cascade of reactions mediated by CaMKs II and IV, and the protein-serine/threonine/tyrosine phosphatase calcineurin.[1,235,236] Activated CaMKs such as CaMKIV mimic cyclic AMP by phosphorylating and activating the CREB (cyclic AMP-responsive element-binding) protein transcription factor that stimulates the expression of those cyclic AMP genes that are configurationally accessible for transcription in granular keratinocytes.[1,237] The membrane becomes a very busy place because the Ca^{2+} surge also stimulates the translocation of phospholipase A2 from the cytosol to plasma membrane where it hydrolyzes membrane phospholipids and releases arachidonic acid, the precursor of prostaglandins and other eicosanoids.[238,239] It is important to note as our cell approaches the granular stage that in one study 91% of the total PLA_2 activity in human skin was found in the cells of the granular and horny layers,[240] which suggests a special role for PLA_2-generated eicosanoids in the later stages of diffpoptosis.

The CaR of murine, but not human, keratinocytes is also specifically coupled to an as yet unknown protein tyrosine kinase that phosphorylates a 62-kDa protein associated with the Ras-GTPase-activating protein when the receptor is activated by Ca^{2+}, nickel, or cobalt.[230,241] This protein kinase is activated within 2 min by the receptor protein, but not by an internal Ca^{2+} surge, because Ca^{2+} ionophores such as the A23187 and X537A that raise the intracellular Ca^{2+} content do not stimulate the protein kinase, while Ni^{2+}, which activates the CaR, stimulates the kinase without significantly raising the internal Ca^{2+} level. The Ca^{2+} surge triggered by the murine CaR activates a protein tyrosine phosphatase (perhaps the protein-serine/threonine/tyrosine protein phosphatase calcineurin?) that activates Fyn, a Src-family protein tyrosine kinase, by a process involving dephosphorylation of the enzyme's Tyr^{531} residue, which causes a configurational change that exposes the enzyme's catalytic site. This surge of protein kinase activity starts 1 h after Ca^{2+} addition and lasts up to 24 h.[1,230]

The basic diffpoptotic plan is first to stop the cell from cycling, then to make keratin cables and plug some of them into desmosomes to link the cell strongly and flexibly to its neighbors, then to load the cell with keratins and the several things needed to make a cornified envelope, and finally to activate the mechanism that will use these things to convert the cell into a hollowed-out corneocyte packed with keratin macrofilaments attached to the inner surface of its lipid-coated horny shell.

On the threshold of the granular layer, the spinous cell has stopped expressing the CLED protein that was triggered by the detachment signal.[139a] It has also stopped displaying large numbers of KGF receptors and is thus no longer protected by the terminal diffpoptosis-suppressing action of KGF from the PTHrP-stimulated dermal fibroblasts: it is now ready to advance to the final, apoptotic part of diffpoptosis.[139] It is much flatter; packed with a network of heavy keratins; riveted strongly to its neighbors by desmosomes; accumulating on its plasma membrane some cornified envelope precursors such as involucrin; still filling up with lipid lamellar bodies; and already loaded with the inactive envelope cross-linking enzyme, TG_K, which, with its fluctuating numbers of myristoyl and palmitoyl adducts, is cycling on and off the cell membrane waiting for the signal to start working.[76,94,154,158–162,242]

What turned on the genes coding for these components in the cell before it reached the top of the spinous layer? Experiments with cultured keratinocytes have shown that a $0.03 \rightarrow 0.12$ mM Ca^{2+} shock shuts down keratins 5 and 14 production and activates the Ca^{2+}-responsive element in the keratin 1 gene's regulatory region, which is followed by the stimulation of the keratin 10 production by the accumulating type 1 molecules (Fig. 3.3).[1,84,85,90–92] A $0.3 \rightarrow 1.2$ mM Ca^{2+} shock shuts down keratin 1 and 10 gene expression and protein production and stimulates the production of "granular cell/corneocyte" components such as filaggrin, involucrin, loricrin, tonofilaments, and TG_K

(Fig. 3.3).[1,84,85,90–92] Exposure to the PKCs activator, 12-*O*-tetradecanoyl phorbol-13 acetate, promotes the appearance of granular cell-specific gene-transactivators, shuts down the production of keratins 1 and 10 mRNAs and proteins while simultaneously stimulating the expression of filaggrin and loricrin mRNAs and proteins by activating the "classical" Ca^{2+}-dependent PKCα as well as transiently stimulating the production of TG_K mRNA in human cells by activating the "novel" Ca^{2+}-independent PKC-η.[85,91,243,244]

One difficulty with labeling components such as involucrin and the cross-linking TG_K as granular stage-specific in the cell culture experiments is that in the real world of the skin involucrin accumulates in the upper *spinous* layer and TG_K *immuno*reactivity *peaks* at the top of the *spinous* layer.[159,160] Another difficulty is the number of different stimulators of Ca^{2+} surges and bursts of membrane-associated PKCs that are known to stimulate the differentiation of cultured human and murine keratinocytes. First, there is the Ca^{2+} surge that occurs when the cell cuts off the flow of β1-integrin signals and detaches from the basement membrane (or culture dish). Then, there is the Ca^{2+}/PKCs signal from the spinous cells' conventional PTH/PTHrP and TRSAW receptors. Finally, there are stronger and longer Ca^{2+}/PKC signals from the CaRs to granular cell-specific nuclear transcription factor complexes when the external Ca^{2+} concentration triples upon entry of the cell into the granular layer and continues rising into the lower horny layer (Fig. 3.1).[222]

How can a tandem sequence of what seem to be the same Ca^{2+}/PKCs signals stimulate the expression of the different sets of genes of keratinocyte diffpoptosis? While the Ca^{2+} signals seem to be the same when they are given at different points in the program, they are in fact not the same. As we learned in the previous chapter, Ca^{2+} released from storage vesicles or flowing into the cell through membrane channels is bound and sequestered very fast and thus cannot diffuse very far from the point of release or portal of entry. Therefore, as the cell differentiates, the Ca^{2+} signals in the forms of quarks, blips, sparks, spikes, repetitive spikes, and (or) waves are differently sited and configured according to the location of their source and their frequency (FM) and amplitude (AM), and they are sent along changing selective intracellular pathways and tunnels toward shifting sets of targets.[245]

When the cell reaches the granular layer, it switches over to another set of transcription factors. For example, it starts accumulating C/EBPα.[140] Among the most important of the many transcription factors involved in diffpoptosis are those that make up the AP1 complexes, which are activated to bind target promoter sequences of DNA by being phosphorylated by PKCs. These complexes are heterodimers of Fos-family proteins (which cannot by themselves bind to AP1 motifs in DNA) and Jun-family proteins (homodimers of which can bind to AP1 motifs but with a much lower affinity than Fos/Jun heterodimers).[246] The genes encoding these components are differently expressed in response to bursts of changing PKC isoform activities as the reconfiguring cells move from layer to layer. In proliferating cultured mouse basal keratinocytes, the DNA-binding activity of AP1 complexes is very low, but is increased by the bursts of PKCs α, δ, and ϵ (but not η) activity triggered directly (PKC-α) or indirectly (PKCs δ and ϵ) by raising the Ca^{2+} concentration in the medium from the proliferation-supporting 0.05 mM to the differentiation-triggering 0.12 mM.[247,248] The AP1 complexes in the Ca^{2+}-stimulated cells consist of Fra-1 or Fra-2 and c-Jun, Jun B, or Jun D, but not c-Fos.[247,248] The PKCs activity also stimulate the expression of the *fra*-2, *jun* B, and *jun* D genes.[247] In the real world of the mouse skin, it appears that AP1 complexes containing Fra-1 and Jun D predominate in spinous cells, while AP1 complexes containing Fra-2 and Jun B predominate in granular cells.[247] Increasing the Ca^{2+} concentration in the growth medium of human keratinocytes from 0.03 to 1.2 mM also increases the expression and nuclear levels of Fos B, Fra 2, and Jun B.[249] In human skin, the c-*fos* gene is not expressed in basal cells but is expressed in the upper spinous and granular layers and lower granular layer; the *fra*-1 gene is also not expressed in basal cells, but it is switched on when the cells reach the first or second spinous layer and is turned off again when the cells reach the upper granular layer; *fos* B and *fra*-2 are expressed in basal cells, but are increasingly expressed when the cells are in the lower and upper spinous layer, respectively, and are then turned off by the time the cells reach the granular layer; *jun* is expressed *only* by granular cells; and *jun* B and D are expressed by cells in all of the layers (Fig. 3.5).[250]

Whether or not, or how, the genes for involucrin, keratins, loricrin, profilaggrin, and transglutaminase TG_K, which have AP1-binding motifs in their promoters,[249,251–254] respond to the Ca^{2+} and PKCs signals triggered by detachment from the basal lamina, activation of PTHrP receptors or activation of the CaRs depends on the responsiveness of their promoters to specific AP1 complexes. For example, PKCs-phosphorylated AP1 heterodimers containing Fra-1 and Jun B or Jun D *stimulate* keratin 1 and involucrin

Fig. 3.5. The same Ca^{2+} and PKCs (particularly PKC-α) signals from integrin detachment, PTHrP, and cation receptors stimulate or inhibit the expression of different sets of genes at different stages of the diffpoptosis program. The different responses discussed in the text to the superficially same signal are due in large part to stage-specific differences in the expression of different components of AP1 transcription complexes that are phosphorylated and activated by PKCs. Here, we see the shifts in AP1 components and their intracellular locations (superscript c indicates cycoplasmic location, while superscript n indicates nuclear location) that occur during human keratinocyte diffpoptosis. The point in the progression from basal cell to granular cell when a component peaks is indicated in bold color-coded letters (e.g., **Fos B^{n}** or **Jun^{nc}**). The dramatic consequences of such shifts can be illustrated with one example. In spinous cells, PKCs-phosphorylated AP1 complexes containing Fra-1, Jun B, and Jun D stimulate the expression of keratin 1 and involucrin genes and at the same time inhibit profilaggrin gene expression. However, when c-Jun appears and peaks in granular cells, it displaces Jun B and the resulting new AP1 complexes shut down the keratin and involucrin genes and turn on the profilaggrin and loricrin genes.

Granular	Fos^{nc}					Jun B^{c}	Jun D^{n}
	Fos^{nc}				**Jun^{nc}**	Jun B^{c}	Jun D^{n}
	Fos^{nc}		Fra-1^{c} Fra-1^{c}			Jun B^{c}	Jun D^{n}
Spinous	Fos^{nc}	Fos B^{n}	Fra-1^{c}	**Fra-2^{nc}**		Jun B^{c}	Jun D^{n}
		Fos B^{n}	Fra-1^{c}	Fra-2^{nc}		Jun B^{c}	Jun D^{n}
		Fos B^{n}		Fra-2^{nc}		Jun B^{c}	Jun D^{n}
		Fos B^{n}		Fra-2^{nc}		Jun B^{c}	Jun D^{n}
Basal		Fos B^{n}		Fra-2^{nc}		Jun B^{c}	**Jun D^{n}**

expression in spinous cells,[254,255] but they *repress* profilaggrin gene expression.[252] However, the appearance and buildup of c-Jun in granular cells would displace Jun B from AP1 sites and thus suppress keratin and involucrin gene expression and stimulate profilaggrin and loricrin expression.[251,252]

Judging from the results of experiments with cultured keratinocytes, our cell probably has a larger complement of various PKCs and greater PKC-stimulated AP1 DNA-binding activity when it reaches the top of the spinous layer than it had when it left the basement membrane.[247,256–258] The strengthening PKCs signals from PTHrP receptors and now the Ca^{2+}-activated polyvalent cation receptors change the composition of AP1 complexes, which shut down the expression of the genes for keratins 1 and 10 (but not the *production* of these proteins), and destabilize the keratin 1 mRNA transcripts.[259] However, as the cell crosses into the granular layer it stops making and processing PTHrP and installing gap junctions (Fig. 3.1).[96,99,100,172,173,176] It continues accumulating the lipid lamellar bodies.[104]

The granular cell's mission is to switch on the rest of the EDC genes in the chromosome 1q21-22 cluster (Fig. 3.4) to make the remaining cornified envelope components. The cell then accumulates the cytoplasmic keratohyalin granules, which give it its name.[104,260] There are three kinds of granule: the P-F-granule, which contains highly phosphorylated profilaggrin; the L-granule, which contains loricrin; and the composite granule, which contains profilaggrin surrounded by loricrin.[260]

The 350-kDa profilaggrin molecule has multiple filaggrin repeats linked by short peptides, a Ca^{2+}-binding S-100 protein-like N-terminal domain, and several phosphorylation sites.[168–170,260] It seems likely that the CaR-triggered Ca^{2+} and PKCs surges and the granular cell-specific Jen transcription factor[261] work together to stimulate the expression of the profilaggrin gene. Then, Ca^{2+} and a group of Ca^{2+}-independent protein serine/threonine kinases (e.g., casein kinase II) collaborate to pack the highly phosphorylated, insoluble protein into the P-F-granules, which are deposited at points of intersection of keratin filament bundles and progressively enlarge as the cell rises to the top of the granular layer.[104,260] The Ca^{2+} surges triggered by the granular cell's CaRs also strongly stimulate the expression of loricrin (from Latin: *lorica*, shell or cover), a very glycine-rich protein which has tandem Gly–Ser repeats separated by glutamine-rich sequences and flanked by glutamine-lysine-rich motifs which will be used by TG_K, and the later-appearing loricrin-selective TG_E, to cross-link the protein into the cornified envelope.[159,160,262] The glycine–serine regions form inverted "Ω-loops," which will be used to dock and anchor keratin macrofilaments to the inner surface of the future horny envelope.[94] Like profilaggrin, loricrin is rendered insoluble by phosphorylation and self-association through disulfide bonding and sequestered in the L-granules.[260] The storage of loricrin and profilaggrin as inactive, insoluble precursors in granules prevents premature and aberrant cornification and keratin filament aggregation. Loricrin as well as the other components of the keratohyalin granules must be shielded from the massive proteolysis by lysosomal cathepsins B and H during the granular→corneal transition. The shield is keratolinin (cystatin A) that began accumulating in and around the granules during the spinous stage.[94,260]

Now we must remember the dramatic, mid-granular buildup of intercellular Ca^{2+} and the increase in the intracellular Ca^{2+} concentration that occurs when cells reach the top of the granular layer and the head of the Ca^{2+} gradient and cross into the horny layer

(Figs. 3.1 and 3.2).[74,75] This intracellular Ca^{2+} surge is probably due to the increasing signaling by the CaRs, responding to the external Ca^{2+} buildup (Figs. 3.1 and 3.2). The cell now enters a transitional state in which it must start processing the accumulated keratohyalin granule components to make the horny envelope. The keratohyalin granules start thinning and their contents disperse throughout the cytoplasm.[104] Profilaggrin is processed in two stages. First, a Ca^{2+}-independent chymotrypsin-like protease chops it into oligomers, which are then chopped into filaggrin monomers by a Ca^{2+}-activated protease.[260] The filaggrin monomers then associate with the keratin filaments to form a temporary scaffolding to orient the filaments until they are locked in place by disulfide bonds. The components of the loricrin aggregates are released (by the reduction of their disulfide linkages) for cornified envelope formation and for docking keratin macrofilaments to their "Ω-loops."[94,260] As various components have been moving to the cell surface to join involucrin and the waiting, but still inactive, TG_K and TG_E, they form a distinct growing band beneath the soon-to-disintegrate plasma membrane.[160]

The final push to cornification and oblivion

The corneocyte is created by a modified form of apoptosis in the component-loaded transitional cell. A feature of this apoptoid process is the use of the specific membrane-associated transglutaminases, TG_K and TG_E, instead of the cytoplasmic "tissue transglutaminase," TG_C, that is used in the non-specialized or general apoptotic mechanism designed simply to get rid of cells as quickly as possible without damaging neighboring cells.[57,94] But the keratinocyte must not be quickly destroyed: its apoptotic transformation to corneocyte is the crowning event of diffpoptosis. The corneocyte must stay around for a while as a part of the body's horny covering.

Apoptosis was invented to enable cells to kill themselves without damaging or killing their normal neighbors by releasing a murderous pack of endonucleases, lipases, phosphatases, and proteases such as the caspase cysteine proteases (ICE [interleukin-1β converting enzyme]-like proteases) during their death throes.[135,263,264] The keratinoctye-specific TG_K and the loricrin-specific TG_E transglutaminases are special guest players in keratinocyte diffpoptosis, which, from their perches on the cell periphery, grab *any* passing large or small proteins with available glutamine and lysine residues and cross-link them with isopeptide bonds into an impervious horny shell while the cell's plasma membrane, nucleus, chromosomes, and other organelles are being shredded by a ravenous horde of hydrolytic enzymes.[1,94,159,160] This part of the diffpoptosis program has been ingeniously adapted to make skin and scales by first having the keratinocyte use Ca^{2+} surges and bursts of various PKCs activities to accumulate polyamines and a pool of large and small glutamine- and lysine-rich proteins that have been specifically designed for making a cornified envelope.[94]

When the granular-transitional cells have made enough of the things needed to become corneocytes and the external Ca^{2+} concentration is high enough, the terminal transition is fast. The death machinery swings into action, probably in response to signals from the CaRs that activate a starter protease such as the SCCE (stratum corneum chymotrypic enzyme) or that stimulate the secretion of an autocrine starter protease or that stimulate a starter caspase (ICE-family protease).[57,133,264–266] It seems likely that Ca^{2+} cycling into and out of the mitochondria induced by the intense Ca^{2+} signaling in these cells opens megachannels in the mitochondrial inner membranes through which the apoptogenic caspase proteases can pass into the cytoplasm and reach key target sites.

The inactive TGs that have been patiently perched on the inner surface of the CaR-armed cell since their appearance and accumulation during the spinous stage now come alive when the CaR signaling peaks at the Ca^{2+}-rich interface of the granular and horny layers (Figs. 3.1 and 3.2). Ca^{2+} activates the full-length 106-kDa TG_K on the cell membrane by stimulating its partial proteolytic processing at specific sites by a CANP(s) (Ca^{2+}-activated neutral protease[s] or calpain[s]) into highly active complexes of 10-, 33-, and 67-kDa fragments that are anchored to the membrane by the 10-kDa fragments.[158,159,267–269] The intense Ca^{2+} signaling also activates the loricrin-specific TG_E. The active TG complexes cross-link cornifin, involucrin, loricrin, pancornulins, and some other proteins (e.g., 36-kDa annexin 1 and a 61-kDa dimer of annexin 1 [driven onto the membrane by Ca^{2+}] or another protein) into the tough, 7–15-nm horny envelope subjacent to the cell membrane with involucrin on the outer surface of the envelope and TG_E selectively cross-linking loricrin into the inner surface of the envelope.[94,159,160] Also somehow involved in regulating this final stage of diffpoptosis is the recently discovered Ca^{2+}-binding CLSP (Calmodulin-Like Skin Protein), which, unlike CaM, binds to TGs.[269a]

As this is happening the keratin-associated filaggrin monomers, which have been serving as a temporary scaffolding for the keratin network, dissociate from the keratin filaments when their positively charged arginine residues are deaminated to citrulline and degraded to free amino acids.[260] The liberated keratin macrofilaments are now attached to the Ω-loops hanging down from the loricrin on the inner surface of the horny envelope.[94] The amino acids of the dissociated, degraded filaggrin are hydrophilic and act as corneocyte moisturizers by binding water.

The Ca^{2+} surges also generate Ca^{2+}·CaM complexes which activate the Ca^{2+}-ATPase pumps in the doomed plasma membrane (Fig. 3.2).[1] While activating the TGs, the Ca^{2+}-activated CANPs may also help switch on the Ca^{2+} pumps by binding via their calmodulin-like domains to the pumps' calmodulin-binding regulatory sites and then activating the pumps by limited proteolysis.[270] This double action makes sure that once it has done its job most of the exchangeable Ca^{2+} is pumped back into intercellular space to prevent a steady leakage of Ca^{2+} from the body through the skin and in the process maintain the head of the transepidermal Ca^{2+} gradient (Fig. 3.2).

During its progression from the spinous stage, the cell has been accumulating lamellar bodies, which may ultimately occupy as much as 10% of the cell's cytoplasmic volume.[76] When the cell arrives at the top of the granular layer, it reconfigures itself into a very special kind of secretory cell.[271] At the border of the horny layer, the lamellar bodies that were formed in and then emerged from the tubular-reticular membrane system of the trans-Golgi network, collected in aggregates and linear arrays in deep invaginations or pockets of the apical surfaces of the cell from which they can easily reach the cell surface.[271] Without these special pathways, the lamellar bodies would get trapped in the mature granular cell's maze of surface keratin filaments, keratohyalin granules, and growing patches of horny envelope. Nested in the invaginations, they are poised for rapid exocytosis. Now, under the normal circumstances of an intact permeability barrier and the head of the Ca^{2+} gradient at the interface between the granular and corneum layers, the cell slowly releases the contents of its lamellar bodies into the intercellular space at the interface of the granular and horny layers. The "accordioned" lipid membranes dumped out of the lamellar bodies uncurl, join end-to-end, and pile up into multilaminar sheets that envelop the cells in the lower two horny layers.[76] The fusion of the envelopes of large numbers of lamellar bodies with the granular cell's membrane greatly increases the cell's surface area : volume ratio as it moves up into the horny corneum.[76] The released lipids are then converted into more apolar ceramides, cholesterol, and free fatty acids by acid-activated hydrolases that were also released from the lamellar bodies.[76] These external layers of processed apolar lipids replace the cornified cells' disintegrated plasma membranes, and the ceramides are covalently linked to glutamine residues of the involucrin on the surface of the cornified envelope to form a lucent hydroxyceramide layer on the outer surface of the envelope.[76,94,160,272–275]

The product of these events is the all-important permeability barrier the invention of which was a sine qua non for the emergence of vertebrates onto dry land.[73,76] Ca^{2+} controls the establishment of the permeability barrier. Removing the barrier (e.g., by repeated tape stripping to remove the corneum) causes a transepidermal water loss and eliminates the Ca^{2+} gradient. This Ca^{2+} depletion and silencing of the upper granular cells' CaRs causes a massive release of lamellar bodies and increases the expression of the genes encoding various enzymes for lipid synthesis (acetyl-CoA carboxylase, farnesyl pyrophosphate synthase, fatty acid synthase, HMG-CoA reductase, serine palmitoyl transferase, squalene synthase) to restore both the lipid barrier and the Ca^{2+} gradient.[73,76] It also triggers the expression and release of a specific group of cytokines (IL-1α, IL-8, IL-10, intercellular adhesion molecule-1 (ICAM-1), interferon-γ, TGF-α, TGF-βs, TNF-α) that stimulate an inflammatory response, lipid synthesis by suprabasal cells, and the proliferation of basal cells.[76,276–280] Particularly important among these is TNF-α, which can stimulate its own expression as well as those of ICAM-1, IL-8, the all-important TGF-α which appears within 1 h and stimulates keratinocyte proliferation, motility, and matrix formation.[277] It seems that TNF-α is located mainly in the granular cells, which, however, do not express their TNF-α gene. But they must have expressed the gene at some earlier stage (probably when they were KGF-stimulated cycling basal cells) and accumulated a store of the cytokine for release in case of barrier disruption.[277] The released TNF-α then spreads to underlying cells, which it stimulates to express their own TNF-α genes as well as those of the other proliferogenic factors.[277,278] Among these underlying target cells may be the dermal fibroblasts, which respond by making more KGF for increasing the basal cells' TGF-α expression and proliferation.[68] However, the proliferogenic response is soon ended (e.g., by 6 h) when the proliferation-phase factors stimulate the expression of the growth-inhibitors such as TGF-β(s) and interferon-γ.[277]

Prematurely restoring signaling from the divalent cation sensors by adding Ca^{2+} (and K^+) to the bathing solution prevents the massive expulsion of lamellar bodies and the burst of lipid synthesis that are urgently needed to promptly restore the permeability barrier.[72,73,76] Evidently, the signals from the granular cells' CaRs do three very important things when the barrier is restored and the gradient is reestablished: signal the presence of the restored permeability barrier; switch off the mechanism that responded to the barrier disruption; and trigger the apoptotic coversion of the cells into lipid-coated corneocytes.

So far, the cells have been bonded together by cholesterol sulfate glue made by the spinous cells' cholesterol sulfotransferase and, of course, the desmosomes. Now, the glue is destroyed by cholesterol sulfatase, which the cells release separately from the lamellar bodies. The acid-activated proteases from the lamellar bodies cut the desmosomal tethers.

Having triggered the key steps of the diffpoptosis program, Ca^{2+} now switches from the constructive Dr. Jekyll to a murderous Mr. Hyde. The phosphatases, phospholipases, proteases, and nucleases such as DNAase I (liberated from its complex with G-actin by the rampaging proteases)[133] that have been unleashed by the strong Ca^{2+} signals from the CaRs and the cascade of events unleashed by caspases leaking out of the disintegrating mitochondria destroy the cell's nucleus and other organelles, leaving a horny shell, purged of its Ca^{2+} and packed with keratin cables anchored to loricrin Ω-loops sticking out of the shell's inner surface. The final products of diffpoptosis are horny cadavers embedded in a water-impermeable lipid, which will eventually be sloughed off into a cellular oblivion.

Summary

The epidermis consists of stacks of upwardly rising keratinocytes with changing transcriptomes for making the things needed for them to end up as dead hard-shelled corneocytes packed with keratin macrofilaments that are sloughed off at the top and replaced at the bottom from clusters of intermittently cycling, clonogenic stem cells and a network of larger, faster cycling amplifying transit cells. The epidermis maintains a steep, upwardly increasing Ca^{2+} gradient (equal and low in the basal and spinous layers, then two and three times higher in the granular and corneum layers, respectively) through which the cells must pass. As the transit amplifying cells proliferate in the low-Ca^{2+} basal layer, their adhesion to the basement membrane by β1 integrins weakens with each cycle and increasing distance from the natal niche. When the adhesion-dependent, β1 integrin-mediated signaling drops below a certain level, the cell no longer makes among other things the Cdk2·cyclin A protein kinase that is needed at the end of the G_1 buildup to trigger and then drive DNA replication. Detachment causes the release of Ca^{2+} from intracellular stores and a surge of Ca^{2+} through membrane channels. The cell starts making the $p21^{Cip/WAF1}$ and $p27^{Kip1}$ inhibitors that prevent the expression of the genes for the principal DNA-replicating enzymes by inhibiting the Cdk2·cyclin E protein kinase. The Ca^{2+} signal triggers the expression and production of keratins 1 and 10 as well as involucrin and SPRR1 (cornifin), the first of the future cornified envelope components, and the TG_K that will collaborate with the later-appearing TG_E to cross-link them and others into the cornified envelope. The downregulation of the basement membrane-binding β1 integrins upregulates the expression and display of the Ca^{2+}-dependent cadherins and assembly of desmosomes to spot-weld the new suprabasal cell to its neighbors and give it the characteristic spinous morphology.

Besides accumulating proliferation-incompatible heavy keratins and the first of several envelope components, the growing spinous cell starts making intracrine PTHrP fragments, which enter the nucleus either directly from the ribosomal factory or in endocytosed receptor complexes and bind to the fibrillar component of the nucleolus where it increases the production of ribosomal RNA for making various spinous-specific components. The PTHrP fragments are also paracrine/autocrine activators of surface receptors, which generate Ca^{2+} surges and bursts of membrane-associated PKCs activity that could drive the spinous program, initiate the granular program, and maintain tissue homeostasis by modulating basal cell proliferation and the production of basement membrane components and other soluble factors by dermal fibroblasts. However, the fragments also control the sizes of the spinous and granular layers and therefore the thickness of the epidermis by delaying the onset of apoptosis by stimulating the expression of the anti-apoptotic Bcl-2 protector protein.

When the cell lifts into the granular layer, it flattens with the inner surface of its plasma membrane studded with involucrin and inactive transglutaminases, stops making gap junctions and PTHrP, and downregulates keratin gene expression using newly appearing, granular cell-specific AP1 complexes and other granular cell-specific transcription factors such as C/EBPα and Jen. It uses these granular cell-specific factors to make the rest of the envelope components such as loricrin as well as the keratin intermediate filament cross-linkers trichohyalin and profilaggrin, which it stores in keratohyalin granules until the Ca^{2+}-triggered signaling from the CaRs reaches a critical intensity. The extracellular Ca^{2+} level does not rise until the cell is halfway up the granular layer. Then, it jumps up dramatically, and the increasingly intense and prolonged signals from the cell's CaRs start driving the final stages of diffpoptosis as the cell rises into the upper granular and lower corneum layers. These signals trigger the breakdown of the keratohyalin granules and the release of loricrin and filaggrin monomers into the cytoplasm. The filaggrin monomers then form a scaffold for orienting and cross-linking the keratin 1·10 heterodimers into a dense macrofibrillar network. The cell membrane starts to disintegrate as the Ca^{2+}-activated, membrane-associated transglutaminases cross-link involucrin, then loricrin and other proteins and peptides into its underlying replacement, a horny envelope with loricrin on its inner surface to attach keratin macrofilaments and involucrin on its outer surface to anchor a lucent hydroxyceramide layer after the disintegration of the cell membrane. At the same time, several enzymes that have been activated, released from mitochondria, or otherwise mobilized by the Ca^{2+} pulses, kill the cell by destroying the nucleus and other organelles. The Ca^{2+} surges also switch on the cell's Ca^{2+} pumps, which dump all of the developing corneocyte's Ca^{2+} load into a large external reservoir, the head of the transepidermal Ca^{2+} gradient, which provides the external Ca^{2+} that will seep down and drive the diffoptosis of the upcoming granule cells. Thus, we have seen Ca^{2+} in all of its roles. We have seen it as a driver of the various stages of the basal cell's growth-division cycle, then as a differentiator that triggered the key steps in the progression from detached basal cell to spinous cell, mature granular cell, and finally as an apoptogenic killer that turned on the machinery that trashed the cell's organelles and then bailed out of the cell on the pump it turned on, leaving a dead, cornified shell packed with a dense network of keratin macrofilaments.

Cancerous keratinocytes

On the road to malignancy

The proliferation of normal keratinocytes in the skin is limited to the basal layer by the need for adhesion-induced signals from β1 integrins and associated receptors for soluble growth factors to activate PtdIns-3K, ILK, PKB, the diffpoptosis-suppressing Id proteins, suppress the expression of the $p21^{Cip/WAF1}$ and $p27^{Kip1}$ cyclin-dependent protein kinase inhibitors, and keep β-catenin flowing into the nucleus to keep the cell cycle genes and the gene for a keratinocyte-specific tumor suppressor active. When these signals stop and the integrin signals from adhesion complexes fade, the cell detaches from the basement membrane, generates a Ca^{2+} surge, shuts down the Id diffpoptosis suppressors, upregulates the p21 inhibitor, shuts down its cell cycle genes, and dismantles its DNA-replicating synthesomes. A second line of defence against hyperproliferation is the restraint imposed by direct contact with neighboring normal cells that does not involve gap junctions. A third line of defence against ectopic, suprabasal proliferation is responsiveness to the higher Ca^{2+} concentrations and anti-proliferation signals from the cell's CaRs in the upper layers. Thus, during the intense competition and ruthless selection leading to the hard-won dominance of a clone that has collected the "full house" of mutations needed to cross the line of no return into invasive skin cancer,[280a] mutant cells *must* appear which

- No longer need attachment to the basal lamina to make the CDKs and replication factories;
- Are not proliferatively restrained by contact with normal neighbors;
- Are not stimulated to diffpoptose by $Ca^{2+}$and signals from CaRs.

When normal cells lift off the basement membrane because of the lowering affinity and density of integrin receptors, they reduce or switch off their cell cycle genes and CDKs but retain their death machinery, which is activated near the top of the stack and the head of the Ca^{2+} gradient. Things change when the carcinogenic clonal selection process leading to papillomas and squamous cell carcinomas is initiated by a carcinogen-induced mutation, most commonly a c-Ha-*ras* gene mutation that results in the production of a constitutively activated Ha-Ras protein.[281–284] (We will learn in the next chapter that it is a mutant Ki-Ras instead of a mutant Ha-Ras that usually drives colon carcinogenesis.) This protein confers an activated phenotype on the cell by enhancing PtdIns(4,5)P_2 breakdown and bypassing soluble growth factors and integrins by directly and continuously stimulating PI-3K, the (PtdIns(3,4,5)P_3) product of which stimulates components of the proliferogenic MAPK protein kinase cascade. This enhances the expression of proliferation-related genes such as c-*myc* and converts the activated keratinocyte into a mini-cytokine factory, which overproduces and constitutively secretes many proliferogenic autocine/paracrine factors among which are TGF-α, the heparin-binding amphiregulin and betacellulin growth factors that also activate EGF/TGF-α receptors,[248] bFGF, IL-1, IL-6, and PDGF, and overexpresses EGF/TGF-α receptors because of an amplified receptor gene or an inability to downregulate the receptors.[285–292] These growth factors would amplify the proliferogenic drive of the activated Ha-Ras protein.

The overproduction of, or hyperresponsiveness to, TGF-α and the other EGF relatives would prevent or at least oppose diffpoptosis and may be responsible for the ability of squamous carcinoma cells to break down the basement membrane and invade the underlying dermis.[293–295] Another effect of overexpressed EGF/TGF-α receptor expression which enables invasiveness and impairs diffpoptosis is downregulation of the expression of the Ca^{2+}-activated, cell–cell binding E-cadherins and redistribution of the cadherin-associated α-, β-, and γ-catenins from the inner surface of the cell membrane to the cytoplasm.[296] This ensures the maintenance of a substantial pool of free β-catenin to sustain c-*myc* expression and other proliferation-related genes.

The initiated cells could shed the need for attachment-dependent integrin signals to initiate DNA replication, and thus free themselves from their proliferative restriction to the basement membrane, by continuing to express integrins in the suprabasal layers. Indeed, psoriasis-like suprabasal proliferation can be induced in mouse skin by attaching the suprabasally (spinous) activated involucrin gene promoter to the genes for α2, α5, and β1 integrin subunits.[297] They also have been found to overproduce $\alpha_6\beta_4$ integrin receptors, some of which contain α6A and α6B splice variants, even after lift-off, which is associated with malignant conversion and expansion of the proliferative compartment into the suprabasal layers.[298,299] Provided they can still organize signaling complexes away from direct contact with basal lamina, these inappropriately expressed integrins would presumably maintain the proliferogenic survival signaling that normally needs matrix adherence. Hauser et al.[88] have also found that unlike normal murine keratinocutes, which drop their cyclin A level when detached from their substrate and are consequently blocked in S phase, the tumorigenic SLC-1 keratinocytes behave as if their integrin signaling does not stop when detached from their substrate because they do not lose cyclin A and Cdk2·cyclin A and do not accumulate the $p27^{Kip1}$ CKI or stop making DNA.

However, the real liberator from the dependence on adhesion and integrin signals for proliferation is the activated Ha-Ras protein.[283] It operates by activating PtdIns-3K by binding directly to the kinase's 110-kDa catalytic component.[28] This keeps ILK and PKB active, and β-catenin flowing into the nucleus to suppress E-cadherin expression, to keep the cyclin D1 gene expressed and to keep the c-*myc* gene and its target Cdk4[32a] gene working to make the Cdk4·D1 cycle-starting CDK and to prevent detachment-triggered programmed death (anoikis).[30,31]

Another reason why neoplastic keratinocytes with their activated Ha-Ras protein do not trigger apoptosis when detached from the basal lamina is the inactivation of the Ca^{2+}-independent PKC-δ.[283] This kinase is inactivated by being phosphorylated by a Src-family protein kinase, which has been stimulated by the EGF/TGF-α receptors activated by the cells' constitutively produced and secreted TGF-α.[283] It seems likely that the 40-kDa fragment of nuclear 79-kDa PKC-δ may play an important role in apoptosis by binding and inactivating the DNA-dependent protein kinase that is involved in DNA repair and by the inclusion of an apoptogenic factor in this chromatin-associated complex.[300]

The key feature of cancerous keratinocytes is a disruption of Ca^{2+} control mechanisms. At first, the initiated cells in mouse skin, for example, are phenotypically normal so application of a Ca^{2+}-ionophore, ionomycin, directly onto the skin before starting promotion reduces the yield of tumors after promotion.[301] However, Ca^{2+}-insensitive mutants ultimately appear during the forced march to cancer during promotion. Indeed, such Ca^{2+}- insensitive mutants must be the first to appear, otherwise the initiated cells would be stimulated to diffpoptose when they leave the low-Ca^{2+} basal layer. Thus, the escape from the clutches of external Ca^{2+} occurs in the early stages of chemical and viral (e.g., human papilloma virus 16) carcinogenesis; initiated but phenotypically semi-normal mutants with basal cell markers may respond to high external Ca^{2+} by only reducing, instead of stopping, proliferation and uncoordinately initiating some, but not all, parts of the differentiation program.[302–314] Eventually, mutants appear, which do not need a small amount of external Ca^{2+} to cycle and do not respond to the normally differentiation-triggering higher concentrations of the ion and may even be stimulated to proliferate by these higher concentrations.[310] As we learned above, Ca^{2+} shock causes normal keratinocytes in culture dishes to express their involucrin and transglutaminase genes, make involucrin and transglutaminase, move the new transglutaminases from the cytosol to the plasma membrane where they are eventually activated, and make cornified envelopes. Human SCC4 squamous carcinoma cells are examples of the loss of responsiveness to Ca^{2+}. These cancer cells produce involucrin and transglutaminase mRNAs, but raising the external Ca^{2+} concentration to 1.2 mM does not increase the transcripts or stimulate their translation into protein.[303] Unfortunately, we do not yet know whether this failure is due to defective CaR signaling or some downstream defect. This loss of responsiveness to Ca^{2+} plus the liberation of the production of CDK components from the need for attachment-induced integrin signaling enables the initiated cells to proliferate in the suprabasal layers. It also facilitates the selective growth of clones of initiated keratinocytes because initiated cells proliferate in a high-Ca^{2+} medium to form colonies while their normal neighbors stop proliferating and differentiate. Thus, after this pivotal phenotypic shift, Ca^{2+} becomes a *tumor promoter* (!) rather than a tumor preventor.

Once the carcinogenic process has been initiated in a niche-anchored basal stem cell or a transit amplifying cell,[25] the descendents of that cell may be *selectively* expanded into a visible papilloma by a PKCs activator such as TPA.[313] TPA could do this by overriding the suppression of the hyperproliferation of initiated, but not yet malignant, keratinocytes by surrounding normal cells, which can be demonstrated both in the culture dish and the skin. Initiated cells that directly contact normal cells are prevented by this contact from expanding into papillomas.[284,315–318] This suppression of the proliferation of initiated cells by contact with their normal neighbors is *not* due to a flow of suppressors from the normal cells or a dilution of stimulators in the initiated cells by transfer through gap junctions because such suppression in vitro requires a high (e.g., 1.2 mM) Ca^{2+} concentration that would shut gap junctions as indicated by inhibition of intercellular transfer of the dye Lucifer Yellow.[315–318] Instead, TPA works by disrupting the proliferatively suppressing attachment of the initiated cells to their normal neighbors by the Ca^{2+}-configured/activated E-cadherins.[150,319] TPA induces the phosphorylation of E-cadherin subunits by PKCs in cell–cell adhesion complexes.[320,321] This causes a redistribution of E-cadherins from the cell membrane to the cytoplasm, which would uncouple the initiated cells from their suppressive normal neighbors.[320,321] This would derepress the proliferogenic mechanism because the surface display of Ca^{2+}-activated/interlinked E-cadherins suppresses proliferation and promotes stratification and diffpoptosis.[152] Specifically, suppressing the Ca^{2+}-dependent, β-catenin-sequestering function of E-cadherin along with the β-catenin-sparing Ha-Ras-activated PtdIns-3K would contribute to the desensitization of the initiated cell to external Ca^{2+} and the creation of a pool of free β-catenin for stimulating c-*myc* and other proliferation-related genes.[30,31,35,283]

TPA and its relatives can also promote the expansion of initiated keratinocytes into papillomas because of the cells' radically altered response to surges of PKCs activities. TPA stimulates keratinocytes to make both TGFα and TGFβ.[289] The TGFα would stimulate proliferation, but the TGFβ overrides TGFα to cause normal cells to stop cycling and differentiate.[14,313] TGFβs stop cells cycling first by inhibiting the expression of cyclin D1, the first of the G_1 cyclins, which is needed to form the Cdk4·cyclin D1 protein kinases that hyperphosphorylate and inactivate the Rb-family pocket proteins and the related p107 protein that block the stimulation of key replication-related genes by the E2F transcription factor.[123,124,322] The TGFβs also stimulate the expression of the $p15^{INK\ 4b}$ protein that selectively binds to Cdk4 protein and thus prevents CAK protein kinase (Cdk7·cyclin H) from binding to Cdk4 and activating it by phosphorylating Cdk4's Thr^{160} residue. The surge of $p15^{INK\ 4b}$ also displaces the $p27^{Kip1}$ inhibitor from its complexes with Cdk4·cyclin D. The liberated $p27^{Kip1}$ can then bind to and prevent the activation of Cdk2·cyclin E, the stimulator of the principal replication genes, and Cdk2·cyclin A, the replication trigger and driver. Thus, the TGFβs stop cells cycling by blocking the G_1-specific protein kinases that hyperphosphorylate and thus inactivate the Rb-family pocket protein transcription-suppressors. The early initiated cells, like their normal neighbors, respond to TPA by increasing TGFβs production and increasing their already constitutive hyperproduction of TGFα, which overrides the inhibitory action of the TGFβ and stimulates the proliferation of the initiated cells. However, in mouse skin the increasingly mutation-loaded clonal descendents of the initiated keratinocyte may lose the ability to produce and (or) respond to autocrine/paracrine TGFβs on their way to becoming squamous cell carcinomas, although they might increase the production and secretion of TGFβ3 late in the carcinogenic clonal selection.[283,289,323,324] The consequence of the loss of TGFβs in mouse skin cells, or the loss of one or both parts of the active type I/type II TGFβ receptor complex, or the inactivation or loss of the TGFβ target gene, $p15^{INK\ 4b}$, the constitutive overexpression of the G_1-specific cyclin-dependent kinases, or the loss of the Rb-family pocket protein suppressors is that TPA, like Ca^{2+}, *stimulates* the proliferation of initiated cells but *stops* the proliferation of the neighboring normal cells.

Some neoplastic keratinocytes overproduce and secrete the PKCs-stimulating, hence TPA-like, PTHrP.[177,325–332] (It is relevant at this point to recall that squamous carcinoma cells display the same unconventional PTH/PTHrP receptor as normal keratinocytes that unlike the conventional receptor, does not stimulate the keratinocytes' adenylyl cyclase.[180–182]) The ability of some carcinoma cells to make and hypersecrete PTHrP depends on the ability of specific transcription factors to stimulate PTHrP gene expression and whether the cell is expressing certain mutant forms of the p53(TP53) protein (i.e., the products of p53(TP53) genes with mutants in codons 248 and 273), which can repress PTHrP gene expression.[330,332] When this PKCs-stimulating protein spills into the blood, it first stimulates bone formation, but when its concentration rises above a threshold value it stimulates Ca^{2+} uptake from the glomerular filtrate by the convoluted tubule cells and demolishes bone to produce the humoral hypercalcemia of malignancy.[177,325–328] Therefore, the PKCs-activating PTHrP may be a TPA-like autocrine/paracrine proliferation stimulator for initiated, but not normal keratinocytes. Thus, the hypersecreted PTHrP, especially its intracellularly processed TRSAW-bearing fragments, might enhance tumor growth by inhibiting the proliferation and stimulating the differentiation of normal neighbors without affecting the proliferation of more advanced tumor cell clones.[188] This promoting action would apply only to squamous cell carcinomas because the PTHrP gene is not expressed in the cells of the highly invasive (but normally not metastasizing) basal carcinomas, which do not differentiate, but is strongly expressed in squamous cell carcinomas which can contain a substantial fraction of differentiated or quasi-differentiated cells.[111]

Speaking of basal cell carcinoma, the most common of the human cancers that arises in both follicular and interfollicular epidermis, it has been found that as might be expected these carcinoma cells have lost their responsiveness to Ca^{2+} signals. It is the mutational disabling or loss of a so-called "gate-keeper" gene (a gene the product of which controls the level of proliferation in a self-renewing system like the epidermis)[24] on a human basal cell's chromosome 9q that starts the clonal expansion needed to accumulate the mutations required to disable the Ca^{2+} response mechanism and produce a basal cell carcinoma. This gene is called "patched" or *Ptch.*[329–339] The *Ptch* product, Ptch, is a large glycoprotein which is plugged into the cell membrane with 12 transmembrane α-helices.[330,333,340] Ptch complexes with, and inactivates, another membrane-inserted protein Smo ("smoothened"), a G-protein-coupled receptor with seven transmembrane α-helices.[330,333,340] The Ptch·Smo combination is really the modular receptor for Shh (sonic hedgehog protein), with Ptch being the ligand binder and Smo the signal generator which is inhibited by Ptch. When Shh binds to Ptch, Smo is derepressed and its signaling sends the Gli transcription factor from the cytoplasm into the nucleus where, like β-catenin, it stimulates the expression of cycle-driving genes (Fig. 3.6).[333,340]

Fig. 3.6. The mechanism that is the basis of basal cell carcinoma, the most frequent of the human cancers. The tumor cells have Smo ("smoothened") and Ptch ("patched") transmembrane proteins plugged into their plasma membranes. By itself, Smo would be active, but normally Smo and Ptch join to form a modular receptor for hedgehog proteins (Hh) in which Ptch silences Smo. When Hh binds to Ptch, the components separate and Smo becomes active. The Smo signaling causes the translocation of Gli1 into the nucleus where it combines with other factors to stimulate the expression of proliferation-driving genes. However, the tumor cells' Ptch genes have been mutationally disabled (symbolized as ***Ptch*** in the drawing) or lost, which unleashes Smo's constitutive proliferogenic activity. It is not yet known whether some or all *normal* basal cells use the Hh-activated Ptch·Smo mechanism to drive their cycles or whether expression or partial expression (e.g., only Smo) of the mechanism is restricted to being a first step in basal cell carcinogenesis.

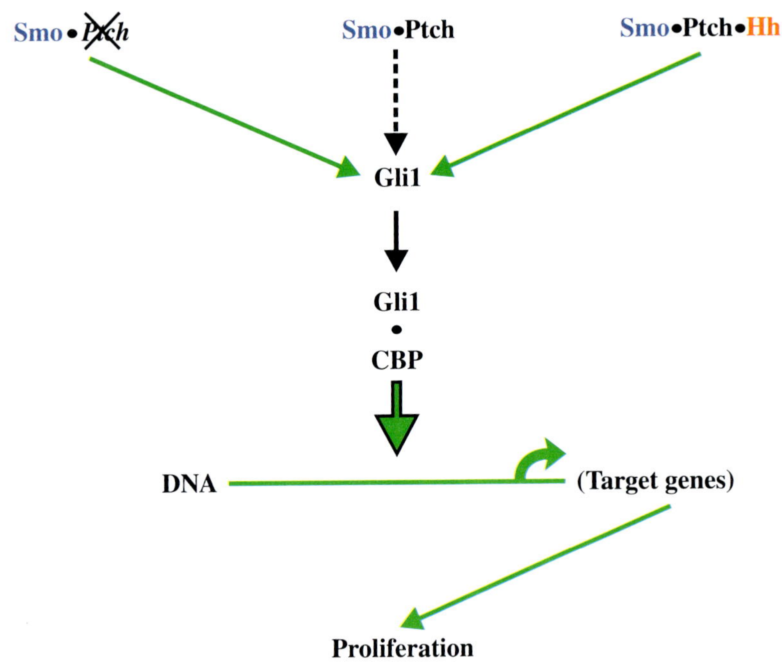

Moreover, human keratinocytes retrovirally transduced with Shh expressed BMP-2, Bcl-2 protector, and other genes characteristic of basal cell carcinoma cells and produced basal cell carcinoma-like growths in nude mice.[341] Indeed, the Gli pathway may connect to the β-catenin pathway by activating the expression of the Wnt gene(s).[342] The loss or disabling of Ptch that occurs in basal cell carcinomas constitutively activates Smo and consequently Gli target genes. This results in the relentless driving of proliferation and a refusal of the cells to respond to Ca^{2+} signaling to stop cycling and diffpoptose.[342a] But the question remains as to what role if any the Shh/Ptch·Smo/Gli mechanism has in normal basal cell proliferation in the mature epidermis apart from hair follicles, or does it appear in whole or in part only as the initial step in basal cell carcinogenesis?

Epidermal homeostasis is partly coordinated and maintained by the diffusion of cycle-promoters, cycle-repressors, and diffpoptosis promoters throughout the basal and suprabasal layers via gap junctions that are closed by the increasingly loud Ca^{2+} signals from the CaRs when the cell hits the high extracellular Ca^{2+} wall in the granular layer. The bulk of the evidence from a wide variety of epithelial and mesenchymal tumors indicates a common strong selection for mutants with disabled gap junctional intercellular communication.[102,343–357] Therefore, such cells cannot be proliferatively restrained by talking to their normal neighbors through gap junctions. However, we have already learned that the TPA-relieved, Ca^{2+}-requiring (*not* Ca^{2+}-inhibited) neighborhood suppression of the expansion of clones of early-stage initiated keratinocytes into papillomas is probably due to Ca^{2+}-dependent, β-catenin-sequestering, attachment of initiated cells to their normal neighbors by type E cadherins rather than gap junctions.[315–318] However, as the clonal selection of initiated keratinocytes advances toward malignancy, there is a progressive change in the structural components of gap junctions and reduction of gap junctional intercellular communication and transjunctional transmission of restraining signals from the normal cellular neighborhood.

The striking loss of gap junctional communication that can occur during the progression from papilloma to metastatic carcinoma is illustrated by the results of experiments by Klann et al.[306] Early, initiated keratinocytes were isolated from a dimethylbenz[*a*]anthracene-treated SENCAR mouse *on the basis of their resistance to proliferative inhibition by incubation in a high-Ca^{2+} medium.* Early- and late-stage papilloma cells and squamous carcinoma cells were generated by a standard in vivo initiation/promotion treatment program. The intercellular transfer of microinjected Lucifer Yellow CH dye between cells in a medium containing 0.05 mM Ca^{2+} dropped progressively and dramatically from a substantial 68 dye-coupled cells per injection for the early Ca^{2+}-resistant cells to a meager 3 dye-coupled cells for metastatic cells. Tumor promoters such as TPA profoundly affect the expression and production of the Cx (connexin) components of gap junctions of these SENCAR mouse keratinocytes.[345–347] Untreated normal cells express Cx31·1 and Cx43, but not Cx26, in vitro. However, tumor promoters switch on the Cx26 gene, transiently stimulate Cx43 gene expression, and suppress Cx31·1 gene expression. In the animal, keratinocytes also normally express Cx31·1 and Cx43, but not Cx26, but initiation by 7,12-dimethylbenz[*a*]anthracene and promotion by TPA results in Cx43 and Cx26 hyperexpression, the initiation of Cx26 hyperexpression, and repression of Cx31·1 expression by the papilloma stage.[340–347] But plasma membrane-associated Cx26 and Cx43 proteins, and presumably the gap junctions they would make, decline and disappear in the late papilloma and squamous carcinoma stages, despite a persistent production of high levels of Cx26 and Cx43 mRNAs.[345] Thus, for some reason the cancerous keratinocyte cannot translate its abundant mRNAs into Cx proteins and install gap junctions in its plasma membrane.

On the other hand, Holden et al.[358] have found no difference in the gap junctional communication between cells from cultures of normal murine keratinocytes, papillomas, or squamous cell carcinomas, but there was a large drop in the junctional communication during the progression from squamous to spindle cell carcinoma.[358] This large drop was not due to abnormal expression of Cx43, but rather to the lack of E-cadherin expression by the spindle cell carcinoma cells, because E-cadherin mediated cell–cell contact is needed to make functional gap junctions.[358]

The end of the road: *malignancy*

Carcinogenesis started with a rare mutation, which, instead of damaging or killing the cell, produces a cell with a radically altered response to tumor promoters and an impaired ability to diffpoptose.[314] Unless it be helped by a tumor promoter, this rare mutant cell will be suppressed and eventually vanish, or it will generate only a slowly growing clone of benign cells. But a tumor promoter will stimulate the cell and its descendants to proliferate into a papilloma. This promoter-driven clonal expansion produces the many rounds of DNA replication and opportunities for more mutations and the large number of cells needed to generate more mutation-loaded clones and eventually a malignant clone(s). While any persisting carcinogens would continue to produce mutations and thus enhance the progression to cancer, the tumor promoters that are selectively driving the expansion of the initial clone *at the expense of its normal neighbors* (which, by contrast, are being stimulated to diffpoptose rather than proliferate),[318] may also promote genetic instability and mutagenesis by triggering a shower of active oxygen radicals via the PKCs-stimulated plasma membrane-associated NAD(P)H oxidase.[359] These radicals together with Ca^{2+}-dependent endonuclease activity break chromosomes and thereby cause unequal chromosome exchanges with resulting gene deletions and rearrangements.[359] At first, the apoptogenic p53(TP53) safety-monitor we met in the last two chapters responds to severe chromosome damage by killing the cell to prevent the emergence of more advanced, and ultimately malignant, mutant clones.[359a] But eventually some cells may suffer a loss-of-function mutation and (or) lose their p53(TP53) gene and with it their safety-check mechanism.[284,360–365] Such cells could not accumulate enough nuclear p53(TP53) to stimulate the genes encoding the cycle-stopping, Cdk-inhibiting $p21^{WAF1/CIP1}$ and the apoptogenic Gadd45 and to organize the repair of damaged genes.[322,362,33] It might be expected that initiated, promoted, and transforming cells might also continue to produce or even over-express the anti-apoptosis Bcl-2 protector protein (which was expressed by their normal basal cell ancestor) to specifically prevent the Ca^{2+}-triggered apoptosis. This would further enhance their resistance to diffpoptosis especially before the loss of p53(TP53).[366] However, the evidence so far indicates that basal cell carcinomas, like normal basal cells, make Bcl-2 protector protein and do so in large amounts, but squamous cell carcinomas *do not*.[131,367] The loss of functional p53(TP53) during clonal progression removes a major obstacle to the ultimate appearance and expansion of a malignant clone, the cells of which ignore, indeed cannot respond to, Ca^{2+}'s iron-fisted command to stop cycling and diffpoptose.

An important requirement for the emergence of a malignant clone of invasive metastasizing cells is a failure to express E-cadherins with a resulting loss of the Ca^{2+}-dependent anchorage to neighboring cells by the cadherins.[368,369] This could be due, for example, to hypermethylation of CpGs around the E-cadherin gene promoter and reconfiguration of the promoter chromatin, which prevents the binding of transcription factors and thus silences the gene.[370,371] However, there may be an intermediate stage in which initiated keratinocytes still expressing E-cadherin receptors in the presence of enough Ca^{2+} can (unlike normal keratinocytes) avoid anoikis and proliferate without signals from adhesion-activated β1 integrins by being included in clusters of Ca^{2+}·cadherin-attached, hence pseudo-adhered, cells.[372] In such clusters, the cells are close enough to each other to get strong survival- and proliferation-sustaining signals from soluble paracrine cytokines and EGF/TGF-α receptors activated by large juxtacrine TGF-α precursors sticking out of their neighbors' membranes.[69,70,372] A *premature* loss of E-cadherins would cause detached, although initiated, keratinocytes to die like detached normal cells by anoikis.[371]

Malignant conversion also needs increased AP1 transcription activity to turn on or turn up the expression of the anti-apoptosis Bcl-2 protector gene, the genes for collagenase and other proteases (e.g., stromelysin), the VEGF (vascular endothelial growth factor) gene and to turn off a novel keratinoctye tumor suppressor gene which is strongly expressed in both normal and papilloma cells.[283] These deadly converts, liberated by their mutant Ha-Ras from a dependence on adhesion-induced integrin signaling, protected from apoptosis by Bcl-2 and "immortalized" at least partially by their ever-active c-Myc stimulating the expression of their telomerase gene, can chop their way through the basement membrane with their proteinases, invade dermal blood vessels, and start new colonies in distant places where they use their VEGF to induce the sprouting of local blood vessels to nourish them.[22,35,283,373] However, these new blood vessels are not just conduits for food and oxygen. Their endothelial cells also promote the proliferation and survival of the tumor cells crowding around them by secreting paracrine growth factors such as bFGF (FGF-2) and anti-apoptosis survival factors such as IGFs.[375] However, they may have a problem making new blood vessels and producing a growing tumor because the metalloproteinases that they used to escape from the mother tumor also chop off the N-terminal parts of plasminogen and collagen XVIII to respectively release the angiostatin and endostatin fragments that block the formation of the colony's blood vessels by overriding the angiogenic actions of FGF and VEGF.[374,375]

A farewell to the skin

We will now visit another example of a cell system where we will see Ca^{2+} exercising its full range of cycle- and diffpoptosis-driving functions — the colon epithelium. But before trying to piece together what happens to a colonocyte, especially from the Ca^{2+} viewpoint, as it glides up the wall of a colon crypt, let us take one final brief look at the diffpoptotic development of a keratinocyte as it rises to the top of the epidermal stack, since we shall see many similarities in the two systems. The epidermal population is maintained by the β1-integrin-primed proliferation on the basement membrane of diffpoptosis-committed transit amplifying cells, which are the descendants of niche-bound stem cells. The stratification and the stepwise differentiation of the cells started by the interruption of β1 integrin signaling are driven by three main Ca^{2+} signals. The first signal is triggered by detachment from the basal lamina and starts the production of keratins, some components of the future cornified envelope, and the expression and display of the Ca^{2+}-activated E-cadherin molecules that tether the cell to its neighbors rather than the basal lamina. The next signal is triggered by autocrine/paracrine factors such as PTHrP in the spinous layer to start and control the granular cell program of production of the rest of the corneocyte components. The final massive signal is triggered by the stimulation of CaRs by the sharply increased external Ca^{2+} level in the upper granular and transitional granular/horny layers. It drives the assembly of the accumulated components by a poised apoptosis mechanism to make a dead, keratin-packed, horny-shelled corneocyte.

References

1. Whitfield JF. Calcium: Cell Cycle Driver, Differentiator and Killer. Austin, R.G. Landes Company, 1997.
2. Dover R. Cell kinetics of keratinocytes. In: Leigh IM, Lane LE, Watt F (editors). The Keratinocyte Handbook. Cambridge, Cambridge University Press, 1994: 203–234.
3. Potten CS. Cell replacement in the epidermis (keratopoiesis) via discrete units of proliferation. Internat Rev Cytol 1981; 69: 271–318.
4. Bata-Csorgo Z, Hammerberg C, Voorhees JJ, et al. Flow cytometric identification of proliferative subpopulations within normal human epidermis and the localization of the primary hyperproliferative population in psoriasis. J Exp Med 1993; 178: 1271–1281.
5. Jones PH. Isolation and characterization of human epidermal stem cells. Clin Sci 1996; 91: 141–146.
6. Jones PH, Harper S, Watt F. Stem cell patterning and fate in human epidermis. Cell 1995; 80: 83–93.
7. Lavker RM, Sun T-T. Heterogeneity in epidermal basal keratinocytes: morphological and functional correlations. Science 1982; 215: 1239–1241.
8. Lavker RM, Sun T-T. Epidermal stem cells. J Invest Dermatol 1983; 81: 121s–127s.
9. Pavlovitch JH, Rizk-Rabin M, Jaffray P, et al. Characteristics of homogeneously small keratinocytes from newborn rat skin: possible stem cells. Am J Physiol 1991; 261: C964–C972.

9*a*. Terskikh VV, Vasiliev AV. Cultivation and transplantation of epidermal keratinocytes. Int Rev Cytol 1999; 188: 41–72.

10. Watt FM. Epidermal stem cells as targets for gene transfer. Human Gene Ther 2000; 11: 2261–2266.
11. Watt FM. Epidermal stem cells: markers, patterning and the control of stem cell fate. Philos Trans R Soc Lond B Biol Sci 1998; 353: 831–837.
12. Whitfield JF. Calcium in Cell Cycles and Cancer. Boca Raton, CRC Press, 1995: 123–151.

13. Parkinson EK, Grabham P, Emmerson A. A subpopulation of cultured human keratinocytes which is resistant to the induction of terminal differentiation-related changes by phorbol-12 myristate, 13-acetate: evidence for an increase in the resistant population following transformation. Carcinogenesis 1983; 4: 857–861.
14. Parkinson EK, Pera MF, Emmerson A, et al. Differential effects of complete and second-stage tumour promoters in normal but not transformed human and mouse keratinocytes. Carcinogenesis 1984; 5: 1071–1077.
15. Hodivala KJ, Watt FM. Evidence that cadherins play a role in the down-regulation integrin expression that occurs during keratinocyte terminal differentiation. J Cell Biol 1994; 124: 589–600.
16. Hotchin NA, Gandarillas A, Watt FM. Regulation of cell surface beta 1 integrin levels during keratinocyte terminal differentiation. J Cell Biol 1995; 128: 1209–1219.
17. Sheppard D. Epithelial integrins. BioEssays 1996; 18: 655–660.
18. Watt FM, Kubler MD, Hotchin NA, et al. Regulation of keratinocyte terminal differentiation by integrin-extracellular matrix interactions. J Cell Sci 1993; 106: 175–182.
19. Watt FM, Hertle MD. Keratinocyte integrins. In: Leigh IM, Lane LE, Watt FM, editors. The Keratinocyte Handbook. Cambridge, Cambridge University Press, 1994: 153–164.
20. Rizk-Rabin M, Pavlovitch JH. Epidermal calcium-binding protein: a marker of early differentiation of basal layer keratinocytes of rats. Cell Tissue Res 1993; 272: 161–168.
21. Muniyappa K, Kironmai KM. Telomere structure, replication and length maintenance. Crit Rev Biochem Mol Biol 1998; 33: 297–336.
22. Ogoshi M, Le T, Shay JW, et al. In situ hybridization analysis of the expression of human telomerase RNA in normal and pathologic conditions of the skin. J Invest Dermatol 1998; 110: 818–823.
23. Sedivy JM. Can ends justify the means?: telomeres and the mechanism of replicative senescence and immortalization in mammalian cells. Proc Natl Acad Sci USA 1998; 95: 9078–9081.
24. Kinzler KW, Vogelstein B. Landscaping the cancer terrain. Science 1998; 280: 1036–1037.
25. Barrandon Y, Morgan JR, Mulligan RC, et al. Restoration of growth potential in paraclones of human keratinocytes by aviral oncogene. Proc Natl Acad Sci USA 1989; 86: 4102–4106.
25*a*. Hullemann E, Boostra J. Regulation of G1 phase progression by growth factors and the extracellular matrix. Cell Mol Life Sci 2001; 58: 80–93.
26. Delcommenne M, Tan C, Gray V, et al. Phosphoinositide-OH kinase-dependent regulation of glycogen synthase kinase 3 and protein kinase B/AKT by the integrin-linked kinase. Proc Natl Acad Sci USA 1998; 95: 11211–11216.
27. Hannigan GE, Leung-Hagesteljn C, Fitz-Gibbon L, et al. Regulation of cell adhesion and anchorage-dependent growth by a new β1-integrin-linked protein kinase. Nature 1996; 379: 91–98.

28. Wymann MP, Pirola L. Structure and function of phosphoinositide3-kinases. Biochim Biophys Acta 1998; 1436: 127–150.
29. Xie W, Li F, Kudlow JE, et al. Expression of the integrin-linked kinase (ILK) in mouse skin. Am J Pathol 1998; 153: 367–372.
30. Khwaja A, Rodriguez-Viciana P, Wennström S, et al. Matrix adhesion and Ras transformation both activate a phosphoinositide 3-OH kinase and protein kinase B/Akt cellular survival pathway. EMBO J 1997; 16: 2782–2793.
31. Shepherd PR, Withers DJ, Siddle K. Phosphoinositide3-kinase: the key switch mechanism in insulin signalling. Biochem J 1998; 333: 471–490.
31*a*. Bouchard C, Thieke K, Maier A, et al. Direct induction of cyclin D2 by Myc contributes to cell cycle progression and sequestration of p27. EMBO J 1999; 18: 5321–5333.
32. He T-C, Sparks AB, Rago C, et al. Identification of c-*myc* as a target of the APC pathway. Science 1998; 281: 1509–1512.
32*a*. Hermeking H, Rago C, Shuhmacher M, et al. Identification of CDK4 as a target of c-MYC. Proc Natl Acad Sci USA 2000; 97: 2229–2234.
33. Shtutman M, Zhurinsky J, Simcha I, et al. The cyclin D1 gene is a target of the β-catenin/LEF-1 pathway. Proc Natl Acad Sci USA 1999; 96: 5522–5527.
34. Novak A, Hsu SC, Leung-Hagersteijn C, et al. Cell adhesion and the integrin-linked kinase regulate the LEF-1 and β-catenin signaling pathways. Proc Natl Acad Sci USA 1998; 95: 4374–4379.
35. Dang CV. cMyc target genes involved in cell growth, apoptosis, and metabolism. Mol Cell Biol 1999; 19: 1–11.
36. Wu C, Keightley SY, Leung-Hagersteijn C, et al. Integrin-linked protein kinase regulates fibronectin matrix assembly, E-cadherin expression, and tumorigenicity. J Biol Chem 1998; 273: 528–536.
37. Coffer PJ, Jin J, Woodgett JR. Protein kinase B (c-Akt): a multifunctional mediator of phosphatidylinositol 3-kinase activation. Biochem J 1998; 335: 1–13.
38. Downward J. Lipid regulated kinases some common themes at last. Science 1998; 279: 673–674.
39. Barth AIM, Näthke IS, Nelson J. Cadherins, catenins and APC protein: interplay between cytoskeletal complexes and signalling pathways. Curr Opin Cell Biol 1997; 9: 683–690.
40. Hatzfeld M. The armadillo family of proteins. In Rev Cytol 1999; 186: 179–224.
41. Behrens J. Cadherins and catenins: role in signal transduction and tumor progression. Cancer and Metastasis Reviews 1999; 18: 15–30.
42. Yokoya F, Immamoto N, Tachibana T, et al. β-Catenin can be transported into the nucleus in a Ran-unassisted manner. Mol Biol Cell 1999; 10: 1119–1131.
43. Werner MH, Burley SK. Architectural transcription factors: proteins that remodel DNA. Cell 1997; 88: 733–736.
44. Pennisi E. How a growth control path takes a wrong turn to cancer. Science 1998; 281: 1438–1440.
44*a*. Roose J, Clevers H. TCF transcription factors: molecular switches in carcinogenesis. Biochim Biophys Acta 1999; 1424: M23–M27.

45. Munemitsu S, Albert I, Souza B, et al. Regulation of intracellular β-catenin levels by the adenomatous polyposis coli (APC) tumor-suppressor protein. Proc Natl Acad Sci USA 1995; 92: 3046–3050.
46. White RL. Colon cancer: molecular biology of the APC protein. Path Biol 1997; 45: 240–244.
47. Hart MJ, los Santos R, Albert IN, et al. Downregulation of β-catenin by human axin and its association with the APC tumor suppressor, β-catenin and GSK3β. Curr Biol 1998; 8: 573–581.
48. Brown JD, Moon RT. Wnt signalling: why is everything so negative? Curr Opin Cell Biol 1998; 10: 182–187.
49. Papkoff J, Rubinfeld B, Schryver B, et al. Wnt-1 regulates free pools of catenins and stabilizes APC-catenin complexes. Mol Cell Biol 1996; 16: 2128–2134.
49*a*. Isshiki M, Anderson RG. Calcium signal transduction from caveolae. Cell Calcium 1999; 26: 201–208.
50. Wary KK, Mainiero F, Isakoff SJ, et al. The adaptor protein Shc couples a class of integrins to the control of cell cycle progression. Cell 1996; 87: 733–743.
51. Chong LD, Traynor-Kaplan A, Bokoch GM, et al. The small GTP-binding protein rho regulates a phosphatidylinositol phosphate 5-kinase in mammalian cells. Cell 1994; 79: 507–513.
52. Guan J-L, Chen H-C. Signal transduction in cell–matrix interactions. Int Rev Cytol 1996; 168: 81–121.
53. Sjaastad MD, Nelson WJ. Integrin-mediated calcium signalling and regulation of cell adhesion by intracellular calcium. BioEssays 1997; 19: 47–55.
54. McNamee HM, Ingber DE, Schwartz MA. Adhesion to fibronectin stimulates inositol lipid synthesis and enhances PDGF-induced inositol lipid breakdown. J Cell Biol 1993; 121: 673–678.
55. Meredith JE, Winitz S, Lewis JM, et al. The regulation of growth and intracellular signaling by integrins. Endocrine Revs 1996; 17: 207–220.
56. Howe A, Aplin E, Alahari S, et al. Integrin signaling and cell growth control. Curr Opin Cell Biol 1998; 10: 220–231.
57. Bowen ID, Bowen SM, Jones AH. Mitosis and Apoptosis. London, Chapman & Hall, 1998.
58. Maruoka Y, Harada H, Mitsuyasu T, et al. Keratinocytes become terminally differentiated in a process involving programmed cell death. Biochem Biophys Res Commun 1997; 238: 886–890.
59. Polakowska RR, Piacentini M, Bartlett R, et al. Apoptosis in human skin development: morphogenesis, periderm, and stem cells. Devel Dynamics 1994; 199: 176–188.
60. Danker K, Gabriel B, Heidrich C, et al. Focal adhesion kinase pp125^{FAK} and the β1 integrin subunit are constitutively complexed HaCaT cells. Exp Cell Res 1998; 239: 326–331.
61. Chou MM, Hou W, Johnson J, et al. Regulation of protein kinase C ζ by PI3-kinase and PDK-1. Curr Biol 1998; 8: 1069–1077.

62. Guadagno TM, Ohtsuka M, Roberts JM, et al. A link between cyclin A expression and adhesion-dependent cell cycle progression. Science 1993; 262: 1572–1575.
63. Symington BE. Fibronectin receptor modulates cyclin-dependent kinase activity. J Biol Chem 1992; 267: 25744–25747.
64. Symington BE. Growth signalling through the $\alpha_5\beta_1$ fibronectin receptor. Biochem Biophys Res Commun 1995; 208: 126–134.
65. Del Sal G, Ruaro EM, Utrera R, et al. Gas1-induced growth suppression requires a transactivation-independent p53 function. Mol Cell Biol 1995; 15: 7152–7160.
66. Varner JA, Emmerson DA, Juliano RL. Integrin alpha 5 beta 1 expression negatively regulates cell growth: reversal by attachment to fibronectin. Mol Cell Biol 1995; 6: 725–740.
66*a*. Langlands K, Down GA, Kealey T. Id proteins are dynamically expressed in normal dermis and dysregulated in squamous cell carcinoma. Cancer Res 2000; 60: 5929–5933.
67. Hunter T. When is a lipid kinase not a lipid kinase? When it is a protein kinase. Cell 1995; 83: 1–4.
68. Werner S. Keratinocyte growth factor: a unique player in epithelial repair processes. Cytokine Growth Factor Revs 1998; 9: 153–165.
69. Brachman R, Lindquist PB, Wagashima M, et al. Transmembrane TGF-α precursors activate EGF/TGF-α receptors. Cell 1989; 56: 691–700.
70. Wong ST, Winchell LF, McCune BK, et al. The TGF-α precursor on the cell surface binds to the EGF receptor on adjacent cells, leading to signal transduction. Cell 1989; 56: 495–506.
71. Grundin TG, Roomans GM, Forslund B, et al. X-ray microanalysis of psoriatic skin. J Invest Dermatol 1985; 85: 378–380.
72. Ahn SK, Hwang SM, Jiang SJ, et al. The changes of epidermal calcium gradient and transitional cells after prolonged occlusion following tape stripping in the epidermis. J Invest Dermatol 1999; 113: 189–195.
73. Elias PM, Nau P, Hanley K, et al. Formation of the epidermal calcium gradient coincides with key milestones of barrier ontogenesis in the rodent. J Invest Dermatol 1998; 110: 399–404.
74. Menon GK, Elias PM. Ultrastructural localization of calcium in psoriatic and normal human epidermis. Arch Dermatol 1991; 127: 57–63.
75. Menon GK, Grayson S, Elias PM. Ionic calcium reservoirs in mammalian epidermis: ultrastructural localization by ion-capture cytochemistry. J Invest Dermatol 1985; 84: 508–512.
76. Elias PM. Stratum corneum architecture, metabolic activity and interactivity with subjacent cell layers. Exp Dermatol 1996; 5: 191–201.
77. Vicanova J, Boelsma E, Mommaas AM, et al. Normalization of epidermal calcium distribution profile in reconstructed human epidermis is related to improvement of terminal differentiation and stratum corneum barrier formation. J Invest Dermatol 1998; 111: 97–106.

78. Sánchez-Mateos P, Cabañas C, Sánchez-Madrid F. Regulation of integrin function. Seminars Cancer Biol 1996; 7: 99–109.
79. Lu KP, Means AR. Regulation of cell cycle by calcium and calmodulin. Endocrine Revs 1993; 14: 40–58.
80. Brown EM, MacLeod RJ. Extracellular calcium sensing and extracellular calcium signaling. Physiol Rev 2001; 81: 239–297.
81. Chattopadhyay N, Mithal A, Brown EM. The calcium-sensing receptor: a window into the physiology and pathophysiology of mineral ion metabolism. Endocrine Revs 1996; 17: 289–307.
82. Oda Y, Tu C-L, Pillai S, et al. The calcium sensing receptor and its alternatively spliced form in keratinocyte differentiation. J Biol Chem 1998; 273: 23344–23352.
83. Barrit GJ. Receptor-activated Ca^{2+} inflow in animal cells: a variety of pathways tailored to meet different intracellular Ca^{2+} signalling requirements. Biochem J 1999; 337: 153–169.
84. Li L, Tennenbaum T, Yuspa SH. Suspension-induced murine keratinocyte differentiation is mediated by calcium. J Invest Dermatol 1996; 106: 254–260.
85. Li L, Tucker RW, Hennings H, et al. Inhibitors of the intracellular Ca^{2+}-ATPase in cultured mouse keratinocytes reveal components that are regulated by distinct intracellular Ca^{2+} compartments. Cell Growth Diff 1995; 6: 1171–1184.
86. Glick AB, Danielpour D, Morgan D, et al. Induction and autocrine receptor binding of transforming growth factor beta-2 during terminal differentiation of primary mouse keratinocytes. Mol Endocrinol 1990; 4: 46–52.
87. Sachsenmeier KF, Sheibani N, Schlosser SJ, et al. Transforming growth factor-beta1 inhibits nucleosomal fragmentation in human keratinocytes following loss of adhesion. J Biol Chem 1996; 271: 5–8.
88. Hauser PJ, Agrawal D, Pledger WJ. Primary keratinocytes have an adhesion dependent S phase checkpoint that is absent in immortalized cell lines. Oncogene 1998; 17: 3083–3092.
89. Medema JP, Sark MWJ, Backendorf C, et al. Calcium inhibits epidermal growth factor-induced activation of $p21^{ras}$ in human primary keratinocytes. Mol Cell Biol 1994; 14: 7078–7085.
90. Pillai S, Bikle DD, Mancianti ML, et al. Calcium regulation of growth and differentiation of normal keratinocytes: modulation of differentiation competence by stages of growth and extracellular calcium. J Cell Physiol 1990; 143: 294–302.
91. Yuspa SH, Kilkenny A, Cheng C, et al. Alterations in epidermal biochemistry as a consequence of stage-specific genetic changes in skin carcinogenesis. Environ Health Perspect 1991; 93: 3–10.
92. Yuspa SH, Kilkenny AE, Steinert PM, et al. Expression of murine epidermal differentiation markers is tightly regulated by restricted extracellular calcium concentrations in vitro. J Cell Biol 1989; 109: 1207–1217.
93. Fuchs E. Epidermal differentiation and keratin gene expression. J Cell Sci 1993; Suppl 17: 197–208.

94. Reichert U, Michel S, Schmidt R. The cornified envelope: a key structure of terminally differentiating keratinocytes. In: Darmon M, Blumenberg M, editors. Molecular Biology of the Skin: The Keratinocyte. San Diego, Academic Press, 1993: 107–150.
94*a*. Lowell S, Jones P, LaRoux I, et al. Stimulation of epidermal differentiation by Delta–Notch signalling at the boundaries of stem-cell clusters. Curr Biol 2000; 10: 491–500.
95. Brissette JL, Kumar NM, Gilula NB, et al. Switch in gap junction protein expression is associated with selective changes in junctional permeability during keratinocyte differentiation. Proc Natl Acad Sci USA 1994; 91: 6453–6457.
96. Cristophers E, Schubert C, Goos M. The epidermis. Hdbk Exper Pharmacol 1982; 81(I): 3–30.
97. Fitzgerald DJ, Fusenig NE, Boukamp P, et al. Expression and function of connexin in normal and transformed human keratinocytes in cultures. Carcinogenesis 1994; 15: 1859–1865.
98. Elias PM. Epidermal lipid membranes and keratinization. Int J Dermatol 1981; 20: 1–19.
99. Kam E, Melville L, Pitts JD. Patterns of junctional communication in skin. J Invest Dermatol 1988; 87: 748–753.
100. Kam E, Watt FM, Pitts JD. Patterns of junctional communication in skin: studies on cultured keratinocytes. Exp Cell Res 1987; 173: 431–438.
101. Salomon D, Masgrau E, Vischer S. Topography of mammalian connexins in human skin. J Invest Dermatol 1994; 103: 240–247.
102. Sawey MJ, Goldschmidt MH, Risek B. Perturbation in connexin 43 and connexin 26 gap- junction expression in mouse skin hyperplasia and neoplasia. Mol Carcinogen 1996; 17: 49–61.
103. Dlugosz AA, Cheng C, Denning MF. Keratinocyte growth factor receptor ligands induce transforming growth factor alpha expression and activate the epidermal growth factor receptor signaling pathway in cultured epidermal keratinocytes. Cell Growth Diff 1994; 5: 1283–1292.
104. Holbrook KA. Ultrastructure of the epidermis. In: Leigh IM, Lane EB, Watt FM, editors. The Keratinocyte Handbook. Cambridge, Cambridge University Press, 1994: 3–39.
104*a*. Guo QM, Malek RL, Kim S, et al. Identification of c-Myc responsive genes using rat. cDNA microarray. Cancer Res 2000; 60: 5922–5928.
105. Kaldis P. The cdk-activating kinase (CAK): from yeast to mammals. Cell Mol Life Sci (CMLS) 1999; 55: 284–296.
106. Shtutman M, Zhurinsky J, Simcha I, et al. The cyclin D1 gene is a target of the β-catenein/LEF-1 pathway. Proc Natl Acad Sci USA 1999; 96: 5522–5527.
107. Barrett J, Lewis BC, Hoang AT. Cyclin A links c-Myc to adhesion-independent proliferation. J Biol Chem 1995; 270: 15923–15925.
108. Boudreau N, Werb Z, Bissell MJ. Suppression by basement membrane requires three-dimensional tissue organization and withdrawal from the cell cycle. Proc Natl Acad Sci USA 1996; 93: 3509–3513.

109. Henriksson M, Lüscher B. Proteins of the Myc network: essential regulators of cell growth and differentiation. Adv Cancer Res 1996; 68: 109–182.
110. Hoang AT, Cohen KJ, Barrett JF. Participation of cyclin A in Myc-induced apoptosis. Proc Natl Acad Sci USA 1994; 91: 6875–6879.
111. Ryan K, Birnie GD. *Myc* oncogenes: the enigmatic family. Biochem J 1996; 314: 713–721.
112. Dou QP, Pardee AB. Transcriptional activation of thymidine kinase, a marker for cell cycle control. Progr Nucl Acid Res 1996; 53: 197–217.
113. Slansky JE, Farnham PJ. Transcriptional regulation of the dihyrofolate reductase gene. BioEssays 1996; 18: 55–62.
114. Jiang CK, Tomic-Canic M, Lucas DJ. TGF beta promotes the basal phenotype of epidermal keratinocytes: transcriptional induction of K#5 and K#14 keratin genes. Growth Factors 1995; 12: 87–97.
115. Glick AB, Kulkarni AB, Tennenbaum T. Loss of expression of transforming growth factor beta in skin and skin tumors is associated with hyperproliferation and a high risk for malignant conversion. Proc Natl Acad Sci USA 1993; 90: 6076–6080.
116. Kato M, Ishizaki A, Hellman U, et al. A human keratinocyte cell line produces two autocrine growth inhibitors, transforming growth factor-beta and insulin-like growth factor binding protein-6, in a calcium and cell density-dependent manner. J Biol Chem 1995; 270: 12373–12379.
117. Tian JQ, Quaroni A. Involvement of p21(WAF1/Cip1) and p27(Kip1) in intestinal epithelial cell differentiation. Am J Physiol 1999; 276: C1245–C1258.
118. Alani RM, Hasskarl J, Munger K, et al. Alterations in cyclin-dependent kinase 2 function during differentiation of primary human keratinocytes. Mol Carcinog 1998; 23: 226–233.
119. DiCunto F, Topley G, Calautti E, et al. Inhibitory function of $p21^{Cip1/WAF1}$ in differentiation of primary mouse keratinocytes independent of cell cycle control. Science 1998; 280: 1069–1072.
120. Missero C, Calautti E, Eckner R, et al. Involvement of the cell-cycle inhibitor Cip1/WAF1 and the E1A-associated p300 protein in terminal differentiation. Proc Natl Acad Sci USA 1995; 92: 5451–5455.
121. Martinez LA, Chen Y, Fischer SM, et al. Coordinated changes in cell cycle machinery occur during keratinocyte terminal differentiation. Oncogene 1999; 18: 397–406.
122. Hauser PJ, Agrawal D, Flanagan M, et al. The role of $p27^{Kip1}$ in the in vitro differentiation of murine keratinocytes. Cell Growth Differ 1997; 8: 203–211.
123. Polyak K. Negative control of cell growth by TGFβ. Biochim Biophys Acta 1996; 1242: 185–199.
124. Reynisdottir I, Polyak K, Massaguie J. Kip/Cip and Ink4 Cdk inhibitors cooperate to induce cell cycle arrest in response to TGF-β. Genes Dev 1995; 19: 1831–1845.
125. Dotto GP, Gilman MZ, Maruyama M. c-myc and c-fos expression in differentiating mouse primary keratinocytes. EMBO J 1986; 5: 2853–2857.

126. Sharpe GR, Fisher C, Redfern CPF. Changes in oncogene mRNA expression during human keratinocyte differentiation. Arch Dermatol Res 1994; 286: 476–480.
127. Younus J, Gilchrest BA. Modulation of mRNA levels during human keratinocyte differentiation. J Cell Physiol 1992; 152: 232–239.
128. Bachs O, Agell N. Calcium and Calmodulin Function in the Cell Nucleus. Austin, RG Landes Company, 1995.
129. Lin S, Hickey R, Malkas L. The biochemical status of the DNA synthesome can distinguish between permanent and temporary cell growth arrest. Cell Growth Diff 1997; 8: 1359–1369.
130. Boudreau N, Sympson CJ, Werb, et al. Suppression of ICE and apoptosis in mammary epithelial cells by extracellular matrix. Science 1995; 267: 891–893.
131. Cerroni L, Kerl H. Aberrant bcl-2 protein expression provides a possible mechanism of neoplastic cell growth in cutaneous basal-cell carcinoma. J Cutan Pathol 1994; 21: 398–403.
132. Hockenbery DM, Zutter M, Hickey W, et al. BCL2 protein is topographically restricted in tissues characterized by apoptotic cell death. Proc Natl Acad Sci USA 1992; 88: 6961–6965.
133. Marthinuss J, Andrade-Gordon P, Seiberg M. A secreted serine protease can induce apoptosis in Pam 212 keratinocytes. Cell Growth Diff 1995; 6: 807–816.
134. Marthinuss J, Lawrence L, Seiberg M. Apoptosis in Pam212, an epidermal keratinocyte cell line: a possible role for bcl-2 in epidermal differentiation. Cell Growth Diff 1995; 6: 239–250.
135. McCall C, Cohen J. Programmed cell death in terminally differentiating keratinocytes: role of endogenous endonucleases. J Invest Dermatol 1991; 97: 111–115.
136. Frisch SM, Francis H. Disruption of epithelial cell–matrix interaction induces apoptosis. J Cell Biol 1994; 124: 619–626.
137. Livesay FJ. Netrins and netrin receptors. Cell Mol Life Sci (CMLS) 1999; 56: 62–68.
138. Mehllen P, Rabizadeh S, Snipas SJ, et al. The DCC gene product induces apoptosis by a mechanism requiring receptor proteolysis. Nature 1998; 395: 801–804.
139. Hines MD, Allen-Hoffmann BL. Keratinocyte growth factor inhibits cross-linked envelope formation and nucleosomal fragmentation in cultured human keratinocytes. J Biol Chem 1996; 271: 6245–6251.
139*a*. Sun L, Sun T-T, Lavker RM. CLED: A calcium-linked protein associated with early differentiation. Exp Cell Res 2000; 259: 96–106.
140. Maytin EV, Habener JF. Transcription factors C/EBPα. C/EBPβ, and CHOP (GADD153) expressed during the differentiation program of keratinocytes in vitro and in vivo. J Invest Dermatol 1998; 110: 238–246.
141. Yoon SJ, LeBlanc-Straceski J, Ward D, et al. Organization of the human keratin type II gene cluster at 12q13. Genomics 1994; 24: 502–508.

142. Huff CA, Yuspa SH, Rosenthal D. Identification of control elements 3′ to the human keratin 1 gene that regulates cell type and differentiation-specific expression. J Biol Chem 1993; 268: 377–384.
143. Rothnagel JA, Greenhalgh DA, Gagne TA, et al. Identification of a calcium-inducible, epidermal-specific regulatory element in the 3′-flanking region of the human keratin 1 gene. J Invest Dermatol 1993; 101: 506–513.
143*a*. Zhu S, Oh HS, Shim M, et al. C/EBPβ modulates the early events of keratinocyte differentiation involving growth arrest and keratin 1 and keratin 10 expression. Mol Cell Biol 1999; 19: 7181–7190.
144. Wegner M, Cao Z, Rosenfeld MG. Calcium-regulated phosphorylation within the leucine zipper of C/EBPβ. Science 1992; 256: 370–373.
145. Yukawa K, Tanaka T, Tsuji S, et al. Expressions of CCAAT/ enhancer-binding proteins β and δ and their activities are intensified by cAMP signaling as well as Ca^{2+}/calmodulin kinases activation in hippocampal neurons. J Biol Chem 1998; 273: 31345–31351.
146. Hodivala KT, Watt FM. Evidence that cadherins play a role in the down-regulation of integrin expression that occurs during keratinocyte terminal differentiation. J Cell Biol 1994; 124: 589–600.
147. Hotchin NA, Gandarillas A, Watt FM. Regulation of cell surface β1 integrin levels during keratinocyte terminal differentiation. J Cell Biol 1995; 128: 1209–1219.
148. Huber O. Nuclear localization of β-catenin by interaction with transcription factor LEF-1. Mech Dev 1996; 59: 3–10.
149. Calautti E, Caboli S, Stein PL, et al. Tyrosine phosphorylation and src family kinases control keratinocyte cell–cell adhesion. J Cell Biol 1998; 141: 1449–1465.
150. Marrs JA, Nelson WJ. Cadherin cell adhesion molecules in differentiation and embryogenesis. Int Rev Cytol 1996; 165: 159–205.
151. Nagar B, Overduin M, Ikura M, et al. Structural basis of calcium-induced E-cadherin rigidification and dimerization. Nature 1996; 380: 360–364.
152. Lewis JE, Jensen PJ, Wheelock MJ. Cadherin function is required for human keratinocytes to assemble desmosomes and stratify in response to calcium. J Invest Dermatol 1994; 102: 870–877.
153. St Croix B, Sheehan C, Rak JW, et al. E-Cadherin-dependent growth suppression is mediated by the cyclin-dependent kinase inhibitor $p27^{KIP1}$. J Cell Biol 1998; 142: 557–571.
154. Fuchs E. Epidermal differentiation and keratin gene expression. J Cell Sci 1993; Suppl 17: 197–208.
155. Kartasova T, Roop DR, Holbrook KA. Mouse differentiation-specific keratins 1 and 10 require a preexisting keratin scaffold to form a filament network. J Cell Biol 1993; 120: 1251–1261.
156. Steinert PM, Marekov LN, Parry DA. Conservation of keratin intermediate filaments: molecular mechanisms by which different keratin molecules integrate into preexisting keratin intermediate filaments during differentiation. Biochemistry 1993; 32: 10046–10056.

157. Kendrew J (editor). The Encyclopedia of Molecular Biology (on CD ROM). Oxford, Blackwell Science Ltd., 1995.
158. Kim SY, Chung SI, Steinert PM. Highly active soluble processed forms of the transglutaminase 1 enzyme in human epidermal keratinocytes. J Biol Chem 1995; 270: 18026–18035.
159. Rice RH, Mehrpouyam M, Qin Q. Transglutaminases in keratinocytes. In: Leigh M, Lane B, Watt F, editors. The Keratinocyte Handbook. Cambridge, Cambridge University Press, 1994: 259–274.
160. Simon M. The epidermal cornified envelope and its precursors. In: Leigh M, Lane B, Watt F, editors. The Keratinocyte Handbook. Cambridge, Cambridge University Press, 1994: 275–292.
161. Steinert PM, Candi E, Kartasova T, et al. Small proline-rich proteins are cross-bridging proteins in the cornified envelopes of stratified squamous epithelia. J Struct Biol 1998; 122: 76–85.
162. Stewart PM, Chung SI, Kim SY. Inactive zymogen and highly inactive proteolytically processed membrane-bound forms of the transglutaminase 1 enzyme in human epidermal keratinocytes. Biochem Biophys Res Commun 1996; 221: 101–106.
163. Lopez-Bayghen E, Vega A, Cadena A. Transcriptional analysis of the 5′-noncoding region of the human involucrin gene. J Biol Chem 1996; 271: 512–520.
163*a*. Ng DC, Shafaee S, Lee D, et al. Requirement of an AP-1 site in the calcium response region of the involucrin promoter. J Biol Chem 2000; 275: 24080–24088.
164. Hohl D, de Viragh PA, Amiguet-Barras F, et al. The small proline-rich proteins constitute a multigene family of differentially regulated cornified envelope precursor proteins. J Invest Dermatol 1995; 104: 902–909.
165. Mische D, Korge BP, Marenholz I, et al. Genes encoding structural proteins of epidermal cornification and S100 calcium-binding proteins form a gene complex (epidermal differentiation complex) on human chromosome 1q21. J Invest Dermatol 1996; 106: 989–992.
166. Ridinger K, Ilg EC, Niggli FK, et al. Clustered organization of S-100 genes in human and mouse. Biochim Biophys Acta 1998; 1448: 254–263.
167. Lee S-C, Kim I-G, Marekov LN, et al. The structure of human trichohyalin: potential multiple function as an EF-hand-like calcium-binding protein, a cornified cell envelope precursor and an intermediate filament-associated (crosslinking) protein. J Mol Biol 1993; 268: 2164–2176.
168. Markova NG, Marekov LN, Chipev CC, et al. Profilaggrin is a major epidermal calcium binding protein. Mol Cell Biol 1993; 13: 613–625.
169. Presland RB, Bassuk JA, Kimball JR, et al. Characterization of two distinct calcium-binding sites in the amino-terminus of human profilaggrin. J Invest Dermatol 1995; 104: 218–223.
170. Presland RB, Haydock PV, Fleckman P, et al. Characterization of the human epidermal profilaggrin gene. Genomic organization and identification of an

S-100-like calcium-binding domain at the amino terminus. J Biol Chem 1992; 267: 23772–23781.
171. Epstein EH, Langston AW, Leung J. Sulfation reactions in the epidermis. Ann NY Acad Sci 1988; 548: 97–101.
172. Danks JA, Ebeling PR, Hayman JA, et al. Parathyroid hormone-related protein: immunohistochemical localization in cancers and in normal skin. J Bone Miner Res 1989; 4: 273–278.
173. Danks JA, McHale JC, Clark SP, et al. In situ hybridization of parathyroid hormone-related protein in normal skin, skin tumors, and gynecological cancers using digoxygenin-labeled probes and antibody enhancement. J Histochem Cytochem 1995; 43: 5–10.
174. Danks JA, Trivett MK, Power DM, et al. Parathyroid hormone-related protein in lower vertebrates. Clin Exper Pharmacol Physiol 1998; 25: 750–752.
174*a*. Enserink M. Fossil opens window on early animal history. Science 1999; 286: 1829.
175. Morris SC. The Crucible of Creation: The Burgess Shale and the Rise of Animals. Oxford, Oxford University Press, 1998.
176. Ingleton PM, Danks JA. Distribution and functions of parathyroid hormone-related protein in vertebrate cells. Internat Rev Cytol 1996; 166: 231–280.
177. Phibrick WM, Wysolmerski JJ, Galbraith S, et al. Defining the roles of parathyroid hormone-related protein in normal physiology. Physiol Rev 1996; 76: 127–173.
178. Whitfield JF, Morley P, Willick G. The Parathyroid Hormone: An Unexpected Bone Builder for Treating Osteoporosis. Austin, RG Landes Company, 1998.
179. Errazahi A, Bouizar Z, Rizk-Rabin M. RT-PCR identification of PTHrp PTH/PTHrp receptor mRNAs during the steps of differentiation pathway of rat newborn keratinocytes: a putative autocrine role of PTHrp. Bone 1998; 23: S248.
180. Orloff JJ, Ganz MB, Ribaudo AE, et al. Analysis of parathyroid hormone-related protein binding and signal transduction mechanisms in benign and malignant squamous cells. Am J Physiol 1992; 262: E599–E607.
181. Orloff JJ, Kats Y, Urena P, Schipani E, et al. Further evidence for a novel receptor for amino-terminal parathyroid hormone-related protein on keratinocytes and squamous carcinoma cell lines. Endocrinology 1995; 136: 3016–3023.
182. Whitfield JF, Chakravarthy BR, Durkin JP, et al. Parathyroid hormone stimulates protein kinase C but not adenylyl cyclase in mouse epidermal keratinocytes. J Cell Physiol 1992; 150: 299–303.
183. Sharpe GR, Dillon JP, Durham B, et al. Human keratinocytes express transcripts for three isoforms of parathyroid hormone-related protein (PTHrP), but not for the parathyroid hormone/PTHrP receptor: effects of 1,25$(OH)_2$ vitamin D3. Br J Dermatol 1998; 138: 944–951.
184. Hanafin NM, Chen TC, Heinrich G, et al. Cultured human fibroblasts and not cultured human keratinocytes express a PTH/PTHrP receptor mRNA. J Invest Dermatol 1995; 105: 133–137.

185. Henderson JE, Kremer R, Rhim JS, et al. Identification and functional characterization of adenylate cyclase-linked receptors for parathyroid hormone-like peptide on immortalized human keratinocytes. Endocrinology 1992; 130: 449–457.
186. Jobert A-S, LeRoy C, Butlen D, et al. Parathyroid hormone-induced calcium release from intracellular stores in a human kidney cell line in the absence of stimulation of cyclic adenosine 3′,5′ monophosphate production. Endocrinology 1997; 138: 5282–5292.
187. Gagnon L, Jouishomme H, Whitfield JF, et al. Protein kinase C-activating domains of parathyroid hormone-related protein. J Bone Miner Res 1993; 8: 497–503.
188. Whitfield JF, Isaacs RJ, Jouishomme H, et al. C-terminal fragment of parathyroid hormone-related protein, PTHrP-(107–111), stimulates membrane-associated protein kinase C activity and modulates the proliferation of human and murine skin keratinocytes. J Cell Physiol 1996; 166: 1–11.
188*a*. Cuthbertson RM, Kemp BE, Barden JA. Structure study of osteostatin PTHrP[Thr107](107–139). Biochim Biophys Acta 1999; 1432: 64–72.
189. Blomme EAG, Werkmeister JR, Zhou H, et al. Parathyroid hormone-related protein expression and secretion in a skin organotypic culture system. Endocrine 1998; 8: 143–151.
190. Amalric F, Bouche G, Bonnet H, et al. Fibroblast growth factor-2 (FGF-2) in the nucleus: translocation process and targets. Biochem Pharmacol 1994; 47: 111–115.
191. Maher DW, Lee BA, Donoghue DJ. The alternatively spliced exon of the platelet-derived growth factor A chain encodes a nuclear targeting signal. Mol Cell Biol 1989; 9: 2251–2253.
192. Aarts MM, Levy D, Stregger S, et al. The nuclear targeting sequence (NTS) of parathyroid hormone-related protein (PTHrP) mediates binding to RNA. Bone 1998: 23: S251.
193. Henderson JE, Amizuka N, Warshawsky H, et al. Nuclear localization of parathyroid hormone-related peptide enhances survival of chondrocytes under conditions that promote apoptotic cell death. Mol Cell Biol 1995; 15: 4064–4075.
194. Nguyen MTA, Karaplis AC. The nucleus: a target site for parathyroid hormone-related peptide (PTHrP) action. J Cell Biochem 1998; 70: 193–199.
194*a*. Nguyen MT, He B, Karaplis AC, Nuclear forms of parathyroid hormone-related peptide are translated from non-AUG sites downstream from the initiator methionine. Endocrinology 2001; 142: 694–703.
195. Moroianu J. Molecular mechanisms of nuclear protein transport. Crit Rev Eukaryot Gene Expr 1997; 7: 61–72.
196. Lam MHC, Briggs LJ, Hu W, et al. Importin β recognizes parathyroid hormone-related protein with high affinity and mediates its nuclear import in the absence of importin α. J Biol Chem 1999; 274: 7391–7398.
197. Moroianu J. Distinct nuclear import and export pathways mediated by members of the karyopherin β family. J Cell Biochem 1998; 70: 231–239.
198. Holt EH, Broadus AE, Brines ML, et al. Parathyroid hormone-related peptide is produced by cultured cerebellar granule cells in response to L-type

voltage-sensitive Ca^{2+} channel flux via a Ca^{2+}/calmodulin-dependent kinase pathway. J Biol Chem 1996; 271: 28105–28111.
199. Merryman JL, DeWille JW, Werkmeister JR, et al. Effects of transforming growth factor-β on parathyroid hormone-related protein production and ribonucleic acid expression by a squamous carcinoma cell line in vitro. Endocrinology 1994; 134: 2424–2430.
200. Werkmeister JR, Merryman JJ, McCauley LK, et al. Parathyroid hormone-related protein production by normal human keratinocytes in vitro. Exp Cell Res 1993; 208: 68–74.
201. Kaiser SM, Sebag M, Rhim JS. Antisense-mediated inhibition of parathyroid hormone-related peptide production in a keratinocyte cell line impedes differentiation. Mol Endocrinol 1994; 8: 139–147.
201*a*. Lam MH, House CM, Tiganis T, et al. Phosphorylation at the cyclin-dependent kinase site (Thr 85) of parathyroid hormone-related protein negatively regulates its nuclear localization. J Biol Chem 1999; 274: 18559–18566.
202. Foley J, Longely BJ, Wysolmerski JJ, Dreyer BE, et al. PTHrP regulates epidermal differentiation in adult mice. J Invest Dermatol 1998; 1122–1128.
203. Vortkamp A, Lee K, Lanske B, et al. Regulation of rate of cartilage differentiation by Indian hedgehog and PTH-related protein. Science 1996; 273: 613–622.
204. Dunbar ME, Young P, Zhang J-P, et al. Stromal cells are critical targets in the regulation of mammary ductal morphogenesis by parathyroid hormone-related protein. Devel Biol 1998; 203: 75–89.
205. Amling M, Neff L, Tanaka S, et al. Bcl-2 lies downstream of parathyroid hormone-related peptide in a signaling pathway that regulates chondrocyte maturation during skeletal development. J Cell Biol 1997; 136: 205–213.
206. Insogna KL, Stewart AF, Morris CF, et al. Native and synthetic analogue of the malignancy-associated parathyroid hormone-like protein have in vivo transforming growth factor-like properties. J Clin Invest 1989; 83: 1057–1060.
207. Wu TL, Insogna KL, Milstone L, et al. Skin-derived fibroblasts respond to human PTH-like adenylate cyclase-stimulating proteins. J Clin Endocrinol Metab 1987; 65: 105–109.
208. Lowik CWGM, Hockman K, Offringa R, et al. Regulation of parathyroid hormone-like protein production in cultured normal and malignant keratinocytes. J Invest Dermatol 1992; 98: 198–203.
209. Blomme EAG, Sugimoto Y, Lin YC, et al. Parathyroid hormone-related protein is a positive regulator of keratinocyte growth factor expression by normal dermal fibroblasts. Mol Cell Endocrinol 1999; 152: 189–197.
210. Shin JH, Ji C, Casinghino S, et al. Parathyroid hormone-related protein enhances insulin-like growth factor-I expression by fetal rat dermal fibroblasts. J Biol Chem 1997; 272: 23498–23502.
211. Barran DT, Ray R, Sorensen AM, et al. Binding characteristics of a membrane receptor that recognizes 1α,25-dihydroxyvitamin D3 and its epimer, 1β-25-dihydroxyvitamin D3. J Cell Biochem 1994; 56: 510–517.
212. Bittiner B, Bleehen SS, MacNeil S. 1α,25 Dihydroxyvitamin D_3 increases intracellular calcium in human keratinocytes. Br J Dermatol 1991; 124: 230–235.

213. Dormanen MC, Bishop JE, Hammond MW, et al. Nonnuclear effects of the steroid hormone 1α,25$(OH)_2$-vitamin D3: analogs are able to functionally differentiate between nuclear and membrane receptors. Biochem Biophys Res Commun 1994; 201: 394–401.
214. MacLaughlin JA, Cantley LC, Holick MF. 1,25 Dihydroxy vitamin D-3 increases calcium and phosphatidylinositol metabolism in differentiating cultured human keratinocytes. J Nutr Biochem 1990; 1: 81–85.
215. Nemere I, Dormanen MC, Hammond MW, et al. Identification of a specific binding protein for 1α,25-dihydroxyvitamin D3 in basal-lateral membranes of chick intestinal epithelium and relationship to transcaltachia. J Biol Chem 1994; 269: 23750–23756.
216. Norman AW, Okamura WH, Farach-Carson MC, et al. Structure-function studies of 1,25-dihydroxyvitamin D3 and the vitamin D endocrine system. 1,25-Dihydroxy-pentadeuterio-previtamin D3 (as a 6-*s-cis* analog) stimulates nongenomic but not genomic biological responses. J Biol Chem 1993; 268: 13811–13819.
217. Wali RK, Baum CL, Sitrin MD, et al. 1,25$(OH)_2$ vitamin D_3 stimulates membrane phosphoinositide turnover, activates protein kinase C, and increases cytosolic calcium in rat colonic epihelium. J Clin Invest 1990; 85: 1296–1303.
218. Bikle D. Vitamin D. A calciotropic hormone regulating calcium-induced keratinocyte differentiation. J Am Acad Dermatol 1997; 37: S42–S52.
219. Ratnam AV, Bikle DD, Cho JK. 1,25 Dihydroxyvitamin D_3 enhances the calcium responses of keratinocytes. J Cell Physiol 1999; 178: 188–196.
220. Rizk-Rabin M, Zineb R, Zhor B, et al. Synthesis of and response to 1,25-dihydroxy cholecalciferol by subpopulations of murine epidermal keratinocytes: existence of a paracrine system for 1,25-dihydroxycholecalciferol. J Cell Physiol 1994; 159: 131–141.
221. Janulis M, Wong MS, Favus MJ. Structure-function requirements of parathyroid hormone for stimulation of 1,25-dihydroxyvitamin D_3 by rat renal proximal tubules. Endocrinology 1993; 133: 713–719.
222. Bikle DD, Ratnam A, Mauro T. Changes in calcium responsiveness and handling during keratinocyte differentiation. Potential role of the calcium receptor. J Clin Invest 1996; 97: 1085–1093.
223. Tu CL, Oda Y, Bikle DD. Effects of a calcium receptor activator on the cellular response to calcium in human keratinocytes. J Invest Dermatol 1999; 113: 340–345.
224. Oda Y, Tu CL, Chang W, et al. The calcium-sensing receptor and its alternatively spliced form in murine epidermal differentiation. J Biol Chem 2000; 275: 1183–1190.
225. Xie Z, Bikle DD. Inhibition of calcium and 1,25-dihydroxyvitamin D_3-induced keratinocyte differentiation by suppression of phospholipase C-γ1. Bone 1998; 23(5S): S366.
226. Xie Z, Bikle DD. Phospholipase C-γ1 is required for calcium-induced keratinocyte differentiation. J Biol Chem 1999; 274: 20421–20424.

227. Chakravarthy BR, Isaacs RJ, Morley P, et al. Stimulation of protein kinase C during Ca^{2+}-induced keratinocyte differentiation. J Biol Chem 1995; 270: 1362–1368.
228. Chakravarthy B, Morley P, Whitfield JF. Ca^{2+}-calmodulin and protein kinase Cs: a hypothetical synthesis of their conflicting convergences on shared substrate domains. Trends Neurosci 1999; 22: 12–16.
229. Filvaroff E, Calautti E, Reiss M. Functional evidence for an extracellular calcium receptor mechanism triggering tyrosine kinase activation associated with mouse keratinocyte differentiation. J Biol Chem 1994; 269: 21735–21740.
230. Whitfield JF, Bird R, Chakravarthy BR, et al. Calcium-cell cycle regulator, differentiator, killer, chemopreventor, and maybe, tumor promoter. J Cell Biochem 1995; Suppl 22: 74–91.
231. Adams MD, Oxender DL. Bacterial periplasmic binding protein tertiary structures. J Biol Chem 1989; 264: 15739–15742.
232. Bockaert J, Pin J-P. Utiliser un recepteur couplé aux protéines G pour communiquer. Un succès évolutif. CR Acad Sci Paris (Life Sci) 1998; 321: 529–551.
233. Conklin BR, Bourne HR. Marriage of the flytrap and the serpent. Nature 1994; 367: 22.
233*a*. Riccardi D. Cell surface, Ca^{2+}(cation)-sensing receptor(s): one or many? Cell Calcium 1999; 26: 77–83.
234. Berridge MJ. Elementary and global aspects of calcium signalling. J Physiol 1997; 499: 291–306.
235. Schwaninger M, Lux G, Blume R. Membrane depolarization and calcium influx induce glucagon gene transcription in pancreatic islet cells through the cyclic AMP-responsive element. J Biol Chem 1993; 268: 5168–5177.
236. Sheng M, Thompson MA, Greenberg ME. CREB: a Ca^{2+}-regulated transcription factor phosphorylated by calmodulin-dependent kinases. Science 1991; 252: 1427–1430.
237. Matthews RP, Guthrie CR, Wailes LM, et al. Calcium/calmodulin-dependent protein kinase types II and IV differentially regulate CREB-dependent gene expression. Mol Cell Biol 1994: 14: 6107–6116.
238. Channon JY, Leslie CC. A calcium-dependent mechanism for associating a soluble arachidonyl-hydrolyzing phospholipase A_2 with membrane in the macrophage cell line RAW264·7. J Biol Chem 1990; 265: 5409–5413.
239. Clark JD, Lin LL, Kriz RW, et al. A novel arachidonic acid-selective cytosolic PLA2 contains a Ca^{2+}-dependent translocation domain with homology to PKC and GAP. Cell 1991; 65: 1043–1051.
240. Bergers MD, Verhagen DR, Jongerius M, et al. A unique phopholipase A2 in human epidermis: its physiologic function and its level in certain dermatoses. J Invest Dermatol 1988; 90: 23–25.
241. Calautti E, Missero C, Stein PL, et al. fyn tyrosine kinase is involved in keratinocyte differentiation control. Genes Devel 1995; 9: 2279–2291.

242. Steinert PM, Kim SY, Chung SI, et al. The transglutaminase 1 enzyme is variably acylated by myristate and palmitate during differentiation in epidermal keratinocytes. J Biol Chem 1996; 271: 26242–26250.
243. Denning MF, Dlugosz AA, Williams EK, et al. Specific protein kinase C isozymes mediate the induction of keratinocyte differentiation markers by calcium. Cell Growth Diff 1995; 6: 149–157.
244. Ueda E, Ohno S, Kuroki T, et al. The η isoform of protein kinase C mediates transcriptional activation of the human transglutaminase 1 gene. J Biol Chem 1996; 271: 9790–9794.
245. Toescu EC, Verkhratsky A. Principles of calcium signaling. In: Verkhratsky A, Toescu EC, editors. Integrative Aspects of Calcium Signalling. New York, Plenum Press, 1998: 1–19.
246. Latchman DS. Eukaryotic Transcription Factors. 3rd ed. London, Academic Press, 1998.
247. Rutberg SE, Saez E, Glick A, et al. Differentiation of mouse keratinocytes is accompanied by PKC-dependent changes in AP-1 proteins. Oncogene 1996; 13: 167–176.
248. Rutberg SE, Saez E, Lo S, et al. Opposing activities of c-Fos and Far-2 on AP-1 regulated transcriptional activity in mouse keratinocytes induced to differentiate by calcium and phorbol esters. Oncogene 1997; 15: 1337–1346.
249. Ng DC, Bikle DD. Calcium increases AP1 protein levels during differentiation. Bone 1998; 23(5S): S562.
250. Welter JF, Eckert R. Differential expression of the fos and jun family members c-fos, fos B, Fra-1, Fra-2, c-jun, jun B and jun D during human epidermal keratinocyte differentiation. Oncogene 1995; 11: 2681–2687.
251. DiSepio D, Jones A, Longley MA, et al. The proximal promoter of the mouse loricrin gene contains a functional AP-1 element and directs keratinocyte-specific but not differentiation-specific expression. J Biol Chem 1995; 270: 10792–10799.
252. Jiang S-I, Steinert PM, Markova NG. Activator protein 1 activity is involved in the regulation of the cell type-specific expression from the proximal promoter of the human profilaggrin gene. J Biol Chem 1996; 271: 24105–24114.
253. Saunders NA, Bernacki SH, Vollberg TM. Regulation of transglutaminase type 1 expression in squamous differentiating rabbit tracheal epithelial cells and human epidermal keratinocytes: effects of retinoic acid and phorbol esters. Mol Endocrinol 1993; 7: 387–398.
254. Welter JF, Crish JF, Agarwal C. Fos-related antigen (*Fra*-1), *jun* B, and *jun* D activate human involucrin promoter transcription by binding to proximal and distal AP1 sites to mediate phorbol ester effects on promoter activity. J Biol Chem 1995; 270: 12614–12622.
255. Deng T, Karin M. Jun B differs from c-Jun in its DNA-binding and dimerization domains, and represses c-Jun by formation of inactive heterodimers. Genes Dev 1993; 7: 479–490.

256. Dlugosz AA, Mischak H, Mushinski JF, et al. Transcripts encoding protein kinase C-alpha, -delta, -epsilon, -zeta, and -eta are expressed in basal and differentiating mouse keratinocytes in vitro and exhibit quantitative changes in neoplastic cells. Mol Carcinog 1992; 5: 286–292.
257. Gherzi R, Sparatore B, Patrone M, et al. Protein kinase C mRNA levels in reconstituted normal epidermis: relationships to cell differentiation. Biochem Biophys Res Commun 1992; 184: 283–291.
258. Matsui MS, Chew SL, DeLeo VA. Protein kinase C in normal human epidermal keratinocytes during proliferation and calcium-induced differentiation. J Invest Dermatol 1992; 99: 565–571.
259. Dlugosz AA, Yuspa SH. Coordinate changes in gene expression which mark the spinous to granular cell transition in epidermis are regulated by protein kinase C. J Cell Biol 1993; 120: 217–225.
260. Dale BA, Resing KA, Presland RB. Keratohyalin granule proteins. In: Leigh IM, Lane LE, Watt F (editors). The Keratinocyte Handbook. Cambridge, Cambridge University Press, 1994: 323–359.
261. Andreoti JM, Jang S-I, Chung E, et al. The expression of a normal epithelium-specific ets transcription factor is restricted to the most differentiated layer in the epidermis. Nucleic Acid Res 1997; 25: 4287–4295.
262. Yoneda K, McBride OW, Korge BP. The cornified cell envelope: loricrin and transglutaminases. J Dermatol 1992; 19: 761–764.
263. Dexter TM, Raff MC, Wyllie AH, editors. Death from inside out: the role of apoptosis in development, tissue homeostasis and malignancy. Phil Trans R Soc B 1994; 345: 231–333.
264. Duke RC, Orjcius DM, Youg J D-E. Cell suicide in health and disease. Sci Amer 1996; 275: 80–87.
265. Hansson L, Stromqvist M, Backman A, et al. Cloning, expression and characterization of stratum corneum chymotryptic enzyme: a skin-specific human serine proteinase. J Biol Chem 1994; 269: 19420–19426.
266. Sondell B, Thornell L-E, Stigbrand T, et al. Immunolocalization of stratum corneum chymotryptic enzyme in human skin and oral epithelium with monoclonal antibodies: evidence of protease specifically expressed in keratinizing squamous epithelia. J Histochem Cytochem 1994; 42: 459–465.
267. Ando Y, Imamura S, Murachi T, et al. Calpain activates two transglutaminases from porcine skin. Arch Dermatol Res 1988; 280: 380–384.
268. Garach-Jehoshua O, Ravid A, Liberman UA, et al. Upregulation of the calcium-dependent protease, calpain, during keratinocyte differentiation. Br J Dermatol 1998; 139: 950–957.
269. Kim SY, Bae CD. Calpain inhibitors reduce the cornified cell envelope formation by inhibiting proteolytic processing of transglutaminase 1. Exp Mol Biol 1998; 30: 257–262.
269*a*. Méhul B, Bernard D, Simonetti L, et al. Identification and cloning of a new calmodulin-like protein from human epidermis. J Biol Chem 2000; 275: 12841–12847.

270. Croall DE, DeMartino GN. Calcium-activated neutral protease (calpain) system: structure, function, and regulation. Physiol Rev 1991; 71: 813–847.
271. Elias PM, Cullander C, Mauro T, et al. The secretory granular cell: the outermost granular cell as a specialized secretory cell. J Invest Dermatol Symp Proc 1998; 3: 87–100.
272. Menon GK, Elias PM, Feingold KR. Integrity of the permeability barrier is crucial for maintenance of the epidermal calcium gradient. Br J Dermatol 1994; 130: 139–147.
273. Menon GK, Ghadially R, Williams ML, et al. Lamellar bodies as delivery systems for hydrolytic enzymes: implications for normal and abnormal desquamation. Br J Dermatol 1992; 126: 337–345.
274. Ponec M. Lipid biosynthesis. In: Leigh IM, Lane EB, Watt F, editors. The Keratinocyte Handbook. Cambridge, Cambridge University Press, 1994: 351–363.
275. Sando GN, Howard EJ, Madison KC. Induction of ceramide glucosyl transferase activity in cultured human keratinocytes. Correlation with culture differentiation. J Biol Chem 1996; 271: 22044–22051.
276. Hauser C, Saurat J-H, Schmitt JA, et al. Interleukin-I is present in normal human epidermis. J Immunol 1986; 136: 3317–3322.
277. Nickoloff BJ, Naidu Y. Perturbation of epidermal barrier function correlates with initiation of cytokine cascade in human skin. J Am Acad Dermatol 1994; 30: 535–546.
278. Tsai J-C, Feingold KR, Crumrine D, et al. Permeability barrier disruption alters the localization and expression of TNF-alpha protein in the epidermis. Arch Dermatol Res 1994; 286: 242–248.
279. Wood LC, Feingold KR, Sequeira-Martine SM, et al. Barrier function coordinately regulates epidermal IL-1 mRNA levels. Exp Dermatol 1994; 3: 56–60.
280. Wood LC, Jackson SM, Elias PM, et al. Cutaneous barrier perturbation stimulates cytokine production in the epidermis of mice. J Clin Invest 1992; 90: 482–487.
280*a*. Greaves M. Cancer. The Evolutionary Legacy. Oxford, Oxford University Press, 2000.
281. Boukamp P, Breitkreutz D, Hülsen A, et al. In vitro transformation and tumor progression. Recent Res Cancer Res 1993; 128: 339–350.
282. Brown K, Kemp CJ, Burns PA, et al. Positive and negative growth control in multistage skin carcinogenesis. Recent Res Cancer Res 1993; 128: 309–321.
283. Yuspa SH. The pathogenesis of squamous cell cancer: lessons learned from studies of skin carcinogenesis. J Dermatol Sci 1998; 17: 1–7.
284. Yuspa SH, Punnonen K, Lee H, et al. The in vitro analysis of biochemical changes relevant to skin carcinogenesis. Recent Res Cancer Res 1993; 128: 299–308.
285. Barnard JA, Graves-Deal R, Pittelkow MR, et al. Auto- and cross-induction within the mammalian epidermal growth factor-related peptide family. J Biol Chem 1994; 269: 22817–22822.

286. Dlugosz AA, Cheng C, Williams EK, et al. Autocrine transforming growth factor alpha is dispensible for v-ras Ha-induced epidermal neoplasia: potential involvement of alternate epidermal growth factor receptor ligands. Cancer Res 1995; 55: 1883–1893.
287. Donnelly MJ, Patel V, Yeudall WA, et al. Autocrine production of TGF-alpha and TGF-beta during tumor progression of rat oral keratinocytes. Carcinogenesis 1993; 14: 981–985.
288. Finzi E, Kilkenny A, Strickland JE, et al. TGF alpha stimulates growth of skin papillomas by autocrine and paracrine mechanisms but does not stimulate neoplastic progression. Mol Carcinog 1988; 1: 7–12.
289. Fürstenberger G, Krieg P, Schnapke R, et al. The role of endogenous factors in skin carcinogenesis. Recent Res Cancer Res 1993; 128: 323–337.
290. Prime SS, Game SM, Matthews JB, et al. Epidermal growth factor and transforming growth factor alpha characteristics of human oral carcinoma cell lines. Br J Cancer 1994; 69: 8–15.
291. Reiss M, Stash EB, Vellucci VF, et al. Activation of the autocrine transforming growth factor alpha pathway in human squamous carcinoma cells. Cancer Res 1991; 51: 6254–6262.
292. Vardy DA, Kari C, Lazarus GS, et al. Induction of autocrine epidermal growth factor receptor ligands in human keratinocytes by insulin/insulin-like growth factor-1. J Cell Physiol 1995; 163: 257–265.
293. Kim K, Akoto-Amanfu E, Cherrick HM, et al. Anchorage-independent growth and the expression of cellular proto-oncogenes in normal human epidermal keratinocytes and in human squamous cell carcinoma cell lines. Oral Surg Oral Med Oral Pathol 1991; 71: 303–311.
294. Reinartz J, Bechtel MJ, Kramer MD, et al. Tumor necrosis factor-alpha-induced apoptosis in a human keratinocyte cell line (HaCaT) is counteracted by transforming growth factor alpha. Exp Cell Res 1996; 228: 334–340.
295. Fligiel SEG, Varani J. In situ epithelial cell invasion in organ culture. Invasion Metastasis 1993; 13: 225–233.
296. Wilding J, Vousden KH, Soutter WP, et al. E-cadherin transfection down-regulates the epidermal growth factor receptor and reverses the invasive phenotype of human papilloma virus-transfected keratinocytes. Cancer Res 1996; 56: 5285–5292.
297. Carroll JM, Romero MR, Watt FM. Suprabasal integrin expression in the epidermis of transgenic mice results in developmental defects and a phenotype resembling psoriasis. Cell 1995; 83: 957–968.
298. Tennenbaum T, Belanger AJ, Glick AB, et al. A splice variant of alpha 6 integrin is associated with malignant conversion in mouse skin tumorigenesis. Proc Natl Acad Sci USA 1995; 92: 7041–7045.
299. Tennenbaum T, Weiner AK, Belanger AJ, et al. The suprabasal expression of alpha 6 beta 4 integrin is associated with a high risk for malignant conversion in mouse skin carcinogenesis. Cancer Res 1993; 53: 4803–4810.

300. Bhardi A, Kraeft S-K, Gounder M, et al. Inactivation of DNA-dependent protein kinase by protein kinase Cδ: implications for apoptosis. Mol Cell Biol 1998; 18: 6714–6728.
301. Yuspa SH, Kilkenny AE, Roop DR, et al. Consequences of exposure to initiating levels of carcinogens in vitro and in vivo. Altered differentiation and growth mutations and transformation. Prog Clin Biol Med 1989; 298: 127–135.
302. Creek KE, Geslani G, Batova A, et al. Progressive loss of sensitivity to growth control by retinoic acid and transforming growth factor-beta at late stages of human papillomavirus type 16-initiated transformation of human keratinocytes. Adv Exp Med Biol 1995; 375: 117–135.
303. Gibson DF, Ratnam AV, Bikle DD. Evidence for separate control mechanisms at the message, protein, and enzyme activation levels for transglutaminase during calcium-induced differentiation of normal and transformed human keratinocytes. J Invest Dermatol 1996; 106: 154–161.
304. Hennings H, Holbrook K. Calcium regulation of cell–cell contact and differentiation of epidermal cells in culture. Exp Cell Res 1983; 143: 127–142.
305. Hennings H, Kruszewski FH, Yuspa SH. Intracellular calcium alterations in response to increased external calcium in normal and neoplastic keratinocytes. Carcinogenesis 1989; 10: 777–780.
306. Klann RC, Fitzgerald DJ, Piccoli C, et al. Gap-junctional intercellular communication in epidermal cell lines from selected stages of SENCAR mouse skin carcinogenesis. Cancer Res 1989; 49: 699–705.
307. Kruszewski FH, Hennings H, Tucker RW, et al. Differences in the regulation of intracellular calcium in normal and neoplastic keratinocytes are not caused by ras gene mutations. Cancer Res 1991; 51: 4206–4212.
308. Rubin HL, Parenteau NL, Rice RH. Coordination of keratinocyte programming in human SCC-13 squamous carcinoma and normal epidermal cells. J Cell Physiol 1989; 138: 208–214.
309. Schneider BL, Kulesz-Martin M, Bowden GT. Induced terminal differentiation and tumorigenic suppression in murine keratinocyte somatic cell hybrids. Mol Carcinog 1995; 13: 6–14.
310. Steele VE, Wyatt GP, Kellof GJ, et al. Differential growth response to exogenous calcium in normal and carcinogen-exposed primary human keratinocyte cell cultures. Anticancer Res 1998; 18: 4067–4070.
311. Yuspa SH, Kilkenny AE, Roop DR, et al. Consequences of exposure to initiating levels of carcinogens in vitro and in vivo: altered differentiation and growth, mutations, and transformation. In: Slaga TJ, Klein-Szanto AJP, Boutwell RK, et al., editors. Skin Carcinogenesis: Mechanisms and Human Relevance. New York, Liss, 1989: 127–135.
312. Yuspa SH, Morgan DL. Mouse stem cells resistant to terminal differentiation associated with initiation of carcinogenesis. Nature 1981; 293: 72–74.
313. Yuspa SH, Morgan D, Lichti U, et al. Cultivation and characterization of cells derived from mouse skin papillomas induced by an initiation-promotion potocol. Carcinogenesis 1986; 7: 949–958.

314. Yuspa SH, Poirier MC. Chemical carcinogenesis from animal models to molecular models in one decade. Adv Cancer Res 1988; 50: 25–70.
315. Hennings H, Lowry DT, Robinson VA. Coculture of neoplastic and normal keratinocytes as a model to study tumor promotion. Skin Pharmacol 1991; 4 Suppl 1: 79–84.
316. Hennings H, Lowry DT, Robinson VA, et al. Activity of diverse tumor promoters in a keratinoctye co-culture model of initiated epidermis. Carcinogenesis 1992; 13: 2145–2151.
317. Hennings H, Robinson VA, Michael DM, et al. Development of an in vitro analogue of initiated mouse epidermis to study tumor promoters and antipromoters. Cancer Res 1990; 50: 4794–4800.
318. Strickland JE, Ueda M, Hennings H, et al. A model for initiated mouse skin: suppression of papilloma but not carcinoma formation by normal epidermal cells in grafts on athymic nude mice. Cancer Res 1992; 52: 1439–1444.
319. Amagai M. Pemphigus. Autoimmunity to epidermal cell adhesion molecules. Adv Dermatol 1996; 11: 319–352.
320. Blum S, Ness W, Petrow W, et al. Localization of protein kinase C in primary cultures of human keratinocytes in relation to cell contact proteins. Cell Signalling 1994; 6: 157–165.
321. Jansen LA, Mesnil M, Jongen WM. Inhibition of gap junctional intercellular communication and delocalization of the cell adhesion molecule E-cadherin by tumor promoters. Carcinogenesis 1996; 17: 1527–1531.
322. MacLachlan T, Sang N, Giordano A. Cyclins, cyclin-dependent kinases and Cdk inhibitors: implications in cell cycle control and cancer. Crit Revs Euk Gene Expression 1995; 5: 127–156.
323. Fowlis DJ, Flanders KC, Duffie E, et al. Discordant transforming growth factor beta 1 RNA and protein localization during chemical carcinogenesis of the skin. Cell Growth Diff 1992; 3: 81–91.
324. Game SM, Huelsen A, Patel V, et al. Progressive abrogation of TGF-beta1 and and EGF growth control is associated with tumour progression in ras-transfected human keratinocytes. Int J Cancer 1992; 52: 461–470.
325. Dunbar ME, Wysolmerski JJ, Broadus AE. Parathyroid hormone-related protein: from hypercalcemia of malignancy to developmental regulatory molecule. Am J Med 1996; 312: 287–294.
326. Foley J, Wysolmerski JJ, Broadus AE, et al. Parathyroid hormone-related protein gene expression in human squamous carcinoma cells is repressed by mutant isoforms of p53. Cancer Res 1996; 56. 4056–4062.
327. Halloran BP, Nissenson RA, editors. Parathyroid Hormone-Related Protein: Normal Physiology and Its Role in Cancer. Boca Raton, CRC Press, 1992.
328. Wysolmerski JJ, Vasavada R, Foley J, et al. Transactivation of the PTHrP gene in squamous cell carcinomas predicts the occurrence of hypercalcemia in athymic mice. Cancer Res 1996; 56: 1043–1049.
329. Aszterbaum M, Rothman A, Johnson R, et al. Identification of mutations in the human *PATCHED* gene in sporadic basal cell carcinomas and in patients with basal cell nevus syndrome. J Invest Dermatol 1998; 110: 885–888.

330. Gailani MR, Bale AE. Developmental genes and cancer: role of patched in basal carcinoma of the skin. J Natl Cancer Inst 1997; 89: 1103–1109.
331. Gailani MR, Stähle-Bäckdahl M, Leffell DJ, et al. The role of the human homologue of *Drosophila patched* in sporadic basal cell carcinoma syndrome. Nature Genet 1996; 14: 78–71.
332. Hahn H, Wicking C, Zaphiropoulos PG, et al. Mutations of the human homologue of Drosophila *patched* in the nevoid basal cell carcinoma syndrome. Cell 1996; 85: 841–851.
333. Ingham PW. The *patched* gene in development and cancer. Curr Opin Genet Devel 1998; 8: 88–94.
334. Kallassy M, Toftgard R, Ueda M, et al. Patched (ptch)-associated preferential expression of smoothened (smoh) in human basal cell carcinoma of the skin. Cancer Res 1997; 57: 4731–4735.
335. Oro AE, Higgines KM, Hu Z, et al. Basal cell carcinoma in mice overexpressing sonic hedgehog. Science 1997; 276: 817–821.
336. Pennisi E. Gene linked to commonest cancer. Science 1996; 272: 1583–1589.
337. Reifenberger J, Wolter M, Weber RG, et al. Missense mutations in sporadic basal cell carcinomas of the skin and primitive neuroectodermal tumors of the central nervous system. Cancer Res 1998; 58: 1798–1803.
338. Unden AB, Zaphiropoulos PG, Bruce K, et al. Human *patched* (PTCH) mRNA is overexpressed consistently in tumor cells of both familial and sporadic basal cell carcinomas. Cancer Res 1997; 57: 2336–2340.
339. Xie J, Murone M, Luoh SM, et al. Activating smoothened mutations in sporadic basal-cell carcinoma. Nature 1998; 391: 90–92.
340. Altaba AR. Catching a Gli-mpse of hedgehog. Cell 1997; 90: 193–196.
341. Fan H, Oro AE, Scott MP, et al. Induction of basal cell carcinoma features in transgenic human skin expressing Sonic Hedgehog. Nat Med 1997; 3: 788–792.
342. Dean M. Cancer as a complex developmental disorder. Cancer Res 1998; 58: 5633–5636.
342*a*. Fan H, Khavari PA. Sonic hedgehog opposes epithelial cell cycle arrest. J Cell Biol 1999; 147: 71–76.
343. Brownell HL, Narsimhan RP, Corbley MJ, et al. Ras is involved in gap junction closure in proliferating fibroblasts or preadipocytes but not in differentiated adipocytes. DNA Cell Biol 1996; 15: 443–451.
344. Brownell HL, Whitfield JF, Raptis L. Cellular Ras partly mediates gap junction closure by the polyoma virus middle tumor antigen. Cancer Lett 1996; 103: 99–106.
345. Budunova IV, Carbajal S, Slaga TJ. The expression of gap junctional proteins during different stages of mouse skin carcinogenesis. Carcinogenesis 1995; 11: 2717–2724.
346. Budunova IV, Carbajal S, Slaga TJ. Effect of diverse tumor promoters on the expression of gap-junctional proteins connexin (Cx)26, Cx31·1, and Cx43 in SENCAR mouse epidermis. Mol Carcinog 1996; 15: 202–214.
347. Budunova IV, Carbajal S, Viaje A, et al. Connexin expression in epidermal cell lines from SENCAR mouse skin tumors. Mol Carcinog 1996; 15: 190–201.

348. Dotto GP, el-Fouly MH, Nelson C, et al. Similar and synergistic inhibition of gap-junctional communication by ras transformation and tumor promoter treatment of mouse primary keratinocytes. Oncogene 1989; 4: 637–641.
349. el-Fouly MH, Trosko JE, Chang CC, et al. Potential role of the human Ha-ras oncogene in the inhibition of gap junctional intercellular communication. Mol Carcinog 1989; 2: 131–135.
350. Esinduy CB, Chang CC, Trosko JE, et al. In vitro growth inhibition of neoplastically transformed cells by non-transformed cells: requirement for gap junctional intercellular communication. Carcinogenesis 1995; 16: 915–921.
351. Kalimi GH, Hampton LL, Trosko JE, et al. Homologous and heterologous gap-junctional intercellular communication in v-raf-, v-myc-, and v-raf/v-myc-transduced rat liver epithelial cell lines. Mol Carcinog 1992; 5: 301–310.
352. Madhukar BV, Oh SY, Chang CC, et al. Altered regulation of intercellular communication by epidermal growth factor, transforming growth factor-beta and peptide hormones in normal human keratinocytes. Carcinogenesis 1989; 10: 13–20.
353. Phipps M, Phipps J, Whitfield JF, et al. Carcinogenic implications of the neighborhood coherence principle (NCP). Med Hypotheses 1990; 31: 289–301.
354. Trosko JE, Chang CC, Madhukar BV, et al. Chemical, oncogene and growth factor inhibition of gap junctional intercellular communication: an integrative hypothesis of carcinogenesis. Pathobiology 1990; 58: 265–278.
355. Trosko JE, Chang CC, Madhukar BV. Modulation of intercellular communication during radiation and chemical carcinogenesis. Radiat Res 1990; 123: 241–251.
356. Trosko JE, Madhukar BV, Chang CC. Endogenous and exogenous modulation of gap junctional intercellular communication: toxicological and pharmacological implications. Life Sci 1993; 53: 1–19.
357. Yamasaki H. Aberrant expression and function of gap junctions during carcinogenesis. Environ Health Perspect 1991; 93: 191–197.
358. Holden PR, McGuire B, Stoler A, et al. Changes in gap junctional intercellular communication in mouse skin carcinogenesis. Carcinogenesis 1997; 18: 15–21.
359. Whitfield JF. Calcium, Cell Cycles, and Cancer. 1st ed. Boca Raton, CRC Press, 1992.
359*a*. Carr AM. Piecing together the p53 puzzle. Science 2000; 287: 1765–1766.
360. Boukamp P, Peter W, Pascheberg U, et al. Step-wise progression in human skin carcinogenesis in vitro involves mutational inactivation of p53, ras H oncogene activation and additional chromosome loss. Oncogene 1995; 11: 961–969.
361. Edington KG, Loughran OP, Berry IJ, et al. Cellular immortality: a late event in the progression of human squamous cell carcinoma of the head and neck associated with p53 alteration and a high frequency of allele loss. Mol Carcinog 1995; 13: 254–265.
362. Missero C, Di Cunto F, Kiyokawa H, et al. The absence of p21CIP1/WAF1 alters keratinocyte growth and differentiation and promotes ras-tumor progression. Genes Dev 1996; 10: 3065–3075.

363. Shin KH, Tannyhill RJ, Liu X, et al. Oncogenic transformation of HPV-immortalized human oral keratinocytes is associated with the genetic instability of cells. Oncogene 1996; 12: 1089–1096.
364. Sugerman PB, Joseph BK, Savage NW. The role of oncogenes, tumor suppressor genes and growth factors in oral squamous cell carcinoma: a case of apoptosis versus proliferation. Oral Dis 1995; 1: 172–188.
365. Weinberg WC, Azzoli CG, Kadiwar N, et al. p53 gene dosage modifies growth and malignant progression of keratinocytes expressing the v-ras Ha oncogene. Cancer Res 1994; 54: 5584–5592.
366. Lu QL, Abel P, Foster CS, et al. bcl-2: role in epithelial differentiation and oncogenesis. Hum Pathol 1996; 2: 102–110.
367. Rodriguez-Vilanueva J, Colome MI, Brisbay S, et al. The expression and localization of bcl-2 protein in normal skin and in non-melanoma skin cancers. Pathol Res Pract 1995; 191: 391–398.
368. Bracke ME, Van Roy FM, Mareel MM. The E-cadherin/catenin complex in invasion and metastasis. Curr Topics Microbiol Immunol 1996; 213: 123–161.
369. Furukawa F, Fujii K, Horiguchi Y, et al. Roles of E- and P-cadherin in the human skin. Microsc Res Tech 1997; 38: 343–352.
370. Henning G, Behrens J, Truss M, et al. Progression of carcinoma cells is associated with alterations in chromatin structure and factor binding at the E-cadherin promoter in vivo. Oncogene 1995; 11: 475–484.
371. Hirohashi S. Inactivation of the E-cadherin-mediated cell adhesion system in human cancer. Am J Pathol 1998; 153: 333–339.
372. Kantak SS, Kramer RH. E-cadherin regulates anchorage-independent growth and survival in oral squamous cell carcinoma cells. J Biol Chem 1998; 272: 16953–16961.
373. Curran AJ, St Denis K, Irish J. Telomerase activity in oral squamous cell carcinoma. Arch Otolaryngol Head Neck Surg 1998; 124: 784–788.
374. Folkman J. Endogenous inhibitors of angiogenesis. The Harvey Lect 1998; 92: 65–82.
375. Cooke R. Dr. Folkman's War. New York, Random House, 2001.

4

Calcium and Colon Cell Diffpoptosis

Colon crypts, stem cells, and a cellular escalator

According to the conventional model, the colonic counterparts of the skin's basal lamina-bound stem cells are a small number of clonogenic cells nestling at the foot of each monoclonally derived crypt shaped like the finger of a glove dipping down into the wall of the organ from the flat mucosal surface (Figs. 4.1 and 4.2).[1–5] A stem cell may divide symmetrically into two stem cells or two differentiation-bound, transit-amplifying daughter cells or it may divide asymmetrically into one stem cell and one differentiation-bound transit cell, but on the average, stem cells produce one stem cell and one differentiation-bound transit cell.[1,3] The first three generations of transit cells are also potentially clonogenic, but the later generations are proliferatively disabled and irreversibly committed to differentiate into three kinds of cell: absorptive cells, EC (enterochromaffin or enteroendocrine) cells, and mucus-producing goblet cells.[1–5] The EC cells are the colon's pressure sensors, which, when squeezed by fecal distension, release 5-HT (serotonin) to stimulate intrinsic sensory neurons with 5-HT_{1P} receptors that signal the motor neurons of the gut's massive "second brain" to start a peristaltic muscular pushing of the fecal mass forward.[6] The absorptive cells are the "work-horses" of the colon. They control the flow of water and electrolytes across the colon epithelium into the blood. They also are the cells with the transporters that limit the response to the sensor signals by taking up the 5-HT squeezed out of the EC cells.[6] (These absorptive cells can also cause a severe hypermotility and diarrhea when their 5-HT transporters are shut off by a SSRI (specific serotonin reuptake inhibitor) such as Prozak.)[6] The mucus cells provide the colon surface with a protective mucus coat and the Ca^{2+}-binding mucopolysaccharides for an intracryptal Ca^{2+} gradient.

These mature, functioning, non-proliferating cells live for only a couple of days before dying and therefore must be replaced by the progeny of the small clone (one to four cells in the mouse) of pluripotent stem cells anchored to a specific niche at the bottom (cell positions 1 and 2) of each one of the millions of monoclonal, finger-like crypts.[3] The proliferative potential of the stem cell is probably maintained, and initiation of the diffpoptosis program suppressed, by signals from the same extracellular matrix-bound clusters of β1 integrins that maintain the proliferative potential and suppress diffpoptosis in its keratinocyte counterpart.[7,8] It is also maintained by signals from receptors for various autocrine, endocrine, juxtacrine, and paracrine factors from blood, neighbors, and nerve terminals. Like all cycling cells, these cells express the c-*myc* gene, the product of which can be either proliferogenic or apoptogenic, depending on the presence of survival signalers.[9–11] The expression of c-*myc* and thus its targets, the cycle-starting–driving

Fig. 4.1. One possible scenario for the control of colon cell proliferation and diffpoptosis by Ca^{2+} and the CaR or CaR variant (Ca^{2+}, or more accurately, divalent cation receptors or sensors; solid triangles). In this version, there is an intracryptal Ca^{2+} gradient (hollow circles) which triggers sensor signaling when the cell reaches a level in the crypt where the Ca^{2+} concentration is at the receptors' set-point. The combination of fading proliferation-driving β_1 intergrin signals, surging CaR signals, emerging E-cadherin cell–cell tethers, and increasing APC expression stop cell cycle initiation, drive the upward gliding, differentiation and finally apoptosis of the absorptive cells on the flat mucosal surface.

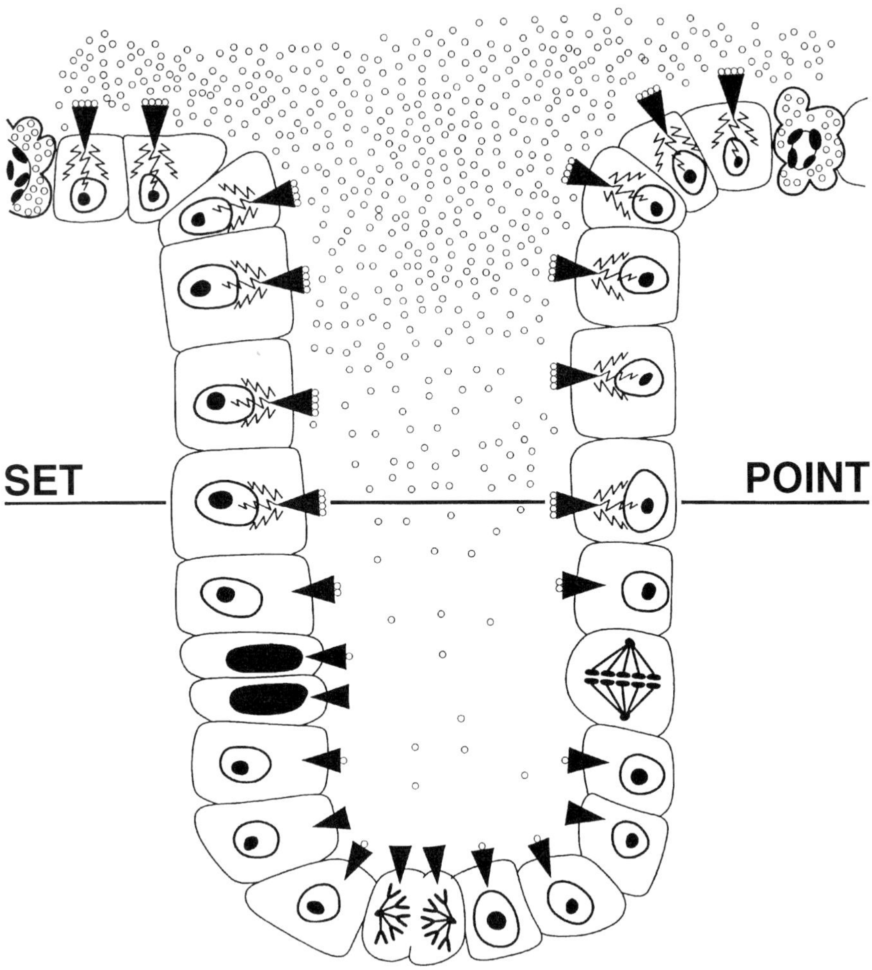

cyclin D1 and CDK 4 genes, are ultimately due to the promotion by β-catenin·Tcf-4 of the formation of transactivator complexes on the c-Myc gene's enhancer region.[12,13] The importance of β-catenin·Tcf-4 for the maintenance crypt cell proliferation and thus the colon epithelium is demonstrated by the absence of proliferating cells in the colon crypts of *Tcf4*$^{-/-}$ mice that die at birth because of a badly depleted, stretched, and broken mucosal epithelium.[12] The proliferation of cells in the early stages of intestinal epithelial development is driven by other factors, but the β-catenin·Tcf-4 mechanism takes over as crypts appear and then remains a major player in the control of crypt cell proliferation.

Fig. 4.2. Another scenario for the control of cell proliferation and diffpoptosis by Ca^{2+} and the Ca^{2+} sensors in the colon crypt. In this scenario, there is no significant intracryptal Ca^{2+} gradient, but at a certain point in their journey up the crypt wall the cells increase the sensor density at which their signaling starts contributing to the cells' differentiation and ultimate apoptosis.

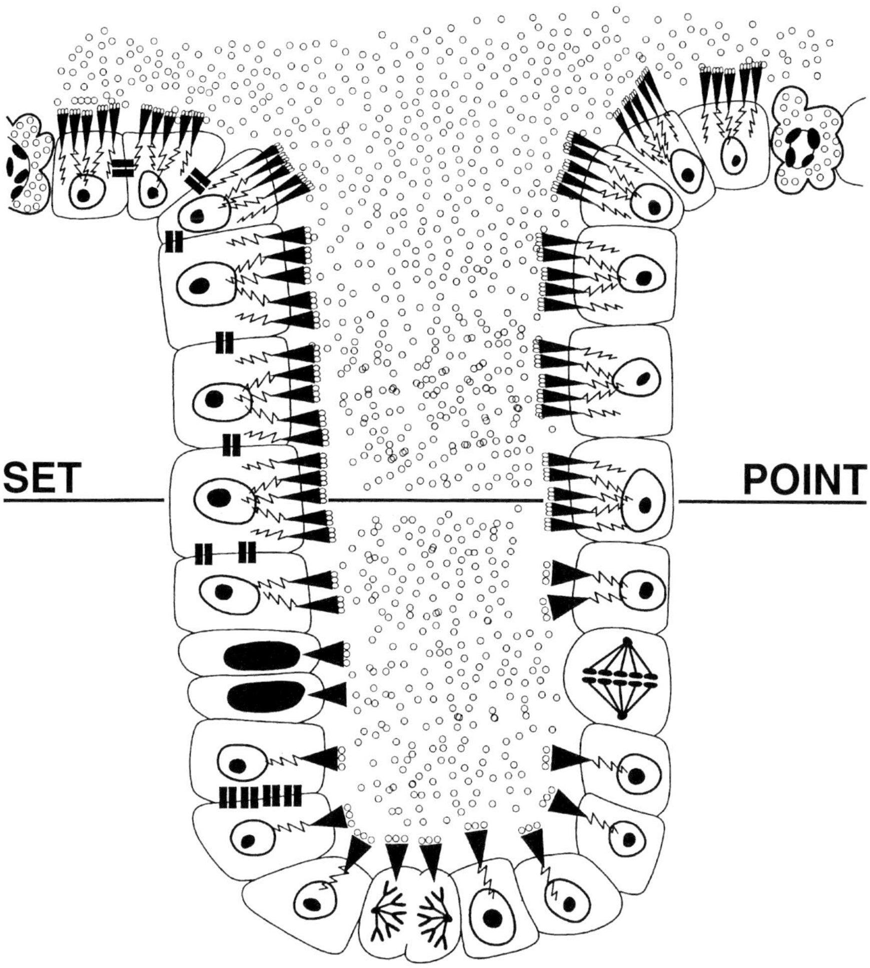

Unlike their counterparts in the benign world of the small intestine, the colon crypt cells must protect themselves from premature death induced by exposure to swarms of fecal bacteria and countless noxious agents, including mutagens. They do this with the anti-apoptotic Bcl-2 protector protein, the expression of which is confined to the stem cell region in the base of the crypt.[3,14–16] While this is essential for keeping a healthy functioning mucosa in a hostile environment, it has the not inconsiderable *disadvantage* of also protecting neoplastic mutants from the apoptogenic weapons of the mutation-detecting gatekeepers such as p53(TP53). This is why there are colon cancers but virtually no small intestine cancers.

Stem cells can somehow detect the loss of one of their members as well as the loss of more distant cells.[3,17] This loss stimulates the stem cells to cycle faster and shift from asymmetric to symmetric divisions, and it stimulates the recruitment of "potential" stem (i.e., clonogenic) cells in the crypt back into the proliferative pool to replace the lost cells.[1,3] How do they detect cell loss? Perhaps cell loss simply increases the available space (cellular Lebensraum) on the basal lamina for integrin-primed proliferation. However, it is likely that it is the intercellular canaliclular or gap junctional communication network that provides the measure of the size of the colon cell population.[18] A loss of cells would change the volume of proliferative modulators flowing through the gap junctional network.[18,19] In murine small intestinal crypts, there is an efficient transfer of microinjected Lucifer Yellow dye between cells at the base of the crypt, but the transfer efficiency drops between the upper middle and the top of the crypt. There is a similar network in the colon,[20] which would account for the spreading of an internal Ca^{2+} surge from cells at the base of a colon crypt to cells in the mid-region of the colon crypt when the M_3 muscarinic receptors of the basal cells are stimulated by acetylcholine.[20] The activated M_3 receptors stimulate the generation of Ins(1,4,5)P_3, which is probably what passes through the gap junctions from cell to cell, releasing Ca^{2+} from internal stores as it goes. Indeed, a similar upwardly spreading Ca^{2+} wave can be started by microinjecting Ins(1,4,5)P_3 into the basal crypt cells.[20] Clearly, a loss of cells would reduce the volume of intercellular traffic of proliferation moderators and increase the accumulation of proliferative promoters in the network's remaining cells.

Interestingly, just such a cryptal gap junctional network has been assigned the central role in a radically new model of the control of intestinal cell proliferation and differentiation.[21] According to this iconoclastic model, *all* of the columnar cells in a crypt have the same proliferative potential as the classical stem cell, and they will proliferate if the concentration of an intraepithelial growth factor that is small enough (1 kD or smaller) to pass through gap junctions exceeds a certain level. Hence, there are no distinctions between actual and potential stem cells, transit amplifying cells or mature cells as there are in the conventional model.[1,3,22] In other words, there would be no asymmetric division of stem cells. The key feature of this model is that the growth factor is produced *only* by "stem" cells in the lower part of the crypt from which it would pass up the chain of cells lining the crypt through gap junctions. At a certain level in the crypt, the concentration of the factor would drop below a threshold level because of dilution and decay. The cells at this level and higher would not be able to proliferate, and the differentiation program would start either by default or be triggered by signals from the CaRs to be discussed further on. Loss of cells from the chain or an increase in the production of the growth factor or the number of growth factor producers would cause a build up of the growth factor and an extension of the proliferative zone to higher levels of the crypt.

As we said above, according to the conventional model a true stem cell on the average (although not necessarily) divides asymmetrically at the end of its cycle into another pluripotent stem cell, which stays in the niche, and another daughter cell, a *potential* clonogenic stem cell, which can still proliferate indefinitely but not unless prodded to do so by something such as a severe loss of cells that requires the recruitment of more stem cells[1,3] or possibly the mutational disabling of a proliferation-monitoring/restraining gatekeeper gene. From the second or third generation onward, a cell cycle counter starts working in the rapidly cycling transit cells, and when the number on the dial reaches "0" they stop cycling. By this time, they have left the proliferative zone crafted by the combined actions of integrins, basal laminar components, and factors from their assigned pericryptal fibroblast escorts.[22] In the mouse, a transit amplifying cell of only limited clonogenic potential, and its descendants will complete eight cycles to produce a clone of

256 terminally differentiated cells. Like a transit amplifying basal keratinocyte, it expresses a set of β1 integrins, the signals from which prime its proliferative response mainly to signals from growth factors such as autocrine/paracrine TGF-α that start the cresting and crashing cell cycle-driving, stage-specific CDKs while moving away from the natal niche.[8,23] The proliferative activity may also be modulated by the circadian ebb and flow of other factors such as EGF and gastrin from the upper gastrointestinal tract and adrenergic, cholinergic, and other agents from the neuronal nets of the gut's "*second brain*" enveloping the crypt.[6,24–28] This precursor cell and its progeny, expressing such proliferation-related genes as cyclin D1, c-*myc*, and c-*myb* and protected from premature apoptosis by the Bcl-2 protein, glide along the basement membrane and up the crypt wall.[3,24–30] However, although the lower crypt is a place of intense cell production, it is also a minor apoptotic death zone, although there is much less spontaneous apoptosis here than in the small intestinal crypts.[3,31–32] While this cell death could be meant to prevent the propagation of cells with damaged or misrepaired DNA, its main purpose is simply to prevent cellular overproduction. Thus, for example, if a stem cell in an undamaged crypt should happen to divide symmetrically into two stem cells there could be a considerable cellular overproduction. This overproduction would shrink the amount of available space on the basal lamina, the cellular *Lebensraum*, which would force an actively cycling cell expressing the potentially apoptogenic c-*myc* to detach from the basement membrane and cut off its integrin-mediated and orchestrated "survival signals."[9–11] This would trigger c-Myc-mediated anoikis (detachment-induced apoptosis)[33] and the prompt disposal of the remnants of the dead cell by its neighbors.[7]

The epithelial cells are assigned a pericryptal fibroblast escort to accompany them up the crypt wall. Indeed, the location of the fibroblasts' proliferation, differentiation, and killing zones and their rate of migration up the crypt wall are the same as those of the epithelial cells.[34–38] There is undoubtedly a continuous exchange of information between the fibroblasts and epithelial cells in the form of factors and collaboratively made changes in basement membrane constituents.

Ca^{2+} and the control of colon cell proliferation, differentiation, and apoptosis

The cellular production in a colon crypt is so enormous that the crypt populations are turned over completely every 3 days in humans and mice.[1,1a,5,31] Each crypt in a mouse's proximal colon contains about 100 proliferating cells, which generate about 6 new cells per hour, while in the distal colon there are about 200 proliferating cells per crypt, which generate about 21 new cells every hour.[1,1a,5] Each of the millions of crypts in the human transverse colon produces 2.7–8.8 cells per hour.[31] Obviously, there has to be a culling mechanism to keep the cell population at a constant size to avoid overgrowth. Until recently, it has been assumed that the population is kept in check by the shedding of cells from the flat surface epithelium of the colon or at the tips of the villi in the small intestine. However, it now appears that colon cells are eliminated by apoptosis and the eating of their remnants by their neighbors at the bottom of the crypt; by apoptosis, cannibalism, and shedding at the top of the crypt; and in the small intestine, by apoptosis and cannibalism by neighbors at the bottom of the crypt, and by apoptosis and then by engulfment and (or) shedding at the villar tip. (The reason why this has not been realized until now is simple — the swift disposal of apoptosing cells by their neighbors and macrophages.) Therefore, differentiation in another continuously renewing epithelium is really differentiation-apoptosis or diffpoptosis. But in this case, the apoptotic mechanism does not make the ultimate functioning cells as it does in the skin by converting component-preloaded keratinocytes into functional corneocytes. Instead, it eliminates the colon cells after they have matured and functionally peaked and thus prevents overpopulation with cellular seniors that have passed their peak.

Does Ca^{2+} affect the proliferation of colon stem and transit-amplifying cells as it does the proliferation of basal keratinocytes? Sadly, there is very little information about Ca^{2+} and colon cell proliferation even in the plastic world of the culture dish at least in part because *normal* colon cells respond badly to the disruption of their β1-integrin basement membrane linkages.[7] Breaking these linkages in order to isolate the cells for culturing triggers anoikis, which kills them within as little as 4 h unless they can rapidly reattach to a collagen substrate and restore signaling from integrin·growth factor receptor clusters.[7] Nevertheless, it seems that Ca^{2+} affects colon cell proliferation in the same way it affects keratinocyte proliferation. Freshly isolated normal human colon cells need about 0.1 mM external Ca^{2+} to proliferate optimally, while their proliferation and the proliferation of rat colon cells is stopped and the cells are induced to differentiate by the high (0.8–2.0 mM) Ca^{2+} concentrations that can be found in the fecal water.[39–42] Ca^{2+} also controls proliferation in the real world of the crypt where the cells need some Ca^{2+} to proliferate maximally but cannot do so with smaller or larger amounts of the ion.[40] Thus, the proliferative activity of cells in crypts from normal rats given a diet containing either 0.5 or 15.0 g of calcium/kg was significantly lower than the optimal activity in the crypts from rats given a diet containing 5.0 g of calcium/kg.[43]

The key, and extremely important, question is whether there is a top-down Ca^{2+} gradient in the colon crypt like the epidermal gradient (Fig. 4.1). The short answer is that we don't know yet. Unlike the air-exposed skin, where the Ca^{2+} gradient has to be produced by keratinocytes bringing the ion up from the blood at the bottom and dumping it out at the top during their conversion to corneocytes, there is already a high Ca^{2+} concentration in the fecal stream flowing by the mouth of the crypt, which enables direct top-loading. Obviously, from the evidence we do have, if the Ca^{2+} concentration at the bottom of the crypt were to equilibrate with the millimolar concentration in the fecal stream by Ca^{2+} diffusing down from the crypt mouth it would raise the external Ca^{2+} well above the optimal level for the proliferation of the stem and transit amplifying cells. And loading the diet with a large amount of Ca^{2+} does shrink the cryptal proliferative zone (i.e., decreases the value of Φ, the % of cycling cells in the upper 40% of the crypt).[44–49] As suggested by Whitfield,[25,26] a Ca^{2+} gradient could be established by a bottom-to-top gradient of Ca^{2+}-binding chondroitin sulfates, heparin sulfates, and particularly the sialo- and sulfomucins secreted by the goblet cells that are concentrated in the lower regions of the crypt.[20,50–56] Indeed, the distribution of mucins (particularly sulfomucins) parallels the proliferative activity in the crypts of human and rat colons.[20,52,55] Lower crypt cells, like basal and spinous keratinocytes, may also help reduce the lumenal and intercellular Ca^{2+} concentration by accumulating and sequestering the ion or dumping it into the blood, while upper crypt cells may oppose downward Ca^{2+} diffusion by actively accumulating the ion. Brenner et al.[57] have recently reported a striking bottom-to-top *intracellular* Ca^{2+} gradient in mouse colon crypts and a crypt level-dependent responsiveness to a change in the extracellular Ca^{2+} concentration and have mentioned that the same gradient has been found in human crypts. Thus, for example, the concentration of Ca^{2+} in the cells of isolated whole murine crypts (determined by loading the cells with the Ca^{2+}-specific fluoroprobe FURA-2) rose from 70 to 83 nM in the lower crypt to 157 nM at the crypt mouth when the crypts were in Ca^{2+}-free buffer. When the crypts were suspended in buffer containing 1 mM Ca^{2+}, the intracellular Ca^{2+} rose from 262 to 297 in the lower crypt to 463 nM at the crypt mouth. The maintenance of this gradient required $1\alpha,25(OH)_2$ vitamin D_3 because it was lost in animals fed a low-vitamin D_3 diet, which indicates the involvement of the transcellular transport of Ca^{2+} by the $1\alpha,25(OH)_2$ vitamin D_3-dependent protein calbindin from the lumen to the blood. It may also indicate the involvement of signals from the CaRs on the cells' apical surfaces,[58–62] because expression of parathyroid gland's CaR is promoted by the seco-steroid (at least it is in keratinocytes).[63]

According to Kállay et al.[63a] and Scheinin et al.[63b], only the chromogranin A-containing enterendocrine cells at the base of the crypts make this particular CaR in the human colon. This suggests that if the Ca^{2+} concentration in the crypt should rise above a certain limit, the signals from the CaRs might cause the enteroendocrine cells to release a proliferative suppressor such as pancreastatin, a cleavage product of their chromogranin A.[63b] Alternatively or additionally, there might be a CaR-suppressible production of a mitogen, such as the diffusible one proposed by Gerike et al.[21], by enteroendocrine cells. However, the specific CaR may not be the only Ca^{2+} sensor. The other crypt cells might have a different sensor (e.g., RyR, the ryanodine-sensitive receptor/channel)[58] because cultured colon cells without their enterendocrine companions are still proliferatively shut down by low Ca^{2+} concentrations.[39–41,63c] Whatever they may be, these Ca^{2+} sensors are most likely the agents through which external Ca^{2+} controls the proliferation and diffpoptosis of the suprabasal colon cells (Figs. 4.1 and 4.2). Perhaps lower crypt cells in the normal man or mouse bearing a minimal number of Ca^{2+} sensors and equipped with calbindin collect Ca^{2+} with their calbindin and take it to their basolateral membranes and dump it into the passing bloodstream. This pumping mechanism plus the mucins could produce a low, proliferation-compatible level of external Ca^{2+} just like the ECaBP-mediated uptake of the ion that we suggested in the previous chapter maintains the low, proliferation-compatible Ca^{2+} level in the basal epidermis. At higher crypt levels, the cell's Ca^{2+} sensor density might increase. The rising flow of PLC-activating, intracellular Ca^{2+}-moblizing signals from the receptors[61] would increase the steady-state intracellular Ca^{2+} level. A variant of this model would be that the cell's Ca^{2+} sensor density does not change during its journey up the crypt, but the cell's internal Ca^{2+} surges when the luminal Ca^{2+} level reaches a specific set point in the upper crypt. In either case, the more intense Ca^{2+} sensor signaling and the rising internal Ca^{2+} level would be enough to stop proliferation and start differentiation.

Integrins, ILK, catenins, and proliferation

The proliferation of the precursor cells in the low-Ca^{2+} base of the colon crypt, like that of the basal keratinocytes in the previous chapter, is driven by adhesion-induced clusters of $\beta 1$ integrins and receptors for growth factors such as EGF and TGF-α. The signals from these clusters stimulate PtdIns-3K, the PtdIns(3,4,5)P_3 product of which stimulates intergrin-associated ILK and membrane-associated PKB.[64–68] ILK is connected to the β1 integrin cytoplasmic domain by the cysteine-rich LIM motifs of the PINCH adaptor protein, which also attach the Nck-2 adaptor protein that connects to various growth factor receptors through its SH2 and SH3 domains.[69,70] These signaling complexes keep the cell in the proliferative mode by increasing the activity of the E2F-family of proliferogenic transcription factors when PKB activates the Cdk4·cyclin D protein kinase that hyperphosphorylates and thus inactivates the E2F-suppressing Rb-family pocket proteins).[71] They also prevent the destruction of β-catenin, which can get into the nucleus to stimulate key proliferation-driving genes such as those for Fra-1 and c-Jun (components of AP-1 transcription complexes), but most importantly the genes for cyclin D1 and the master transcription factor for several replication genes, c-*myc*.[12,72–74] The c-Myc product heterodimerizes with Max protein to form Myc·Max complexes, which, among other things, stimulate the expression of the gene for the Cdc25A protein phosphatase that also activates the G_1 buildup, starting Cdk4·cyclin D1 protein kinase.[9] The complexes also stimulate the expression of TERT, the telomerase catalytic subunit, which if c-Myc expression should persist and TERT should exceed a certain level would contribute to cellular immortalization and tumorigenesis.[75] Premature apoptosis is prevented by the cells making sure that c-Myc, the E2F1 expression it stimulates, the apoptosis-triggering p53(TP53) that excessive stimulation of E2F1 can in turn stimulate, and Cdc25A are kept within the proliferogenic realm using survival signals from autocrine IGF-I, which stimulate the expression of the anti-apoptosis Bcl-2 protector protein, and by suppressing the expression of pro-apoptosis proteins such as BAD.[9,67,72,76,77] ILK and PKB also disable GSK-3β,[67] which prevents the protein kinase from marking β-catenin for ubiquitination and consignment to the proteasome shredder and downregulating the cycle-driving cyclin D1.[73,78–81] The signals from the β1-integrin-assembled complexes might also suppress the expression of the APC (adenomatous polyposis coli) protein that joins with another protein, axin or conductin (axil), to bring β-catenin into contact with GSK-3β and ultimately the shredder.[81,82] They may also suppress the expression and display of the Ca^{2+}-dependent E-cadherins, which reduce the availability of β-catenin for gene stimulation by using it to link the E-cadherin C-terminal tails to α-catenin, vinculin, α-actinin, and F-actin to make the intracellular parts of the cables with which the Ca^{2+}- binding cadherins tie neighboring cells together.[83–85]

The proliferation of the colon stem and transit amplifying cells also requires cAMP signals, a predominance of the proliferation-promoting PKAI (type I cAMP-activated protein kinase) over the differentiation-promoting PKAII,[86,87] and expression of the 320–350 kDa Ki-67 protein, which is a universal marker for all proliferating cells, including colon cells.[48,88–90] The proliferation-driving signals from the integrin·growth factor receptor complexes increase the expression of the RI (type I regulatory) subunit gene and reduce the expression of the RII subunit gene.[86,87] We shall soon see that a reversal of the expressions of the RI and RII genes is part of the colon cell's diffpoptosis program.

Ca^{2+}, E-cadherins, APC, and diffpoptosis

As an epidermal basal cell moves away from the natal cluster, its β1 integrin complement drops and its attachment to the basal lamina weakens. Eventually, the integrin signaling drops below a critical level, which triggers a Ca^{2+} signal that triggers diffpoptosis. We know that colon cells reduce the expression of their β1 integrins as they move toward the top of the crypt and the colon surface,[7,8] but without a reliable in vitro model system we do not know whether there is a Ca^{2+} surge associated with the integrin shift like that which occurs in keratinocytes. The fading signaling from β1 integrins and their associated growth factor receptors, and the PtdIns-3K/ILK/PKB activities they stimulate, derepresses E-cadherin gene expression and increases GSK-3β activity, which, respectively, sequester and destroy β-catenin and thereby shut down the expression of the catenin's cycle-driving target genes. However, colon cells do not detach from the proliferative zone and lift into the differentiation and killing zones, as do keratinocytes. Instead, colonocytes stay attached and glide along the basement membrane in a chain (a cellular escalator) as they go from the proliferative zone in the lower third of the crypt to the differentiation zone and finally the killing zone. The cell can no longer make cycle-driving c-Myc·Max transcription complexes, which start the events leading ultimately to the waves of cyclins E and A for the Cdk2·cyclin protein kinases needed to make the replication factories. But as if this were not enough to stop cycling, the upwardly moving colon cell, like a detaching keratinocyte, also abruptly stops expressing the Ki67 protein, reduces the expression of the RI subunits of PKAI, increases the expression of the RII subunits for the differentiation-promoting PKAII,[86,87] and possibly because of an increasing volume of signals from the β4 subunits of emerging differentiation-specific integrins (such as α6β4),[91] starts expressing the $p21^{Cip1/WAF1}$ CDK inhibitor, which prevents the G_1 buildup of materials for the replication factories by inhibiting Cdk4·cyclin D and Cdk2·cyclin E protein kinases as well as any existing Cdk2·cyclin A protein kinase.[88–97] This p53(TP53)-independent (like that in keratinocytes) burst of $p21^{Cip1/WAF1}$ expression may be used to shut down the cell cycle without triggering apoptosis, which would allow the cell to differentiate and function until it reaches the killing ground.[95–97] p21's job is to stop the colon cell cycling without being involved in starting differentiation.[97] After it has done its job of silencing the Cdk·cyclins, it drops down to a lower level,[97] which is still high enough to prevent any resurgence of the Cdk·cyclins without interfering with differentiation as its does in keratinocytes.[98] The surge of p21 is followed by a surge of another multifunctional CKI, $p27^{Kip1}$, which, despite it being a CDK inhibitor, is probably involved in differentiation rather than cell cycle suppression[97] as it is in keratinocytes.[98]

The developing Ca^{2+}-dependent, E-cadherin-mediated cell–cell adhesion might be responsible for the enhanced expression of TGF-β1 and TGF-β2 and the type II component of the TGF-βs' Type I·Type II receptor complex in the upper part of the crypt.[99–101] The autocrine/paracrine TGF-βs are responsible for the producer cells expressing the $p15^{INK4b}$, $p21^{Cip1/WAF1}$, and $p27^{Kip1}$ CKIs and suppressing the expression and activities of CDKs and shutting down c-*myc* in the upper, differentiation zone of the crypt.[96–102] In other words, the TGF-βs may throw the switch that turns off proliferation when the cells reach the differentiation zone. They also stimulate their producer cells to make differentiation-specific matrix-binding integrins, and these factors diffuse over to the escorting fibroblasts and cause them to make the matrix components that bind and activate these integrin receptors.[103–105]

The development of a proliferatively disabled functioning polarized columnar epithelial cell with the expression and placing of enzymes, channels, and other components at appropriate places in the apical micovillar and basolateral membranes requires the continued actions of different integrins plus several emerging cell–cell and cell–matrix adhesion components, starting with E-cadherins. As in the keratinocytes of the previous chapter, extracellular Ca^{2+} puts the emerging E-cadherins into their active configuration. Their cytoplasmic domains combine with β- (or γ-) and α-catenins to form complexes that link them to α-actin, ankyrin, and other components of the cytoskeleton as well as to enzymes such as Na^+/K^+-ATPase.[84,106,107] These complexes include receptor kinases such as the EGF/TGF-α receptor that binds to the conserved core region of β-catenin[108] and from this vantage point can drive various functions, including the expression of differentiation-specific genes. Using the establishment of functional polarity in Madin–Darby canine kidney cells as a model, it is likely that the formation of tight junctions and desmosomes by the differentiating colon cells is started and driven by global and localized Ca^{2+} surges triggered by the cell–cell contact and the Ca^{2+} sensors.[109–112] These surges drive the translocation of junctional components to appropriate assembly sites on the cell membrane to form tight junctions and desmosomes. The result is tight, Ca^{2+}-driven, cell–cell adhesion by cytoskeleton-linked signaling complexes, which reconfigures and polarizes the now-proliferatively silenced cell and puts clusters of key enzymes at particular regions of the cell surface such as disaccharidases, peptidases, Na^+-dependent glucose co-transporters in the apical brush border and α_2- and β-adrenergic receptors, VIP (vasoactive intestinal peptide) receptors, Na^+/K^+-ATPase, and outwardly directed Na^+-independent glucose transporters on the basolateral border.[109–112] Other stem cell progeny become goblet cells, most of which stay near home in the lower crypt to make mucins for the Ca^{2+}-binding shield.

A major player in colon cell maturation is the APC (adenomatous polyposis coli) tumor suppressor protein, the gene for which is mutationally disabled or eliminated during the progression of a colon cell's progeny to carcinoma. As colon cells mature, they express and accumulate the 312-kDa APC phosphoprotein.[113] APC seems to have at least three roles in diffpoptosis, two of which involve association with the cytoskeleton and cell–cell adhesion.[114–120] First, as we have learned, APC associates with, and is phosphorylated by, GSK-3β on an axin (or conductin) platform, which enables it to bind β-catenins in the lateral cytoplasm and compete with, and thus directly modulate, cell-tethering by the E-cadherins.[66,81–83,106,107,117–119] APC also strongly promotes microtubule assembly with its C-terminal region and collects in clusters near the ends of cytoplasmic microtubules.[115,116] There it promotes the formation of junctional complexes and signaling as well as various microtubule-mediated functions, which include moving the cell up the crypt wall toward the killing zone. Obviously, APC's cell-moving and Ca^{2+}·E-cadherin's cell-tethering actions must be balanced to produce the moving chain of tethered functional cells. Indeed, upsetting this balance by hyperexpressing E-cadherin would, and does, jam the "escalator." Finally, the APC-bound "excess" β-catenin on the axin platform is marked for ubiquitination by the GSK-3β kinase and sent to the proteasome shredder.[81] In this way, APC keeps β-catenin at a low level.[121–125] This is needed to balance cell production and diffpoptosis because any free β-catenin would enter the nucleus where it would displace the Groucho and CtBP transcription co-repressors and bind to enhancer-bound Tcf-4 factors to form active transcription complexes on the enhancers of cell cycle-driving genes, which would extend the proliferative zone and oppose differentiation and apoptosis.[12,72,76,121–125]

Another key player in colon cell maturation is DCC (Deleted in Colon Cancer), which, as its name indicates, is lost during the later stages of the progression to malignant colon cancer. DCC has been considered as a cell–cell adhesion protein belonging to the N-CAM (Neural Cell Adhesion Molecule) family.[126–128] However, from the outset there was something wrong with this idea because there is not enough of it to be a cell adhesion factor. We now know that it is a *receptor*, which, when activated by contacting the laminin-like netrin-1 in the extracellular migration trackways with its Ig-like and fibronectin-type III extracellular motifs, sends signals (that involve G-proteins and Ca^{2+}) into the colon cell through its cytoplasmic domain, which suppress anoikis and somehow promote differentiation (possibly by inhibiting CDK activity) and guide the differentiating cell's migration up the crypt wall just as neurons use their DCC receptors to steer growing neurites toward their targets.[126–130] However, DCC is also part of an important "fail-safe" mechanism that guards against a cell being able to detach from the basal lamina with impunity and metastasize to distant places.[131] Thus, the apoptogenic mechanism is silent when the cell is attached to the basal lamina and DCC is attached to its ligand and presumably steering the cell along the trackway and emitting an anoikis-suppressing signal. Should the colon cell try to detach itself from the trackway, there would be an increase in caspase-3 activity, which, by cleaving it at Asp*1290*, converts DCC into a potent apoptosis driver.[131]

Once again, a Ca^{2+} surge has been cited as the common mediator of the action of another differentiation factor. However, we must take the opportunity at this point to remind the reader that in our current primitive state of knowledge that all Ca^{2+} surges are not the same. These several regulators and adhesion complexes trigger different space–time–frequency-modulated Ca^{2+} transients of various shapes and sizes in different parts of the cell. These specifically patterned streams of transients are the words and phrases of the complex Ca^{2+} language that different agents use to tell cells how to reconfigure themselves into mature absorptive or goblet cells.[132]

As the DCC-directed, upwardly gliding cell is proliferatively shut down by the appearance of TGF-βs and their receptor complexes, the fading of proliferation-related integrins and the cadherin-triggered/mediated adhesion to its neighbors, it of course shuts off its c-*myc* and c-*myb* genes.[27,28,133] It might be expected that as it moves up the crypt and stops cycling, the differentiating colon cell would also stop making its principal autocrine proliferogen, TGF-α. But it doesn't! It keeps making TGF-α and the EGF/TGF-α receptors.[134] However, these receptors are probably now activated at least in part by the EGF-related amphiregulin, which is expressed only by differentiated columnar epithelial and goblet cells.[135] Because the receptor signals may be different and the signals' old targets have been replaced in the functionally reconfigured cell by new diffpoptosis-related ones, neither the emerging amphiregulin nor the TGF-α made at the top of the crypt can make the maturing cell start cycling.[134]

In the skin, basal cells express the PTHrP gene at a low level and display PTHrP receptors, but they do not translate any PTHrP messages into PTHrP holoprotein or produce their characteristic selection of PTHrP fragments. However, when a cell enters the spinous layer, PTHrP gene expression surges and the cell now translates the messages into PTHrP proteins, only to stop again when it enters the granular layer. Rat colon cells also express their PTHrP receptor gene(s) and probably make receptors and produce PTHrP mRNA at all crypt levels, but they translate these messages into PTHrP proteins only at the top of the crypt and on the flat epithelial surface.[136] Similarly, PTHrP mRNA is made all along the crypt and villus in rat jejunum, but the messages are translated into PTHrP proteins only when the cells leave the crypt and start moving up the villus.[136] The functions of PTHrP and its fragments in the cells of the colon's surface epithelium and the small intestinal villi are unknown, except that PTH/PTHrP receptor signals can stimulate Ca^{2+} transport by chicken small intestine cells.[137] However, while the keratinocyte's particular PTH/PTHrP receptors are coupled to phospholipase-Cβ, but not to adenylyl cyclase, the colon receptors are more conventionally coupled to adenylyl cyclase[138] and almost certainly to phospholipase-Cβ. PTHrP and its fragments are undoubtedly very important autocrine/paracrine factors in the mature colon epithelium just as it is in many other producer tissues. Significantly, the translation of its mRNA coincides with the entry of the cell into the apoptotic or killing zone[31,32] and the activity of the Ca^{2+} sensors as they meet the very high Ca^{2+} levels at the top of the crypt. It seems likely that PTHrP's job is to hold off terminal apoptosis as it does in cartilage by stimulating the expression of the Bcl-2 protector protein.[139]

When the colon cell nears the mouth of the crypt and the edge of the Ca^{2+}-rich fecal stream, it, like the granular keratinocyte, almost certainly hits a wall of external Ca^{2+} and is battered by signals from its apical Ca^{2+} sensors (Fig. 4.1). This signaling would expel Ca^{2+} from internal stores, which in turn would open SOCs (stores-operated channels) and start an inward flow of external Ca^{2+}.[61,140] This could be why upper crypt cells (like the upper granular keratinocytes in the previous chapter) have the largest amount of intracellular Ca^{2+}.[57] These signals would complete the maturation of the cell. As in the granular layer of the epidermis, the rising intensity of signaling from Ca^{2+} sensors could also be responsible for the large drop in gap junctional intercellular communication in the upper crypt.[20,21] But these signals also prepare the cell for something else — apoptosis.

Now we must revisit the APC tumor-suppressor protein, which will be at the center of the stage in the next section. So far, this protein and E-cadherin have been keeping the free β-catenin pool at a low level to promote cell maturation and the chain of mutually tethered cells gliding along the basal lamina without re-starting the proliferogenic genes that would disrupt the chain. When APC builds up beyond a certain critical level as the cell nears the top of the crypt,[113] it could weaken the E-cadherin-mediated linkage of the cell to its neighbors by dropping the β-catenin content below the level needed to make and maintain the tethering cables. The cell would soon break away from the chain and start the population-limiting apoptogenic mechanism.[31] Indeed, Morin et al.[141] have found that introducing a normal, active APC gene in colorectal carcinoma cells with disabled APC genes kills the cells by apoptosis. At this advanced stage of its life history, the cell starts getting ready to kill itself. It stops making the anti-apoptogenic Bcl-2 protein that protected its ancestors from premature death during their youth when they were at the bottom of the crypt and suicidally starts making the pro-apoptosis BAD, Bax, and $Bcl\text{-}X_S$ proteins.[3,13–15]

Now the end is near and the aging cell starts accumulating DNase I[142] and transglutaminase, which will eventually be used to convert it into a hardened corneocyte-like squame that is either swiftly engulfed by its neighbors or falls off into the passing fecal stream.[143,144] However, we must not forget the fibroblasts that have faithfully escorted the epithelial cells and helped nourish and drive their growth and differentiation during the journey up the crypt. They too, like the wives and retainers of an ancient King, must apoptose and be swiftly eliminated by macrophages to balance the intense cell production going on in the lower crypt.[31]

Colon cancer

The beginning of the road to cancer — aberrant crypts, polyps, and a loss of Ca^{2+} control

The steady battering of the colon with carcinogens over the years results in as many as 50% of western people harboring adenomas in their colons by 70 years of age. These carcinogens produce a genomic variability that a cell and its progeny can draw on to evolve by classical Darwinian natural selection strategies that enable them to survive and flourish despite damaged genomes and the best efforts of checkpoint guardians, cycle suppressors, and apoptogens.[145] Colon carcinogenesis, like carcinogenesis in other tissues, starts with genomic damage. In the proximal colon, bile acids are principal carcinogens. The N-nitrosylated conjugated bile acids are analogs of methylating carcinogens such as MNNG (*N′*-methyl-*N′*-nitrosoguanine) and MNU (*N*-nitrosourea), which start the carcinogenic ball rolling by methylating cytosines, which when deaminated to thymine causes mismatching G/C→G/Me-C→G/T mutations that triggers mismatch repair.[145,146] In the distal colon, BAFs (bulk adduct formers) are the principal carcinogens, which cause chromosomal rearrangements, loss or translocation of parts of chromosomes or loss of entire chromosomes, i.e., chromosomal instability and aneuploidy.[145] The cell tries to remove the adducts by cutting them out (nucleotide excision repair), but these molecular scissors can make things worse by too much chromosome cutting! Since Darwinian success is measured only by the enhanced ability of an individual cell to survive and replicate its genome, the intense selection pressure in a transforming population of cells will favor those clones that can override the growth arresting/repair mechanisms and proliferate despite their damaged chromosomes and aneuploidy. The pressure will then be on inactivating or getting rid of the cycle-suppressing pocket proteins and Cdk4/6·cyclin D-suppressing CKIs by disabling their genes by mutation or shutting them down by methylating the cytosines in the CpG/GpC islands in their 5′-regulatory regions, and hyperexpressing key cycle drivers.[145] And last, but not least, will be the disabling or elimination of the apoptogenic p53 (TP53) and mitotic checkpoint mechanisms. The result of this successful evolution by natural selection is a colon tumor and ultimately carcinoma in the cells of which about 500 of the 300 000 gene transcripts from about 45 000 different genes are expressed at significantly different levels from those in normal colon cells.[146]

However, while the world's focus has been on gene mutations as the agents of carcinogenic progression, there is something else which is probably equally important. This is the methylation of the cytosines in CpG islands in the promoters of the genes for growth suppressors and differentiation promoters.[145,147] As we learned in Chap. 1, cytosine methylation by DNA-MTase (DNA methytransferase) is a kind of molecular "not-to-be-read" Post-it® note that is put on to genes just after replication to duplicate their mother cell's functional characteristics. Since the genes for growth (tumor) suppressors and differentiation promoters have CpG islands in their promoters, they can be as permanently and as replicably silenced by hypermethylation as by a disabling mutation. Thus, as we shall soon see colon cancers are the products of both gene mutation and epimutation (methylation).[146]

As a person ages, the DNA-Mtase activity in various tissues including the colon rises.[146] One of the effects of this in the colon is the hypermethylation of the CpG islands, and consequent silencing, of the estrogen receptor (ER).[146] An initiating mutation in cells with their silenced, epimutated ERs triggers their clonal expansion into the preneoplastic lesions to be described below.

The initiating genomic change caused by a N-nitrosylated conjugated bile acid in the proximal colon or a BAF carcinogen in the distal colon drifting down into the crypt from the fecal stream must happen in one of the actual stem cells or in a cell of the first two or three generations of still uncommitted, still clonogenically competent transit cells, but not in later rapidly cycling transit amplifying cells with their cycle counters running down, which eventually apoptose although they complete more cycles before doing so. Cancer is common in the colon, but rare in the small intestine.[3] Why? Mutagen-damaged cells in both colon and small intestine may respond to this chromosomal damage with p53(TP53)-triggered apoptosis.[95] It happens that the most apoptosis-prone cells in the mouse small intestine crypts are the clonogenic stem cells in position 4 where they do not contain the anti-apoptogenic Bcl-2 protein, but in the colon the most apoptosis-prone cells are only the non-clonogenic transit amplifying cells in the mid-crypt region (cell positions 11 to 12) far from the well-protected, Bcl-2-loaded clonogenic stem cells in positions 1 and 2.[3,16,17,148] The importance of Bcl-2 is demonstrated by the fact that in Bcl-2-null mice the level of spontaneous apoptosis rises significantly in colon, while there is no increased apoptosis in small intestine crypts.[3,16] Therefore, in the small intestine, mutagen-damaged clonogenic cells, unprotected by Bcl-2, kill themselves, while mutagen-damaged clonogenic colon cells protected by Bcl-2 survive as dangerous, self-renewing cells in which the clonal selection process has been started and are primed to respond to tumor promoters by proliferating rather than differentiating.

According to conventional wisdom, the mutated and epimutated human colorectal cancer clones appear in hyperplastic adenomatous polyps.[4,149,150] However, a large fraction of early lesions as well as advanced colorectal cancers have arisen from flat mucosa rather than polyps. In fact, the smallest detectable precursor lesions are clusters or foci of hyperplastic, distended, flask-shaped, branching, bifurcated (dividing) aberrant crypts.[150–177] These ACFs (Aberrant Crypt Foci) are the products an abnormal crypt cycle and stem cell expansion. An ACF starts with excessive clonogenic cell proliferation at the base of a crypt, which causes a branching or fission that moves up until there are two crypts, which in turn branch and divide to produce a crypt cluster.[2,155,158]

Some hyperplastic ACFs turn into, or perhaps start out as, the dangerous dysplastic adenomatous ACFs. The post-hyperplasia-generating mutations that start the carcinogenic process disrupt the differentiation program and extend the proliferative zone. In normal human colon crypts, the nuclear, proliferation-marking Ki67 and the hMSH2 mismatch-repair proteins are intensely expressed in the lower third of the crypt, but they are sharply downregulated and the $p21^{Cip1/WAF1}$ CKI upregulated when the tethered cells glide out of the proliferative zone. Even in the largest of the hyperplastic ACFs, proliferatively active cells expressing Ki67 and hMSH2 proteins were still restricted to the lower third of the crypts, while p21 was expressed only in the upper part of the crypts and the surface epithelium.[96] However, in the carcinoma-hatching dysplastic ACFs, the mechanism(s) that shuts off Ki67 and hMSH2 expression and turns on the cycle-stopping $p21^{Cip1/WAF1}$ at the border of the diffpoptosis zone has been disabled, and the strict compartmentalization of cycling cells as indicated by the expression of Ki67 and hMSH2 is lost. Ki67 and hMSH2 are now expressed all the way up to the top of the crypt and the surface epithelium and overlapped with the expression of $p21^{Cip1/WAF1}$ in the upper regions.[96]

This extension of the proliferative zone into the differentiation and killing zones is the result of the initiated cells persistently producing autocrine/paracrine growth factors such as PDGF-B (platelet-derived growth factor-B), IGF-I, IGF-I variants, TGF-α, partly processed TGF-α precursors, and the EGF-like Cripto, which is not expressed by normal colon cells.[25,135,178,179] Because they can stimulate themselves with their hyperexpressed growth factors, the proliferation of the initiated/activated cells is not limited by the need for signals from adhesion-assembled complexes of β1 integrins and expogenous growth factors and their receptors. An example of the importance of a TGF-α-autocrine/paracrine loop for the proliferation of at least some colon carcinoma cells is the ability of a 23-base antisense oligonucleotide that recognizes TGF-α mRNA to reduce TGF-α secretion and proliferation of LIM1215 colon carcinoma cells.[180] Later, mutant cells appear which have either lost the Type II TGF-β receptor and are no longer susceptible to proliferative inhibition by TGF-β, or may be stimulated by TGF-βs or a paracrine factor such as PDGF-B induced by TGF-β.[99a,181–186] The mutants increase PKC-βII expression and reduce PKCs α and βI expression.[186a] They are also likely to start constitutively making PTHrP, which is known to stimulate adenylyl cyclase and proliferation of at least some initiated cells such as human LoVo colon carcinoma cells.[138] Indeed, an increased production of PTHrP is one of the earliest changes on the road to cancer and peaks in adenocarcinoma cells.[187] The possibility of PTHrP being an autocrine/paracrine enhancer of tumor progression that operates through the PTH/PTHrP receptor is suggested by an association of hyperparathyroidism with colon tumors.[188–191]

Excessive growth of normal, uninitiated cells is limited by anoikis[33] as well as cryptal branching and fission. An excess of cells would reduce the available basal lamina space, which would reduce or stop the flow anti-apoptosis signals from adhesion-induced β1-clusters. This would trigger anoikis and the swift disposal of the cadavers by their neighbors.[7,9] However, the carcinogen-initiated cells don't need signals from β1 integrins because of the persistent production of autocrine, PtdIns-3K-activating growth factors such as Cripto, TGF-α, IGF-I, and the resulting hyperexpression of cell cycle drivers such as the c-*myc* gene and inhibitors of apoptosis.[9,67,133,192] Another likely contributor to initiated cells' resistance to anoikis is the surging PTHrP, which may shield the cells by increasing the expression of the Bcl-2 protector protein.[139]

At the heart of colon carcinogenicity are the failures to effectively assemble the proliferation-suppressing, cell–cell/cell–matrix adhesion systems and express the TGF-βs and TGF-β receptors that switch on CKI genes and silence cell cycle genes such as those encoding cyclin D1 and c-Myc and to produce functioning columnar cells that ultimately apoptose. A common first step (in about 70% of all non-hereditary, non-germ line colorectal carcinomas) is a mutation in the first half of the coding region of the Apc^+ gene on the long (q) arm of human chromosome 5.[101,123] This starts neoplastic development.[149–193] It is a disabling of the *Apc* gene that is responsible for the hereditary adenomatous polyposis coli from which the gene and its product got their name.[149–193] In this disease, the colon surface becomes studded with polyps, some of which spawn carcinomas.[149–195] In hereditary non-polyposis colon carcinoma (HNPCC) and some sporadic colon carcinomas, the *Apc* mutation is one of the many resulting from the disabling of the hMSH2 gene (on chromosome 2p 15-16) of the proofreading/editing mismatch repair system that scans newly replicated DNA for any single base-pair mismatches and multiple-base insertion–deletion loop-type misalignments, which it cuts out and replaces with correctly matched DNA.[22,196,197] However, in the rest of the non-germ line sporadic colorectal cancers the *Apc* gene is disabled even with a functioning mismatch repair mechanism.

Some ACFs, the relatively innocuous non-dysplastic ACFs, owe their hyperplasia to a mutant, constitutively hyperactive Ki-Ras protein, but they still have normal Apc^{+} genes and are not neoplastic.[155,170–172,174,193] On the other hand, clonogenic cells in the dysplastic crypt foci are either homozygous $Apc^{-/-}$ or they have one normal Apc^{+} and one disabled, Apc^{-}, gene. At least one APC mutant, specifically Apc^{1309} in the central part of the gene, is a dominant negative, the product of which blocks the activity of the normal wild-type protein and is associated with a severe phenotype of an early appearance of thousands of polyps.[194] But despite this, $Apc^{-/-}$ crypts replicate faster into ACFs than $Apc^{\pm}$ crypts.[156]

Obviously, it is the loss of APC function that puts most clonogenic colon cells on the road to cancer? Why? How? Apc^{+} is what Kinzler and Vogelstein[195] have called a *gatekeeper gene*, the product of which (like the basal keratinocyte's *Ptch* gene discussed in the previous chapter) prevents runaway proliferation. The answer lies in APC's ability to form a complex with axin (or conductin), the GSK-3β protein kinase, and β-catenin.[12,66,73,81,82,121–124,198] β-Catenin in this complex is phosphorylated by GSK-3β and thus marked for ubiquitination and proteasomal shredding, which stops it from getting into the nucleus to combine with the Tcf-4 architectural transcription factor and stimulate cell cycle drivers such as the c-*myc* gene.[12,64–86,121–124,198] However, if APC is disabled or lost, there will be enough β-catenin to promote proliferation and delay or prevent differentiation and apoptosis.[72–74,76,121–124,199,200] And diacylglycerol- / bile acid- / growth factor-stimulated upregulated PKC-βII in the initiated cells promotes the accumulation of β-catenin by phosphorylating and inhibiting GSK-3β.[186a] Indeed, while overexpressing β-catenin stimulates the expression of cycle-driving genes, it reduces the expression of a principal differentiation-related gene, the gene for ZO-1, the zonula occludens of the apical tight junctions that are among the key components of polarized functioning colon cells, and it would also result in a c-Myc-mediated hyperexpression of TERT, the catalytic subunit of telomerase, which could contribute to tumorigenesis by downregulating $p19^{ARF}$ and p53(TP53) and immortalizing the mutant cells.[75,186a,201] Thus, the cause of hyperproliferation and the expansion of the proliferative zone is too much β-catenin, only a fraction of which, at least in this early stage, can still be locked into cell-tethering cables by Ca^{2+}·E-cadherin. Indeed, some colon cancer cells still express a perfectly normal, functional APC protein and GSK-3β kinase, but they make a mutant β-catenin which cannot be phosphorylated and thus marked for the proteasome shredder by GSK-3β.[12,73,122]

However, there is a negative feedback mechanism that can mitigate the consequences of APC loss and excess β-catenin. β-Catenin·Tcf-4 also stimulates the expression of Tcf-1, a close relative of Tcf-4 that shares gene targets with Tcf-4, but which cannot bind β-catenin. Because of this a surge of Tcf-1 can negatively feed back onto β-catenin·Tcf-4, compete with β-catenin·Tcf-4 for target genes, and limit the complex's ability to stimulate proliferation.[125] It follows from this that a loss of Tcf-1 during carcinogenesis could enhance the hyperproliferative consequence of APC loss.[125]

Although hyperexpression of β-catenin and its accumulation in the nucleus was first found to be responsible for starting colon carcinomas, it is now being found in other tumors. Thus, mutant non-phosphorylatable β-catenins are particularly prevalent in anaplastic thyroid carcinoma; and a large nuclear accumulation of β-catenin has now been found in the proliferating cells of 43% of human hepatocellular carcinomas, but not in the cells of adjacent non-tumorous (normal, cirrhotic nodular, dysplastic) liver tissue.[81,202] Deregulated β-catenin signaling has also been found in endometrial cancers, melanomas, ovarian cancers, pilomatricomas, and prostate cancers.[73]

One thing seems to be missing that could greatly help β-catenin drive proliferation and impede differentiation. As we learned in Chap. 1, the key to starting a cycle rather than shifting out of the proliferative mode is the expression of Cdk4·cyclin D, which hyperphosphorylates and inactivates the Rb-family pocket proteins that suppress the expression of replication-driving genes. Indeed, *most* cancers of continuously self-renewing epithelia have disabled their Cdk4·cyclin D1-inhibiting p16$^{INK\ 4a}$ protein, a consequence of which would be increased inactivation of Rb-family proteins.[202a,203] Another result of the inactivation or deletion of the INK 4a locus would be a loss of the locus's alternative product p19^{ARF}, which, by blocking the action of the p53(TP53)-destroying Mdm2, stops excessive oncogene-driven proliferation by stabilizing the cycle-stopping, apoptogenic p53(TP53).[202a] Therefore, these cells would tend to proliferate excessively because they cannot effectively restrain the G_1-driving Cdk4·cyclin D1/D2 protein kinases, which causes them to resist shifting into the differentiation mode. However, p16$^{INK\ 4a}$ mutations have rarely been observed in colorectal cancers.[145,203,204] This rarity of disabled mutant p16$^{INK\ 4a}$/p19^{ARF} genes in colorectal cancers is unusual in view of the fact that their frequency is second only to that of p53(TP53). Are colorectal cancer cells really so different from others? The exciting answer is NO! One of the results of the surge of DNA-MTase (DNA methyltransferase) activity that starts along with the loss of APC and the appearance of ACFs is the permanent epimutational silencing of the otherwise normal p16$^{INK\ 4a}$ gene in the cells of 30–40% of colon tumors by the hypermethylation of the CpG islands in the gene's promoter.[145,147]

As in skin carcinogenesis, a striking feature of colon carcinogenesis is a loss of the cells' responsiveness to Ca^{2+}.[25,26,41,49,205] Cells in the very earliest stages of carcinogenesis still express the E-cadherin tethers and are still responsive to Ca^{2+}. Indeed, at this stage ACF formation in carcinogen (azoxymethane)-treated rats can still be reduced by feeding the animals a diet containing either a suboptimal or superoptimal amount of calcium.[43] Also in humans, the abnormally expanded proliferation zone in the colon crypts of relatives of familial colon cancer patients shrinks back to the lower crypt region when the diet is supplemented with Ca^{2+}.[42] However, there are indications of things to come in the *Apc*$^{-/-}$ familial adenomatous polyposis patients. Incubation in a medium containing 2.2–5.0 mM Ca^{2+} reduces the abnormal proliferation of colon cells taken from 13 out of 14 persons *at risk* of developing colon cancer (because of previous colonic tumors or familial association), but it reduces the growth of colon cells from only 3 out of 10 persons *actually with* familial polyposis.[206] In other words, the colon cells of those persons with familial polyposis had already lost their Ca^{2+}-dependent proliferative control mechanism.

As we learned above, there is a striking *intracellular* Ca^{2+} gradient in the normal colon crypts of humans and mice.[57] This gradient collapses *before* the appearance of tumors in carcinogen (e.g., 1,2-dimethylhydrazine)-treated mice.[207] Moreover, the gradient was found by Brenner et al.[208] to have collapsed even in the crypts from the seemingly normal regions of the mucosa of carcinogen-treated mice with colon tumors, but it had not collapsed in the crypts of the carcinogen-treated mice that did not eventually develop tumors. Thus, the intracellular Ca^{2+} concentration ranged between 212 ± 51 and 228 ± 47 nM in the crypts from normal mucosal regions of tumor-bearing colons, but the intracellular Ca^{2+} concentration was a normal 179 ± 29 nM at the base and 297 ± 62 nM at the mouth of crypts from the colons of mice that had not developed tumors. Therefore, the initiator of tumor development seems to disable the cryptal Ca^{2+} gradient maker mechanism, but we do not know whether there is also a loss of Ca^{2+}'s ability to affect proliferation.

The loss of functional APC and the initial loss of responsiveness to Ca^{2+} are associated with a radical shift in the cellular response to a surge of PKC activity specifically that of the overexpressed PKC-βII.[186a] PKC stimulators, such as long-chain diacylglycerols (particularly those with $C_{18:8}$ oleic acid residues), the diacylglycerol-like TPA, and secondary bile acids, inhibit the proliferation of normal colon epithelial cells, but do not affect, or actually *stimulate*, the proliferation of initiated colon cells starting with the $Apc^{-/-}$ familial adenomatous polyposis cells.[209,210] (However, it must be noted that *over-expressed* PKC-β1 inhibits the proliferation of some colon cancer cells.[211]) The reasons for this dramatic conversion of PKC activity from proliferative inhibitor to stimulator at the onset of carcinogenesis are the overexpression of PKC-βII and the downregulation of PKCs α and βI, the ability of PKC-βII stimulated by TPA or the other agents to trigger β-catenin accumulation and thus proliferation and to stimulate another protein kinase that catalyzes the phosphorylation of the Tyr residues of a 63-kDa protein in premalignant and malignant colon cells but *not* in normal cells or those colon carcinoma cells that are still induced to differentiate by PKCs.[186a,212] This other protein kinase is probably c-Src, which is known to be activated early in carcinogenesis, can be stimulated by membrane-associated PKCs, and when activated, decreases the differentiation-promoting Ca^{2+}·E-cadherin-mediated cell–cell tethering.[106,213–215] Therefore, the PKCs-stimulating tumor promoters may stimulate the proliferation and expansion of the initiated $Apc^{-/-}$colon cells by stimulating β-catenin buildup and overcoming the Ca^{2+}·E-cadherin-mediated suppression of their proliferation by contacting neighboring normal cells just as they overcome the suppression of the expansion of initiated keratinocytes into papillomas by contacting normal keratinocytes.[216–218]

Another important factor in the growing unresponsiveness of progressive generations of initiated colon cells to differentiation signals may be the overexpression of the proliferation-related, cell cycle-driving PKAI with respect to the differentiation-promoting PKAII.[219,220] Indeed, Bradbury et al.[219,220] have found that human colon cancer cells, especially those from poor histological grade tumors, have no detectable RII subunits, while "normal" cells from tumor-adjacent and distant mucosa have both RI and RII subunits. The importance of this declining expression of the differentiation-driving

PKAII is indicated by the ability of RIα-specific antisense oligodeoxynucleotides to suppress, without cytotoxicity, the anchorage-independent proliferation of LS-174T human colon cancer cells in vitro.[221] Moreover, it is further indicated by the ability of subcutaneously injected antisense oligodeoxynucleotides to reduce the growth of these LS-147T tumor cells by as much as 50% in athymic nude mice.[221] Overexpressing RIIβ, but not a disabled RIIβ, in these cells almost completely suppressed PKAI and reduced cell proliferation in monolayer cultures and suppressed their abilities to proliferate in suspension and in a serum-free medium.[222] Thus, the regulatory subunits of the PKAs are targets for a gene-therapeutic strategy to knock colon cancer cells out of the RI-upregulating proliferation mode and into the RII-upregulating differentiation mode.[223]

In about 50–60% of dysplastic ACFs (adenomas), there are $Apc^{-/-}$ cells which also had previously started making a mutant, constitutively signaling, hyperactive Ha-Ras, or much more commonly a Ki-Ras protein with a mutated codon-12 or -13.[149,155,163,170–172,193] These hyperactive Ras proteins can block the differentiation of epithelial cells, are mitogenic, and eliminate the cell's need for external Ca^{2+} to proliferate.[224,225] They drive the proliferogenic MEK/MAP kinase cascade by activating the Raf-1 protein kinase (see Chap. 2). As we learned in the previous chapter, activated Ras proteins such as Ha-Ras can stimulate PtdIns-3K, which in turn stimulates the ILK and PKB protein kinases that suppress E-cadherin expression, inhibit GSK-3β, and thus keep the cycle-driving β-catenin flowing into the nucleus.[12,67] It would also appear that Ras-activated Raf-1 kinase phosphorylates BAD protein, which releases it from inactive BAD·Bcl-2 complexes and at the same time frees Bcl-2 to dimerize with itself or heterodimerize with Bax protein to promote cell survival.[9,226] But Ki-Ras is rather special. The mutant Ki-Ras, *unlike* a mutant Ha-Ras, can prevent the colon cells' apicobasal polarization and upregulate CEA (carcinoembryonic antigen), which prevents the formation of normal, laterally adhering polarized colon cells.[227] Thus, the hyperactive Ki-Ras mutant would disrupt differentiation and collaborate with an abnormally expressed Bcl-2 protein and a surging β-catenin to further enhance the abilities of adenoma and carcinoma cells to continue proliferating even in the killing zone of the crypt and on the mucosal surface despite the increasingly intense, and normally apoptogenic, signaling from the CaRs. Ultimately, a hyper-signaling Ki-Ras mutant, an upregulated Bcl-2, a disabled or lost β-catenin-destroying APC, and an increasingly overexpressed PKAI combine to stamp out traces of normal differentiation and drive the adenomatous development to the threshold of malignancy.

As the mass of competing mutant clones progresses from polyp to premalignant adenoma, there is an increasing loss of responsiveness to Ca^{2+}. The cells still need external Ca^{2+} to proliferate, but even this will be lost with the transition from adenoma to carcinoma. However, the proliferation-stopping, differentiation-triggering Ca^{2+}-switch no longer works. Thus, for example, the supernormal proliferative activity of primary human colon adenoma cells in the presence of 0.05–0.1 mM Ca^{2+} can be increased even further by raising the Ca^{2+} concentration to a high physiological level such as 2.2 mM, which would stop normal colon cells proliferating.[39] Therefore, the cells of mutant clones can proliferate, indeed be induced to proliferate more rapidly, when confronted by the high Ca^{2+} concentration in the fecal stream.

Loading the diet with Ca^{2+} (e.g., with 2 g/day as calcium carbonate or phosphate) reduces the compensatory, tumor-promoting hyperproliferation of colon crypt cells caused by the destruction of cells on the flat mucosal surface by cytolytic surfactant secondary bile acids and long-chain fatty acids in the fecal water of persons consuming a typical high-fat, low-fiber, Western diet.[48,49] Since it prevents the appearance of clones with mutant Ras proteins and reduces the size and number of carcinogen-induced tumors in rats by forming inactive complexes with the cytolytic secondary bile acids and long-chain fatty acids in the fecal water, it should reduce the risk of colorectal cancer in humans.[48,49,228,229] Indeed, Ca^{2+} reduces the hyperproliferative activity in the colons of high-risk humans and normal mice mainly by promoting apoptosis and thus shrinking the cryptal proliferative zone rather than inhibiting proliferation.[44–49,229a] However, a key question remains as to how accessible the proliferative zones in the depths of the crypts are to Ca^{2+} in the fecal water. Loading the crypts with Ca^{2+} would directly reduce the proliferation and promote the differentiation of normal cells as well as still phenotypically but early initiated cells. However, this loading would act like the tumor-promoting TPA by suppressing normal cells without affecting or even stimulating the proliferation of initiated cells.[49] This possibility is supported by the observation that loading the diet of rats treated with the carcinogen azoxymethane with Ca^{2+} when carcinogenesis was in full swing (as indicated by extensive ACF formation) actually greatly *increased* the number of tumors located 12–16 cm from the rectal end.[49] It is also supported by reports that dietary Ca^{2+} supplementation *stimulates* the proliferation of epithelial cells in the colons of patients with adenomatous polyps.[230,231]

There is likely to be another serious drawback to loading colons containing initiated cells with Ca^{2+} above a certain concentration. Cells in the early post-initiation stages are probably proliferatively restrained by things flowing into them from their normal neighbors through their still functional gap junctions.[18] However, gap junctions are designed to be slammed shut by Ca^{2+}.[18] This is meant to prevent the draining of components from the cellular network through a damaged dying cell.[18] Ca^{2+} was chosen as the damage detector and drain plugger simply because its accumulation is a universal consequence of cell injury. The closing of gap junctions by supra-optimal Ca^{2+} loading would release the initiated colon cells from the restraining influences of their normal neighbors.

Yet another drawback of dietary Ca^{2+} loading is of particular concern to men. There is a report that Ca^{2+} from food sources and dietary supplements increases the risk of advanced prostate cancer.[232]

The end of the road — a lucky winning clone and colon carcinoma

As the tumor promoters in the fecal stream and the increasingly endogenously unrestrained CDKs relentlessly drive the proliferation of mutating clones, a lucky winner of the mutation lottery will emerge with all the mutations (Greaves's "Full House")[232a] it needs to escape from advanced adenoma to the ultimate freedom of carcinoma and be resistant to apoptosis.[232b] Adenoma cells had increased their expression of the Bcl-2 protector protein, but during the transition to carcinoma, Bcl-2 is downregulated and its anti-apoptogenic relative Bcl-w is upregulated to maintain the protection.[1,233] Another anti-apoptosis change, which occurs in about 50–75% of large adenomas on the threshold of adenocarcinoma, is a loss-of-function mutation of the p53(TP53) gene on the short (p) arm of human chromosome 17 (17p13 to be precise).[123] The disabled p53(TP53) mutant protein has a prion-like "infectiousness": it can associate with the product of the normal allele and cause the normal protein to reconfigure itself to the inactive mutant form.[234] In other words, the G_1/S checkpoint of a p53(TP53)$^{+/-}$ heterozygote is as useless as that of a p53(TP53)$^{-/-}$ homozygote. But eventually cells appear which have also lost their normal allele. The loss of p53(TP53) function coincides with the appearance of carcinoma in the polyp or adenoma. It has eliminated a major safety checkpoint guardian, which when activated by DNA damage would normally stop cells proliferating until the damage is repaired, or failing this, trigger apoptosis.[9,92,95,235] Therefore, the loss of functional p53(TP53) enables the survival and expansion of dangerous criminal clones. Obviously, the combination of a defective mismatch repair mechanism due to mutation or methylation of the hMLH1 gene promoter, such as is found in some sporadic colon carcinomas,[196] and the loss of p53(TP53) would give a very strong push to the accumulation of mutations and epimutations, which are responsible for the emergence of carcinoma clones. And the upregulation of Bcl-w would help protect them from the ravages of potent chemotherapeutic agents or the dangers of metastasis.

The ultimate appearance of carcinoma (malignant) cells is tightly coupled to the loss of TGF-β signaling, which is most often due to the disabling mutation of the Type II receptor, but can also be due to the loss of functional Smad proteins (Smad 2 or Smad 4) that mediate the proliferation-stopping response of normal colon cells to TGF-βs.[99a] Also, in about 70–90% of colon carcinoma cells, but not adenoma cells, the DCC gene (the smad 4 (DPC4) gene's neighbor on the short arm of human chromosome 18)[235a] is disabled. This is an important step toward malignancy because the DCC receptor, a differentiation promoter and migration guide, triggers apoptosis if a normal cell, in a misguided attempt to escape from its niche, should pull it away from its laminin-like netrin-1 ligand in the migration trackway.[126–131] In other words, DCC signaling is part of a tumor-suppressing, fail-safe anti-metastasis mechanism, without which the cell can safely lift anchor and metastasize to distant places and ultimately kill its owner.[131,193,235a]

Up to now, the roiling mass of struggling but still tightly tethered clones has been safely confined to a "carcinoma in situ." Eventually, breakaway cells will emerge which cannot make the Ca^{2+}·E-cadherins with which their progenitors tied themselves to their neighbors and sequestered the cycle-promoting, differentiation-restraining β-catenin.[81,236–241] But how can this be, since the emerging malignant mutants' E-cadherin genes seem to be perfectly normal?[146] What has happened? As in the case of the loss of $p16^{INK\ 4a}$ function, the failure to make the protein appears to be one of the consequences of a rising level of DNA-MTase, which replicably and permanently silences the still normal E-cadherin gene by hypermethylating the cytosines of a CpG island that spans the gene's transcription start site.[81,145,147]

Yet another mutation involved in the enabling of metastasis is an upregulation of the expression of the Ca^{2+}-binding S100A4 protein.[242] This is in addition to the earlier replacement of the normal cell's expression of S100A6 isoform IV with the expression of S100A6 isoforms I and III.[243]

With these several mutations and epimutations and their strong Bcl-w shield, the carcinoma cells have completed the divorce from control by external Ca^{2+} that began when their carcinogen-damaged ancestor lost its functional APC. The extent of this divorce is indicated by the fact that while the cells in the early hyperplastic stage up to the well-differentiated adenocarcinoma stage still express CaRs, these receptors, and with them the cells' responsiveness to proliferation-restraining signals from external Ca^{2+}, vanish in the later stages.[63a,b]

As we have learned, the loss of functional APC results in the over-expression and a buildup in the nucleus of the cycle-driving β-catenin. However, it seems that β-catenin is most strongly expressed by the cells on the tumor's invasion front, while cells in the core of the tumor have little or no β-catenin in their nuclei, but have it on their cell membranes as do normal colon cells.[244] This raises the possibility that normal cells somehow tell the cells in the oncoming tumor invasion front to load their nuclei with β-catenin.[244]

References

1. Bach S, Renehan AG, Potten S. Stem cells: the intestinal stem cell as a paradigm. Carcinogenesis 2000; 21: 469–476.

1*a*. Bjerkness M, Cheng H. Clonal analysis of mouse intestinal epithelial progenitors. Gastroenterology 1999; 116: 7–14.

2. Gordon JL, Schmidt GH, Roth KA. Studies of intestinal stem cells using normal, chimeric, and transgenic mice. FASEB J 1992; 6: 3039–3050.
3. Potten CS. Stem cells in gastrointestinal epithelium: numbers, characteristics and death. Philos Trans R Soc Lond B Biol Sci 1998; 353: 821–830.
4. Lev R. Adenomatous Polyps of the Colon. New York, Springer-Verlag, 1990.
5. Sunter JP, Appleton DR, DeRodriguez MSB, et al. A comparison of cell proliferation at different sites within the large bowel of the mouse. J Anat 1979; 129: 833–842.
6. Gershon MD. The Second Brain. New York, Harper–Collins, 1998.
7. Sträter J, Wedding U, Barth TFE, et al. Rapid onset of apoptosis in vitro follows disruption of β_1-integrin/matrix interactions in human colonic crypt cells. Gastroenterology 1996; 110: 1776–1784.
8. Zutter MM, Santoro SA. Widespread histologic distribution of the $\alpha_2\beta_1$ integrin cell-surface collagen receptor. Am J Pathol 1990; 137: 113–120.
9. Bowen ID, Bowen SM, Jones AH. Mitosis and Apoptosis. London, Chapman & Hall, 1998.
10. Henriksson M, Lüscher B. Proteins of the Myc network: essential regulators of cell growth and differentiation. Cancer Res 1996; 68: 109–182.
11. Ryan KM, Birnie GD. *Myc* oncogenes: the enigmatic family. Biochem J; 1996: 713–721.
12. Roose J, Clevers H. TCF transcription factors: molecular switches in carcinogenesis. Biochim Biophys Acta 1999; 1424: M23–M37.
13. Bernards R. CDK-independent activities of D type cyclins. Biochim Biophys Acta 1999; 1424: M17–M22.
14. Hockenbery DM, Zutter M, Hickey W, et al. BCL2 protein is topographically restricted in tissues characterized by apoptotic cell death. Proc Natl Acad Sci USA 1992; 88: 6961–6965.
15. Korsmeyer SJ, Shutter JR, Veis DJ, et al. BCL-2/Bax: a rheostat that regulates an anti-oxidant pathway and cell death. Semin Cancer Biol 1993; 4: 327–332.
16. Merritt AJ, Potten CS, Watson AJ, et al. Differential expression of bcl-2 in intestinal epithelia. Correlation with attenuation of apoptosis in colonic crypts and the incidence of colonic neoplasia. J Cell Sci 1995; 108: 2261–2271.

17. Potten CS. Regeneration in epithelial proliferative units as exemplified by small intestinal crypts. Ciba Found Symp 1991; 160: 54–71.
18. Loewenstein W. The touchstone of life. New York, Oxford University Press, 1999.
19. Bjerkness M, Cheng H. Functional gap junctions in mouse small intestinal crypts. Anat Rec 1985; 212: 364–367.
20. Lindqvist SM, Sharp P, Johnson IT, et al. Acetylcholine-induced calcium signaling along the rat colonic crypt axis. Gastroenterology 1998; 115: 1131–1143.
21. Gerike TG, Paulus U, Potten CS, et al. A dynamic model of proliferation and differentiation in the intestinal crypt based on a hypothetical intraepithelial growth factor. Cell Prolif 1998; 31: 93–110.
22. Shibata D. The dynamics of early intestinal tumour proliferation: to be or not to be. Cancer Surveys 1998; 32: 181–200.
23. Giancotti FG, Mainiero F. Integrin-mediated adhesion and signalling in tumorigenesis. Biochim Biophys Acta 1994; 1198: 47–64.
24. Goodlad RA, Wright NA. Epidermal growth factor. Baillière's Clin Gastroenterology 1996; 10: 33–47.
25. Whitfield JF. Calcium signals and cancer. Crit Rev Oncogenesis 1992; 3: 55–90.
26. Whitfield JF. Calcium signals in cell proliferation, differentiation, and death. In: Foà PP, Walsh MF, editors. Ion Channels and Ion Pumps. New York, Springer-Verlag, 1994: 19–58.
27. Melani C, Rivoltini L, Parmiani G, et al. Inhibition of proliferation by c-myb antisense oligodeoxynucleotides in colon adenocarcinoma cell lines that express c-myb. Cancer Res 1991; 51: 2897–2901.
28. Ramsay RG, Thompson MA, Hayman JA, et al. Myb expression is higher in malignant human colonic carcinoma and premalignant adenomatous polyps than in normal mucosa. Cell Growth Diff 1992; 3: 723–730.
29. Saville MK, Watson RJ. B-Myb: a key regulator of the cell cycle. Adv Cancer Res 1998; 72: 109–140.
30. Thompson MA, Ramsay RG. Myb: an old oncoprotein with new roles. BioEssays 1995; 17: 341–350.
31. Hall PA, Coates PJ, Ansari B, et al. Regulation of cell number in the mammalian gastrointestinal tract: the importance of apoptosis. J Cell Sci 1994; 107: 3569–3577.
32. Sträter J, Koretz K, Günthert AR, et al. In situ detection of enterocytic apoptosis in normal colonic mucosa and in familial adenomatosis polyposis. Gut 1995; 37: 819–825.
33. Frisch SM, Francis H. Disruption of epithelial cell–matrix interactions induces apoptosis. J Cell Biol 1994; 124: 619–626.
34. Marsh MN, Trier JS. Morphology and cell proliferation of subepithelial fibroblasts in adult mouse jejunum. I. Structural features. Gastroenterology 1974; 67: 622–635.
35. Marsh MN, Trier JS. Morphology and cell proliferation of subepithelial fibroblasts in adult mouse jejunum. II. Radioautographic studies. Gastroenterology 1974; 67: 636–645.

36. Parker FG, Barnes EN, Kaye GI. The pericryptal fibroblast sheath. IV. Replication, migration and differentiation of the subepithelial fibroblasts of the crypt and villus of the rat jejunum. Gastroenterology 1974; 67: 607–621.
37. Pascal RR, Kaye GI, Lane N. Colonic pericryptal fibroblast sheath: replication, migration, and cytodifferentiation of a mesenchymal cell system in adult tissue. I. Autoradiographic studies of normal rabbit colon. Gastroenterology 1968; 54: 835–851.
38. Pascal RR, Kaye GI, Lane N. Colonic pericryptal fibroblast sheath: replication, migration and cytodifferentiation of a mesenchymal cell system in adult tissue. II. Fine structural aspects of normal rabbit and human colon. Gastroenterology 1968; 54: 852–865.
39. Buset M, Lipkin M, Winawer S, et al. Inhibition of human colonic epithelial cell proliferation in vivo and in vitro by calcium. Cancer Res 1986; 46: 5426–5430.
40. Friedman EA. A multistage model of human colon carcinoma development integrating cell culture studies with pathology. Cancer Invest 1985; 3: 453–461.
41. Friedman EA. Use of tissue culture of human colonic epithelial cells to study mechanisms of calcium protection. In: Lipkin M, Newmark HL, Kelloff G, editors. Calcium, Vitamin D, and the Prevention of Colon Cancer. Boca Raton, CRC Press, 1991: 159–168.
42. Lipkin M, Friedman E, Winawer SJ, et al. Colonic epithelial cell proliferation in responders and non-responders to supplemental dietary calcium. Cancer Res 1989; 49: 248–254.
43. Li H, Kramer PM, Lubet RA, et al. Effect of calcium on azoxymethane-induced aberrant crypt foci and cell proliferation in the colon of rats. Cancer Lett 1998; 124: 39–46.
44. Bostick RM. Human studies of calcium supplementation and colorectal epithelial cell proliferation. Cancer Epidemiol Biomarkers 1997; 6: 971–979.
45. Garay CA, Engstrom PF. Chemoprevention of colorectal cancer: dietary and pharmacological approaches. Oncology 1999; 13: 89–97.
46. Holt PR, Atillasoy EO, Gilman J, et al. Modulation of abnormal colonic epithelial cell proliferation and differentiation by low-fat dairy foods: a randomized controlled trial. J Amer Med Assoc (JAMA) 1998; 280: 1074–1079.
47. Holt PR, Lipkin M, Newmark H. Calcium intake and colon cancer biomarkers. J Amer Med Assoc (JAMA) 1999; 281: 1172–1173.
48. Lipkin M. Calcium modulation of intermediate biomarkers in the gastrointestinal tract. In: Lipkin M, Newmark HL, Kelloff G, editors. Calcium, Vitamin D, and the Prevention of Colon Cancer. Boca Raton, CRC Press, 1991: 377–425.
49. Whitfield JF, Bird RP, Chakravarthy BR, et al. Calcium-cell cycle regulator, differentiator, killer, chemopreventor, and maybe, tumor promoter. J Cell Biochem 1995; Suppl 22: 74–91.
50. Chiarugi VP, Vannuchi S. Surface heparan sulphate as a control element in eukaryotic cells: a working model. J Theor Biol 1976; 61: 61–71.

51. Dawson PA, Filipe MI. An ultrastructural and histochemical study of the mucous membrane adjacent to and remote from carcinoma of the colon. Cancer 1976; 37: 2388–2398.
52. Filipe MI. Mucins in the human gastrointestinal epithelium: a review. Invest Cell Pathol 1979; 2: 195–216.
53. Forstner JF, Forstner GG. Calcium binding to intestinal goblel cell mucin. Biochim Biophys Acta 1975; 368: 283–292.
54. Forstner JF, Forstner GG. Effects of calcium on intestinal mucin: implications for cystic fibrosis. Pediatr Res 1976; 10: 609–613.
55. Sunter JP, Wright NA, Appleton DR. Cell population kinetics in the epithelium of the colon of the male rat. Virchow's Arch B Cell Pathol 1978; 26: 275–287.
56. Whitfield JF, Perris AD, Youdale T. The calcium-mediated promotion of mitotic activity in rat thymocyte populations by growth hormone, neurohormones, parathyroid hormone and prolactin. J Cell Physiol 1969; 73: 203–212.
57. Brenner BM, Russell N, Albrecht S, et al. The effect of dietary vitamin D_3 on the intracellular calcium gradient in mammalian colonic crypts. Cancer Lett 1998; 127: 43–53.
58. Brown EM, MacLeod RJ. Extracellular calcium sensing and extracellular calcium signaling. Physiol Revs 2001; 81: 240–297.
59. Butters RR, Chattopadhyay N, Nielsen F, et al. Cloning and characterization of a calcium-sensing receptor from the hypercalcemic New Zealand white rabbit reveals unaltered responsiveness to extracellular calcium. J Bone Miner Res 1997; 12: 568–579.
60. Chattopadhyay N, Cheng I, Rogers K, et al. Identification and localization of extracellular Ca^{2+}-sensing receptor in the rat intestine. Am J Physiol 1998; 274: G22–G130.
61. Gama L, Baxendale-Cox LM, Breitwieser GE. Ca^{2+}-sensing receptors in intestinal epithelium. Am J Physiol 1997; 273: C1168–C1175.
62. Kállay E, Kifor O, Chattopadhyay N, et al. Calcium-dependent c-*myc* proto-oncogene expression and proliferation of CACO-2 cells: a role for a luminal extracellular calcium-sensing receptor. Biochem Biophys Res Commun 1997; 232: 80–83.
63. Bikle D. Vitamin D. A calciotropic hormone regulating calcium-induced keratinocyte differentiation. J Am Acad Dermatol 1997; 37: S42–S55.
63*a*. Kállay E, Bajna E, Wrba F, et al. Dietary calcium and growth modulation of human colon cancer cells: role of the extracellular calcium-sensing receptor. Cancer Detect Prev 2000; 24: 127–136.
63*b*. Scheinin Y, Kállay E, Wrba F. Immunocytochemical localization of the extracellular calcium-sensing receptor in normal and malignant large intestinal mucosa. J Histochem Cytochem 2000; 48: 595–601.
63*c*. Buras RR, Shabahang M, Dovoodi F, et al. The effect of extracellular calcium on colonocytes: evidence for differential responsiveness based upon degree of cell differentiation. Cell Prolif 1995; 28: 245–262.

64. Delcommene M, Tan C, Gray V, et al. Phosphoinositide-OH kinase-dependent regulation of glycogen synthase kinase 3 and protein kinase B/AKT by the integrin-linked kinase. Proc Natl Acad Sci USA 1998; 95: 1121–1126.
65. Hannigan GE, Leung-Hagesteijn C, Fitz-Gibbon L, et al. Regulation of cell adhesion and anchorage-dependent growth by a new β1-integrin-linked protein kinase. Nature 1996; 379: 91–98.
66. Pennisi E. How a growth control path takes a wrong turn to cancer. Science 1998; 281: 1438–1440.
67. Sheperd PR, Withers DJ, Siddle K. Phosphoinositide 3-kinase: the key switch mechanism in insulin signaling. Biochem J 1998; 333: 471–490.
68. Wymann MP, Pirola L. Structure and function of phosphoinositide 3-kinases. Biochim Biophys Acta 1998; 1436: 127–150.
69. Tu Y, Li F, Goicoechea S, et al. The LIM-only protein PINCH directly interacts with integrin-linked kinase and is recruited to integrin-rich sites in spreading cells. Mol Cell Biol 1999; 19: 2425–2434.
70. Tu Y, Li F, Wu C. Nck-2, a novel Src homology 2/3 containing adaptor protein that interacts with the LIM-only protein PINCH and components of growth factor signaling pathways. Mol Biol Cell 1998; 9: 3367–3382.
71. Brennan P, Babbage JW, Burgering BMT, et al. Phosphatidyl 3-kinase couples the interleukin-2 receptor to the cell cycle regulator E2F. Immunity 1997; 7: 679–689.
72. HeT-C, Sparks AB, Raga C, et al. Identification of c-*myc* as a target of the APC pathway. Science 1998; 281: 1509–1512.
73. Morin PJ. β-Catenin signaling and cancer. BioEssays 1999; 21: 1021–1030.
74. Shtutman M, Zhurinsky J, Simcha I, et al. The cyclin D1 gene is a target of the β-catenin/LEF-1 pathway. Proc Natl Acad Sci USA 1999; 96: 5522–5527.
75. Greider CW. Telomerase activation. Trends Genet 1999; 15: 109–112.
76. Mann B, Gelos M, Siedow A, et al. Target genes of β-catenin–T-cell factor / lymphoid-enhancing factor signaling in human colorectal carcinomas. Proc Natl Acad Sci USA 1999; 96: 1603–1608.
77. Yamasaki L. Balancing proliferation and apoptosis in vivo: the Goldilocks theory of E2F/DP action. Biochim Biophys Acta 1999; 1423: M9–M15.
78. Aberle H, Bauer A, Stappert J, et al. β-Catenin is a target for the ubiquitin–proteasome pathway. EMBO J 1997; 16: 3797–3804.
79. Hinds P. The cell cycle Cold Spring Harbor, 20–24 May 1998. Biochim Biophys Acta 1999; 1423: R63–R67.
80. Morin PJ, Sparks AB, Korinek V, et al. Activation of β-catenin–Tcf signaling in colon cancer by mutations in β-catenin or APC. Science 1997; 275: 1787–1790.
81. Behrens J. Cadherins and catenins: role in signal transduction and tumor progression. Cancer and Metastasis Reviews 1999; 18: 15–30.
82. Hart MJ, los Santos R, Albert IN, et al. Downregulation of β-catenin by human axin and its association with the APC tumor suppressor, β-catenin and GSK3β. Curr Biol 1998; 8: 573–581.

83. Barth AIM, Näthke IS, Nelson J. Cadherins, catenins and APC protein: interplay between cytoskeletal complexes and signaling pathways. Curr Opin Cell Biol 1997; 9: 683–690.
84. Hatzfield M. The armadillo family of proteins. Int Rev Biol 1999; 186: 179–224.
85. Novak A, Hsu SC, Leung-Hagersteijn C, et al. Cell adhesion and the integrin-linked kinase regulate the LEF-1 and β-catenin signaling pathways. Proc Soc Natl Acad Sci USA 1998; 95: 4374–4379.
86. Cho-Chung YS, Clair T. The regulatory subunit of cAMP-dependent protein kinase as a target for chemotherapy of cancer and other cellular dysfunctional-related diseases. Pharm Ther 1993; 60: 265–288.
87. Cho-Chung YS, Clair T, Yokozai H. Role of site-selective cAMP analogs in the control and reversal of malignancy. Pharm Ther 1991; 50: 1–33.
88. Biggs JR, Kraft AS. Inhibitors of cyclin-dependent kinase and cancer. J Mol Med 1995; 73: 509–514.
89. Duchrow M, Schluter C, Key G. Cell proliferation-associated nuclear antigen defined by antibody Ki-67: a new kind of cell cycle-maintaining proteins. Arch Immunol Ther Exp (Warsz) 1995; 43: 117–121.
90. El-Deiry WS, Tokino T, Waldman T, et al. Topological control of $p21^{WAF/CIP1}$ expression in normal and neoplastic tissues. Cancer Res 1995; 55: 2910–2919.
91. Clarke AS, Lotz MM, Chao C, et al. Activation of the p21 pathway of growth arrest and apoptosis by the beta 4 integrin cytoplasmic domain. J Biol Chem 1995; 270: 22673–22676.
92. Gartel AL, Tyner AL. The growth-regulatory role of p21(WAF1/CIP1). Progr Mol Subcell Biol 1998; 20: 43–71.
93. Heidebrecht HJ, Buck F, Haas K, et al. Monoclonal antibodies Ki-S3 and Ki-S5 yield new data on the “Ki-67” proteins. Cell Prolif 1996; 29: 413–425.
94. MacLachlan TK, Sang N, Giordano A. Cyclins, cyclin-dependent kinases and cdk inhibitors: implications in cell cycle control and cancer. Crit Rev Eukaryot Gene Express 1995; 5: 127–156.
95. Oren M, Prives C. p53: upstream, downstream and offstream. Review of the 8th p53 workshop (Dundee, July 5–9, 1996). Biochim Biophys Acta 1996; 1288: R13–R19.
96. Polyak K, Hamilton SR, Vogelstein B, et al. Early alteration of cell-cycle-regulated gene expression in colorectal neoplasia. Am J Pathol 1996; 149: 381–387.
97. Tian JQ, Quaroni A. Involvement of p21(WAF1/Cip1) and p27(Kip1) in intestinal epithelial cell differentiation. Am J Physiol 1999; 276: C1245–C1258.
98. DiCunto F, Topley G, Calautti E, et al. Inhibitory function of $p21^{Cip1/WAF1}$ in differentiation of primary mouse keratinocytes independent of cell cycle control. Science 1998; 280: 1069–1072.
99. Avery A, Paraskeva C, Hall P, et al. TGF-beta expression in the human colon: differential immunostaining along crypt epithelium. Br J Cancer 1993; 68: 137–139.

99*a*. Markovitz S. TGF-β receptors and DNA repair genes, coupled targets in a pathway of human colon carcinogenesis. Biochim Biophys Acta 2000; 1470: M13–M20.

100. Sun L, Wu S, Coleman K, et al. Autocrine transforming growth factor-beta 1 and beta-2 expression is increased by cell crowding and quiescence in colon carcinoma cells. Exp Cell Res 1994; 214: 215–224.

101. Winesett MP, Ramsey GW, Barnard JA. Type II TGF(beta) receptor expression in intestinal cell lines and the intestinal tract. Carcinogenesis 1996; 17: 989–995.

102. Li CY, Suardet L, Little JB. Potential role of WAF1/Cip1/p21 as a mediator of TGF-beta cytoinhibitory effect. J Biol Chem 1995; 270: 4971–4974.

103. Pignatelli M, Bodmer WF. Genetics and biochemistry of collagen binding-triggered glandular differentiation in a human colon carcinoma cell line. Proc Natl Acad Sci USA 1988; 85: 5561–5565.

104. Pignatelli M, Bodmer WF. Integrin-receptor-mediated differentiation and growth inhibition are enhanced by transforming growth factor-beta in colorectal tumor cells grown in collagen gel. Int J Cancer 1989; 44: 518–523.

105. Richman PI, Bodmer WF. Control of differentiation in human colorectal carcinoma cell lines: epithelial–mesenchymal interactions. J Pathol 1988; 156: 197–211.

106. Marrs JA, Nelson WJ. Cadherin cell adhesion molecules in differentiation and embryogenesis. Int Rev Cytol 1996; 165: 159–205.

107. Jawhari A, Farthing M, Pignatelli M. The importance of the E-cadherin–catenin complex in the maintenance of intestinal epithelial homeostasis: more than intercellular glue? Gut 1997; 41: 581–584.

108. Hoschuetzky H, Aberele H, Kemler R. Beta-catenin mediates the interaction of the cadherin–catenin complex with epidermal growth factor receptor. J Cell Biol 1994; 127: 1375–1380.

109. Nigam SK, Rodriguez-Boulan E, Silver RB. Changes in intracellular calcium during the development of epithelial polarity and junctions. Proc Natl Acad Sci USA 1992; 89: 6162–6166.

110. Stuart RO, Nigam SK. Regulated assembly of tight junctions by protein kinase C. Proc Natl Acad Sci USA 1995: 92: 6072–6076.

111. Stuart RO, Sun A, Bush KT, et al. Dependence of epithelial intercellular junction biogenesis on thapsigargin-sensitive intracellular calcium stores. J Biol Chem 1996; 271; 13636–13641.

112. Stuart RO, Sun A, Panichas M, et al. Critical role for intracellular calcium in tight junction biogenesis. J Cell Physiol 1994; 159: 423–433.

113. Smith KJ, Johnson KA, Bryan TM, et al. The APC gene product in normal and tumor cells. Proc Natl Acad Sci USA 1993; 90: 2846–2850.

114. Bhattacharya G, Boman BM. Phosphorylation of the adenomatous polyposis coli protein and its possible regulatory effects in cells. Biochem Biophys Res Commun 1995; 208: 103–110.

115. Munemitsu S, Souza B, Muller O, et al. The APC gene product associates with microtubules in vivo and promotes their assembly in vitro. Cancer Res 1994; 54: 3676–3681.

116. Nathke IS, Adams CL, Polakis P, et al. The adenomatous polyposis coli tumor suppressor protein localizes to plasma membrane sites involved in active cell migration. J Cell Biol 1996; 134: 165–179.
117. Inomata M, Ochiai A, Akimoto S, et al. Alteration of beta-catenin expression in colonic epithelial cells of familial adenomatous polyposis patients. Cancer Res 1996; 56: 2213–2217.
118. Rubinfield B, Albert I, Porfiri E, et al. Binding of GSK3β to the APC–β-catenin complex and regulation of complex assembly. Science 1996; 272: 1023–1026.
119. Senda T, Miyashiro I, Matsumine A, et al. The tumor suppressor protein APC colocalizes with beta-catenin in the colon epithelial cells. Biochem Biophys Res Commun 1996; 223: 329–334.
120. Waddell WR, Miesfeld RL. Adenomatous polyposis coli, protein kinases, protein tyrosine phosphatase; the effect of sulindac. J Surg Oncol 1995; 58: 252–256.
121. Korinek V, Barker N, Morin PJ, et al. Constitutive transcriptional activation by a β-catenin–Tcf complex in $APC^{-/-}$ colon carcinoma. Science 1997; 275: 1784–1787.
121*a*. Miller JR, Hocking A, Brown J, et al. Mechanism and function of signal transduction by the Wnt/β-catenin and Wnt/Ca^{2+} pathways. Oncogene 1999; 18: 7860–7872.
122. Morin PJ, Sparks AB, Korinek V, et al. Activation of β-catenin–Tcf signaling in colon cancer by mutations in β-catenin or APC. Science 1997; 275: 1787–1790.
123. Peifer M. β-Catenin as oncogene: the smoking gun. Science 1997; 275: 1752–1753.
124. Rubinfeld B, Robbins P, El-Gamil M, et al. Stabilization of β-catenin by genetic defects in melanoma cell lines. Science 1997; 275: 1790–1792.
125. Roose J, Clevers H. TCF transcription factors: molecular switches in carcinogenesis. Biochim Biophys Acta 1999; 1424: M23–M37.
126. Fearon ER. DCC; Is there a connection between tumorigenesis and cell guidance? Biochim Biophys Acta 1996; 1288: M17–M23.
127. Fearon ER, Piercall WE. The deleted in colorectal cancer (DCC) gene: a candidate tumor suppressor gene encoding a cell surface protein with similarity to neural cell adhesion molecules. Cancer Surveys 1995; 24: 3–17.
128. Livesey FJ. Netrins and netrin receptors. Cell Mol Life Sci (CMLS) 1999; 56: 62–68.
129. Chen YQ, Hsieh JT, Yao F, et al. Induction of apoptosis and G2/M cell cycle arrest by DCC. Oncogene 1999; 18: 2747–2754.
130. Song H, Poo M. Signal transduction underlying growth cone guidance by diffusible factors. Curr Opin Neurobiol 1999; 9: 355–363.
131. Mehlen P, Rabizadeh S, Snipas SJ, et al. The DCC gene product induces apoptosis by a mechanism requiring receptor proteolysis. Nature 1998; 395: 801–804.

132. Berridge MJ, editor. Microdomains and elemental events in calcium signalling. Cell Calcium 1996; 20: 95–226.
133. Torelli G, Venturelli D, Colo A, et al. Expression of c-myb protooncogene and other cell cycle-related genes in normal and neoplastic human colonic mucosa. Cancer Res 1987; 47: 5266–5269.
134. Huang F, Sauma S, Yan Z, et al. Colon absorptive epithelial cells lose their proliferative response to TGF-α as they differentiate. Exp Cell Res 1995; 219: 8–14.
135. Saeki T, Stronberg K, Qi CF, et al. Differential immunohistochemical detection of amphiregulin and cripto in human normal colon and colorectal tumors. Cancer Res 1992; 52: 3467–3473.
136. Li H, Seitz PK, Thomas ML, et al. Widespread expression of the parathyroid hormone-related peptide and PTH/PTHrP receptor genes in intestinal epithelial cells. Lab Invest 1995; 73: 864–870.
137. Zhou L-X, Nemere L, Norman AW. A parathyroid related peptide induces transcaltachia (the rapid, hormonal stimulation of intestinal Ca^{2+} transport). Biochem Biophys Res Commun 1992; 186: 69–73.
138. Yu D, Seitz PK, Selvanayagam P, et al. Effects of parathyroid hormone-related peptide on adenosine 3′,5′-monophosphate and ornithine decarboxylase in a human colonic cell line. Endocrinology 1992; 130: 1993–2000.
139. Amling M, Neff L, Tanaka S, et al. Bcl-2 lies downstream of parathyroid hormone-related peptide in a signaling pathway that regulates chondrocyte maturation during skeletal development. J Cell Biol 1997; 136: 205–213.
140. Barritt GJ. Receptor-activated Ca^{2+} inflow in animal cells: a variety of pathways tailored to meet afferent intracellular signalling requirements. Biochem J 1999: 337: 153–169.
141. Morin PJ, Vogelstein B, Kinzler KW. Apoptosis and APC in colorectal tumorigenesis. Proc Natl Acad Sci USA 1996; 93: 7950–7954.
142. Polzar B, Zanotti S, Stephan H, et al. Distribution of deoxyribonuclease I in rat tissues and its correlation to cellular turnover and apoptosis (programmed cell death). Eur J Cell Biol 1994; 64: 200–210.
143. Albaugh GP, Iyengar V, Lohani A, et al. Isolation of exfoliated colonic epithelial cells, a novel, non-invasive approach to the study of cellular markers. Int J Cancer 1992; 52: 347–350.
144. Iyengar V, Albaugh GP, Lohani A, et al. Human stools as a source of viable colonic epithelial cells. FASEB J 1991; 5: 2856–2859.
145. Breivik J, Gaudernack G. Carcinogenesis and natural selection: a new perspective to the genetics and epigenetics of colorectal cancer. Adv Cancer Res 1999; 76: 185–212.
146. Zhang L, Zhou W, Velculescu VE, et al. Gene expression profiles in normal and cancer cells. Science 1997; 276: 1268–1272.
147. Baylin SB, Herman JG, Graff JR, et al. Alterations in DNA metabolism: a fundamental aspect of neoplasia. Adv Cancer Res 1998; 72: 141–196.
148. Potten CS. Significance of spontaneous and induced apoptosis in gastrointestinal tract of mice. Cancer Metastasis Rev 1992; 11: 179–195.

149. Fearon ER. Molecular abnormalities in colon and rectal cancer. In: Mendelsohn J, Howley PM, Israel MA, et al., editors. The Molecular Basis of Cancer. Philadelphia, W B Saunders, 1995: 340–357.
150. Shamsuddin AM. Diagnostic Assays for Colon Cancer. Boca Raton, CRC Press, 1992.
151. Shimoda T, Ikegami M, Fujisaki J, et al. Early colorectal carcinoma with special reference to its development de novo. Cancer 1989; 64: 1138–1146.
152. Archer MC, Bruce WR, Chan CC, et al. Aberrant crypt foci and microadenoma as markers for colon cancer. Environ Health Perspect 1992; 98: 195–197.
153. Bird RP. Observation and quantification of aberrant crypts in the murine colon treated with a colon carcinogen: preliminary findings. Cancer Lett 1987; 37: 147–151.
154. Bird RP, McLellan EA, Bruce WR. Aberrant crypts, putative precancerous lesions, in the study of the role of diet in the aetiology of colon cancer. Cancer Surv 1989; 8: 189–200.
155. Bjerknes M. Expansion of mutant stem cell populations in the human colon. J Theor Biol 1996; 178: 381–385.
156. Bjerknes M, Cheng H, Hay K, et al. APC mutation and the crypt cycle in murine and human intestine. Am J Pathol 1997; 150: 833–839.
157. Caderni G, Giannini A, Lancioni L, et al. Characterization of aberrant crypt foci in carcinogen-treated rats: association with intestinal carcinogenesis. Br J Cancer 1995; 71: 763–769.
158. Fujimitsu Y, Nakanishi H, Inada K, et al. Development of aberrant crypt foci involves a fission mechanism as revealed by isolation of aberrant crypts. Jpn J Cancer Res 1996; 87: 1199–1203.
159. Magnuson BA, Carr I, Bird RP. Ability of aberrant crypt foci characteristics to predict colonic tumor incidence in rats fed cholic acid. Cancer Res 1993; 53: 4499–4504.
160. Magnuson BA, Shirtliff N, Bird RP. Resistance of aberrant crypt foci to apoptosis induced by azoxymethane in rats chronically fed cholic acid. Carcinogenesis 1994: 15: 1459–1462.
161. Minkin S. Statistical analysis of aberrant crypt assays for colonic cancer promotion studies. Biometrics 1994; 50: 279–288.
162. Moen CJ, van der Valk MA, Bird RP, et al. Different genetic susceptibility to aberrant crypts and colon adenoma in mice. Cancer Res 1996; 56: 2382–2386.
163. Nucci MR, Robinson CR, Longo P, et al. Phenotypic and genotypic characteristics of aberrant crypt foci in human colorectal mucosa. Hum Pathol 1997; 28: 1396–1407.
164. Pretlow TP, O'Riordan MA, Pretlow TG, et al. Aberrant crypts in human colonic mucosa: putative preneoplastic lesions. J Cell Biochem Suppl 1992; 16G: 55–62.
165. Roncucci L. Early events in human colorectal carcinogenesis. Aberrant crypts and microadenoma. Ital J Gastroenterol 1992; 24: 498–501.
166. Roncucci L, Modica S, Pedroni M, et al. Aberrant crypt foci in patients with colorectal cancer. Br J Cancer 1998; 77: 2343–2348.

167. Rosenberg DW, Liu Y. Induction of aberrant crypts in murine colon with varying sensitivity to colon carcinogenesis. Cancer Lett 1995; 92: 209–214.
168. Shpitz B, Hay K, Medline A, et al. Natural history of aberrant crypt foci. A surgical approach. Dis Colon Rectum 1996; 39: 763–767.
169. Siu IM, Robinson DR, Schwartz S, et al. The identification of monoclonality in human aberrant crypt foci. Cancer Res 1999; 59: 63–66.
170. Smith AJ, Stern HS, Penner M, et al. Somatic APC and K-ras codon 12 mutations in aberrant crypt foci from human colons. Cancer Res 1994; 54: 5527–5530.
171. Stopera SA, Bird RP. Expression of ras oncogene mRNA and protein in aberrant crypt foci. Carcinogenesis 1992; 13: 1863–1868.
172. Stopera SA, Murphy LC, Bird RP. Evidence for a ras gene mutation in azoxymethane-induced colonic aberrant crypts in Sprague-Dawley rats: earliest recognizable precursor lesions of experimental colon cancer. Carcinogenesis 1992; 13: 2081–2085.
173. Sutherland LA, Bird RP. The effect of chenodeoxycholic acid on the development of aberrant crypt foci in the rat colon. Cancer Lett 1994; 76: 101–107.
174. Uchida K, Kado S, Onoue M, et al. Relationship between the nature mucus and crypt multiplicity in aberrant crypt foci in the rat colon. Jpn J Cancer Res 1997; 88: 807–814.
175. Vaccina F, Scorcioni F, Pedroni M, et al. Scanning electron microscopy of aberrant crypt foci in human colorectal mucosa. Anticancer Res 1998; 18: 3451–3456.
176. Wargovich MJ, Chen CD, Jimenez A, et al. Aberrant crypts as a biomarker for colon cancer: evaluation of potential chemopreventive agents in the rat. Cancer Epidemiol Biomarkers Prev 1996; 5: 355–360.
177. Yao K, Latta M, Bird RP. Modulation of colonic aberrant crypt foci and proliferative indexes in colon and prostate glands of rats by vitamin E. Nutr Cancer 1996; 26: 99–109.
178. Anzano MA, Rieman D, Prichett W, et al. Growth factor production by human colon carcinoma cell lines. Cancer Res 1989; 49: 2898–2904.
179. Shirai H, Ueno E, Osaki M, et al. Expression of growth factors and their receptors in human early colorectal carcinogenesis: immunohistochemical studies. Anticancer Res 1995; 15: 2889–2894.
180. Sizeland AM, Burgess AW. Anti-sense transforming growth factor α oligonucleotides inhibit autocrine stimulated proliferation of a colon carcinoma cell line. Mol Biol Cell 1992: 3: 1235–1243.
181. Hsu S, Huang F, Friedman E. Platelet-derived growth factor-B increases colon cancer cell growth in vivo by a paracrine effect. J Cell Physiol 1995; 165: 239–245.
182. Huang F, Hsu S, Yan Z, et al. The capacity for growth stimulation by TGF-β1 seen only in advanced colon cancers cannot be ascribed to mutations in APC, DCC, p53 or ras. Oncogene 1994; 9: 3701–3706.

183. Hsu S, Huang F, Hafez M, et al. Colon carcinoma cells switch their response to transforming growth factor β1 with tumor progression. Cell Growth Diff 1994; 5: 267–275.
184. MacKay SL, Yaswen LR, Tarnuzzer RW, et al. Colon cancer cells that are not growth inhibited by TGF-β lack functional type I and type II TGF-β receptors. Ann Surg 1995; 221: 767–777.
185. Wang J, Han W, Zoborowska E, et al. Reduced expression of transforming growth factor β type I receptor contributes to the malignancy of human colon carcinoma cells. J Biol Chem 1996; 271: 17366–17371.
186. Wang J, Sun L, Myeroff L, et al. Demonstration that mutation of the type II transforming growth factor β receptor inactivates its tumor suppressor activity in replication error-positive colon carcinoma cells. J Biol Chem 1995; 270: 22044–22049.
186*a*. Gökmen-Polar Y, Murray NR, Velasco MA, et al. Elevated protein kinase CβII is an early promotive event in colon carcinogenesis. Cancer Res 2001; 61: 1375–1381.
187. Malakouti S, Asadi FK, Kukreja SC, et al. Parathyroid hormone-related protein expression in the human colon: immunohistochemical evaluation. Am Surg 1996; 62: 540–545.
188. Conteas CN, Desai TK, Arlow FA. Relationship of hormones and growth factors to colon cancer. Gastroenterol Clin North Am 1988; 17: 761–772.
189. Farr HW. Hyperparathyroidism and cancer. CA 1976; 26: 66–74.
190. Feig DS, Gottesman IS. Familial hyperparathyroidism in association with colonic carcinoma. Cancer 1987; 60: 429–432.
191. Sharma S, Longo WE, Baniadam B, et al. Colorectal manifestations of endocrine disease. Dis Colon Rectum 1995; 38: 318–323.
192. Barrett JF, Lewis BC, Hoang AT, et al. Cyclin A links c-Myc to adhesion-independent cell proliferation. J Biol Chem 1995; 270: 15923–15925.
193. Carethers JM. The cellular and molecular pathogenesis of colorectal cancer. Gastroenterol Clin North Am 1996; 25: 737–754.
194. Dihlmann S, Gebert J, Siermann A, et al. Dominant negative effect of the *APC*^{1309} mutation: a possible explanation for genotype–phenotype correlations in familial adenomatous polyposis. Cancer Res 1999; 59; 1857–1860.
195. Kinzler KW, Vogelstein B. Landscaping the cancer terrain. Science 1998; 280: 1036–1037.
196. Marra G, Boland CR. DNA repair and colorectal cancer. Gastroenterol Clin North Am 1996; 25: 755–771.
197. Peltomäki P, de la Chapelle A. Mutations predisposing to hereditary non-polyposis colorectal cancer. Adv Cancer Res 1997; 71: 93–119.
198. Polakis P. The oncogenic activation of β-catenin. Curr Opin Genet Devel 1999; 9: 15–21.
199. Hickman JA, Potten CS, Merritt AJ, et al. Apoptosis and cancer chemotherapy. Philos Trans R Soc Lond B Biol Sci 1994; 345: 319–325.

200. Ayhan A, Yasui W, Yokozaki H, et al. Loss of heterozygosity at the bcl-2 gene locus and expression of bcl-2 in human gastric and colorectal carcinomas. Jpn J Cancer Res 1994; 85: 584–591.
201. Wagenaar RA, Crawford HC, Matrisian LM. Stabilized β-catenin immortalizes colonic epithelial cells. Cancer Res 2001; 61: 2097–2104.
202. Nhieu JTV, Renard CA, Wei Y, et al. Nuclear accumulation of mutated β-catenin in hepatocellular carcinoma is associated with increased cell proliferation. Am J Pathol 1999; 155: 703–710.
202*a*. Sherr CJ. Tumor surveillance via the ARF-p53 pathway. Genes Dev 1998; 12: 2984–2991.
203. Sherr CJ. Cancer cell cycles. Science 1996; 274: 1672–1677.
204. Ruas M, Peters G. The $p16^{INK\ 4a}$/CDKN2A tumor suppressor and its relatives. Biochim Biophys Acta 1998; 1378: F115–F177.
205. Whitfield JF. Calcium in Cell Cycles and Cancer. Boca Raton, CRC Press, 1995.
206. Friedman E, Lipkin M, Winawer S, et al. Heterogeneity in the response of familial polyposis epithelial cells to increasing levels of calcium in vitro. Cancer 1989; 63: 2486–2491.
207. Brenner BM, Albrecht S, Davies RJ. Abnormal calcium regulation in colonic crypts following DMH-induced carcinogenesis. Gastroenterology 1996; 110: A496.
208. Brenner BM, Albrecht S, Russell N, et al. The colonic crypt intracellular calcium gradient predicts susceptibility to colorectal cancer. Abstract 95th Ann Convention Am Soc Colon Rectal Surg, June 9–14, 1996.
209. Friedman EA, Gillin S, Lipkin M. 12-*O*-Tetradecanoyl-phorbol-13-acetate stimulation of DNA synthesis in cultured preneoplastic familial polyposis colonic epithelial cells but not in normal colonic cells. Cancer Res 1984; 44: 4078–4086.
210. Friedman EA, Urmacher C, Winawer S. A model for human colon carcinoma evaluation based on the differential response of cultured preneoplastic, premalignant and malignant cells to 12-*O*-tetradecanoylphorbol-13-acetate. Cancer Res 1984; 44: 1568–1578.
211. Phan SC, Morotomi M, Guillem JG, et al. Decreased levels of 1,2-*sn* diacylglycerol in human colon tumors. Cancer Res 1991; 51: 1571–1573.
212. Lee H, Winawer S, Friedman E. Signal-transduction through pp63 tyrosine phosphorylation in human colon carcinoma cells. Proc Am Assoc Cancer Res 1991; 51: 450.
213. Bolen JB, Veillette A, Schwartz AM, et al. Activation of $pp60^{c\text{-}src}$ protein kinase activity in human colon carcinoma. Proc Natl Acad Sci USA 1987; 84: 2251–2255.
214. Cartwright CAA, Meister AI, Eckhart W. Activation of the $pp60^{c\text{-}src}$ protein kinase is an early event in colonic carcinogenesis. Proc Natl Acad Sci USA 1990; 87: 558–562.
215. Weber TK, Steele G, Summerhayes IC. Differential $pp60^{c\text{-}src}$ activity in well and poorly differentiated human colon carcinomas and cell lines. J Clin Invest 1992; 90: 815–821.

216. Hennings H, Lowry DT, Robinson VA. Co-culture of neoplastic and normal keratinocytes as a model to study tumor promotion. Skin Pharmacol 1991; 4 Suppl 1: 79–84.
217. Hennings H, Lowry DT, Robinson VA, et al. Activity of diverse tumor promoters in a keratinocyte co-culture model of initiated epidermis. Carcinogenesis 1992; 13: 2145–2151.
218. Strickland JE, Ueda M, Hennings H, et al. A model for initiated mouse skin: suppression of papilloma but not carcinoma formation by normal epidermal cells in grafts on athymic nude mice. Cancer Res 1992; 52: 1439–1444.
219. Bradbury AW, Carter DC, Miller WR, et al. Protein kinase A (PK-A) regulatory subunit expression in colorectal cancer and related mucosa. Br J Cancer 1994; 69: 738–742.
220. Bradbury AW, Miller WR, Clair T, et al. Overexpressed type I regulatory subunit (RI) of cAMP-dependent protein kinase (PKA) as tumor marker in colorectal cancer. Proc Am Assoc Cancer Res 1990; 31: 172 (abstract).
221. Clair T, Yokozaki H, Tortora G, et al. An antisense oligodeoxynucleotide targeted against the type I regulatory subunit (Riα) mRNA of cAMP-dependent protein kinase (PKA) inhibits the growth of LS-147T human colon carcinoma in athymic nude mice. Proc Am Assoc Cancer Res 1991; 51: 1645 (abstract).
222. Nesterova M, Yokozaki H, McDuffie E, et al. Overexpression of RII β regulatory subunit of protein kinase A in human colon carcinoma cells induces growth arrest and phenotypic changes that are abolished by site-directed mutation of RIIβ. Eur J Biochem 1996; 235: 486–494.
223. Bold RJ, Warren RE, Ishizuka J, et al. Experimental gene therapy of human colon cancer. Surgery 1994; 116: 189–195.
224. Boynton AL, Whitfield JF. Different calcium requirements for proliferation of conditionally and unconditionally tumorigenic mouse cells. Proc Natl Acad Sci USA 1976; 73: 1651–1654.
225. Durkin JP, Whitfield JF. Characterization of G_1 transit induced by the mitogenic–oncogenic viral Ki-ras gene product. Mol Cell Biol 1986; 6: 1386–1392.
226. Eastman A. Survival factors, intracellular signal transduction, and the inactivation of endonucleases in apoptosis. Sem Cancer Biol 1995; 6: 45–52.
227. Yan Z, Deng X, Chen M, et al. Oncogenic c-Ki-ras but not oncogenic c-Ha-ras up-regulates CEA expression and disrupts basolateral polarity in colon epithelial cells. J Biol Chem 1997; 272: 27902–27907.
228. Bartram HP, Kasper K, Liebscher DG, et al. Effects of calcium and deoxycholic acid on human colonic cell proliferation in vitro. Ann Nutr Metab 1997; 41: 315–323.
229. Garay CA, Engstrom PF. Chemoprevention of colorectal cancer: dietary and pharmacological approaches. Oncology 1999; 13: 89–97.
229*a*. Penman ID, Liang QL, Bode J, et al. Dietary calcium supplementation increases apoptosis in the distal murine colonic epithelium. J Clin Pathol 2000; 53: 302–307.

230. Bostick RM, Potter JD, Fosdick L, et al. Calcium and colorectal epithelial cell proliferation: a preliminary randomized, double-blinded, placebo-controlled clinical trial. J Natl Cancer Inst 1993; 85: 132–141.
231. Kleibeuker JH, Welberg JW, Mulder NH, et al. Epithelial cell proliferation in the sigmoid colon of patients with adenomatous polyps increases during oral calcium supplementation. Br J Cancer 1993; 67: 500–503.
232. Giovannucci E, Rimm EB, Wolk A, et al. Calcium and fructose intake in relation to risk of prostate cancer. Cancer Res 1998; 58: 442–447.
232*a*. Greaves M. Cancer. The Evolutionary Legacy. Oxford, Oxford University Press, 2000.
232*b*. Lifshitz S, Lamprecht S, Benharroch D, et al. Apoptosis (programmed cell death) in colonic cells: from normal to transformed stage. Cancer Lett 2001; 163: 229–238.
233. Wilson JW, Nostro MC, Balzi M, et al. Bcl-w expression in colorectal adenocarcinoma. Br J Cancer 2000; 82: 178–185.
234. Prion Diseases. Available at: http://www-micro.msb.le.ac.uk/335/Prions.html. Accessed 1999.
235. Meek DW. Multisite phosphorylation and integration of stress signals at p53. Cell Signal 1998; 10: 159–166.
235*a*. Chen YQ, Hsieh JT, Yao F, et al. Induction of apoptosis and G2/M cell cycle arrest by DCC. Oncogene 1999; 18: 2747–2754.
236. Dimanche-Boitrel MT, Vakaet L Jr, Pujuguet P, et al. In vivo and in vitro invasiveness of a rat colon-cancer cell line maintaining E-cadherin expression: an enhancing role of tumor-associated myofibroblasts. Int J Cancer 1994; 56: 512–521.
237. Nigam AK, Savage FJ, Boulos PB, et al. Loss of cell–cell and cell–matrix adhesion molecules in colorectal cancer. Br J Cancer 1993; 68: 507–514.
238. Pignatelli M. Models of colorectal tumor differentiation. Cancer Surv 1993; 16: 3–13.
239. Pignatelli M, Liu D, Nasim MM, et al. Morphoregulatory activities of E-cadherin and beta-1 integrin in colorectal tumor cells. Br J Cancer 1992; 66: 629–634.
240. Takayama T, Shiozaki H, Shibamoto S, et al. Beta-catenin expression in human cancers. Am J Pathol 1996; 148: 39–46.
241. Vermeulen SJ, Bruyneel EA, Bracke ME, et al. Transition from the noninvasive to the invasive phenotype and loss of alpha-catenin in human colon cancer cells. Cancer Res 1995; 55: 4722–4728.
242. Takenaga K, Nakanishi H, Wada K, et al. Increased expression of S100A4, a metastasis-associated gene, in human colorectal adenocarcinomas. Clin Cancer Res 1997; 3: 2309–2316.
243. Stulik J, Österreicher J, Koupilova K, et al. Differential expression of the $Ca^{2=}$ binding S100A6 protein in normal, preneoplastic and neoplastic colon mucosa. Eur J Cancer 2000; 36: 1050–1059.
244. Brabletz T, Jung A, Hermann K, et al. Nuclear overexpression of the oncoprotein β-catenin in colorectal cancer is localized predominantly at the invasion front. Pathol Res Pract 1998; 194: 701–704.

5
Calcium and Neurons

We have seen Ca^{2+} and its assistants orchestrating key parts of the cell cycle and terminal differentiation or diffpoptosis, that ingenious amalgam of differentiation and apoptosis, of keratinocytes and colon cells. Now we travel from the gut to the brain (but we must not forget that the gut has its own large 'brain' [Gershon's second brain][1]) where Ca^{2+} truly reigns supreme in a role that has expanded from its first job of being the ionic eyes and ears of the first eukaryotic cells to being one of the central figures in the sensory systems of multicellular animals.

Although not experimentally confirmed, we can be sure that Ca^{2+} spikes and oscillations and cresting and crashing waves of CDKs drive the proliferation of the multipotent ependymal progenitor cells in the embryonic brain and the persistent subventricular generative zones of the lateral ventricle and hippocampal dentate gyrus of the adult brain that provide neurons and glial cells for dentate gyrus and olfactory bulb, just as they drive the cycling of basal keratinocytes and colon crypt cells.[2–8] Here too we find the CaRs with as yet undefined functions, although they may function in modulating cation channel-dependent establishment of LTP (long-term potentiation) in hippocampal neurons and help drive the proliferation of glial cells and neuroblasts.[8a,9,10] But there is something special about mature neurons. They are proliferatively disabled like post-mitotic keratinocytes and colon cells, maybe because of unlicensed replication origins or the locking up of the genes for key parts of replication factories. Unlike the very short-lived keratinocytes and colon cells, they are very long-lived and can still grow. Through activity-driven growth without proliferation, called *plasticity*, they continuously shape and reshape the brain to produce a structural/functional configuration as characteristic of the individual as the fingerprints.

To build the basic 'starting' brain, neurons must climb along supporting glial cells or other neurons and send out processes (neurites) to synaptically contact other neurotropin-emitting neurons and peripheral target cells. Ca^{2+} and its now familiar assistants such as CaM and CaMKII, and undoubtedly the CaRs, are at the helms of the axonal growth cones. Indeed, Ca^{2+} transients modulate growth and the expression of differentiation-related genes according to their frequency and distribution.[10a] The advancing tip of a neurite, the autonomously driven growth cone,[11,11a] moves along a track marked by a set of matrix components or 'bricks' such as the netrins, which it grabs with the DCC (deleted in colon cancer) receptors that we met in the previous chapter.[12–15] Some of the matrix guideposts say 'go forward!' while others say 'stop and turn!,' depending on the kinds of signals from the various guidepost receptors and the gradient of diffusible factors emanating from the target.[13] Thus, for example, the repellant MAG (myelin-associated glycoprotein) and the attractant netrin-1 attract if the cellular cyclic AMP level be high, but repel if the level be low.[16] The growth cone has a large finger(s), a filopodium with a bundle of microfilaments, that grabs the trackway's 'bricks' with its receptors and pulls the process along the trackway.

Ca^{2+}'s role in neurite production and growth cone navigation depends on the location, frequency, and amplitude of Ca^{2+} spikes and waves.[10a,16–18] It seems low-frequency Ca^{2+} surges drive neurite growth, but when a rapidly advancing growth cone with a few or no Ca^{2+} transients reaches a turning point on the trackway it slams its brakes on with a burst of Ca^{2+} transients (from both the smooth endoplasmic reticulum and probably voltage-gated membrane channels at the extreme leading edge) that stop migration by causing the dispersal of the finger's bundle of microfilaments, the withdrawal of the finger, the expansion of the cone, and the generation of multiple protrusions. The frequency of the Ca^{2+} surges then drops. Depending on the position of the peripheral→central Ca^{2+} gradient in the growth cone, a new filopodium(-ia) protrudes and the cone sets off again in a new direction with a much reduced frequency of Ca^{2+} transients.[10a,11a,16a,19]

It seems likely that Ca^{2+} somehow works through a molecular switch, a growth-associated protein known as neuromodulin or GAP-43, the overexpression of which enhances memory in mice.[11a,b] This switch is concentrated on the inner surface of the plasma membranes of axons and growth cones and the growth cones' cortical cytoskeleton where it can increase actin filament length, hence filopodium/lamellipodium protrusion and growth cone migration.[11a] The Ca^{2+}·CaM produced by a Ca^{2+} transient binds to the switch's IQ (isoleucine-glutamine) site that also contains Ser^{41}, which

Fig. 5.1. A neuromodulin (NM) switch with which Ca^{2+} and PKC might control the movements of axonal growth cones. An activated navigation factor receptor sends a Ca^{2+} signal that produces Ca^{2+}·CaM that binds to NM's IQ motif (•), which also contains PKC's target residue, Ser^{41}. When the Ca^{2+} signals are infrequent and Ca^{2+}·CaM production is low, PKC can get at NM's IQ domain to make the phospho-NM (P•NM) needed for filopodial production and growth cone advance. However, if Ca^{2+} signaling and Ca^{2+}·CaM production exceed a critical level, Ca^{2+}·CaM will block the IQ domain and prevent PKC from phosphorylating NM. The Ca^{2+}·CaM also stimulates the phosphatase calcineurin that dephosphorylates existing phospho-NM. The conflicting convergence of Ca^{2+}·CaM and PKC on NM's IQ domain and the Ca^{2+}·CaM-stimulated calcineurin thus combine to cause the filopodia to retract and stop the growth cone advance.

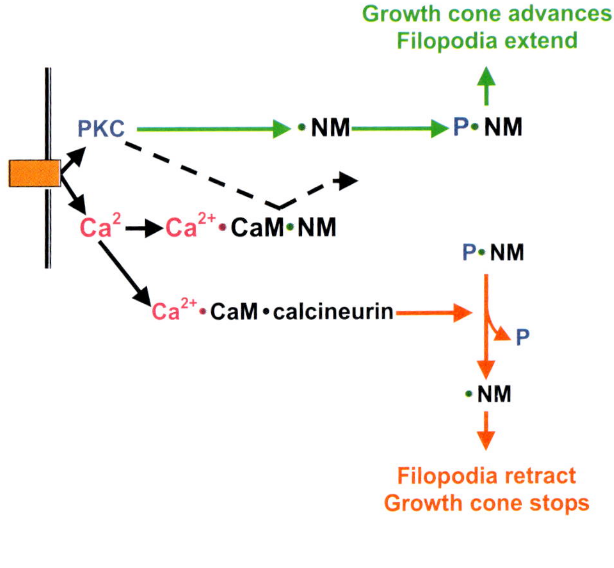

happens to be the target for phosphorylation by PKC (Fig. 5.1). This is how Ca^{2+} controls the switch, because it is PKC- phosphorylated neuromodulin that is required for the formation of filopodia and the moving of the growth cone toward its target.[11a] The subcritical amount of Ca^{2+}·CaM produced by a low-frequency of well-spaced Ca^{2+} signals from the surface receptors for a growth factor or a trackway component would not prevent activated, membrane-associated PKC from phosphorylating neuromodulin's Ser^{41}. The PKC-phosphorylated neuromodulin then promotes filopodia extension and growth cone migration. However, if the frequency of Ca^{2+} transients and the resulting Ca^{2+}·CaM production should exceed a critical level, it would promptly bind to the IQ site before the more slowly responding PKCs could get to it and phosphorylate Ser^{41}. At the same time, the Ca^{2+}·CaM will also stimulate the phosphatase calcineurin, which would dephosphorylate existing phospho-neuromodulin. The conflicting convergence of Ca^{2+}·CaM and PKC on the switch's IQ domain and the surge of Ca^{2+}·CaM·calcineurin activity combine to cause filopodia to retract and stop the growth cone's advance.

In the last two chapters, we learned that PTHrP plays a key role in the differentiation of cartilage, colon, and skin by delaying the final apoptotic demise of the differentiating cells. It is beginning to appear that the peptide also prevents apoptosis in at least some neurons. Depolarizing cerebellar granule cells with excessive external K^+ or activating *N*-methyl-D-aspartate (NMDA) receptor/channels results in the opening of L-type Ca^{2+} channels and a surge of Ca^{2+}·CaM, which stimulates the expression of PTHrP that enhances the survival of the neurons making and secreting it.[20,21] The autocrine/paracrine PTHrP does this by stimulating the producer cells to express the Bcl-2 anti-apoptotic protector protein.[22] However, PTHrP does something else — it prevents an excessive or prolonged Ca^{2+} surge through L-type Ca^{2+} channels in kainic acid-treated granule cells from killing the cells through a cyclic AMP/PKA-induced phosphorylation and inactivation of the channels.[20] Thus, it seems worth speculating that the intense signaling of neurons and their neurites when interacting normally with the various 'bricks' in their trackways or from excessively sustained glutamate receptor activation and depolarization following injury may produce the Ca^{2+}·CaM needed to produce, or at least try to produce, enough PTHrP to protect themselves from excitotoxic and apoptotic death.

Eventually, successful neurons establish synaptic contact with their targets. Complex, cross-talking systems are set up to cause the release of neurotransmitters and drive signal transmission and the memory of transmitted signals. Now we must look at this wonderful synaptic world from the viewpoint of Ca^{2+}.

The growing axons with their probing growth cones eventually meet and form connections — synapses — with other neurons to form a neural net. These synaptic connections can be strengthened or weakened — a property known as "plasticity."[23–25] Plasticity is an essential feature of synaptic transmission, and depending on the nature or pattern of the preceding signal, the synaptic activity can either be enhanced or suppressed, a phenomenon called "activity-dependent synaptic plasticity" or ADSP.[24,26–28] ADSP is considered to be extremely important for many neuronal functions including learning and memory. It is a fundamental property of various types of synapse in both vertebrate and invertebrate central and peripheral nervous systems.[29–33] Several forms of plasticity at both presynaptic and postsynaptic sites have been described which briefly or prolongedly enhance or suppress synaptic transmission.[26–30] This enhancement or suppression might last from milliseconds to several days.[34] Studies over the years have clearly demonstrated that all forms of plasticity are linked to a rise in the intracellular Ca^{2+} concentration ($[Ca^{2+}]_i$). Although it is not quite clear as to how something as seemingly simple as a change in the $[Ca^{2+}]_i$ could induce such diverse and specific forms of plasticity, it is becoming apparent that the origin of the Ca^{2+} and where in the cell it surges and crashes determine the kind of plasticity.[24,26,34] Moreover, the amplitude and the frequency of individual Ca^{2+} signals, which occur in the spatial and the temporal domains, respectively, also contribute to the diversity and the specificity of the plasticity.[24,26,34]

Intercellular Ca^{2+} waves and signaling networks

Before going down to the individual neuron, we must look at the whole picture of the flow of Ca^{2+} signals throughout the brain.[34a] Neurons are surrounded by, and talk to, glial cells — there is a reciprocal exchanging of information in the forms of ions and the neurotransmitter glutamate.[34b,c] K^+ and glutamate released from neurons can reach adjacent glial cells where they activate Ca^{2+} channels and receptors, which results in receptor-activated PLC (phospholipase-C) activity that generates $InsP_3$ (inositol-1,4,5-trisphosphate) and an $InsP_3$-induced release of Ca^{2+} from internal stores, which produces Ca^{2+}·CaM and stimulates various enzymes such as type 1 adenylyl cyclase, protein kinases, and phosphatases. Each glial cell is in turn connected to a network of other glial cells and neurons by as many as 30 000 gap junctions through which $InsP_3$ can flow and trigger the release of Ca^{2+} from more internal stores. The passage of the Ca^{2+} wave with its wake of Ca^{2+}·CaM complexes through the glial cell toward the gap junctional passageways to other glial cells or neurons is amplified at "way stations" along the endoplasmic reticulum consisting of clusters of associated mitochrondria.[34a] The mitochondria-amplified Ca^{2+} waves also stimulate NO synthase that makes the NO that causes the vascular dilation needed to increase the blood supply to active neurons, which is an indicator used in functional brain scanning.

Presynaptic Ca^{2+} signaling and short-term plasticity

Although long-term synaptic plasticity has been the main focus of research because of its importance in learning and memory, recent research has focussed on short-term synaptic enhancement because of its importance in information processing, a principal function of neurons.[26,34] In most synapses, the action potential arriving at the nerve terminal triggers a transient increase in $[Ca^{2+}]_i$, which can induce the expression of various forms of short-term synaptic plasticity at the presynaptic site (Fig. 5.2).[34] The major presynaptic entry points of Ca^{2+} are N-type and P/Q-type voltage-sensitive Ca^{2+} channels (VSCCs), which cluster around the transmitter release sites on the axon terminals.[35–36a] The transient surge of Ca^{2+} through these channels is responsible for the release of neurotransmitters, which is seen as changes in the postsynaptic or junctional potential. If the $[Ca^{2+}]_i$ is still elevated when the next action potential arrives, the preexisting calcium ions combine with the incoming ions to produce a longer-lasting Ca^{2+} surge. This progressive, activity-driven Ca^{2+} buildup in the presynaptic terminal is believed to be the basis of the short-term enhancement (STE) of synaptic transmission.

It was initially thought that STE is a single event,[34] but it is now clear that it consists of three main phases, facilitation (lasting less than a second), augmentation (lasting

Fig. 5.2. How CaM and PKCs may affect the Ca^{2+}-induced release of a neuron's axonal neuromodulin-bound and its dendritic neurogranin-bound CaM stores and the release of the neurotransmitter glutamate into the synaptic space. The emptying of these CaM stores provides the CaM for making the Ca^{2+}·CaM needed for the presynaptic release of glutamate and the postsynaptic stimulation of various Ca^{2+}·CaM-dependent enzymes. The chain of events, started by a surge of Ca^{2+} through depolarization(∿↗)-opened VSCCs (voltage-sensitive Ca^{2+} channels) in the axon terminus, which lead to the release of CaM stores, glutamate secretion, stimulation of postsynaptic NMDA and mGluR receptors, and the postsynaptic accumulation of Ca^{2+}·CaM are described in the text. CaMAC, Ca^{2+}·CaM-dependent adenylyl cyclase; CaMKII, Ca^{2+}·CaM-dependent protein kinase II; GLU, glutamic acid; mGluR, metabotropic glutamate receptor; NG, neurogranin; NM, neuromodulin; NMDAR, NMDA glutamate receptor; NO, nitric oxide; NOS, nitric oxide synthase; DP, depolarization. Solid lines, stimulation; broken lines, inhibition/blockage.

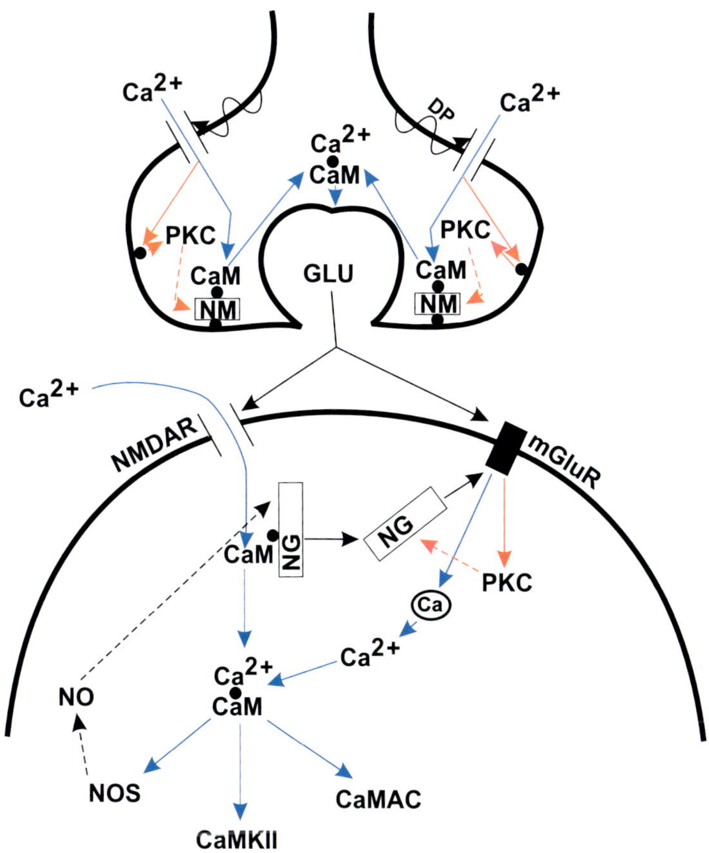

several seconds), and post-tetanic potentiation (PTP, lasting a few minutes).[37,38] Facilitation, in turn, may have a fast-decaying (F1) and slow-decaying (F2) component.[29,31,34] All these phases depend on a rise in $[Ca^{2+}]_i$. When an action potential reaches the nerve terminal and opens VSCCs, the Ca^{2+} streaming through these channels creates microdomains or ephemeral hot spots where the $[Ca^{2+}]_i$ could be as high as 100 μM,[39–41] and it is at these hot spots where Ca^{2+} drives transmitter release by binding to low-affinity sites.[42] The length of time the VSCCs are open is directly related to the duration of the action potential. When the action potential subsides, the VSCCs close, Ca^{2+} flow stops, and the Ca^{2+} hot spots quickly vanish as their Ca^{2+} diffuses into the core of the nerve terminal where it is sequestered by intracellular buffers.[26,34,43] However, a small amount of Ca^{2+} remains free to modestly increase the residual [Ca^{2+}]. Katz and Miledi[44] were the first to propose a "residual Ca^{2+}" hypothesis to explain facilitation of neurotransmission at the neuromuscular junction in frogs. In their experiments, removing extracellular Ca^{2+} during the initial stimulation (conditioning) eliminated the facilitation of transmitter release during a subsequent "test" stimulation when Ca^{2+} had been put back in the medium. Therefore, they proposed that a certain fraction of the Ca^{2+} that surged into the presynaptic terminal during the initial stimulation stays around in the terminal to supplement the Ca^{2+} surge triggered by the next action potential and to enhance transmitter release. Thus, each action potential would be expected to, and in fact does, add to the residual Ca^{2+} pool. For example, it has been estimated that in crayfish terminals the Ca^{2+} load from each action potential increases the amount of Ca^{2+} in the residual pool by 10 nM.[45] In small granule cell-Purkinje cell synaptic terminals, a single action potential can increase the residual Ca^{2+} by as much as 200–300 nM, which subsequently drops with a time constant of 100–200 ms.[46,47] When this terminal is bombarded with a high-frequency train of action potentials, the residual Ca^{2+} level can rise as high as 1000 nM and then slowly decline with a time constant of several seconds to minutes.[48–50]

It is now well accepted that the residual Ca^{2+} affects all three phases of presynaptic enhancement: facilitation, augmentation, as well as the potentiation.[34] But how does the level of residual Ca^{2+} affect these enhancement phases each with its own decay kinetics? To understand this, one must consider the spatiotemporal dynamics of Ca^{2+} distribution.[26,34,38] As we learned above, there are ephemeral high-Ca^{2+} hot spots clustered around, or very near, the channels through which the ion flows. These hot spots are very different from the slow, low-concentration, and diffusely spread residual Ca^{2+} in the core of the terminal. Therefore, the Ca^{2+} binding sites of these two compartments are exposed to very different levels of Ca^{2+}. If the binding sites are directly involved in exocytosis (transmitter release), then their location determines how fast the enhancement of transmitter release occurs. Thus, the Ca^{2+} in the hot spots can more rapidly trigger transmitter release than in the core of the terminal. The hot-spot Ca^{2+} transients may contribute to the facilitation (F1 and F2) phase of the enhancement, while the residual core Ca^{2+} may contribute to longer lasting augmentation and potentiation phases of synaptic enhancement.[34,44,48,49,51,52]

While residual Ca^{2+} theory generally holds, it is not clear whether bound or free Ca^{2+} plays a critical role in establishing STE. Some studies indicate that the residual unbound Ca^{2+} in the hot spots on or near the sites directly linked to exocytosis is responsible for F1 and F2 components of facilitation.[44,51,52] However, in some other studies Ca^{2+} chelators, at concentrations high enough to block transmitter release, did not affect facilitation, which suggests only a minimal role of residual unbound Ca^{2+} in this process.[53] It is likely that Ca^{2+}, bound to targets that are either directly or indirectly linked to transmitter release, might play a role. It is known that the exocytosis-target sites must bind at least 4 Ca^{2+} ions to trigger transmitter release.[41,54] If Ca^{2+} stays bound to one or more of these sites after the passing of an action potential, it would facilitate the triggering of transmitter release by a subsequent action potential when the remaining sites are occupied by the incoming Ca^{2+}.[34]

Experimental results have suggested that different phases of synaptic enhancement correspond to various time constants of the decay of residual Ca^{2+} and also depend on the action of Ca^{2+} on at least two different sites.[24,26,47–49,55] Thus, the Ca^{2+} bound to low-affinity binding sites in the hot spots near the Ca^{2+} channels rapidly fans out into the core where the induction of the F1 and F2 facilitation phases is believed to occur.[44,51,52] The residual, diffuse Ca^{2+} in the terminal's core drops relatively slowly, and it is here where augmentation and post-tetanic potentiation are thought to occur.[48,49] Generally, the residual Ca^{2+} is pumped out of the cell by ATP-driven Ca^{2+} pumps in the plasma membrane and (or) is sequestered in endoplasmic reticulum, mitochondria, or endoplasmic reticulum–mitochondrial clusters.[34*a*,43,56,57] Thus, the speed of the removal of residual Ca^{2+}, which determines the duration of the $[Ca^{2+}]_i$ transients, is strictly a function of the efficiency with which these Ca^{2+}-handling mechanisms operate and which may vary from terminal to terminal. AUG and PTP require a large number presynaptic action potentials for induction.[58,59] The buildup of Ca^{2+} by a rapid train of APs will result in a rapid loading of mitochondria, which then slowly release their Ca^{2+}.[34*a*,50,60–62] Augmentation subsides as the unbound Ca^{2+} in the terminal core is pumped out of the cell. However, the slow extrusion of Ca^{2+} from the mitochondria causes a brief surge of core Ca^{2+}, which is probably responsible for the potentiation before it is ejected from the neuron by the membrane Ca^{2+} pumps.[50] Indeed, blocking mitochondrial Ca^{2+} release abolishes PTP.[50] Although the uptake and release of Ca^{2+} by the endoplasmic reticulum do not seem to play a role in presynaptic activity in some synapses,[50] both mitochondria and endoplasmic reticulum seem to play a role at other synapses.[63,64]

While the induction of presynaptic STE appears to be largely due to Ca^{2+} entry through the VSCCs during an action potential, recent studies indicate that other sources of Ca^{2+} may also contribute to this process. A Na^+-dependent increase in $[Ca^{2+}]_i$ has been implicated in short-term enhancement of synaptic transmission, particularly PTP.[34,48,65] This Ca^{2+} surge could be due to Na^+-induced increase in the release of intracellular Ca^{2+} or the blockade of Ca^{2+} extrusion by the Na^+/Ca^{2+} exchanger.[34] Both ionotropic and metabotropic glutamate receptors, which have also been found in the presynaptic sites, appear to be involved in the establishment STE.[66–68] It is suggested that the facilitation of transmission occurs when metabotropic glutamate receptors activate Ca^{2+}-dependent release of Ca^{2+} from internal stores.[68,69] Furthermore, presynaptic Ca^{2+} channel activity can be modulated by activated G-protein coupled metabotropic receptors,[70] and Ca^{2+} currents affected by these receptors can modulate presynaptic transmitter release.[68,69] Similarly, ionotropic receptors such as the NMDA, AMPA, and kainate receptors, may alter the level of presynaptic Ca^{2+} before the arrival of an action potential and thereby contribute to enhanced presynaptic transmission.[66,67,67*a*,67*b*] Such a mechanism has been proposed for nicotinic receptors at the hippocampal mossy fiber terminal[71] and for glutamate receptors at the lamprey reticulospinal axons.[66] The action of these receptors could be direct by allowing Ca^{2+} through their channels or indirect by depolarizing the terminal to activate presynaptic VSCCs.[66–71] Although the site of STE is predominantly presynaptic, recent studies in *Aplysia* neurons have indicated that preventing a postsynaptic Ca^{2+} rise or inducing hyperpolarization can actually block the induction of STE,[72–74] suggesting that postsynaptic cell may also contribute to STE, at least during the initiation stage.

Postsynaptic Ca^{2+} signaling and long-term plasticity

It is clear from the preceding section that presynaptic Ca^{2+} plays a pivotal role in synaptic plasticity and the dumping of neurotransmitters out of presynaptic storage vesicles. When these transmitters activate their receptors on postsynaptic membranes, they modulate Ca^{2+} dynamics in the associated postsynaptic dendrites, neuronal soma (body), or axon. As in the presynaptic site, an increase in postsynaptic $[Ca^{2+}]_i$ also gives rise to various forms of plasticity. However, the impact of Ca^{2+} signaling at the postsynaptic site is much more complex than at the presynaptic sites because of the variety of Ca^{2+} sources and the greater number of intracellular compartments and targets.[24,75,76] Unlike presynaptic Ca^{2+}, postsynaptic Ca^{2+} generates longer lasting plasticity, the duration of which persists well after the inducing-signal and Ca^{2+} surges have disappeared.[24,77,78] Thus, Ca^{2+} triggers rather than maintains postsynaptic plasticity. This long-lasting effect of Ca^{2+} on postsynaptic plasticity forms the basis of long-term potentiation (LTP) that was first discovered in mammalian hippocampus[79,80] and has subsequently been shown in other brain structures.[81–82a] Recent studies[83] indicate that this longer lasting Ca^{2+} signal also induces another form of plasticity, long-term depression (LTD).

How could the same $[Ca^{2+}]_i$-elevating postsynaptic event on one hand strengthen (LTP) and on the other hand weaken (LTD) synaptic transmission in the same neuron? As we will see further on, the amplitude and the spatiotemporal distribution of the Ca^{2+} signals may play a role in this process, with LTD having a lower Ca^{2+}-induction threshold than LTP.[26,84–86] Such long-term effects of Ca^{2+} on synaptic plasticity appears to be crucial for learning and memory. Since most of the work on LTP and LTD has employed the excitatory synapses of hippocampal glutamatergic neurons, most of the following discussion is based on these studies. However, it is believed that the mechanisms of LTP and possibly LTD are the same in all the glutamatergic neurons.

While the Ca^{2+} involved in presynaptic plasticity comes into the cell through action-potential-opened VSCCs (Fig. 5.2), there are various postsynaptic sources of Ca^{2+} that may contribute to different forms of plasticity.[26,87] The main sources of postsynaptic Ca^{2+} are the VSCCs, ligand-gated calcium channels (LGCCs) such as the NMDA glutamate receptor channel and internal Ca^{2+} stores, such as endoplasmic reticulum and mitochondria, all of which have unique spatial distributions and properties.[24,34a,87–89] Although it is now virtually certain that Ca^{2+} from all these postsynaptic sources either independently or together is involved in some form of synaptic plasticity,[88] the mechanism by which the Ca^{2+} signals are processed and integrated in enhancing/potentiating postsynaptic transmission is not clearly understood.

The major source of Ca^{2+} for the induction of LTP is believed to be the widely distributed postsynaptic glutamate-type NMDA receptor channel (Figs. 5.2 and 5.3).[88–91] Indeed, preventing Ca^{2+} from entering through this channel with selective inhibitors, like APV and MK-801, or the intracellular injection of Ca^{2+} chelators inhibit most forms of LTP and LTD.[92–99] NMDA receptor channels are heavily concentrated in postsynaptic densities often along with AMPA-type glutamate receptor channels.[25,100–102] A unique property of the NMDA receptor channel is its voltage dependence because under resting conditions it is plugged by Mg^{2+}, which can only be expelled by a positive voltage shift.[103–105] This means that two things must occur if Ca^{2+} is to flow through an NMDA receptor — the receptor must bind glutamate as it is dumped out of presynaptic vesicles,

Fig. 5.3. The influence of conflicting convergences of Ca^{2+}·CaM and PKCs on the postsynaptic responses to the stimulation of NMDA receptors only (**A**), or NMDA and mGluR receptors (**B**). When NMDA receptor/channels alone are activated (**A**), the response is brief because the Ca^{2+} flow through the channels is switched off by Ca^{2+}·CaM feeding back and binding to the receptors' R1 subunits, and the Ca^{2+} that has entered the cell is expelled when Ca^{2+}·CaM stimulates the membrane Ca^{2+} pumps. Ca^{2+}·CaM also displaces PKCs from their binding site on AKAP79 in postsynaptic densities. The liberated PKCs bind to the cell membrane, but whether they will be activated depends on the availability of Dgs (diacylglycerols). If mGluR receptors are also activated (**B**), DGs from the PtdIns(4,5)P_2 breakdown by receptor-activated PLCs will directly enhance NMDA receptor-channel activity by phosphorylating the R1 subunit's Ca^{2+}·CaM binding site as well as preventing Ca^{2+}·CaM from binding to this site and inactivating the receptor. The PKCs will prevent the Ca^{2+} entering the cell from being expelled by blocking the activation of the Ca^{2+} pump by phosphorylating the Ca^{2+}·CaM binding/activation site. They will also prevent the neurogranins from re-binding CaM by phosphorylating the CaM binding site. As symbolized by the heavy lettering, the result of the conflicting convergence of CaM, Ca^{2+}·CaM, and the mGluR-activated PKCs is to increase and prolong the buildup of Ca^{2+}·CaM and thus the activity of LTP-establishing Ca^{2+}·CaM-dependent enzymes CaMKIV and CaMAC. Except for CaMKIV (Ca^{2+}·CaM-dependent protein kinase IV), the symbols are the same as in Fig. 5.2.

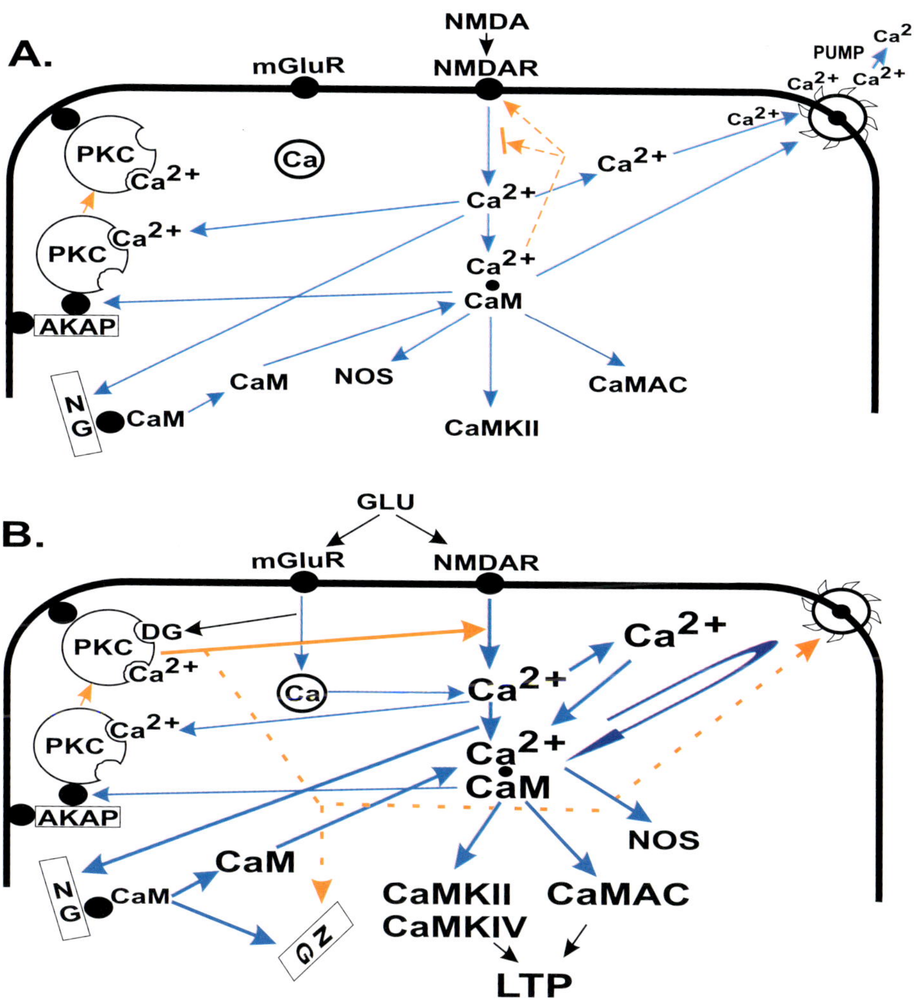

and the postsynaptic membrane must be depolarized to unplug the channel. During low-frequency synaptic transmission, the released glutamate binds to both NMDA and AMPA receptors, but it is the activated AMPA receptors that generate the postsynaptic response by opening their gates for Na^+, which depolarizes the membrane.[105–107] The NMDA receptors don't contribute much to the response under these conditions because of their Mg^{2+} plugs.[105–107] However, during high-frequency synaptic transmission the depolarization of the postsynaptic membrane by Na^+ flowing through the AMPA receptor channels is great enough to cause the NMDA receptors to spit out their Mg^{2+} plugs and allow Ca^{2+} to flow into the cell to drive LTP.[105–109]

Thus, the AMPA receptor is an important member of the LTP-inducing team. A number of studies have indicated increased AMPA channel function, due to modification of the receptor or to an increase in the number of the receptors at the postsynaptic density during LTP.[110–112] It is believed that while NMDA receptor activation is involved in the induction of LTP, AMPA receptor activation is also required for its expression.[111] Although the AMPA receptor channels are generally not permeable to Ca^{2+}, under certain conditions their subunits can be altered to make them permeable to Ca^{2+} and thus able to induce LTP.[111–114] In fact, certain types of neuron express a kind of AMPA receptor that is highly permeable to Ca^{2+}, the entry of Ca^{2+} through which can induce LTP.[115] However, the relationship between AMPA receptor-induced LTP (AMPAR-LTP) and NMDA receptor-induced LTP (NMDAR-LTP) is not yet defined.

While NMDA receptor-mediated Ca^{2+} entry is needed to induce most forms of LTP in many types of neurons, Ca^{2+} flowing in through VSCCs also induces some form of postsynaptic plasticity.[24] This has been suggested by the results of experiments in which the activation of VSCCs, while NMDA receptor channels were blocked, was enough to induce LTP in hippocampal CA1 and CA3 neurons.[26,116–118] This LTP is generally termed VSCC-LTP. In many cases, VSCC-induced LTP in CA1 pyramidal and other neurons has been blocked specifically by L-type channel blockers, indicating that Ca^{2+} influx through this kind of channel can trigger LTP induction.[119,120] Activation of L-type channels appears to be necessary for generating Ca^{2+} spikes,[121] and such spikes are believed to be needed for VSCC-LTP.[120–123] It has been suggested that these L-type channels may not by themselves be a good enough Ca^{2+} source for inducing VSCC-LTP, but rather a source of depolarization to activate other VSCCs.[24] Besides L-type channels, other VSCCs, P/Q-, R-type, and low-voltage-sensitive T-type channels have also been implicated in LTP induction.[24]

Yet, another source of Ca^{2+} for triggering synaptic plasticity is the internal stores (Fig. 5.2). Several studies have indicated that presynaptic activity could raise $[Ca^{2+}]_i$ in postsynaptic neurons by triggering the release of Ca^{2+} from internal stores[23] and that this Ca^{2+} could stimulate gene expression and induce synaptic plasticity.[23] Two major mechanisms, $InsP_3$ (inositol-1,4,5- trisphosphate)-mediated Ca^{2+} release and CaCR (Ca^{2+}- induced Ca^{2+} release), are believed to be involved in the emptying of the internal Ca^{2+} stores.[23] Activation of postsynaptic G-protein-linked, type 1 metabotropic receptors (mGluR1s) by glutamate can activate a specific phospholipase-C (PLC) that releases $InsP_3$ and diacylglycerol from membrane phosphoinositides.[124] $InsP_3$, acting on its receptors on endoplasmic reticulum, mobilizes intracellular Ca^{2+}, and diacyglycerol can activate PKCs, which, as we will see below, are also implicated in LTP induction (Figs. 5.2 and 5.3).[23,124–126] This internally stored Ca^{2+} can also be mobilized directly by CaCR either by Ca^{2+} interacting with $InsP_3$ receptors or RyRs (ryanodine-sensitive receptors).[23,124] Several studies have indicated that Ca^{2+} spikes not only replenish the internal Ca^{2+} stores, but can also enhance $InsP_3$-induced release of Ca^{2+} from the endoplasmic reticulum.[23,127,128] This coincident increase could occur if postsynaptic depolarization is paired with the synaptic activation of mGluR1 to generate $InsP_3$. Indeed, it has been shown with hippocampal CA1 pyramidal neurons that repetitive mGluR1 receptor activation along with back-propagating action potentials synergistically cause release of Ca^{2+} from $InsP_3$-sensitive stores (Figs. 5.2 and 5.3),[128] and the increase in $[Ca^{2+}]_i$ is a very large (multi-micromolar) surge in apical dendrites. It is suggested that a small increase in the $InsP_3$ level due to metabotropic receptor activation (or for that matter other G-protein coupled receptors linked to PLC) could prime or sensitize the $InsP_3$ receptors on the endoplasmic reticulum to respond to subthreshold Ca^{2+} signals from VSCCs or other receptor-operated channels.[23] A similar sensitization could be obtained by activating the metabotropic receptor-like CaRs by Ca^{2+} dumped into the synaptic space along with the glutamate.[8a,129] It has been shown that parallel fiber inputs into Purkinje cells evoke a localized increase in postsynaptic $[Ca^{2+}]_i$, which is in part due to AMPA receptor-induced depolarization and the other due to mGlu→$InsP_3$-dependent release of Ca^{2+} from endoplasmic reticulum.[130] Such metabotropic receptor-mediated synaptic plasticity has been described in hippocampus,[131,132] cerebellum,[133,134] and visual cortex.[135] mGluR-induced increase in $[Ca^{2+}]_i$ has also been implicated in LTD induction.

As mentioned above, one of explanations of how a rise in $[Ca^{2+}]_i$ can induce the diametrically opposite LTP and LTD is based on the different Ca^{2+}-sensitivities of these processes. More Ca^{2+} is needed to induce LTP than LTD, which agrees with the observation that strong high-frequency stimulation induces LTP whereas weak, prolonged low-frequency stimulation induces LTD.[24,84–86] Using potent caged Ca^{2+}-chelators, which can be released to modulate $[Ca^{2+}]_i$ surges, it has been shown that a seconds-long postsynaptic elevation of $[Ca^{2+}]_i$ to high levels induces LTP whereas a modest, minute(s)-long elevation induces LTD, suggesting that the amplitude and duration of the Ca^{2+} signal determines the kind of synaptic modification.[136] This is also reflected in the fact that the several Ca^{2+}-dependent enzymes implicated in these processes (see below) also have differential Ca^{2+} sensitivity. For example, Ca^{2+}/CaM-dependent kinase II (CaMKII), which is believed to be involved in LTP formation, has a lower affinity for Ca^{2+}/CaM compared to protein phosphatase 1 that is believed to be involved in LTD formation.[84,137]

What is the molecular basis of Ca^{2+}-induced long-term synaptic plasticity? Initial clues from early studies suggested a role for protein kinases in this process. Thus, LTP-inducing stimuli in hippocampus slices enhanced the phosphorylation of specific proteins, while there was no such enhancement by a stimulation that did not induce LTP.[138] This was soon supported by the observation that induction of LTP could be blocked by protein kinase inhibitors.[139,140] It is now clear that a number of Ca^{2+}-dependent protein kinases (CaMKII, CaMKIV, MAPK, PKA, PKC, and Src kinases) and phosphatases (calcineurin, PP1) are activated or suppressed during the induction of LTP/LTD and are believed to participate in this process.[104,106,141–145] Understanding how these signaling cascades are interlinked and how the signals they generate collaborate to enhance synaptic transmission has been the major challenge.

NMDA receptor channels, which are the critical sources of Ca^{2+} for LTP induction, are concentrated in the postsynaptic density where they interact with a number of signaling molecules, including Ca^{2+}, CaM, and several protein kinases, through cytoskeletal scaffolding consisting of actin and associated proteins such as postsynaptic density-95 (PSD95) and spinophilin.[106] Similarly, AMPA receptors and VSCCs also interact, either directly or through the cytoskeletal scaffolding, with various signaling molecules which modulate channel activity and link the surface Ca^{2+} signal to protein syntheses and gene expressions which appear to be needed for some forms of long-term synaptic plasticity.

One of the protein kinases that is strongly implicated in enhanced postsynaptic transmission is CaMKII,[106,146] which is found in high concentrations in post synaptic densities (Figs. 5.2 and 5.3).[147–148] This is an oligomeric enzyme that is activated upon binding of Ca^{2+}·CaM,[147] which causes the enzyme to autophosphorylate into a constitutively active, Ca^{2+}·CaM-independent form.[147] The active autophosphorylated enzyme then associates with postsynaptic densities, at least partly by binding to NMDA receptors.[149–151] Thus, CaMKII activity stays high long after the Ca^{2+} signal has passed, which fits with Ca^{2+} triggering rather than maintaining long-term synaptic plasticity. It has been shown in cultured hippocampal and other neurons that increasing $[Ca^{2+}]_i$ by activating NMDA receptors causes CaMKII to constitutively activate itself by autophosphorylating its Thr^{286}.[106,150,152] Indeed, CaMKII inhibitors or CaM antagonists block LTP-induction in CA1 pyramidal neurons.[139,153–155a] Furthermore, the Thr^{286}→Ala mutation, which eliminates CaMKII's ability to phosphorylate itself, inhibits NMDAR-induced LTP in CA1 and other hippocampal neurons.[156]

How could CaMKII activity affect postsynaptic plasticity? Among CaMKII's targets are the AMPA receptors that are often found alongside NMDA receptors in the postsynaptic densities.[100,147,148] The phosphorylation of the GluR1 subunit of AMPA receptors by CaMKII has been seen in cultured hippocampal neurons[157] and CA1 pyramidal neurons,[114] and it is known to increase the ion conductance of the receptors' channels.[106,158,159] Activating NMDA receptors has been shown to stimulate the phosphorylation of the AMPA receptors' GluR1 subunit at its specific CaMKII phosphorylation site in cultured hippocampal cells.[157] The same phosphorylation has been observed following the induction of LTP in area CA1,[114] suggesting a critical role for CaMKII-induced phosphorylation of AMPA receptor in LTP induction. The importance of the GluR1 subunit and its phosphorylation in LTP has been confirmed by the absence of LTP in the CA1 pyramidal cells of transgenic mice without GluR1 subunits in their AMPA receptors.[111]

It is suggested that with low-frequency stimulation, there is only a modest increase in $[Ca^{2+}]_i$, which is believed to induce LTD by generating just enough Ca^{2+}·CaM complexes to activate the protein phosphatase, calcineurin, but not enough to activate CaMKII.[84,137] Calcineurin in turn activates protein phosphatase 1 (PP1) by dephosphorylating and thus inactivating the PP1-inhibiting I-1.[137] PP1 could then inhibit LTD induction by switching off CaMKII by dephosphorylating the kinase's phospho-Thr^{286}. So how does the larger Ca^{2+} influx needed to activate CaMKII activity and drive LTP suppress PP1 activity? The answer is cyclic AMP: it plays a critical part in determining the nature of long-term synaptic plasticity. Cyclic AMP swings into action when the Ca^{2+} influx through NMDA receptors activates the type 1 adenylyl cyclase that makes it. Its job is to activate the protein kinase, PKA, which we have met often in other parts of our tour.[160] PKA then phosphorylates I-1, which then binds to, and inhibits, PP1.[161] This promotes LTP by preventing the dephosphorylation and inactivation of autophosphorylated CaMKII.[161] This link between PKA, PP1, and LTP has been suggested by experiments in which PKA inhibitors were found to block LTP and that this block was lifted by inhibiting PP1 or calcineurin.[161–164] Thus, there is a delicately poised balance between protein kinase and phosphatase activities, the tipping of which in one or the other direction can determine what kind of synaptic plasticity is induced.

Of course, this is not all! Members of another family of kinases — the conventional Ca^{2+}-dependent PKCs — are also very important players in LTP.[106,165–167] The results from several studies have indicated that these PKCs are persistently activated and phosphorylate AMPA and NMDA receptors during LTP induction.[165] The NMDA receptor's R1 (NR1) subunit, like several other proteins,[168] has a Ca^{2+}·CaM binding domain that is also a PKC target site (Fig. 5.3).[169,170] Ca^{2+}·CaM binding and phosphorylation of this shared target domain by PKCs are mutually exclusive — binding of Ca^{2+}·CaM to this domain prevents PKC from phosphorylating it while PKC phosphorylation prevents Ca^{2+}·CaM from binding to it.[168–170] Phosphorylation of the R1 subunit by PKCs increases the probability of the NMDA receptor's channel opening and thus the influx of Ca^{2+} and the generation of Ca^{2+}·CaM by promoting the ejection of the Mg^{2+} plug.[168–170] The persistently activated PKCs associated with LTP induction would keep the Ca^{2+} flowing through the channels by preventing the Ca^{2+}·CaM it makes from feeding back and closing the NMDA channels. The PKCs do this by phosphorylating the Ca^{2+}·CaM binding site on the NMDA receptors' NR1 subunits and preventing Ca^{2+}·CaM from binding to the sites and closing the channels.[168] The PKCs would also home in on and phosphorylate the Ca^{2+}-ejecting pump's inhibitory domain and prevent Ca^{2+}·CaM from binding onto it.[171–174] This would further increase the size of the Ca^{2+} surge by preventing the Ca^{2+}·CaM complexes from lifting the block and turning on the pump to expel the excess Ca^{2+}.[168] Another PKC substrate that may be involved in LTP is neurogranin, a postsynaptic protein which binds CaM instead of Ca^{2+}·CaM in the PKC target domain.[175,176] Neurogranin promptly releases its CaM as Ca^{2+}·CaM when Ca^{2+} comes surging in through the NMDA receptors or VSCCs (Figs. 5.2 and 5.3).[175] A persistently enhanced phosphorylation of neurogranin has been seen in hippocampal neurons during LTP, which correlates to the persistent PKCs activation.[177,178] Presumably, the Ca^{2+} surging through the glutamate-activated NMDA receptors in these cells releases CaM from neurogranin and the resulting Ca^{2+}·CaM spreads out to stimulate key players in LTP induction such as CaMKII, CaMKIV, and adenylyl cyclase (Fig. 5.3).[168] By homing in on and phosphorylating neurogranin's CaM binding domain, the persistently active PKCs would keep the level of free CaM up to feed Ca^{2+}·CaM formation by the Ca^{2+} flowing through the NMDA channels which the kinases have locked open. Clearly, the conflicting convergence of PKCs, CaM, and Ca^{2+}·CaM on their shared domains in key enzymes, pumps, sequestering proteins, and receptor channels has a very powerful, although as yet inadequately appreciated, impact on the response of various mechanisms, including gene transcription to the Ca^{2+} signal surges that drive LTP formation.[168]

Beside adenylyl cyclase, CaMKII, PKCs, and PKA, tyrosine kinases and MAP kinases have also been implicated in postsynaptic plasticity.[106,142,165] The NR2A and 2B subunits of NMDA receptors are known to be phosphorylated by tyrosine kinases.[142,179,180] This phosphorylation is believed to relieve the Zn^{2+}-inhibition of the receptors and increase the current flowing through them.[181] Elevated tyrosine phosphorylation of NMDA receptors by Src kinase has been observed during LTP.[182,183] Indeed, selectively inhibiting Src kinase blocks LTP.[184,185] Src kinase is certainly strategically placed to affect LTP. It is located in the postsynaptic densities, and it must somehow be tethered to NMDA receptors because it can be immunoprecipitated along with them by anti-NMDA receptor antibody.[186] Another Src-like kinase implicated in NMDA receptor phosphorylation is Fyn tyrosine kinase.[186–188] It is generally believed that the enhancement of NMDA receptor conductance by tyrosine phosphorylation increases Ca^{2+} influx into the postsynaptic dendrites, which in turn potentiates the AMPA-receptor-mediated current leading to enhanced synaptic activity.[184] There is evidence that activation of AMPA receptors can induce MAP kinase activity, which may play a role in LTP.[189] It is conceivable that Ca^{2+}-dependent activation of this motley band of kinases ultimately homes in on one target, the CREB (cAMP-response element binding protein) family of transcription factors, which stimulate the expressions of specific genes, the products of which are required for long-term synapse modification.[190–192]

Calcium and neurotransmitter release

In the preceding section, we saw how Ca^{2+} regulates synaptic plasticity. Now we will see how a rise in $[Ca^{2+}]_i$ can modulate the presynaptic release of transmitters that trigger synaptic plasticity. A large rise in the presynaptic $[Ca^{2+}]_i$ from a basal 0.1 mM to as much as 200 μM[24,40,41,193–195] is a major part of the complex mechanism that regulates neurotransmitter release. As we have just learned, the major presynaptic ports of entry of Ca^{2+} are the VSCCs. Both N- and P/Q-type Ca^{2+} channels are involved in triggering neurotransmitter release from synaptic vesicles in both the central and peripheral nervous systems.[35–36a,42,196,197] The finding that these channels interact with neurotransmitter-loaded synaptic vesicles with their α-subunits in nerve terminals as opposed to dendritic spines strongly supports this involvement.[42,198–202] Ca^{2+}, surging through these channels when they are opened by the arrival of the action potential at the terminal, boots up a complex mechanism that pulls the synaptic vesicles down onto the presynaptic membranes, initiates their fusion with the membrane and finally dumps their load of transmitters into the synaptic cleft.[199–204]

In most eukaryotic cells, molecular products addressed for export are packaged into membrane vesicles which fuse with the plasma membrane and dump their contents into the extracellular milieu — a process known as exocytosis.[203,204] Regulated exocytosis is the basis of neurotransmission which includes the release of neurotransmitters and neuromodulators.[205–209] In neurons, these transmitters and modulators are packaged in at least two different organelles — small synaptic vesicles (SSVs), which generally contain the so called 'fast' neurotransmitters (e.g., glutamate, GABA, glycine, aspartate, serotonin, and acetylcholine) and large dense-core vesicles (LDCVs), which contain the 'slow' transmitters (neuropeptides).[203] The contents of these vesicles are released by Ca^{2+}-driven exocytosis, with the SSVs needing a lot of Ca^{2+} and the LDCVs requiring much less to be dumped.[203] In neurons, most of the synaptic vesicles, the 'reserve pool,' are held back from the synaptic membrane in a filamentous cage and only a small proportion, the RRP ('readily releasable pool'), is kept near the release sites where they are very near or actually attached to the terminal's Ca^{2+} channels where the Ca^{2+} hot spots can form when the ions start flowing through the channels.[205,210] It is these frontline vesicles that promptly release their loads of transmitter (e.g., amino acids and acetylcholine) upon the arrival of the channel-opening action potential. Neurotransmitter dumping is the work of a highly specialized signal-booted membrane trafficking cycle, which includes vesicle formation, recruitment, targeting to release sites, docking, fusion, and recycling, many parts of which it shares with constitutive membrane trafficking.[205,210]

First, we must look at Ca^{2+}'s roles in vesicle formation and recruitment to the RRP. Ca^{2+} seems not to be involved in vesicle formation, but it is involved in a big way in vesicle recruitment. In order to avoid RRP depletion at the terminal, synaptic vesicles in the 'reserve pool' must be released from their cage and sent to the RRP. How is this done? As we learned above, synaptic vesicles are tethered to cytoplasmic microfilaments (e.g., actin). Although the details of this tethering are unknown, a vesicle-associated protein, synapsin, is somehow involved in it.[211–214] There are five synapsin isoforms (Ia, Ib, IIa, IIb, and IIIa), many of which are substrates of Ca^{2+}-regulated protein kinases.[210,212,215–217] Phosphorylation and dephosphorylation are believed to control the release of synaptic vesicles from the reserve pool and their transfer to the RRP.[210,212] A number of kinases, CaMKI and II, PKA, PKCs, and MAPK are known to phosphorylate the N- and C-termini of the synapsins.[212,214,218–220] Dephosphorylated synapsin I associates with the cytoskeleton and ties the reserve-pool synaptic vesicles to the filamentous cage.[210,212,221] Ca^{2+} causes synapsin I to change its shape, reduce its affinity for synaptic vesicles as well as cytoskeletal actin filaments. This reconfiguration releases the vesicle from the filaments.[210,222] But how might Ca^{2+} do this? Ca^{2+} surging through the terminal channels upon the arrival of an action potential makes Ca^{2+}·CaM, which activates CaMKII, which is also attached to the synaptic vesicles. CaMKII liberates the vesicles from the reserve pool by phosphorylating synapsin's C-terminus.[210,223] Several studies support this mechanism. For example, presynaptic injection of CaMKII into a squid's giant axon synapse enhances neurotransmitter release.[224] Furthermore, injecting unphosphorylated synapsin I inhibited transmitter release, which was reversed by phosphorylating synapsin with CaMKII, but not with CaMKI or PKA.[224] Although CaMKII-mediated phosphorylation may play the biggest role in the release of vesicles from the reserve pool, recent studies have suggested that synapsins may also be phosphorylated by other kinases. A role for PKA and CaMKI in the phosphorylation of synapsin is suggested by the ability of the phosphorylation of the protein's N-terminal PKA and CaMKI phosphorylation sites to release synaptic vesicles. Indeed, a Ca^{2+} surge into the terminal can activate Ca^{2+}-dependent adenylate cyclase and thus PKA as well as CaMKI, both of which could release vesicles from the 'reserve pool'.[212] Then, of course, there are the PKCs. Activation of the Ca^{2+}-dependent conventional PKCs may increase both the size and rate of RRP replenishment in hippocampal cultures.[225] Some studies indicate that synapsins are also PKC substrates, but it is not certain whether this PKC-stimulated RRP replenishment is mediated through synapsin phosphorylation. Indeed, it is likely that there may be a parallel mechanism.

After their release from the 'reserve pool' by Ca^{2+}-stimulated phosphorylation, the synaptic vesicles have to be targeted, docked, and primed to join the RRP at the active sites on the presynaptic membranes. Nature is notoriously conservative, so neurons use the same, but much speeded up, mechanism that drives specific membrane fusion and exocytosis in other eukaryotic cells. A key part of the fusion machinery is the so-called

20S-particle or complex, a product of the specific interactions of the proteins on the vesicles and the target membranes. It includes NSFs (*N*-ethylmaleimide-sensitive fusion proteins)[226–229] and SNAPs (soluble NSF-attachment proteins).[227,228,230] SNAPs bind to receptors on the vesicles and the target membranes that are collectively called SNAREs (soluble NSF-attachment protein receptors).[229–231] Depending on where they are, the SNAREs on synaptic vesicles are called v(vesicle)SNAREs (found on synaptic vesicles) and the SNAREs on the plasma membranes are called t(target)SNAREs.[229,231a–233] The original SNARES were VAMP (synaptolbrevin), syntaxin, and SNAP-25.[234–238] According to the SNARE hypothesis,[238a–240] the v(vesicle)-SNARE VAMP (synaptobrevin), attached to the vesicle with its C-terminus inserted into the vesicle membrane, binds to the t(terminal)SNARE consisting of the axon terminal membrane-bound (by its C-terminus inserted into the membrane) syntaxin and its associated SNAP-25 to produce a stable 7S "docking" complex. Although syntaxin can by itself associate with VAMP(synaptobrevin), SNAP-25 substantially enhances the association.[241,242] This 7S vSNARE·tSNARE complex provides a high-affinity receptor for SNAPs, which in turn bind NSFs to make the 20S fusion complex.[231,231a,243] By itself, NSF is a weak ATPase but its activity rises when it binds SNAPs.[244,245] The SNAP-activated NSF-mediated ATP hydrolysis primes the docked vesicles for Ca^{2+}-triggered fusion and transmitter dumping,[244] possibly by causing the disassembly of the VAMP·syntaxin·SNAP complex.[245] Besides being attached to VAMP(synaptobrevin), the SNAP-25·syntaxin complex also interacts with a N-type Ca^{2+} channel's synprint (synaptic protein interaction) site.[246–248] This interaction puts the membrane-tethered synaptic vesicles alongside the Ca^{2+} entry port where the ephemeral high-Ca^{2+} hot spots appear.

How does Ca^{2+} persuade the now 'docked and primed' vesicles to fuse with the plasma membrane and dump their transmitters? Many Ca^{2+}-sensor proteins are implicated in this process.[249] Among them are the four (I, II, IV) synaptotagmin isoforms, most of which are highly expressed in neurons.[250,251] These are low-affinity Ca^{2+} binding proteins found associated with synaptic vesicles,[250,251] hence they are grouped with the vSNAREs. They bind to membrane phospholipids and the tSNARE syntaxin in a Ca^{2+}-dependent manner.[251–254] Synaptotagmins have two Ca^{2+}-binding C2 domains, C2A (at the membrane) and C2B (away from the membrane).[232,254–255a] At the resting Ca^{2+} concentration, synaptotagmin tethers are believed to hold the vesicles in the RRP close to the cell membrane by binding to the vesicle's SV2 protein with their C2B domains and to the tSNARE proteins on the cell membrane with their C2A domains.[200,251,256] When Ca^{2+} surges through the adjacent channel, it binds to the C2A domain and sharply shoves the protein deeper into the membrane phospholipid bilayer,[200] as happens with all other proteins carrying C2 domains. On the other vesicle's end, the C2B domain dissociates from SV2,[257,258] and the membrane-insertion of C2A domain induces synaptotagmins to cluster through the Ca^{2+}-dependent interaction between their C2B domains.[259–261] This clustering of SNARE-linked synaptotagmins is believed to organize a vSNARE complex at a single locus on the vesicle, which enables better fusion of the vesicle with the membrane.[200] The Ca^{2+}-driven plunge of the C2A domain into the membrane is believed to facilitate rapid fusion of synaptic vesicles in a number of ways. It pulls vesicles to the synaptic membrane by facilitating the "zippering" together of the vesicle-bound VAMP(synaptobrevin) and the cell membrane's tSNARE (syntaxin·SNAP-25) complex.[207,238a] The vesicle is pulled down to the axon terminal membrane by the simultaneous shortening of the C-terminal regions of syntaxin and synaptobrevin.[238a] Furthermore, the insertion of the C2A domain into the membrane may promote vesicle fusion either by perturbing the membrane lipid bilayer and altering the physical relationship between SNAREs and membrane lipids or by inducing conformational/structural changes in SNAREs.[200] Although the precise mechanism of Ca^{2+}-induced vesicle fusion is not clear, recent kinetic studies have clearly established that synaptotagmin C2 domains are designed to detect and respond rapidly to both rises and falls in the Ca^{2+} levels at the synaptic terminals and thus are ideally suited to regulate synaptic vesicle exocytosis and endocytosis.

Ca^{2+} and neuronal cell death

While Ca^{2+} is essential for neuroblast proliferation, driving axon growth cones, and neurotransmitter secretion, it can also kill neurons just as we saw it killing colon cells and keratinocytes during the normal course of diffpoptosis. However, the neurons left after the extensive culling in the developing brain are meant to live a long time, which requires a careful control of the intracellular Ca^{2+} level.[262–264] Since the early observation that amputated axons must have extracellular Ca^{2+} to die,[265] much evidence of the ion playing the lead role in the killing of neurons under various conditions has been collected. Indeed, the glutamate-driven excitotoxicity, oxidative stress, mitochondrial dysfunction, and apoptosis that have been proposed to kill neurons in stroked brains, and the degenerating brains of Alzheimer, Huntington, and Parkinson patients are all Ca^{2+}-dependent.[266–277] This Ca^{2+}-dependence of neuronal killing has given rise to the 'Ca^{2+} hypothesis' of neurodegeneration,[267,278] which simply means that Ca^{2+} overload kills neurons. Although there are some exceptions,[279–282] the "Ca^{2+} hypothesis" appears to apply to most, if not all, kinds of neurodegeneration. So how are neurons overloaded with Ca^{2+} and how does it kill them?

Normally, the $[Ca^{2+}]_i$ is held around 100 nM in resting neurons.[23,263,283] As we have just seen, local rises in $[Ca^{2+}]_i$ during neuronal stimulation drive neurotransmitter release and synaptic plasticity. An increase in the $[Ca^{2+}]_i$, which can reach μM levels,[30,40,41] is usually only transient because of prompt binding to intracellular Ca^{2+}-buffers, pumping into storage vesicles and mitochondria, and expulsion from the cell by Ca^{2+}-pumps or transporters in the plasma membrane.[34a,262,284] Thus, internal Ca^{2+} homeostasis is achieved by strictly balancing Ca^{2+} influx with expulsion and internal sequestration. It is believed that a serious disturbance of this balance leads to Ca^{2+}-overload, which is the prime suspect in the murder of neurons in many neurodegenerative disorders. Although how Ca^{2+} overload kills neurons is far from being defined, we know that it triggers the abnormal activity of a host of Ca^{2+}-dependent enzymes and cellular processes. It is believed that a multipronged attack by surging Ca^{2+} on many signaling enzymes, including endonucleases, nitric oxide synthase, protein kinases, protein phosphatases, phospholipases, and proteases, generates lethal ROS (reactive oxygen species) and switches on the "death genes."

As we learned in the preceding section, the major sources of Ca^{2+} in neurons are membrane Ca^{2+}-channels, such as VSCCs and LGCCs, and the endoplasmic reticulum stores.[23,262,285,286] Neurons, like other cells, have many ways of ensuring that the Ca^{2+} signal lasts no longer than needed to trigger the physiological response. As we have seen earlier, Ca^{2+} influx through the channels creates Ca^{2+} hot spots which quickly vanish as the ions diffuse away.[24,34,43,287] This rapid disappearance of the hot spots is the work of Ca^{2+} binding proteins such as calbindin, calmodulin, and parvalbumins, as well as many cytoskeletal proteins.[43,288–291a] These buffering proteins have two important jobs — to put the Ca^{2+} surges where they are needed and then to keep them brief.[23,288–292] It is not clear whether a defective functioning of these buffers contributes to the Ca^{2+}-overloading of neurons. Some studies have indicated that Ca^{2+} binding proteins such as calbindin actually protect neurons from being killed by excessive excitation (excitotoxicity).[293] Other studies have found that neurons expressing Ca^{2+} buffers such as parvalbumin or calbindin are more susceptible to injury than neurons without them.[294,295]

Ca^{2+} binding proteins have only limited buffering capacities and cannot reduce a Ca^{2+} surge below a certain level. Therefore, to reduce a large Ca^{2+} surge, neurons must rely on their endoplasmic reticula and mitochondria, which can hold large amounts of Ca^{2+}.[34a,263,284,296–298] These organelles have two jobs in a neuron — buffering Ca^{2+} by sequestering it and releasing it when needed for various functions.[284,296] Ca^{2+} is pumped into the endoplasmic reticulum by a Ca^{2+}/Mg^{2+}-dependent ATPase known as SERCA (sarco/endoplasmic reticulum Ca^{2+} ATPase).[23,263,284,299] There are many isoforms of this enzyme, two (SERCA2b and SERCA3) of which are highly expressed in brain.[300] After SERCA has pumped it into the endoplasmic reticulum, the Ca^{2+} is kept there by binding to calbindin and calreticulin.[291,301] In addition to SERCA, the endoplasmic reticulum has the channels of $InsP_3$ receptors through which the stored Ca^{2+} can be released when the neuron needs it to respond to stimulants.[23,284] As we have seen in the preceding section, Ca^{2+} escapes from the endoplasmic reticulum through these channels when they are opened by $InsP_3$ from the hydrolysis of membrane phosphoinositides by the phospholipase-C stimulated by glutamate-activated metabotropic receptors.[124] The Ca^{2+} released from the endoplasmic reticulum (or for that matter flowing in through plasma membrane Ca^{2+} channels) can trigger a further release of stored Ca^{2+} (the so-called CICR — Ca^{2+}-induced Ca^{2+} release) by activating RyRs (ryanodine receptors).[23,284]

A cytoplasmic Ca^{2+} surge above a certain level is sensed by the SERCAs, which pump the excess Ca^{2+} back into the endoplasmic reticulum.[299] It is not clear how the SERCAs detect Ca^{2+} surges in neurons, but CaMKII is likely to be the sensor. Heart muscle SERCA is turned on either when it is phosphorylated or when its suppressor, the protein phospholamban, is inactivated by phosphorylation.[302–305] The brain's SERCA2b is likely activated when phosphorylated by CaMKII, of which neurons have a lot.[146–152] Thus, it seems likely, although it remains to be proven, that when the $[Ca^{2+}]_i$ surges upwards the CaMKII surge sensor turns on SERCA2b, which pumps the excess Ca^{2+} into the endoplasmic reticulum where it binds to calbindin and calreticulin.[299]

The value of the SERCA pumps for regulating neuronal $[Ca^{2+}]_i$ has been demonstrated by the fact that switching them off with thapsigargin prolongs and increases Ca^{2+}-induced depolarization.[284,306,307] Moreover, thapsigargin can empty a neuron's endoplasmic reticulum Ca^{2+} stores and trigger apoptosis.[308–311] However, it is not known whether it is the emptying of the endoplasmic reticulum Ca^{2+} stores or the rise in $[Ca^{2+}]_i$ that follows the silencing of the SERCAs that damages the neuron. There is some evidence to suggest that it is the depletion of endoplasmic reticulum Ca^{2+} that is toxic because dantrolene, an inhibitor of ryanodine receptor-triggered release of endoplasmic reticulum Ca^{2+}, can prevent thapsigargin-induced apoptosis in neurons.[309] In fact, dantrolene has been shown to reduce ischemia-induced damage in neurons under several experimental conditions.[312,313] Furthermore, stroke-induced ATP shortage empties the endoplasmic reticulum Ca^{2+} by shutting down the ATP-fueled SERCAs. Such a depletion of the neuron's endoplasmic reticulum Ca^{2+} stores inhibits protein synthesis,[298] which can produce a severely damaging stress response like that caused by stroke.[298] According to the results of some studies, parts of the brain that are especially vulnerable to damage (such as the CA1 region of the hippocampus) have many $InsP_3$ receptor channels and few SERCAs.[298] Consequently, an $InsP_3$-mediated release of Ca^{2+} following ischemia would efficiently and prolongedly empty the Ca^{2+} stores because of a lack of SERCAs to refill them.[298] In support of this, the results from some studies have indicated that activators of the $InsP_3$-generating mGluR1/5 metabotropic glutamate receptors kill neurons, while their inactivation them protects them.[314–316] Moreover, injecting a mGluR1 agonist into the cerebral ventricles triggers seizures and kills neurons.[317] Thus, regardless of whether it be the depletion of the Ca^{2+} stores per se or the resulting increase in $[Ca^{2+}]_i$ that is important, endoplasmic reticulum Ca^{2+} appears to play an important role in the killing of neurons.

The neuron's mitochondria can also hold a lot of Ca^{2+}. It is becoming increasingly evident that mitochondrial uptake and release of Ca^{2+} plays an important role in neuron function.[284,297,318,319] We will see below that mitochondrial Ca^{2+}, $Ca^{2+}{}_m$, plays a big part in the killing of neurons in several neurodegenerative disorders. Mitochondria, like the endoplasmic reticulum, can sequester Ca^{2+}, although their affinity for Ca^{2+} is lower and they require a higher $[Ca^{2+}]_i$ to induce them to load up with the ion.[297,318,320–322] The inner mitochondrial membrane is involved in the selective uptake and release of ions through specific channels and transporters. Ca^{2+} is brought into the mitochondrion by a specific low-affinity, high-capacity Ca^{2+} uniporter whose activity is highly dependent on the extramitochondrial Ca^{2+} concentration. Thus, when the cytoplasmic $[Ca^{2+}]$ rises above a threshold or set-point level , the uniporter can carry much of this Ca^{2+} into the mitochondria.[297–323] This Ca^{2+} uptake is largely driven by the mitochondrial membrane potential, $\Delta\psi$.[297,318,320–322] Calcium is expelled from a mitochondrion mainly by the Na^+/Ca^{2+} exchanger.[319,324] The activity of Na^+/Ca^{2+} exchanger is in turn coupled to the Na^+/H^+ exchanger that extrudes Na^+ from the mitochondrion at the expense of the respiration-generated pH gradient, ΔpH.[297] Thus, the flow of protons into the mitochondrial matrix to make ATP creates a counterflow of Na^+ out of the matrix via the Na^+/H^+ exchanger, which in turn creates a counterflow of Ca^{2+} out of the matrix as the ejected Na^+ flows back into the matrix via the Na^+/Ca^{2+} exchanger. The pH gradient across the inner mitochondrial membrane is maintained by respiration and drives ATP production by the ATP synthase. The ΔpH and the $\Delta\psi$ together constitute the mitochondrial proton electrochemical gradient, Δp.[319,323,325,326] ATP synthesis requires a flow of positively charged protons (that have been pumped into the intermembrane space by the four respiratory complexes to create the ΔpH) through the rapidly spinning ADP-Pi-linking F1 heads of the ATP synthase generators into the more negative matrix.[326a,b] The proton flow (i.e., its 'amperage') through the generators is a function of the transmembrane proton (pH) gradient and the overall transmembrane charge gradient (potential) $\Delta\psi$. Should the proton flow stop, the synthase changes into an ATP-eating ATPase. Obviously, $\Delta\psi$ determines mitochondrial Ca^{2+} uptake and release. As we shall see below, factors affecting

$\Delta\psi$ and mitochondrial energy metabolism can cause mitochondrial Ca^{2+} overload, which triggers a series of biochemical events that kill the neuron.

The plasma membrane has the major ports of Ca^{2+} entry, such as the VSCCs and LGCCs, as well as the machinery to expel Ca^{2+} from the cytosol to keep the neuron's $[Ca^{2+}]_i$ within safe limits. There are two major Ca^{2+}-expelling devices — the ATP-fueled Ca^{2+} pump and the Na/Ca^{2+} exchanger, which depends on the normally steep inwardly directed Na^+ gradient to drive a counterflow of Ca^{2+} out of the cell.[327–329] The plasma membrane Ca^{2+} pump is a Ca^{2+} ATPase[328,329] like the endoplasmic reticulum's SERCAs, but it differs from the SERCAs in being activated by Ca^{2+}·CaM.[328,329] The plasma membrane pump uses one ATP to eject one Ca^{2+}, while a SERCA uses one ATP to pump two Ca^{2+} ions into the endoplasmic reticulum.[262] The activity of the plasma membrane pump is also regulated by phosphorylation by protein kinases such as PKA and PKC, as well as a number of fatty acids.[262,330,331] Ca^{2+} flowing through opened membrane channels or released from internal stores produces Ca^{2+}·CaM complexes that switch on (or more accurately de-inhibit) the Ca^{2+} pumps.[168] The excess cytosolic Ca^{2+} is pumped out and the resting Ca^{2+} level is restored. The Na^+/Ca^{2+} exchanger also senses a rise in $[Ca^{2+}]_i$ and exploits the inwardly directed Na^+ gradient to expel one Ca^{2+} ion for every 2–3 Na^+ flowing into the cell.[266,327] The incoming Na^+ is then detected and expelled by the Na^+/K^+ ATPase pumps if and when it reaches a critical level.[262,327] Since ATP fuels both of the neuron's Ca^{2+}-expelling devices, a reduction of ATP production can produce a lethal Ca^{2+}-overload.

The neuron killer in many neurodegenerative diseases is excitotoxicity — the lethality of too much stimulation by a neurotransmitter such as glutamate.[262,267,268] Glutamate is a major neurotransmitter in the central nervous system and is present in millimolar concentrations in the gray matter.[262] The concentration of glutamate in the extracellular space is normally kept at a very low level.[332] However, when neurons are stimulated, glutamate is dumped into the synaptic space in boluses or quanta (quantal release), which bind to various postsynaptic receptors that trigger diverse neuronal responses, most of which, including synaptic plasticity and transmitter release, are Ca^{2+}-dependent. The released glutamate must not be allowed to stay around for too long and persistently activate receptors. It is quickly sucked back up into glial cells and neurons by their glutamate transporters that are driven by the same transmembrane Na^+ gradient that drives the Na^+/Ca^{2+} exchanger.[332–334] However, under certain conditions, such as the ischemia in a stroked brain, enough glutamate is dumped out of the presynaptic neurons to overwhelm the transporters and stay outside the cell where they overload the postsynaptic neurons with Ca^{2+} by persistently stimulating their various glutamate receptors.[262] It is this persistent glutamate signaling that kills neurons. How does Ca^{2+} overload occur under these conditions and how does it kill neurons?

As we have learned above, glutamate activates its ionotropic NMDA, AMPA, and kainate receptors (so called because they admit Na^+, K^+, and Ca^{2+} ions into the cell) and opens their ion channels,[335–337] and it activates the G-protein-coupled, $InsP_3$-generating metabotropic receptors that are related to the CaRs we met so often in the previous chapters and, as we have seen, mobilize Ca^{2+} from the endoplasmic reticulum.[23,262] While the Ca^{2+}-admitting NMDA receptor channels are the principal instruments of glutamate-induced neurotoxicity,[335] other glutamate receptors, such as the AMPA receptors, become accessories to the crime.[267,338,339] For example, AMPA receptors, which normally admit only Na^+, can admit Ca^{2+} under certain conditions such as the ischemia of stroke where their subunit composition changes.[340–342] Besides the failure to sweep up released glutamate to stop the flow of Ca^{2+} through the NMDA receptor channels, the neuron suffers a from a growing fuel shortage during ischemia.[272,297,343] This energy shortage, which is due to the depletion of the cellular ATP pool, also contributes to the Ca^{2+} overload. The ATP shortage shuts down the plasma membrane's Na^+/K^+ ATPase, which causes the Na^+ gradient to collapse and thus depolarize the neuron's plasma membrane.[262,327,344] (It should be noted that the Na^+/K^+ ATPase is a major fuel consumer that uses about 40% of a neuron's ATP to maintain the ionic gradient and the plasma membrane potential.)[344] The collapse of the Na^+ gradient disables the gradient-driven glutamate transporter, the consequence of which is a buildup of extracellular glutamate and its signaling to a lethal level. The depolarization stimulates Ca^{2+} influx in two ways — it opens VSCCs and blows the Mg^{2+} plugs out of NMDA receptors, which enables Ca^{2+} to flow through them when they are activated by the accumulating glutamate.[103–105] The Ca^{2+}·CaM made by the rising $[Ca^{2+}]_i$ should turn on the plasma membrane's Ca^{2+} pumps to expel the excess Ca^{2+},[168,328,329] but the cell is running out of ATP fuel to operate them. As if this were not enough, the fuel shortage and the collapse of the Na^+ gradient cause the Na^+/Ca^{2+} exchanger to work in reverse and bring more Ca^{2+} into the increasingly stressed neuron! The excessive glutamate also simultaneously activates the G-protein-linked metabotropic glutamate receptors that stimulate protein kinases such as PKCs,[124] which can phosphorylate and keep the plasma membrane Ca^{2+} pumps from working by preventing Ca^{2+}·CaM from binding to them and turning them on.[168] The PKCs have another dangerous capability — they can keep the glutamate-activated NMDA receptors open and passing Ca^{2+} by phosphorylating the Ca^{2+}·CaM binding site on the receptors' NR1 subunits, which prevents them from being shut down by the Ca^{2+}·CaM complexes produced by the incoming Ca^{2+}.[168] As we have seen earlier, a rise in $[Ca^{2+}]_i$ can also activate other protein kinases such as PKA, CaMKII, and tyrosine kinases, which are known to activate both VSCCs and LGCCs.[104,106,141–145] Although it is far from clear as to how efficiently protein kinases can phosphorylate and activate ion channels when the ATP level is low, it is obvious that a combination of open Ca^{2+} channels and disabled Ca^{2+} extrusion systems results in Ca^{2+} overloading of neurons.

One of the major consequences of the disruption of Ca^{2+} homeostasis is the induction of oxidative stress by the excessive generation of ROS, the various reactive oxygen species, which kill neurons in many neurodegenerative disorders.[319,322,345–347] ROSs modulate the opening of many ion channels, including Ca^{2+} channels, which allows the entry of more Ca^{2+}, which in turn generates more ROS, thus completing a lethal vicious cycle.[267,348,349] Although mitochondria are by far the largest sources of ROS, the cyclooxygenase (COX) pathway, which makes prostaglandins, and the microsomal NADPH cytochrome P450 reductase system also generate them.[350,351] The sustained elevation of Ca^{2+} can overactivate cytosolic phospholipase A2 (cPLA2), which releases arachidonic acid (AA) from membrane phospholipids.[347,350,352] Arachidonic acid can generate ROS during its conversion to prostaglandin and leukotrienes by the COX pathway.[350] Elevated levels of free arachidonate have been found in ischemic/anoxic brain[262,353] and in glutamate-treated slices of hippocampus where it is believed to be due to the activation of cPLA2 by Ca^{2+} entering through the open NMDA receptor channels.[354–357] Arachidonic acid also uncouples mitochondrial oxidative phosphorylation[358] and causes the opening of the mitochondrial permeability transition pore, a megachannel we will soon meet below which can unleash a lethal cascade of events. Indeed, the brains of mice lacking cPLA2 are less susceptible to ischemic injury.[359] A number of studies have indicated that a glutamate-induced $[Ca^{2+}]_i$ surge causes the translocation to membranes and activation of PKCs which are known to be involved in ROS generation.[360,361] Thus, the ROS that damage cellular proteins, nucleic acids, and membrane lipids collaborate with the excitotoxic glutamate to severely damage and kill neurons.

As if all of this were not enough, an excessively elevated $[Ca^{2+}]_i$ also generates NO (nitric oxide). NO is not highly toxic by itself, but the peroxynitrites produced when it reacts with oxygen radicals such as superoxide are very toxic.[362,363] In neurons, NO is produced by the neuron-specific NO synthase (nNOS), which is a Ca^{2+}·CaM-activated enzyme.[363] The surge of Ca^{2+} through continuously stimulated NMDA receptor channels and the resulting Ca^{2+}·CaM complexes should and indeed do result in an excessive production of NO in glutamate-treated cultured neurons.[364–369] However, the NO story is rather confusing because NO has been implicated in both neuronal death and survival.[362,370] In cultured neurons, glutamate-induced excitotoxicity has been linked to NO production.[367–372] Furthermore, it has been shown that mice lacking nNOS are resistant to NMDA-induced neuronal damage in vivo,[373] and the neurons in cortical cultures from these mice are resistant to glutamate-induced killing.[373] Other studies could not link NO to neuron killing.[373] NO toxicity could result from its effect on the electron transport system and inhibition of mitochondrial respiration.[363,374,375] NO can reversibly interact with mitochondrial cytochrome oxidase (in the proton-pumping complex IV) and react with superoxide anion to produce peroxynitrite.[363,376–378] Peroxynitrite is known to inhibit both the Mn- and Cu/Zn-superoxide dismutases that sweep up ROS.[379,380] High concentrations of peroxynitrites also inhibit cytochrome c reductase in the proton-pumping complex III of the respiratory chain.[381,382] Moreover, it can also inhibit the plasma membrane glutamate transport mechanism.[383] Thus, NO could kill neurons mainly by inhibiting mitochondrial complexes III and IV and with this the stopping of the spinning-inducing flow of protons through the ATP generators and the consequent conversion of the ATP generators into ATP-destroyers.[384]

The involvement of mitochondria in many neurodegenerative disorders has attracted much research interest lately. Most of the attention is focused on the relationship between the glutamate-induced increase in $[Ca^{2+}]_i$ and mitochondrial $\Delta\psi$. Prolonged activation of NMDA receptors massively increases $[Ca^{2+}]_i$ and kills cultured hippocampal, cortical, and cerebellar neurons.[385–388] The involvement of mitochondria in this neuron killing is indicated by the ability of inhibitors of the electron transport chain to increase the sensitivity of the cells to NMDA-induced Ca^{2+} overload . Thus, hippocampal cells exposed to antimycin A (a complex III inhibitor) and cyanide (a complex IV inhibitor) were more vulnerable to killing by NMDA.[389–390] Inhibitors of the mitochondrial proton-pumping complex I also increased the vulnerability of neurons to murder by NMDA.[391,392] As mentioned above, one of the earliest events in the killing of neurons is the collapse of cytosolic Ca^{2+} homeostasis, also referred to as delayed Ca^{2+} deregulation, DCD, which occurs in neurons, following the overstimulation of NMDA receptors.[321,322,325] We also saw that much of the Ca^{2+} flowing into the cell through the NMDA receptor channels starts being taken up by mitochondria once the rising level of cytosolic Ca^{2+} reaches reaches a set point. This in-rushing Ca^{2+} depolarizes mitochondria and somehow triggers the production of ROSs,[321,322,325] which, as mentioned above, do many damaging things such as disable the Ca^{2+} pumps in the plasma membrane.[393–396] Ca^{2+} pump failure further increases the cytosolic [Ca^{2+}], which in turn increases mitochondrial Ca^{2+} loading and depolarization, ending in the collapse of the mitochondrial $\Delta\psi$.[321,322,325] The combination of $\Delta\psi$ collapse, free radical generation, NO production, and increased ADP/ATP ratio (resulting from the collapse of the $\Delta\psi$ and ATP production) trigger the opening of a megachannel at the contact points between the inner and outer mitchondrial membranes — PTP, the inner mitochondrial permeability transition pore.[321,322,325] PTP is probably the key mediator of the mitochondria's involvement in cell death.[321,322,325] The opening of the PTP allows ions and molecules such as cytochrome c to leak into the cytoplasm and trigger apoptosis.[397–399] Indeed, inhibiting the opening of the PTP megachannel with the immunosuppressent, cyclosporin A, has been shown to prevent or delay mitochondrial depolarization and promote the recovery of $\Delta\psi$ in cortical and cerebellar granule cells.[400,401] This is probably how cyclosporin A prevents or delays neuronal apoptosis in stroke and other neurodegenerative conditions.[402–406]

Mitochondria are believed to play a role in both necrotic and apoptotic deaths of neurons.[407] Necrotic death occurs when the neuronal injury is severe enough to completely collapse mitochondrial function and shut off the ATP supply.[322] When the injury is less severe, although still fatal, mitochondrial depolarization and opening of the PTP will release a flood of apoptogenic agents into the cytoplasm, but the mitochondria will still be able to make enough ATP to fuel the apoptogenic mechanism.[322] Ca^{2+} is involved in both kinds of killing. The big PTP pore is made up of inner membrane proteins such as the adenine nucleotide translocator (ANT) and outer membrane proteins like porin, which is a voltage-dependent anion channel (VDAC) and its associated protein cyclophilin D.[398,408] In fact, the inhibition of pore-opening and the above-mentioned neuroprotective effect of cyclosporin A is due to it preventing the pore from forming by binding to the pore's cyclophilin D component.[398,408] The Ca^{2+}-triggered opening of a huge hole in the mitochondrial inner membrane collapses the ΔpH, which uncouples mitochondrial respiration from ATP synthesis, induces mitochondrial swelling, and releases apoptogenic agents such as cytochrome c, AIF (apoptosis-inducing factor), and procaspases into the cytoplasm.[398,408–410] The mitochondrial Ca^{2+} loading, oxidative stress, and ATP shortage that open the big channel appear to collaborate in the killing of neurons in the ischemic brains. There are also strong indications that the collapse of $\Delta\psi$ recruits the apoptogenic, channel-forming Bax protein into the initial pore complex along with a number of other proteins, and that the Bax channel is the conduit through which various apoptogenic molecules escape into the cytosol during apoptosis.[411] Both Bcl-2 and Bcl-X_L protector proteins can bind Bax to prevent megachannel formation and thus prevent Ca^{2+}-induced neuronal apoptosis.[411,412] While the cytochrome c flowing out of the mitochondria through the PTP megachannel triggers protein degradations in the cytosol by activating proteases known as caspases, AIF flowing out along with it continues on to the nucleus where it induces the apoptotic destruction of chromatin.[398]

The apoptogenic factors flowing out of the big channels in dysfunctional mitochondria can affect a number of signaling molecules, including protein kinases and transcription factors, many of which are implicated in cell death.[398,413–415] Thus, proteolysis by caspases activates kinases such as PKC-δ, MEKK1, FAK, and PAK2, which are known to promote apoptosis. Furthermore, the cleaved kinase domain of MEKK1 can itself stimulate caspase activity to further promote apoptogenesis.[416] Caspases may increase neuronal susceptibility to apoptotic death by destroying 'cell survival' proteins such as Akt-1 and Raf-1 kinases.[417,418] The destruction of Raf-1 is serious because it is taken to mitochondria by Bcl-2 where it would normally phosphorylate BAD and thus facilitate the anti-apoptotic function of Bcl-2.[419] Consequently, the destruction of Raf-1 by caspases will prevent BAD phosphorylation and inhibit the protective function of Bcl-2. Suppressing mitochondrial Raf-1 enzymatic activity has also been shown to inhibit the anti-apoptotic activity of Akt.[420] Caspases also hobble the DNA-repair machinery by destroying the DNA-dependent protein kinase, (DNA-PK), and poly (ADP-ribose) polymerase (PARP).[421] A number of cytoskeletal and nuclear proteins such as tau, GAS-2, gelsolin, fodrin, nuclear lamins and the inhibitory subunit of DNA fragmentation factor, U1-70K and NuMA, have been shown to be destroyed by caspases,[421,422] and their destruction is likely to be involved in apoptogenesis. Activation of calcineurin, a Ca^{2+}-dependent protein phosphatase (PP2B), by the Ca^{2+} influx can dephosphorylate BAD, which then translocates to mitochondria and promotes apoptosis by forming a complex with Bcl-X_L.[423] Indeed, it has been shown that the glutamate-induced $[Ca^{2+}]_i$ increase in hippocampal neurons can activate PP2B and trigger the translocation of BAD to mitochondria and apoptosis.[423]

Many transcription factors are implicated in apoptosis. The transcription factor NF-κB is known to play a role in apoptogenesis in many cells by controlling the expression of genes involved in inhibiting cell death.[424] Thus, inhibition of transcriptional activity of NF-κB has been shown to promote cell death in many cases.[424] Recent studies have indicated that the proteolytic cleavage of NF-κB or its suppressor protein I-κBα by caspases produces dominant-negative fragments that block NF-κB activity.[425,426] Consequently, caspases can promote cell death by blocking the protective effect of NF-κB. Another transcription factor, p53(TP53), appears to play a key role in apoptosis induced by various stimuli including ROS-induced DNA damage and hypoxia.[427] p53(TP53) has been shown to stimulate pro-apoptotic genes such as Bax and silence anti-apoptotic genes such as Bcl-2 thereby shifting the balance towards apoptosis.[427] p53(TP53) also induces p85, a regulator of PI-3 kinase, which has been shown to participate in oxidative stress-induced apoptosis.[427] The activity of p53(TP53) is suppressed by the mdm-2 oncoprotein, which, as we learned in Chap. 2, binds p53(TP53), prevents it from stimulating gene expression, and delivers it to the proteasome shredder.[427] Recently, it has been shown that mdm-2 is cleaved by a specific caspase which stops p53(TP53) degradation and thus stabilizes its apoptogenic function.[428] Caspases also destroy cell-cycle-suppressor proteins such as the E2F1 binding/suppressing retinoblastoma (pRb) 'pocket' protein. The destruction of pRb unleashes E2F1, which stimulates the expression of the apoptogenic p53(TP53), which feeds back and promotes the further destruction of RB by the caspases.[427] Clearly, the creation of the mitochondrial megachannel triggered by an uncontrolled Ca^{2+} surge sets off an impressive cascade of lethal events.[429]

Besides indirectly activating caspases, the surging Ca^{2+} directly activates other proteases — the calpains we met in Chaps. 2 and 3.[430] There are two isoforms of this enzyme, μ-calpain activated by micromolar Ca^{2+} and m-calpain activated by millimolar Ca^{2+}, both of which are highly expressed in all parts of the brain.[430,434,435] μ-Calpain appears to predominate in neurons.[430–435] Calpain targets a number of cellular proteins, which includes cytoskeletal proteins (spectrin, MAP2, tubulin), membrane proteins such as

Ca^{2+}-channel components, the neural N-CAM cell-adhesion molecule, the EGF and PDGF receptors, and enzymes such as Ca^{2+}-ATPase, the CaMKs, and the PKCs,[430,431] which are variously involved in neuronal survival and death. Moreover, both calpains and caspases share many substrates.[431] Unlike the apoptogenic caspases, the calpains appear to be involved in both necrosis and apoptosis.[430,431] There is growing evidence to suggest that Ca^{2+}-activated calpain plays a key part in excitotoxicity-induced neuron killing.[431,436] A burst of calpain activity follows glutamate or NMDA treatment of cultured hippocampal and cortical neurons,[437,438] and blocking calpain activity with specific inhibitors protects neurons from excitotoxicity.[438–440] Calpain blockers have been shown to protect neurons against hypoxia-induced death.[441,442] There is some evidence to suggest that only when Ca^{2+} enters through certain ports, such as the NMDA receptor channel, can it activate calpain, which then helps to murder the neuron.[437]

Our tour of the world of brains and neurons must come to an end. Once again, we have seen the many faces of Ca^{2+}. We have seen it building brains by driving neuroblast proliferation and the migration of axonal growth cones, operating and individually shaping brains by triggering release of neurotransmitters, but we have also seen it killing neurons and destroying brains.

Now we must go on to the last part of our tour of Calcium Land. We will find this last place very different — yet maybe not so different.

References

1. Gershon MD. The Second Brain. New York, Harper–Collins, 1998.
2. Alonso G. Neuronal progenitor-like cells expressing polysialylated neural cell adhesion molecule are present on the ventricular surface of the adult rat brain and spinal cord. J Comp Neurol 1999; 414: 149–166.
3. Doe CQ, Fuerstenberg S, Peng CY. Neural stem cells: from fly to vertebrates. J Neurobiol 1998; 36: 111–127.
4. Marmur R, Mabie PC, Gokhan S, et al. Isolation and developmental characterization of cerebral cortical multipotent progenitors. Dev Biol 1998; 204: 577–591.
5. Peretto P, Merighi A, Fasolo A, et al. The subependymal layer in rodents: a site of structural plasticity and cell migration in the adult brain mammalian brain. Brain Res Bull 1999; 49: 221–243.
6. Takagi Y, Nozaki K, Takahashi J, et al. Proliferation of neuronal precursor cells in the dentate gyrus is accelerated after transient forebrain ischemia in mice. Brain Res 1999; 831: 283–287.
7. Vescovi AL, Snyder EY. Establishment and properties of neural stem cell clones: plasticity in vitro and in vivo. Brain Pathol 1999; 9: 569–598.
8. Wagner JP, Black IB, DiCicco-Bloom E. Stimulation of neonatal and adult brain neurogenesis by subcutaneous injection of basic fibroblast growth factor. J Neurosci 1999; 19: 6006–6016.

8*a*. Brown EM, MacLeod RJ. Extracellular calcium sensing and extracellular calcium. Physiol Rev 2000; 81: 239–297.

9. Chattopadhyay N, Legradi G, Bai M, et al. Calcium-sensing receptor in the rat hippocampus: a developmental study. Brain Res Dev Brain Res 1997; 100: 13–21.

10. Chattopadhyay N, Ye CP, Yamaguchi T, et al. Extracellular calcium-sensing receptor induces cellular proliferation and activation of a nonselective cation channel in U373 human astrocytoma cells. Brain Res 1999; 851: 116–124.
11. Rehder V, Cheng S. Autonomous regulation of growth cone filopodia. J Neurobiol 1998; 34: 179–192.
12. Culotti JG, Merz DC. DCC and netrins. Curr Opin Cell Biol 1998; 10: 609–613.
12*a*. Dickson BJ. Moving on. Science 2001; 291: 1910–1911.
13. Keino-Masu K, Masu M, Hinck L, et al. Deleted in colorectal cancer (DCC) encodes a netrin receptor. Cell 1996; 175–185.
14. Mueller BK. Growth cone guidance. Annu Rev Neurosci 1999; 22: 351–388.
15. Song H, Poo M. Signal transduction underlying growth cone guidance by diffusible factors. Curr Opin Neurobiol 1999; 9: 355–363.
16. Goldberg DJ, Grabham PW. Braking news: calcium in the growth cone. Neuron 1999; 22: 423–425.
16*a*. Gomez TM, Robles E, Poo M. Filopodial calcium transients promote substrate-dependent growth cone turning. Science 2001; 291: 1983–1987.
17. Amato A, Al-Mohanna F, Bolsover SR. Spatial organization of calcium dynamics in growth cones of secretory neurons. Dev Brain Res 1996; 92: 101–110.
18. Kostyuk P. Plasticity in Nerve Cell Function. Oxford, Oxford University Press, 1999.
19. Zheng, JQ, Poo MM, Connor JA. Calcium and chemotropic turning of nerve growth cones. Perspect Dev Neurobiol 1996; 4: 205–213.
19*a*. Routtenberg A, Cantallops I, Zaffuto S, et al. Enhanced learning after genetic overexpression of a brain growth protein. Proc Natl Acad Sci USA 2000; 97: 7657–7662.
20. Brines ML, Ling Z, Broadus AE. Parathyroid hormone-related protein protects against kainic acid excitotoxicity in rat cerebellar granule cells by regulating L-type channel calcium flux. Neurosci Lett 1999; 274: 13–16.
21. Holt EH, Broadus AE, Brines ML. Parathyroid hormone-related peptide is produced by cultured cerebellar granule cells in response to L-type voltage-sensitive Ca^{2+} channel flux via a Ca^{2+}/calmodulin-dependent kinase pathway. J Biol Chem 1996; 271: 28105–28111.
22. Amling M, Neff L, Tanaka S, et al. Bcl-2 lies downstream of parathyroid hormone-related peptide in a signaling pathway that regulates chondrocyte maturation during skeletal development. J Cell Biol 1997; 136: 205–213.
23. Berridge MJ. Neuronal calcium signaling. Neuron 1998; 21: 13–26.
24. Hanse E and Konnerth A. Calcium and activity-dependent synaptic plasticity. In: Verkhratsky A, Toescu EC, editors. Integrative Aspects of Calcium Signalling. New York, Plenum Press, 1998: 333–358.
25. Lee SH, Sheng M, Lee SH, et al. Development of neuron–neuron synapses. Curr Opin Neurobiol 2000; 10: 125–131.
26. Zucker RS. Calcium- and activity-dependent synaptic plasticity. Curr Opin Eurobiol 1999; 9: 305–313.

27. Fischer TM, Carew TJ. Activity-dependent regulation of neural networks: the role of inhibitory synaptic plasticity in adaptive gain control in the siphon withdrawal reflex of Aplysia. Biol Bull 1997; 192: 164–166.
28. Clark GA, Hawkins RD, Kandel ER. Activity-dependent enhancement of presynaptic facilitation provides a cellular mechanism for the temporal specificity of classical conditioning in Aplysia. Learn Mem 1994; 1: 243–257.
29. Zucker RS. Short-term synaptic plasticity. Annu Rev Neurosci 1989; 12: 13–31.
30. Atwood HL, Dixon D, and Wojtowicz JM. Rapid introduction of long-lasting synaptic changes at crustacean neuromuscular junctions. J Neurobiol 1989; 20: 373–385.
31. Magleby K. Short-term changes in synaptic efficacy. In: Edleman GM, Gall LE, Maxwell W, Cowan WM, editors. Synaptic Function. New York, Wiley Press, 1987: 21–56.
32. Atwood HL and Wojtowicz JM. Short-term and long-term plasticity and physiological differentiation of crustacean motor synapses. Int Rev Neurobiol 1986; 28: 275–360.
33. Zengel J, Magleby K, Horn J, et al. Facilitation, augmentation and potentiation of synaptic transmission at the superior cervical ganglion of the rabbit. J Gen Physiol 1980; 76: 213–231.
34. Fisher SA, Fisher TM, and Carew TJ. Multiple overlapping processes underlying short-term synaptic enhancement. Trends Neurosci 1997; 20: 170–177.
34*a*. Simpson PB. The local control of cytosolic Ca^{2+} as a propagator of CNS communication — integration of mitochondrial transport mechanisms and cellular responses. J Bioenerg Biomembranes 2000; 32: 5–13.
34*b*. Carmignoto G. Reciprocal communication systems between astrocytes and neurones. Progr Neurobiol 2000; 62: 561–581.
34*c*. Araque A, Carmignoto G, Haydon PG. Dynamic signaling between astrocytes and neurons. Annu Rev Physiol 2001; 63: 795–813.
35. Reuter H. Diversity and function of presynaptic calcium channels in the brain. Curr Opin Neurobiol 1996; 6: 331–337.
36. Robitaille R, Adler EM, Charlton MP. Strategic location of calcium channels at transmitter release sites of frog neuromuscular synapses. Neuron 1990; 5: 773–779.
36*a*. Rahamimoff R. Ion channels in presynaptic nerve terminals and control of transmitter release. Physiol Rev 1999; 79: 1019–1088.
36*b*. Lee A, Scheuer T, Catterall WA. Ca^{2+}/calmodulin-dependent facilitational and inactivation of P/Q-type Ca^{2+} channels. J Neurosci 2000; 20: 6830–6838.
37. Magleby KL, Zengel JE. A quantitative description of stimulation-induced changes in transmitter release at the frog neuromuscular junction. J Gen Physiol 1982; 80: 613–638.
38. Zengel JE, Magleby KL. Augmentation and facilitation of transmitter release. A quantitative description at the frog neuromuscular junction. J Gen Physiol 1982; 80: 583–611.

39. Matthews G. Neurotransmitter release. Annu Rev Neurosci 1996; 19: 219–233.
40. Llinas R, Sugimori M, Silver RB. Microdomains of high calcium concentration in a presynaptic terminal. Science 1992; 256: 677–679.
41. Heidelberger R, Heinemann C, Neher E, Matthews G, et al. Calcium dependence of the rate of exocytosis in a synaptic terminal. Nature 1994; 371: 513–515.
42. Bennett MR. The concept of a calcium sensor in transmitter release. Prog Neurobiol 1999; 59: 243–277.
43. Blaustein MP. Calcium transport and buffering in neurons. Trends Neurosci 1988; 11: 438–443.
44. Katz B, Miledi R. The role of calcium in neuromuscular facilitation. J Physiol (Lond) 1968; 195: 481–492.
45. Tank DW, Regehr WG, Delaney KR. A quantitative analysis of presynaptic calcium dynamics that contribute to short-term enhancement. J Neurosci 1995; 15: 7940–7952.
46. Regehr WG, Atluri PP. Calcium transients in cerebellar granule cell presynaptic terminals. Biophys J 1995; 68: 2156–2170.
47. Atluri PP, Regehr WG. Determinants of the time course of facilitation at the granule cell to Purkinje cell synapse. J Neurosci 1996; 16: 5661–5671.
48. Delaney KR and Tank DW. A quantitative measurement of the dependence of short-term synaptic enhancement on presynaptic residual calcium. J Neurosci 1994; 14: 5885–5902.
49. Regehr WG, Delaney KR, Tank DW. The role of presynaptic calcium in short-term enhancement at the hippocampal mossy fiber synapse. J Neurosci 1994; 14: 523–537.
50. Tang Y, Zucker RS. Mitochondrial involvement in post-tetanic potentiation of synaptic transmission. Neuron 1997; 18: 483–491.
51. Rahamimoff R, Yaari Y. Delayed release of transmitter at the frog neuromuscular junction. J Physiol (Lond) 1973; 228: 241–257.
52. Van der Kloot W. Facilitation of transmission at the frog neuromuscular junction at 0 degrees C is not maximal at time zero. J Neurosci 1994; 14: 5722–5724.
53. Winslow JL, Duffy SN, Charlton MP. Homosynaptic facilitation of transmitter release in crayfish is not affected by mobile calcium chelators: implications for the residual ionized calcium hypothesis from electrophysiological and computational analyses. J Neurophysiol 1994; 72: 1769–1793.
54. Lando L, Zucker RS. Ca^{2+} cooperativity in neurosecretion measured using photolabile Ca^{2+} chelators. J Neurophysiol 1994; 72: 825–830.
55. Kamiya H, Zucker RS. Residual Ca^{2+} and short-term synaptic plasticity. Nature 1994; 371: 603–606.
56. Garcia ML, Strehler EE. Plasma membrane calcium ATPases as critical regulators of calcium homeostasis during neuronal cell function. Front Biosci 1999; 4: D869–D882.
57. Verkhratsky AJ, Petersen OH. Neuronal calcium stores. Cell Calcium 1998; 24: 333–343.

58. Zengel JE, Magleby KL. Augmentation and facilitation of transmitter release. A quantitative description at the frog neuromuscular junction. J Gen Physiol 1982; 80: 583–611.
59. Magleby KL, Zengel JE. A quantitative description of tetanic and post-tetanic potentiation of transmitter release at the frog neuromuscular junction. J Physiol (Lond) 1975; 245: 183–208.
60. Werth JL, Thayer SA. Mitochondria buffer physiological calcium loads in cultured rat dorsal root ganglion neurons. J Neurosci 1994; 14: 348–356.
61. Bleakman D, Roback JD, Wainer BH, et al. Calcium homeostasis in rat septal neurons in tissue culture. Brain Res 1993; 600: 257–267.
62. Lin YQ, Brain KL, Bennett MR. Calcium in sympathetic boutons of rat superior cervical ganglion during facilitation, augmentation, and potentiation. J Auton Nerv Syst 1998; 73: 26–37.
63. Peng Y-Y. Ryanodine-sensitive component of calcium transients evoked by nerve firing at presynaptic nerve terminals. J Neurosci 1996; 16: 6703–6712.
64. Peng Y-Y. Effects of mitochondria on calcium transients at intact presynaptic terminals depend on frequency of nerve firing. J Neurophysiol 1998; 80: 186–195.
65. Mulkey RM, Zucker RS. Posttetanic potentiation at the crayfish neuromuscular junction is dependent on both intracellular calcium and sodium ion accumulation. J. Neurosci 1992; 12: 4327–4336.
66. Cochilla AJ and Alford S. NMDA receptor-mediated control of presynaptic calcium and neurotransmitter release. J Neurosci 1999; 19: 193–205.
67. Rohrbough J, Spitzer NC. Ca^{2+}-permeable AMPA receptors and spontaneous presynaptic transmitter release at developing excitatory spinal synapses. J Neurosci 1999; 19: 8528–8541.
67*a*. Emptage NJ, Reid CA, Fine A. Calcium stores in hippocampal synaptic boutons mediate short-term plasticity, store-operated Ca^{2+} entry, and spontaneous transmitter release. Neuron 2001; 29: 197–208.
67*b*. Schmitz D, Mellor J, Nicoll RA. Presynaptic kainate receptor mediation of frequency facilitation at hippocampal mossy fiber synapses. Science 2001; 291: 1972–1976.
68. Anwyl R. Metabotropic glutamate receptors: electrophysiological properties and role in plasticity. Brain Res Rev 1999; 29: 83–120.
69. Cochilla AJ and Alford S. Metabotropic glutamate receptor-mediated control of neurotransmitter release. Neuron 1998; 20: 1007–10016.
70. Takahashi T, Forsythe ID, Tsujimoto T, et al. Presynaptic calcium current modulation by a metabotropic glutamate receptor. Science 1996; 274: 594–597.
71. Gray R, Rajan AS, Radcliffe KA, et al. Hippocampal synaptic transmission enhanced by low concentrations of nicotine. Nature 1996; 383: 713–716.
72. Lin XY, Glanzman DL. Long-term potentiation of Aplysia sensorimotor synapses in cell culture: regulation by postsynaptic voltage. Proc R Soc Lond B Biol Sci 1994; 255: 113–188.
73. Bao JX, Kandel ER, Hawkins RD. Involvement of presynaptic and postsynaptic mechanisms in a cellular analog of classical conditioning at Aplysia sensory-motor neuron synapses in isolated cell culture. J Neurosci 1998; 18: 458–466.

74. Bao JX, Kandel ER, Hawkins RD. Involvement of pre- and postsynaptic mechanisms in posttetanic potentiation at Aplysia synapses. Science 1997; 275: 969–973.
75. Eilers J, Konnerth A. Dendritic signal integration. Curr Opin Neurobiol 1997; 7: 385–390.
76. Denk W, Yuste R, Svoboda K, et al. Imaging calcium dynamics in dendritic spines. Curr Opin Neurobiol 1996; 6: 372–378.
77. Perkel DJ, Petrozzino JJ, Nicoll RA, Connor JA, et al. The role of Ca^{2+} entry via synaptically activated NMDA receptors in the induction of long-term potentiation. Neuron 1993; 11: 817–823.
78. Regehr WG, Tank DW. Postsynaptic NMDA receptor-mediated calcium accumulation in hippocampal CA1 pyramidal cell dendrites. Nature 1990; 345: 807–810.
79. Dunwiddie TV, Lynch G. The relationship between extracellular calcium concentrations and the induction of hippocampal long-term potentiation. Brain Res 1979; 169: 103–110.
80. Regehr WG, Tank DW. Dendritic calcium dynamics. Curr Opin Neurobiol 1994; 4: 373–382.
81. Kirkwood A, Dudek SM, Gold JT, et al. Common forms of synaptic plasticity in the hippocampus and neocortex in vitro. Science 1993; 260: 1518–1521.
82. Bear MF, Kirkwood A. Neocortical long-term potentiation. 1993; 3: 197–202.
82*a*. Maren S. Long-term potentiation in the amygdala: a mechanism for emotional learning and memory. Trends Neurosci 1999; 22: 561–567.
83. Daniel H, Levenes C, Crepel F. Cellular mechanisms of cerebellar LTD. Trends Neurosci 1998; 21: 401–407.
84. Lisman JA. Mechanism for the Hebb and the anti-Hebb processes underlying learning and memory. Proc Natl Acad Sci USA 1989; 86: 9574–9578.
85. Artola A, Singer W. Long-term depression of excitatory synaptic transmission and its relationship to long-term potentiation. Trends Neurosci 1993; 16: 480–487.
86. Hansel C, Artola A, Singer W. Relation between dendritic Ca^{2+} levels and the polarity of synaptic long-term modifications in rat visual cortex neurons. Eur J Neurosci 1997; 9: 2309–2322.
87. Paulsen O, Sejnowski TJ. Natural patterns of activity and long-term synaptic plasticity. Curr Opin Neurobiol 2000; 10: 172–179.
88. Chittajallu R, Alford S, Collingridge GL. Ca^{2+} and synaptic plasticity. Cell Calcium 1998; 24: 377–385.
89. Collingridge GL, Bliss TV. Memories of NMDA receptors and LTP. Trends Neurosci 1995; 18: 54–56.
90. Bliss TV, Collingridge GL. A synaptic model of memory: long-term potentiation in the hippocampus. Nature 1993; 361: 31–39.
91. Wilson MA, Tonegawa S. Synaptic plasticity, place cells and spatial memory: study with second generation knockouts. Trends Neurosci 1997; 20: 102–106.
92. Nicoll RA, Oliet SH, Malenka RC. NMDA receptor-dependent and metabotropic glutamate receptor-dependent forms of long-term depression

coexist in CA1 hippocampal pyramidal cells. Neurobiol Learn Mem 1998; 70: 62–72.
93. Escobar ML, Chao V, Bermudez-Rattoni F. In vivo long-term potentiation in the insular cortex: NMDA receptor dependence. Brain Res 1998; 779: 314–319.
94. Bear MF, Press WA, Connors BW. Long-term potentiation in slices of kitten visual cortex and the effects of NMDA receptor blockade. J Neurophysiol 1992; 67: 841–851.
95. Dudek SM, Bear MF. Homosynaptic long-term depression in area CA1 of hippocampus and effects of *N*-methyl-D-aspartate receptor blockade. Proc Natl Acad Sci USA 1992; 89: 4363–4367.
96. Frankiewicz T, Potier B, Bashir ZI, et al. Effects of memantine and MK-801 on NMDA-induced currents in cultured neurones and on synaptic transmission and LTP in area CA1 of rat hippocampal slices. Br J Pharmacol 1996; 117: 689–697.
97. Webber TJ, Green EJ, Winters RW, Schneiderman N, McCabe PM, et al. Contribution of NMDA and non-NMDA receptors to synaptic transmission from the brachium of the inferior colliculus to the medial subdivision of the medial geniculate nucleus in the rabbit. Exp Brain Res 1999; 124: 295–303.
98. Rauschecker JP, Egert U, Kossel A. Effects of NMDA antagonists on developmental plasticity in kitten visual cortex. Int J Dev Neurosci 1990; 8: 425–435.
99. Collingridge GL, Kehl SJ, McLennan H. Excitatory amino acids in synaptic transmission in the Schaffer collateral-commissural pathway of the rat hippocampus. J Physiol (Lond) 1983; 334: 33–46.
100. Nicoll RA, Malenka RC, Kauer JA. Functional comparison of neurotransmitter receptor subtypes in mammalian central nervous system. Physiol Rev 1990; 70: 513–565.
101. Bekkers JM, Stevens CF. NMDA and non-NMDA receptors are co-localized at individual excitatory synapses in cultured rat hippocampus. Nature 1989; 341: 230–233.
102. Jones KA, Baughman RW. Both NMDA and non-NMDA subtypes of glutamate receptors are concentrated at synapses on cerebral cortical neurons in culture. Neuron 1991; 7: 593–603.
103. McBain CJ, Mayer ML. *N*-Methyl-D-aspartic acid receptor structure and function. Physiol Rev 1994; 74: 723–760.
104. Bennett MR. The concept of long term potentiation of transmission at synapses. Prog Neurobiol 2000; 60: 109–137.
105. Malenka RC. Synaptic plasticity in the hippocampus: LTP and LTD. Cell 1994; 78: 535–538.
106. Soderling TR, Derkach VA. Postsynaptic protein phosphorylation and LTP. Trends Neurosci 2000; 23: 75–80.
107. Malenka RC, Nicoll RA. Long-term potentiation-A decade of progress. Science 1999; 285: 1870–1874.
108. Isaac JT, Nicoll RA, Malenka RC. Silent glutamatergic synapses in the mammalian brain. Can J Physiol Pharmacol 1999; 77: 735–737.

109. Linden DJ. The return of the spike: postsynaptic action potentials and the induction of LTP and LTD. Neuron 1999; 22: 661–666.
110. Braithwaite SP, Meyer G, Henley JM. Interactions between AMPA receptors and intracellular proteins. Neuropharmacology 2000; 39: 919–930.
111. Zamanillo D, Sprengel R, Hvalby O, et al. Importance of AMPA receptors for hippocampal synaptic plasticity but not for spatial learning. Science 1999; 284: 1805–1811.
112. Shi SH, Hayashi Y, Petralia RS, et al. Rapid spine delivery and redistribution of AMPA receptors after synaptic NMDA receptor activation. Science 1999 11; 284: 1811–1816.
113. Jia Z, Agopyan N, Miu P, et al. Enhanced LTP in mice deficient in the AMPA receptor GluR2. Neuron 1996; 17: 945–956.
114. Barria A, Muller D, Derkach V, et al. Regulatory phosphorylation of MPA-type glutamate receptors by CaM-KII during long-term potentiation. Science 1997; 276: 2042–2045.
115. Gu JG, Albuquerque C, Lee CJ, et al. Synaptic strengthening through activation of Ca^{2+}-permeable AMPA receptors. Nature 1996; 381: 793–796.
116. Kapur A, Yeckel MF, Gray R, et al. L-Type calcium channels are required for one form of hippocampal mossy fiber LTP. J Neurophysiol 1998; 79: 2181–2190.
117. Izumi Y, Zorumski CF. LTP in CA1 of the adult rat hippocampus and voltage-activated calcium channels. Neuroreport 1998; 9: 3689–3691.
118. Grover LM, Teyler TJ. *N*-Methyl-D-aspartate receptor-independent long-term potentiation in area CA1 of rat hippocampus: input-specific induction and preclusion in a non-tetanized pathway. Neuroscience 1992; 49: 7–11.
119. Kullmann DM, Perkel DJ, Manabe T, et al. Ca^{2+} entry via postsynaptic voltage-sensitive Ca^{2+} channels can transiently potentiate excitatory synaptic transmission in the hippocampus. Neuron 1992; 9: 1175–1183.
120. Huang YY, Malenka RC. Examination of TEA-induced synaptic enhancement in area CA1 of the hippocampus: the role of voltage-dependent Ca^{2+} channels in the induction of LTP. J Neurosci 1993; 13: 568–576.
121. Miura M, Yoshioka M, Miyakawa H, et al. Properties of calcium spikes revealed during GABAA receptor antagonism in hippocampal CA1 neurons from guinea pigs. J Neurophysiol 1997; 78: 2269–2279.
122. Cormier RJ, Greenwood AC, Connor JA. Bidirectional synaptic plasticity correlated with the magnitude of dendritic calcium transients above a threshold. J. Neurophysiol 2001; 85: 399–406.
123. Petrozzino JJ, Connor JA. Dendritic Ca^{2+} accumulations and metabotropic glutamate receptor activation associated with an *N*-methyl-D-aspartate receptor-independent long-term potentiation in hippocampal CA1 neurons. Hippocampus 1994; 4: 546–558.
124. Fagni L, Chavis P, Ango F, et al. Complex interactions between mGluRs, intracellular Ca^{2+} stores and ion channels in neurons. Trends Neurosci 2000; 23: 80–88.

125. Angenstein F, Riedel G, Reyman KG, et al. Transient translocation of protein kinase Cgamma in hippocampal long-term potentiation depends on activation of metabotropic glutamate receptors. Neuroscience 1999; 93: 1289–1295.
126. Sweatt JD, Atkins CM, Johnson J, et al. Protected-site phosphorylation of protein kinase C in hippocampal long-term potentiation. J Neurochem 1998; 71: 1075–1085.
127. Hernandez-Cruz A, Escobar AL, Jimenez N. Ca^{2+}-induced Ca^{2+} release phenomena in mammalian sympathetic neurons are critically dependent on the rate of rise of trigger Ca^{2+}. J Gen Physiol 1997; 109: 147–167.
128. Nakamura T, Barbara JG, Nakamura K, et al. Synergistic release of Ca^{2+} from IP3-sensitive stores evoked by synaptic activation of mGluRs paired with backpropagating action potentials. Neuron 1999; 24: 727–737.
129. Flower DR. Modelling G-protein-coupled receptors for drug design. Biochim Biophys Acta 1999; 1422: 207–234.
130. Takechi H, Eilers J, Konnerth A. A new class of synaptic response involving calcium release in dendritic spines. Nature 1998; 396: 757–760.
131. Reyes M, Stanton PK. Induction of hippocampal long-term depression requires release of Ca^{2+} from separate presynaptic and postsynaptic intracellular stores. J Neurosci 1996; 16: 5951–5960.
132. Riedel G, Reymann KG. Metabotropic glutamate receptors in hippocampal long-term potentiation and learning and memory. Acta Physiol Scand 1996; 157: 1–19.
133. Inoue T, Kato K, Kohda K, et al. Type 1 inositol 1,4,5-trisphosphate receptor is required for induction of long-term depression in cerebellar Purkinje neurons. J Neurosci 1998; 18: 5366–5373.
134. Kohda K, Inoue T, Mikoshiba K. Ca^{2+} release from Ca^{2+} stores, particularly from ryanodine-sensitive Ca^{2+} stores, is required for the induction of LTD in cultured cerebellar Purkinje cells. J Neurophysiol 1995; 74: 2184–2188.
135. Komatsu Y. GABAB receptors, monoamine receptors, and postsynaptic inositol trisphosphate-induced Ca^{2+} release are involved in the induction of long-term potentiation at visual cortical inhibitory synapses. J Neurosci 1996; 16: 6342–6352.
136. Yang SN, Tang YG, Zucker RS. Selective induction of LTP and LTD by postsynaptic $[Ca^{2+}]_i$ elevation. J Neurophysiol 1999; 81: 781–787.
137. Mulkey RM, Endo S, Shenolikar S, et al. Involvement of a calcineurin/inhibitor-1 phosphatase cascade in hippocampal long-term depression. Nature 1994; 369: 486–488.
138. Bar PR, Schotman P, Gispen WH, et al. Changes in synaptic membrane phosphorylation after tetanic stimulation in the dentate area of the rat hippocampal slice. Brain Res 1980; 198: 478–484.
139. Malinow R, Schulman H, Tsien RW. Inhibition of postsynaptic PKC or CaMKII blocks induction but not expression of LTP. Science 1989; 245: 862–866.

140. Malenka RC, Kauer JA, Perkel DJ, et al. An essential role for postsynaptic calmodulin and protein kinase activity in long-term potentiation. Nature 1989; 340: 554–557.
141. Micheau J, Riedel G. Protein kinases: which one is the memory molecule? Cell Mol Life Sci 1999; 55: 534–548.
142. Gurd JW. Protein tyrosine phosphorylation: implications for synaptic function. Neurochem Int 1997; 31: 635–649.
143. Foster TC. Involvement of hippocampal synaptic plasticity in age-related memory decline. Brain Res Brain Res Rev 1999; 30: 236–249.
144. Norman ED, Thiels E, Barrionuevo G, et al. Long-term depression in the hippocampus in vivo is associated with protein phosphatase-dependent alterations in extracellular signal-regulated kinase. J Neurochem 2000; 74: 192–198.
145. Huang LQ, Rowan MJ, Anwyl R. Role of protein kinases A and C in the induction of mGluR-dependent long-term depression in the medial perforant path of the rat dentate gyrus in vitro. Neurosci Lett 1999; 274: 71–74.
146. Hayashi Y, Shi SH, Esteban JA, et al. AMPA receptors into synapses by LTP and CaMKII: requirement for GluR1 and PDZ domain interaction. Science 2000; 287: 2262–2267.
147. Braun AP, Schulman H. The multifunctional calcium/calmodulin-dependent protein kinase: from form to function. Annu Rev Physiol 1995; 57: 417–445.
147*a*. Gardoni F, Caputi A, Cimino M, et al. Calcium/calmodulin-dependent protein kinase II is associated with NR2A/B subunits of NMDA receptor in postsynaptic densities. J Neurochem 1998; 71: 1733–1741.
148. Kennedy MB. The postsynaptic density at glutamatergic synapses. Trends Neurosci 1997; 20: 264–268.
149. Strack S, McNeill RB, Colbran RJ. Mechanism and regulation of calcium/calmodulin-dependent protein kinase II targeting to the NR2B subunit of the *N*-methyl-D-aspartate receptor. J Biol Chem 2000; 275: 23798–23806.
150. Strack S, Colbran RJ. Autophosphorylation-dependent targeting of calcium/calmodulin-dependent protein kinase II by the NR2B subunit of the *N*-methyl-D-aspartate receptor. J Biol Chem 1998; 273: 20689–20692.
151. Leonard AS, Lim IA, Hemsworth DE, et al. Calcium/calmodulin-dependent protein kinase II is associated with the *N*-methyl-D-aspartate receptor. Proc Natl Acad Sci USA 1999; 96: 3239–3244.
152. Shen K, Meyer T. Dynamic control of CaMKII translocation and localization in hippocampal neurons by NMDA receptor stimulation. Science 1999; 284: 162–166.
153. Wang JH, Kelly PT. Postsynaptic injection of CA^{2+}/CaM induces synaptic potentiation requiring CaMKII and PKC activity. Neuron 1995; 15: 443–452.
154. Hvalby O, Hemmings HC Jr, Paulsen O, Raastad M, Storm JF, Andersen P, et al. Specificity of protein kinase inhibitor peptides and induction of long-term potentiation. Proc Natl Acad Sci USA 1994; 91: 4761–4765.
155. Schnabel R, Palmer MJ, Kilpatrick IC, et al. CaMKII inhibitor, KN-62, facilitates DHPG-induced LTD in the CA1 region of the hippocampus. Neuropharmacology 1999; 38: 605–608.

155*a*. Otmakhov N, Griffith LC, Lisman JE. Postsynaptic inhibitors of calcium/calmodulin-dependent protein kinase type II block induction but not maintenance of pairing-induced long-term potentiation. J Neurosci 1997; 17: 5357–5365.
156. Giese KP, Fedorov NB, Filipkowski RK, et al. Autophosphorylation at Thr286 of the alpha calcium-calmodulin kinase II in LTP and learning. Science 1998; 279: 870–873.
157. Tan SE, Wenthold RJ, Soderling TR. Phosphorylation of AMPA-type glutamate receptors by calcium/calmodulin-dependent protein kinase II and protein kinase C in cultured hippocampal neurons. J Neurosci 1994; 14: 1123–1129.
158. Derkach V, Barria A, Soderling TR. Ca^{2+}/calmodulin-kinase II enhances channel conductance of alpha-amino-3-hydroxy-5-methyl-4-isoxazolepropionate type glutamate receptors. Proc Natl Acad Sci USA 1999; 96: 3269–3274.
159. McGlade-McCulloh E, Yamamoto H, Tan SE, et al. Phosphorylation and regulation of glutamate receptors by calcium/calmodulin-dependent protein kinase II. Nature 1993; 362: 640–642.
160. Blitzer RD, Wong T, Nouranifar R, et al. Postsynaptic cAMP pathway gates early LTP in hippocampal CA1 region. Neuron 1995; 15: 1403–1414.
161. Blitzer RD, Connor JH, Brown GP, et al. Grating of CaMKII by cAMP-regulated protein phosphatase activity during LTP. Science 1998; 280: 1940–1942.
162. Elgersma Y, Silva AJ. Molecular mechanisms of synaptic plasticity and memory. Curr Opin Neurobiol 1999; 9: 209–213.
163. Huang YY, Li XC, Kandel ER. cAMP contributes to mossy fiber LTP by initiating both a covalently mediated early phase and macromolecular synthesis-dependent late phase. Cell 1994; 79: 69–79.
164. Storm DR, Hansel C, Hacker B, et al. Impaired cerebellar long-term potentiation in type I adenylyl cyclase mutant mice. Neuron 1998; 20: 1199–1210.
165. Swope SL, Moss SI, Raymond LA, et al. Regulation of ligand-gated ion channels by protein phosphorylation. Adv Second Messenger Phosphoprotein Res 1999; 33: 49–78.
166. Ramakers GM, Pasinelli P, Hens JJ, et al. Protein kinase C in synaptic plasticity: changes in the in situ phosphorylation state of identified pre- and postsynaptic substrates. Prog Neuropsychopharmacol Biol Psychiatry 1997; 21: 455–486.
167. Feng TP. The involvement of PKC and multifunctional CaM kinase II of the postsynaptic neuron in induction and maintenance of long-term potentiation. Prog Brain Res 195; 105: 55–63.
168. Chakravarthy B, Morley P, Whitfield J. Ca^{2+}-calmodulin and protein kinase Cs: a hypothetical synthesis of their conflicting convergences on shared substrate domains. Trends Neurosci 1999; 22: 12–16.
169. Tingley WG, Ehlers MD, Kameyama K, et al. Characterization of protein kinase A and protein kinase C phosphorylation of the *N*-methyl-D-aspartate

receptor NR1 subunit using phosphorylation site-specific antibodies. J Biol Chem 1997; 272: 5157–5166.
170. Hisatsune C, Umemori H, Inoue T, et al. Phosphorylation-dependent regulation of *N*-methyl-D-aspartate receptors by calmodulin. J Biol Chem 1997; 272: 20805–20810.
171. Verma AK, Paszty K, Filoteo AG, et al. A Protein kinase C phosphorylates plasma membrane Ca^{2+} pump isoform 4a at its calmodulin binding domain. J Biol Chem 1999; 274: 527–531.
172. Penniston JT, Enyedi A. Modulation of the plasma membrane Ca^{2+} pump. J Membr Biol 1998; 165: 101–109.
173. Enyedi A, Elwess NL, Filoteo AG, et al. Protein kinase C phosphorylates the "a" forms of plasma membrane Ca^{2+} pump isoforms 2 and 3 and prevents binding of calmodulin. J Biol Chem 1997; 272: 27525–27528.
174. Hofmann F, Anagli J, Carafoli E, Vorherr T, et al. Phosphorylation of the calmodulin binding domain of the plasma membrane Ca^{2+} pump by protein kinase C reduces its interaction with calmodulin and with its pump receptor site. J Biol Chem 1994; 269: 24298–24303.
175. Gerendasy DD, Sutcliffe JG. RC3/neurogranin, a postsynaptic calpacitin for setting the response threshold to calcium influxes. Mol Neurobiol 1997; 15: 131–163.
176. Gerendasy DD, Herron SR, Jennings PA, Sutcliffe JG, et al. Calmodulin stabilizes an amphiphilic alpha-helix within RC3/neurogranin and GAP-43/neuromodulin only when Ca^{2+} is absent. J Biol Chem 1995; 270: 6741–6750.
177. Ramakers GM, De Graan PN, Urban IJ, et al. Temporal differences in the phosphorylation state of pre- and postsynaptic protein kinase C substrates B-50/GAP-43 and neurogranin during long-term potentiation. J Biol Chem 1995; 270: 13892–13898.
178. Chen SJ, Sweatt JD, Klann E. Enhanced phosphorylation of the postsynaptic protein kinase C substrate RC3/neurogranin during long-term potentiation. Brain Res 1997; 749: 181–187.
179. Menegoz M, Lau LF, Herve D, et al. Tyrosine phosphorylation of NMDA receptor in rat striatum: effects of 6-OH-dopamine lesions. Neuroreport 1995; 7: 125–128.
180. Lau LF, Huganir RL. Differential tyrosine phosphorylation of *N*-methyl-D-aspartate receptor subunits. J Biol Chem 1995; 270: 20036–20041.
181. Zheng F, Gingrich MB, Traynelis SF, et al. Tyrosine kinase potentiates NMDA receptor currents by reducing tonic zinc inhibition. Nat Neurosci 1998; 1: 185–191.
182. Rostas JA, Brent VA, Voss K, et al. Enhanced tyrosine phosphorylation of the 2B subunit of the *N*-methyl-D-aspartate receptor in long-term potentiation. Proc Natl Acad Sci USA 1996; 93: 10452–10456.
183. Rosenblum K, Dudai Y, Richter-Levin G. Long-term potentiation increases tyrosine phosphorylation of the *N*-methyl-D-aspartate receptor subunit 2B in rat dentate gyrus in vivo. Proc Natl Acad Sci USA 1996; 93: 10457–10460.

184. Lu YM, Roder JC, Davidow J, et al. Src activation in the induction of long-term potentiation in CA1 hippocampal neurons. Science 1998; 279: 1363–1367.
185. Huang CC, Hsu KS. Protein tyrosine kinase is required for the induction of long-term potentiation in the rat hippocampus. J Physiol (Lond) 1999; 520 (Pt. 3): 783–796.
186. Suzuki T, Okumura-Noji K. NMDA receptor subunits epsilon 1 (NR2A) and epsilon 2 (NR2B) are substrates for Fyn in the postsynaptic density fraction isolated from the rat brain. Biochem Biophys Res Commun 1995; 216: 582–588.
187. Lu YF, Kojima N, Tomizawa K, et al. Enhanced synaptic transmission and reduced threshold for LTP induction in fyn-transgenic mice. Eur J Neurosci 1999; 11: 75–82.
188. Grant SG, O'Dell TJ, Karl KA, et al. Impaired long-term potentiation, spatial learning, and hippocampal development in fyn mutant mice. Science 1992; 258: 1903–1910.
189. Perkinton MS, Sihra TS, Williams RJ. Ca^{2+}-permeable AMPA receptors induce phosphorylation of cAMP response element-binding protein through a phosphatidylinositol 3-kinase-dependent stimulation of the mitogen-activated protein kinase signaling cascade in neurons. J Neurosci 1999; 19: 5861–5874.
190. Curtis J, Finkbeiner S. Sending signals from the synapse to the nucleus: possible roles for CaMK, Ras/ERK, and SAPK pathways in the regulation of synaptic plasticity and neuronal growth. J Neurosci Res 1999; 58: 88–95.
191. Schulz S, Siemer H, Krug M, et al. Direct evidence for biphasic cAMP responsive element-binding protein phosphorylation during long-term potentiation in the rat dentate gyrus in vivo. J Neurosci 1999; 19: 5683–5692.
192. Silva AJ, Kogan JH, Frankland PW, et al. CREB and memory. Annu Rev Neurosci 1998; 21: 127–148.
193. Augustine GJ, Neher E. Neuronal Ca^{2+} signalling takes the local route. Curr Opin Neurobiol 1992; 2: 302–307.
194. Zucker RS. Calcium and transmitter release. J Physiol Paris 1993; 87: 25–36.
195. Burgoyne RD, Morgan A. Calcium sensors in regulated exocytosis. Cell Calcium 1998; 24: 367–376.
196. Fossier P, Baux G, Tauc L. Role of different types of Ca^{2+} channels and a reticulum-like Ca^{2+} pump in neurotransmitter release. J Physiol Paris 1993; 87: 3–14.
197. Huang CC, Hsu KS, Gean PW. Isoproterenol potentiates synaptic transmission primarily by enhancing presynaptic calcium influx via P- and (or) Q-type calcium channels in the rat amygdala. J Neurosci 1996; 16: 1026–1033.
198. Catterall WA. Interactions of presynaptic Ca^{2+} channels and snare proteins in neurotransmitter release. Ann N Y Acad Sci 1999; 868: 144–159.
199. Catterall WA. Structure and function of neuronal Ca^{2+} channels and their role in neurotransmitter release. Cell Calcium 1998; 24: 307–323.
200. Tsien RW, Lipscombe D, Madison DV, et al. Multiple types of neuronal calcium channels and their selective modulation. Trends Neurosci 1988; 11: 431–438.

201. Rettig J, Sheng ZH, Kim DK, et al. Isoform-specific interaction of the alpha1A subunits of brain Ca^{2+} channels with the presynaptic proteins syntaxin and SNAP-25. Proc Natl Acad Sci USA 1996; 93: 7363–7368.
202. Kim DK, Catterall WA. Ca^{2+}-dependent and -independent interactions of the isoforms of the alpha1A subunit of brain Ca^{2+} channels with presynaptic SNARE proteins. Proc Natl Acad Sci USA 1997; 94: 14782–14786.
203. Langley K, Grant NJ. Are exocytosis mechanisms neurotransmitter specific? Neurochem Int 1997; 31: 739–757.
204. Zucker RS. Exocytosis: a molecular and physiological perspective. Neuron 1996; 17: 1049–1055.
205. Martin TF. Stages of regulated excocytosis. Trends Cell Biol 1997; 271–276.
206. Hartmann J. Calcium and exocytosis. In: Verkhratsky A, Toescu EC, editors. Integrative Aspects of Calcium Signalling. New York, Plenum Press, 1998: 199–229.
207. Zheng X, Bobich JA. A sequential view of neurotransmitter release. Brain Res Bull 1998; 47: 117–128.
208. Calakos N, Scheller RH. Synaptic vesicle biogenesis, docking, and fusion: a molecular description. Physiol Rev 1996; 76: 1–29.
209. Bajjalieh SM. Synaptic vesicle docking and fusion. Curr Opin Neurobiol 1999; 9: 321–328.
210. Turner KM, Burgoyne RD, Morgan A. Protein phosphorylation and the regulation of synaptic membrane traffic. Trends Neurosci 1999; 22: 459–464.
211. Hilfiker S, Pieribone VA, Czernik AJ, et al. Synapsins as regulators of neurotransmitter release. Philos Trans R Soc Lond B Biol Sci 1999; 354: 269–279.
212. Hosaka M, Hammer RE, Sudhof TC. A phospho-switch controls the dynamic association of synapsins with synaptic vesicles. Neuron 1999; 24: 377–387.
213. Nicol S, Rahman D, Chan KM, et al. Interaction of synapsin I with actin and SSVs: differential regulation by calmodulin. Biochem Soc Trans 1998; 26: S110.
214. Greengard P, Valtorta F, Czernik AJ, et al. Synaptic vesicle phosphoproteins and regulation of synaptic function. Science 1993; 259: 780–785.
215. Sudhof TC, Czernik AJ, Kao H-T, et al. Synapsins: mosaics of shared and individual domains in a family of synaptic vesicle phosphoproteins. Science 1989; 245: 1474–1480.
216. Rosahl TW, Soillane D, Missler M, et al. Essential functions of synapsins I and II in synaptic vesicle regulation. Nature 1995; 375: 488–493.
217. Hosaka M, Sudhof TC. Synapsin III, a novel synapsin with an unusual regulation by Ca^{2+}. J Biol Chem 1998; 273: 13371–13374.
218. Matsubara M, Kusubata M, Ishiguro K, et al. Site-specific phosphorylation of synapsin I by mitogen-activated protein kinase and Cdk5. J Biol Chem 1996; 271: 21108–21113.
219. Coffey ET, Sihra TS, Nicholls DG, et al. Phosphorylation of synapsin I and MARCKS in nerve terminals is mediated by Ca^{2+} entry via an Aga-GI sensitive Ca^{2+} channel which is coupled to gluatamate exocytosis. FEBS Lett 1994; 353: 264–268.

220. Czernik AJ, Pang DT, Greengard P. Amino acid sequences surrounding the cAMP- dependent and calcium/calmodulin-dependent phosphorylation sites in rat and bovine synapsin I. Proc Natl Acad Sci USA 1987; 84: 7518–7522.
221. Ceccaldi PE, Grohovaz F, Benfenati F, et al. Dephosphorylated synapsin I anchors synaptic vesicles to actin cytoskeleton: an analysis by videomicroscopy. J Cell Biol 1995; 128: 905–912.
222. Torri-Tarelli F, Bossi M, Fesce R, et al. Synapsin I partially dissociates from synaptic vesicles during exocytosis induced by electrical stimulation. Neuron 1992; 9: 1143–1153.
223. Llinas R, Gruner JA, Sugimori M, et al. Regulation by synapsin I and Ca^{2+}-calmodulin- dependent protein kinase II of the transmitter release in squid giant synapse. J Physiol (Lond) 1991; 436: 257–282.
224. Llinas R, McGuinness TL, Leonard CS, et al. Intraterminal injection of synapsin I or calcium/calmodulin-dependent protein kinase II alters neurotransmitter release at the squid giant synapse. Proc Natl Acad Sci USA 1985; 82: 3035–3039.
225. Stevens CF, Sullivan JM. Regulation of the readily releasable vesicle pool by protein kinase C. Neuron 1998; 21: 885–893.
226. Haas A. NSF — fusion and beyond. Trends Cell Biol 1998; 8: 471–473.
227. Burgoyne RD, Morgan A. Analysis of regulated exocytosis in adrenal chromaffin cells: insights into NSF/SNAP/SNARE function. Bioessays 1998; 20: 328–335.
228. Hohl TM, Parlati F, Wimmer C, et al. Arrangement of subunits in 20 S particles consisting of NSF, SNAPs, and SNARE complexes. Mol Cell 1998; 2: 539–548.
229. Hay JC, Scheller RH. SNAREs and NSF in targeted membrane fusion. Curr Opin Cell Biol 1997; 9: 505–512.
230. Wilson DW, Whiteheart SW, Wiedmann M, et al. A multisubunit particle implicated in membrane fusion. J Cell Biol 1992; 117: 531–538.
231. Sollner T, Whiteheart SW, Brunner M, et al. SNAP receptors implicated in vesicle targeting and fusion. Nature 1993; 362: 318–324.
231*a*. Sollner T, Bennett MK, Whiteheart SW, et al. A protein assembly-disassembly pathway in vitro that may correspond to sequential steps of synaptic vesicle docking, activation, and fusion. Cell 1993; 75: 409–418.
232. Davis AF, Bai J, Fasshauer D, et al. ER Kinetics of synaptotagmin responses to Ca^{2+} and assembly with the core SNARE complex onto membranes. Neuron 1999; 24: 363–376.
233. Walch-Solimena C, Blasi J, Edelmann L, et al. The t-SNAREs syntaxin 1 and SNAP-25 are present on organelles that participate in synaptic vesicle recycling. J Cell Biol 1995; 128: 637–645.
234. Trimble WS, Cowan DM, Scheller RH. VAMP-1: a synaptic vesicle-associated integral membrane protein. Proc Natl Acad Sci USA 1988; 85: 4538–4542.
235. Baumert M, Maycox PR, Navone F, et al. Synaptobrevin: an integral membrane protein of 18,000 daltons present in small synaptic vesicles of rat brain. EMBO J 1989; 8: 379–384.

236. Bennett MK, Calakos N, Scheller RH. Syntaxin: a synaptic protein implicated in docking of synaptic vesicles at presynaptic active zones. Science 1992; 257: 255–259.
237. Oyler GA, Polli JW, Higgins GA, et al. Distribution and expression of SNAP-25 immunoreactivity in rat brain, rat PC-12 cells and human SMS-KCNR neuroblastoma cells. Brain Res Dev Brain Res 1992; 65: 133–146.
238. Oyler GA, Higgins GA, Hart RA, et al. The identification of a novel synaptosomal-associated protein, SNAP-25, differentially expressed by neuronal subpopulations. J Cell Biol 1989; 109: 3039–3052.
238*a*. Brunger A. Structural insights into the molecular mechanism of calcium-dependent vesicle-membrane fusion. Curr Opin Struct Biol 2001; 11: 63–173.
239. Rothman JE. Mechanisms of intracellular protein transport. Nature 1994; 372: 55–63.
240. Rothman JE, Orci L. Molecular dissection of the secretory pathway. Nature 1992; 355: 409–415.
241. Pevsner J, Hsu SC, Braun JE, et al. Specificity and regulation of a synaptic vesicle docking complex. Neuron 1994; 13: 353–361.
242. Chapman ER, An S, Barton N, et al. SNAP-25, a t-SNARE which binds to both syntaxin and synaptobrevin via domains that may form coiled coils. J Biol Chem 1994; 269: 27427–27432.
243. Mehta PP, Battenberg E, Wilson MC. SNAP-25 and synaptotagmin involvement in the final Ca^{2+}-dependent triggering of neurotransmitter exocytosis. Proc Natl Acad Sci USA 1996; 93: 10471–10476.
244. Haynes LP, Barnard RJ, Morgan A, et al. Stimulation of NSF ATPase activity during t-SNARE priming. FEBS Lett 1998; 436: 1–5.
245. Barnard RJ, Morgan A, Burgoyne RD. Stimulation of NSF ATPase activity by alpha-SNAP is required for SNARE complex disassembly and exocytosis. J Cell Biol 1997; 139: 875–883.
246. Burgoyne RD, Morgan A. Calcium sensors in regulated exocytosis. Cell Calcium 1998; 24: 367–376.
247. Mochida S, Yokoyama CT, Kim DK, et al. Evidence for a voltage-dependent enhancement of neurotransmitter release mediated via the synaptic protein interaction site of N-type Ca^{2+} channels. Proc Natl Acad Sci USA 1998; 95: 14523–14528.
248. Mochida S, Sheng ZH, Baker C, et al. Inhibition of neurotransmission by peptides containing the synaptic protein interaction site of N-type Ca^{2+} channels. Neuron 1996; 17: 781–788.
249. Bennett MR. The concept of a calcium sensor in transmitter release. Prog Neurobiol 1999; 59: 243–277.
250. Schiavo G, Osborne SL, Sgouros JG. Synaptotagmins: more isoforms than functions? Biochem Biophys Res Commun 1998; 248: 1–8.
251. Mikoshiba K, Fukuda M, Ibata K, et al. A role of synaptotagmin, a Ca^{2+} and inositol polyphosphate binding protein, in neurotransmitter release and neurite outgrowth. Chem Phys Lipids 1999; 98: 59–67.

252. Li C, Ullrich B, Zhang JZ, et al. Ca^{2+}-dependent and -independent activities of neural and non-neural synaptotagmins. Nature 1995; 375: 594–599.
253. Shao X, Li C, Fernandez I, et al. Synaptotagmin–syntaxin interaction: the C2 domain as a Ca^{2+}-dependent electrostatic switch. Neuron 1997; 18: 133–142.
254. Mochida S, Fukuda M, Niinobe M, et al. Roles of synaptotagmin C2 domains in neurotransmitter secretion and inositol high-polyphosphate binding at mammalian cholinergic synapses. Neuroscience 1997; 77: 937–943.
255. Chapman ER, Davis AF. Direct interaction of a Ca^{2+}-binding loop of synaptotagmin with lipid bilayers. J Biol Chem 1998; 273: 13995–14001.
255*a*. Fernández-Chacón R, Königstorfer A, Gerber SH, et al. Synaptotagmin I functions as a calcium regulator of release probability. Nature 2001; 410: 41–49.
256. Chapman ER, Jahn R. Calcium-dependent interaction of the cytoplasmic region of synaptotagmin with membranes. Autonomous function of a single C2-homologous domain. J Biol Chem 1994; 269: 5735–5741.
257. Schivell AE, Batchelor RH, Bajjalieh SM. Isoform-specific, calcium-regulated interaction of the synaptic vesicle proteins SV2 and synaptotagmin. J Biol Chem 1996; 271: 27770–27775.
258. Pyle RA, Schivell AE, Hidaka H, Bajjalieh SM, et al. Phosphorylation of synaptic vesicle protein 2 (SV2) modulates binding to synaptotagmin. J Biol Chem 2000;
259. Sugita S, Sudhof TC. Specificity of Ca^{2+}-dependent protein interactions mediated by the C2A domains of synaptotagmins. Biochemistry 2000; 39: 2940–2949.
260. Sugita S, Hata Y, Sudhof TC. Distinct Ca^{2+}-dependent properties of the first and second C2-domains of synaptotagmin I. J Biol Chem 1996; 271: 1262–1265.
261. Osborne SL, Herreros J, Bastiaens PI, et al. Calcium-dependent oligomerization of synaptotagmins I and II. Synaptotagmins I and II are localized on the same synaptic vesicle and heterodimerize in the presence of calcium. J Biol Chem 1999; 274: 59–66.
262. Sattler R, Tymianski M. Calcium and cellular death. In: Verkhratsky A, Toescu EC, editors. Integrative Aspects of Calcium Signalling. New York, Plenum Press, 1998: 333–358.
263. Sattler R, Tymianski M. Molecular mechanisms of calcium-dependent excitotoxicity. J Mol Med 2000; 78: 3–13.
264. Trump BF, Berezesky IK. Calcium-mediated cell injury and cell death. FASEB J 1995; 9: 219–228.
265. Schlaepfer WW, Bunge RP. Effects of calcium ion concentration on the degeneration of amputated axons in tissue culture. J Cell Biol 1973; 59: 456–470.
266. Zipfel GJ, Lee JM, Choi DW. Reducing calcium overload in the ischemic brain. N Engl J Med 1999; 341: 1543–1544.
267. Choi DW. Calcium: still center-stage in hypoxic-ischemic neuronal death. Trends Neurosci 1995; 18: 58–60.
268. Choi DW. Calcium and excitotoxic neuronal injury. Ann N Y Acad Sci 1994; 747: 162–171.

269. Usachev YM, Thayer SA. Controlling the urge for a Ca^{2+} surge: all-or-none Ca^{2+} release in neurons. Bioessays 1999; 21: 743–750.
270. Nicholls DG, Budd SL, Castilho RF, et al. Glutamate excitotoxicity and neuronal energy metabolism. Ann N Y Acad Sci 1999; 893: 1–12.
271. Mattson MP, Pedersen WA, Duan W, et al. Cellular and molecular mechanisms underlying perturbed energy metabolism and neuronal degeneration in Alzheimer's and Parkinson's diseases. Ann N Y Acad Sci 1999; 893: 154–175.
272. Nicholls DG, Budd SL. Mitochondria and neuronal survival. Physiol Rev 2000; 80: 315–360.
273. Budd SL, Nicholls DG. Mitochondria in the life and death of neurons. Essays Biochem 1998; 33: 43–52.
274. Montal M. Mitochondria, glutamate neurotoxicity and the death cascade. Biochim Biophys Acta 1998; 1366: 113–126.
275. Toescu EC. Apoptosis and cell death in neuronal cells: where does Ca^{2+} fit in? Cell Calcium 1998; 24: 387-403.
276. Jayaraman T, Marks AR. Calcium-dependent signalling in apoptosis. In Verkhratsky A, Toescu EC, editors. Integrative Aspects of Calcium Signalling. New York, Plenum Press, 1998: 333–358.
277. Mattson MP, Rydel RE, Lieberburg I, et al. Altered calcium signaling and neuronal injury: stroke and Alzheimer's disease as examples. Ann N Y Acad Sci 1993; 679: 1–21.
278. Choi DW. Calcium-mediated neurotoxicity: relationship to specific channel types and role in ischemic damage. Trends Neurosci 1988; 11: 465–469.
279. Price MT, Olney JW, Samson L, et al. Calcium influx accompanies but does not cause excitotoxin-induced neuronal necrosis in retina. Brain Res Bull 1985; 14: 369–376.
280. Collins F, Schmidt MF, Guthrie PB, et al. Sustained increase in intracell calcium promotes neuronal survival. J Neurosci 1991; 11: 2582–2587.
281. Franklin JL, Johnson EM Jr. Suppression of programmed neuronal death by sustained elevation of cytoplasmic calcium. Trends Neurosci 1992; 15: 501–508.
282. Johnson EM Jr, Koike T, Franklin J. A "calcium set-point hypothesis" of neuronal dependence on neurotrophic factor. Exp Neurol 1999; 115: 163–166.
283. Berridge MJ, Bootman MD, Lipp P. Calcium — a life and death signal. Nature 1998; 395: 645–648.
284. Verkhratsky AJ, Petersen OH. Neuronal calcium stores. Cell Calcium 1998; 24: 333–343.
285. Berridge MJ. The AM and FM of calcium signalling. Nature 1997; 386: 759–760.
286. Putney JW Jr. Calcium signaling: up, down, up, down...what's the point? Science 1998; 279: 191–192.
287. Augustine GJ, Neher E. Neuronal Ca^{2+} signalling takes the local route. Curr Opin Neurobiol 1992; 2: 302–307.

288. Chard PS, Bleakman D, Christakos S, et al. Calcium buffering properties of calbindin D28k and parvalbumin in rat sensory neurones. J Physiol (Lond) 1993; 472: 341–357.
289. Nowycky MC, Pinter MJ. Time courses of calcium and calcium-bound buffers following calcium influx in a model cell. Biophys J 1993; 64: 77–91.
290. Kasai H, Petersen OH. Spatial dynamics of second messengers: IP3 and cAMP as long-range and associative messengers. Trends Neurosci 1994; 17: 95–101.
291. Sala F, Hernandez-Cruz A. Calcium diffusion modeling in a spherical neuron. Relevance of buffering properties. Biophys J 1990; 57: 313–324.
291*a*. Baimbridge KG, Celio MR, Rogers JH. Calcium-binding proteins in the nervous system. Trends Neurosci 1992; 15: 303–308.
292. Thayer SA, Miller RJ. Regulation of the intracellular free calcium concentration in single rat dorsal root ganglion neurones in vitro. J Physiol (Lond) 1990; 425: 85–115.
293. Mattson MP, Rychlik B, Chu C, et al. Evidence for calcium-reducing and excito-protective roles for the calcium-binding protein calbindin-D28k in cultured hippocampal neurons. Neuron 1991; 6: 41–51.
294. Freund TF, Buzsaki G, Leon A, et al. Relationship of neuronal vulnerability and calcium binding protein immunoreactivity in ischemia. Exp Brain Res 1990; 83: 55–66.
295. Weiss JH, Koh J, Baimbridge KG, et al. Cortical neurons containing somatostatin- or parvalbumin-like immunoreactivity are atypically vulnerable to excitotoxic injury in vitro. Neurology 1990; 40: 1288–1292.
296. Bennett MR, Huxlin KR. Neuronal cell death in the mammalian nervous system: the calmortin hypothesis. Gen Pharmacol 1996; 27: 407–419.
297. Siesjo BK, Hu B, Kristian T. Is the cell death pathway triggered by the mitochondrion or the endoplasmic reticulum? J Cereb Blood Flow Metab 1999; 19: 19–26.
298. Paschen W, Doutheil J. Disturbances of the functioning of endoplasmic reticulum: a key mechanism underlying neuronal cell injury? J Cereb Blood Flow Metab 1999; 19: 1–18.
299. Misquitta CM, Mack DP, Grover AK. Sarco/endoplasmic reticulum Ca^{2+} (SERCA)-pumps: link to heart beats and calcium waves. Cell Calcium 1999; 25: 277–290.
300. Baba-Aissa F, Raeymaekers L, Wuytack F, et al. Distribution and isoform diversity of the organellar Ca^{2+} pumps in the brain. Mol Chem Neuropathol 1998; 33: 199–208.
301. Nori A, Villa A, Podini P, et al. P Intracellular Ca^{2+} stores of rat cerebellum: heterogeneity within and distinction from endoplasmic reticulum. Biochem J 1993; 291: 199–204.
302. James P, Inui M, Tada M, et al. Nature and site of phospholamban regulation of the Ca^{2+} pump of sarcoplasmic reticulum. Nature 1989; 342: 90–92.
303. Xu A, Narayanan N. Ca^{2+}/calmodulin-dependent phosphorylation of the Ca^{2+}-ATPase, uncoupled from phospholamban, stimulates Ca^{2+}-pumping in native

cardiac sarcoplasmic reticulum. Biochem Biophys Res Commun 1999; 258: 66–72.
304. Xu A, Hawkins C, Narayanan N. Phosphorylation and activation of the Ca^{2+}-pumping ATPase of cardiac sarcoplasmic reticulum by Ca^{2+}/calmodulin-dependent protein kinase. J Biol Chem 1993; 268: 8394–8397.
305. Narayanan N, Xu A. Phosphorylation and regulation of the Ca^{2+}-pumping ATPase in cardiac sarcoplasmic reticulum by calcium/calmodulin-dependent protein kinase. Basic Res Cardiol 1997; 92 Suppl 1: 25–35.
306. Shmigol A, Kostyuk P, Verkhratsky A. Role of caffeine-sensitive Ca^{2+} stores in Ca^{2+} signal termination in adult mouse DRG neurones. Neuroreport 1994; 5: 2073–2076.
307. Markram H, Helm PJ, Sakmann B. Dendritic calcium transients evoked by single back-propagating action potentials in rat neocortical pyramidal neurons. J Physiol (Lond) 1995; 485 (Pt. 1): 1–20.
308. Wei H, Wei W, Bredesen DE, et al. Bcl-2 protects against apoptosis in neuronal cell line caused by thapsigargin-induced depletion of intracellular calcium stores. J Neurochem 1998; 70: 2305–2314.
309. Nath R, Raser KJ, Hajimohammadreza I, et al. Thapsigargin induces apoptosis in SH-SY5Y neuroblastoma cells and cerebrocortical cultures. Biochem Mol Biol Int 1997; 43: 197–205.
310. Takadera T, Ohyashiki T. Apoptotic cell death and CPP32-like activation induced by thapsigargin and their prevention by nerve growth factor in PC12 cells. Biochim Biophys Acta 1998; 1401: 63–71.
311. Silverstein FS, Nelson C. The microsomal calcium-ATPase inhibitor thapsigargin is a neurotoxin in perinatal rodent brain. Neurosci Lett 1992; 145: 157–160.
312. Berg M, Bruhn T, Frandsen A, et al. Kainic acid-induced seizures and brain damage in the rat: role of calcium homeostasis. J Neurosci Res 1995; 40: 641–646.
313. Zhang L, Andou Y, Masuda S, et al. Dantrolene protects against ischemic, delayed neuronal death in gerbil brain. Neurosci Lett 1993; 158: 105–108.
314. Bruno V, Battaglia G, Kingston A, et al. Neuroprotective activity of the potent and selective mGlu1a metabotropic glutamate receptor antagonist, (+)-2-methyl-4 carboxyphenylglycine (LY367385): comparison with LY357366, a broader spectrum antagonist with equal affinity for mGlu1a and mGlu5 receptors. Neuropharmacology 1999; 38: 199–207.
315. Bruno V, Copani A, Knopfel T, et al. Activation of metabotropic glutamate receptors coupled to inositol phospholipid hydrolysis amplifies NMDA-induced neuronal degeneration in cultured cortical cells. Neuropharmacology 1995; 34: 1089–1098.
316. McDonald JW, Fix AS, Tizzano JP, et al. Seizures and brain injury in neonatal rats induced by 1*S*,3*R*-ACPD, a metabotropic glutamate receptor agonist. J Neurosci 1993; 13: 4445–4455.
317. Camon L, Vives P, de Vera N, Martinez E, et al. Seizures and neuronal damage induced in the rat by activation of group I metabotropic glutamate receptors

with their selective agonist 3,5-dihydroxyphenylglycine. J Neurosci Res 1998; 51: 339–348.
318. Murphy AN. Ca^{2+}-mediated mitochondrial dysfunction and the protective effects of Bcl-2. Ann N Y Acad Sci 1999; 893: 19–32.
319. Greene JG. Mitochondrial function and NMDA receptor activation: mechanisms of secondary excitotoxicity. Funct Neurol 1999; 14: 171–184.
320. Blass JP. Mitochondria, neurodegenerative diseases, and selective neuronal vulnerability. Ann N Y Acad Sci 1999; 893: 434–439.
321. Nicholls DG, Budd SL, Castilho RF, et al. Glutamate excitotoxicity and neuronal energy metabolism. Ann N Y Acad Sci 1999; 893: 1–12.
322. Nicholls DG, Budd SL. Mitochondria and neuronal survival. Physiol Rev 2000; 80: 315–360.
323. Nicholls DG, Budd SL. Mitochondria and neuronal glutamate excitotoxicity. Biochim Biophys Acta 1998; 1366: 97–112.
324. Gunter TE, Gunter KK, Sheu SS, et al. Mitochondrial calcium transport: physiological and pathological relevance. Am J Physiol 1994; 267: C313–C339.
325. Nicholls DG, Ward MW. Mitochondrial membrane potential and neuronal glutamate excitotoxicity: mortality and millivolts. Trends Neurosci 2000; 23: 166–174.
326. Simpson PB, Russell JT. Role of mitochondrial Ca^{2+} regulation in neuronal and glial cell signalling. Brain Res Brain Res Rev 1998; 26: 72–81.
326*a*. Sambongi Y, Ueda T, Wada Y, et al. A biological molecular motor, proton-translocating ATP synthase: multidisciplinary approach for a unique membrane enzyme. J Bioenerg Biomembranes 2000; 32: 441–447.
326*b*. Pedersen PL, Ko YH, Hong S. ATP synthases in the year 2000: defining the different levels of mechanism and getting a grip on each. J Bioenerg Biomembranes 2000; 32: 423–432.
327. Blaustein MP, Lederer WJ. Sodium/calcium exchange: its physiological implications Physiol Rev 1999; 79: 763–854.
328. Carafoli E, Brini M. Calcium pumps: structural basis for and mechanism of calcium transmembrane transport. Curr Opin Chem Biol 2000; 4: 152–161.
329. Garcia ML, Strehler EE. Plasma membrane calcium ATPases as critical regulators of calcium homeostasis during neuronal cell function. Front Biosci 1999; 4: D869–D882.
330. Verma AK, Paszty K, Filoteo AG, et al. Protein kinase C phosphorylates plasma membrane Ca^{2+} pump isoform 4a at its calmodulin binding domain. J Biol Chem 1999; 274: 527–531.
331. Zylinska L, Guerini D, Gromadzinska E, et al. Protein kinases A and C phosphorylate purified Ca^{2+}-ATPase from rat cortex, cerebellum and hippocampus. Biochim Biophys Acta 1998; 1448: 99–108.
332. Nicholls D, Attwell D. The release and uptake of excitatory amino acids. Trends Pharmacol Sci 1990; 11: 462–468.
333. He Y, Janssen WG, Rothstein JD, et al. Differential synaptic localization of the glutamate transporter EAAC1 and glutamate receptor subunit GluR2 in the rat hippocampus. J Comp Neurol 2000; 418: 255–269.

334. Gundersen V, Danbolt NC, Ottersen OP, et al. Demonstration of glutamate/aspartate uptake activity in nerve endings by use of antibodies recognizing exogenous D-aspartate. Neuroscience 1993; 57: 97–111.
335. Sattler R, Charlton MP, Hafner M, et al. Distinct influx pathways, not calcium load, determine neuronal vulnerability to calcium neurotoxicity. J Neurochem 1998; 71: 2349–2364.
336. Michaelis EK. Molecular biology of glutamate receptors in the central nervous system and their role in excitotoxicity, oxidative stress and aging. Prog Neurobiol 1998; 54: 369–415.
337. Ozawa S, Kamiya H, Tsuzuki K. Glutamate receptors in the mammalian central nervous system. Prog Neurobiol 1998; 54: 581–618.
338. Pellegrini-Giampietro DE, Gorter JA, Bennett MV, et al. The GluR2 (GluR-B) hypothesis: Ca^{2+}-permeable AMPA receptors in neurological disorders. Trends Neurosci 1997; 20: 464–470.
339. Feldmeyer D, Kask K, Brusa R, et al. Neurological dysfunctions in mice expressing different levels of the Q/R site-unedited AMPAR subunit GluR-B. Nat Neurosci 1999; 2: 57–64.
340. Yamaguchi K, Yamaguchi F, Miyamoto O, et al. The reversible change of GluR2 RNA editing in gerbil hippocampus in course of ischemic tolerance. J Cereb Blood Flow Metab 1999; 19: 370–375.
341. Rump A, Sommer C, Gass P, et al. Editing of GluR2 RNA in the gerbil hippocampus after global cerebral ischemia. J Cereb Blood Flow Metab 1996; 16: 1362–1365.
342. Bennett MV, Pellegrini-Giampietro DE, Gorter JA, et al. The GluR2 hypothesis: Ca^{++}-permeable AMPA receptors in delayed neurodegeneration. Cold Spring Harb Symp Quant Biol 1996; 61: 373–384.
343. Coyle JT, Puttfarcken P. Oxidative stress, glutamate, and neurodegenerative disorders. Science 1993; 262: 689–695.
344. Calabresi P, De Murtas M, Pisani A, et al. Vulnerability of medium spiny striatal neurons to glutamate: role of Na^+/K^+ ATPase. Eur J Neurosci 1995; 7: 1674–1683.
345. Beal MF. Mitochondrial dysfunction in neurodegenerative diseases. Biochim Biophys Acta 1998; 1366: 211–223.
346. Beal MF. Mitochondria, free radicals, and neurodegeneration. Curr Opin Neuroiol 1996; 6: 661–666.
347. Simonian NA, Coyle JT. Oxidative stress in neurodegenerative diseases. Annu Rev Pharmacol Toxicol 1996; 36: 83–106.
348. Lafon-Cazal M, Pietri S, Culcasi M, et al. NMDA-dependent superoxide production and neurotoxicity. Nature 1993; 364: 535–537.
349. Dykens JA. Isolated cerebral and cerebellar mitochondria produce free radicals when exposed to elevated Ca^{2+} and Na^+: implications for neurodegeneration. J Neurochem 1994; 63: 584–591.
350. Schmidley JW. Free radicals in central nervous system ischemia. Stroke 1990; 21: 1086–1090.

351. Montal M. Mitochondria, glutamate neurotoxicity and the death cascade. Biochim Biophys Acta 1998; 1366: 113–126.
352. Bonventre JV. Roles of phospholipases A2 in brain cell and tissue injury associated with ischemia and excitotoxicity. J Lipid Mediat Cell Signal 1997; 171: 71–79.
353. Bazan NG, Rodriguez de Turco EB, et al. Mediators of injury in neurotrauma: intracellular signal transduction and gene expression. J Neurotrauma 1995; 12: 791–814.
354. Farooqui AA, Litsky ML, Farooqui T, et al. Inhibitors of intracellular phospholipase A2 activity: their neurochemical effects and therapeutical importance for neurological disorders. Brain Res Bull 1999; 49: 139–153.
355. Farooqui AA, Horrocks LA. Excitotoxicity and neurological disorders: involvement of membrane phospholipids. Int Rev Neurobiol 1994; 36: 267–323.
356. Dumuis A, Sebben M, Haynes L, et al. NMDA receptors activate the arachidonic acid cascade system in striatal neurons. Nature 1988; 336: 68–70.
357. Weichel O, Hilgert M, Chatterjee SS, et al. Bilobalide, a constituent of Ginkgo biloba, inhibits NMDA-induced phospholipase A2 activation and phospholipid breakdown in rat hippocampus. Naunyn Schmiedebergs Arch Pharmacol 1999; 360: 609–615.
358. Cocco T, Di Paola M, Papa S, et al. Arachidonic acid interaction with the mitochondrial electron transport chain promotes reactive oxygen species generation. Free Radic Biol Med 1999; 27: 51–59.
359. Bonventre JV, Huang Z, Taheri MR, et al. Reduced fertility and postischaemic brain injury in mice deficient in cytosolic phospholipase A2. Nature 1997; 390: 622–625.
360. Buchner K, Adamec E, Beermann ML, et al. Isoform-specific translocation of protein kinase C following glutamate administration in primary hippocampal neurons. Brain Res Mol Brain Res 1999; 64: 222–235.
361. Vaccarino FM, Liljequist S, Tallman JF. Modulation of protein kinase C translocation by excitatory and inhibitory amino acids in primary cultures of neurons. J Neurochem 1991; 57: 391–396.
362. Lipton SA, Choi YB, Pan ZH, et al. A redox-based mechanism for the neuroprotective and neurodestructive effects of nitric oxide and related nitrosocompounds. Nature 1993; 364: 626–632.
363. Heales SJ, Bolanos JP, Stewart VC, et al. Nitric oxide, mitochondria and neurological disease. Biochim Biophys Acta 1999; 1410: 215–228.
364. Bredt DS Endogenous nitric oxide synthesis: biological functions and pathophysiology. Free Radic Res 1999; 31: 577–596.
365. Oh S. The generation of nitric oxide and its roles in neurotransmission and neurotoxicity. Keio J Med 1995; 44: 53–61.
366. Chalimoniuk M, Strosznajder J. NMDA receptor-dependent nitric oxide and cGMP synthesis in brain hemispheres and cerebellum during reperfusion after transient forebrain ischemia in gerbils: effect of 7-nitroindazole. J Neurosci Res 1998; 54: 681–690.

367. Kiedrowski L, Costa E, Wroblewski JT. Glutamate receptor agonists stimulate nitric oxide synthase in primary cultures of cerebellar granule cells. J Neurochem 1992; 58: 335–341.
368. Gunasekar PG, Kanthasamy AG, Borowitz JL, et al. NMDA receptor activation produces concurrent generation of nitric oxide and reactive oxygen species: implication for cell death. J Neurochem 1995; 65: 2016–2021.
369. Liu DM, Wu JN, Chiou AL, et al. NMDA induces NO release from primary cell cultures of human fetal cerebral cortex. Neurosci Lett 1997; 223: 145–148.
370. Lipton SA. Neuronal protection and destruction by NO. Cell Death Differ 1999; 6: 943–951.
371. Sattler R, Xiong Z, Lu WY, et al. Specific coupling of NMDA receptor activation to nitric oxide neurotoxicity by PSD-95 protein. Science 1999; 284: 1845–1848.
372. Kiedrowski L, Costa E, Wroblewski JT. Glutamate receptor agonists stimulate nitric oxide synthase in primary cultures of cerebellar granule cells. J Neurochem 1992; 58: 335–341.
373. Grunewald T, Beal MF. NOS knockouts and neuroprotection. Nat Med 1999; 5: 1354–1355.
374. Brown GC. Nitric oxide as a competitive inhibitor of oxygen consumption in the mitochondrial respiratory chain. Acta Physiol Scand 2000; 168: 667–674.
375. Brown GC. Nitric oxide and mitochondrial respiration. Biochim Biophys Acta 1999; 1411: 351–369.
376. Nicholls P, Sharpe M, Torres J, et al. Nitric oxide as an effector and a substrate for cytochrome c oxidase. Biochem Soc Trans 1998; 26: S323.
377. Almeida A, Heales SJR, Bolanos JP, et al. Glutamate neurotoxicity is associated with nitric oxide-mediated mitochondrial dysfunction and glutathione depletion. Brain Res 1998; 790: 209–216.
378. Cleeter MW, Cooper JM, Darley-Usmar VM, et al. Reversible inhibition of cytochrome c oxidase, the terminal enzyme of the mitochondrial respiratory chain, by nitric oxide. Implications for neurodegenerative diseases. FEBS Lett 1994; 345: 50–54.
379. MacMillan-Crow LA, Crow JP, Thompson JA. Peroxynitrite-mediated inactivation of manganese superoxide dismutase involves nitration and oxidation of critical tyrosine residues. Biochemistry 1998; 37: 1613–1622.
380. Ischiropoulos H, Zhu L, Chen J, et al. Peroxynitrite-mediated tyrosine nitration catalyzed by superoxide dismutase. Arch Biochem Biophys 1992; 298: 431–437.
381. Bolanos JP, Heales SJ, Land JM, Clark JB. Effect of peroxynitrite on the mitochondrial respiratory chain: differential susceptibility of neurones and astrocytes in primary culture. J Neurochem 1995; 64: 1965–1972.
382. Lizasoain I, Moro MA, Knowles RG, et al. Nitric oxide and peroxynitrite exert distinct effects on mitochondrial respiration which are differentially blocked by glutathione or glucose. Biochem J 1996; 314: 877–880.
383. Trotti D, Rossi D, Gjesdal O, et al. Peroxynitrite inhibits glutamate transporter subtypes. J Biol Chem 1996; 271: 5976–5979.

384. Kinosita K, Yasuda R, Noji H. F_1-ATPase: a highly efficient rotary ATP machine. Essays Biochem 2000; 35: 3–18.
385. Budd SL, Nicholls DG. Mitochondria, calcium regulation, and acute glutamate excitotoxicity in cultured cerebellar granule cells. J Neurochem 1996; 67: 2282–2291.
386. Isaev NK, Zorov DB, Stelmashook EV, et al. Neurotoxic glutamate treatment of cultured cerebellar granule cells induces Ca^{2+}-dependent collapse of mitochondrial membrane potential and ultrastructural alterations of mitochondria. FEBS Lett 1996; 392: 143–147.
387. Ankarcrona M, Dypbukt JM, Orrenius S, et al. Calcineurin and mitochondrial function in glutamate-induced neuronal cell death. FEBS Lett 1996; 394: 321–324.
388. Ankarcrona M, Dypbukt JM, Bonfoco E, et al. Glutamate-induced neuronal death: a succession of necrosis or apoptosis depending on mitochondrial function. Neuron 1995; 15: 961–973.
389. Schinder AF, Olson EC, Spitzer NC, et al. Mitochondrial dysfunction is a primary event in glutamate neurotoxicity. J Neurosci 1996; 16: 6125–6133.
390. Dubinsky JM, Rothman SM. Intracellular calcium concentrations during "chemical hypoxia" and excitotoxic neuronal injury. J Neurosci 1991; 11: 2545–2551.
391. Marey-Semper I, Gelman M, Levi-Strauss MA. Selective toxicity toward cultured mesencephalic dopaminergic neurons is induced by the synergistic effects of energetic metabolism impairment and NMDA receptor activation. J Neurosci 1995; 15: 5912–5918.
392. Saporito MS, Heikkila RE, Youngster SK, et al. Dopaminergic neurotoxicity of 1-methyl-4-phenylpyridinium analogs in cultured neurons: relationship to the dopamine uptake system and inhibition of mitochondrial respiration. J Pharmacol Exp Ther 1992; 260: 1400–1409.
393. Gao J, Yin DH, Yao Y, et al. Loss of conformational stability in calmodulin upon methionine oxidation. Biophys J 1998; 74: 1115–1134.
394. Zaidi A, Michaelis ML. Effects of reactive oxygen species on brain synaptic plasma membrane Ca^{2+}-ATPase. Free Radic Biol Med 1999; 27: 810–821.
395. Yao Y, Yin D, Jas GS, et al. Oxidative modification of a carboxyl-terminal vicinal methionine in calmodulin by hydrogen peroxide inhibits calmodulin-dependent activation of the plasma membrane Ca-ATPase. Biochemistry 1996; 35: 2767–2787.
396. Rohn TT, Hinds TR, Vincenzi FF. Inhibition of Ca^{2+}-pump ATPase and the Na^+/K^+-pump ATPase by iron-generated free radicals. Protection by 6,7-dimethyl-2,4-DI-1-pyrrolidinyl-7*H*-pyrrolo[2,3-*d*] pyrimidine sulfate (U-89843D), a potent, novel, antioxidant/free radical scavenger. Biochem Pharmacol 1996; 51: 471–476.
397. Green D, Kroemer G. The central executioners of apoptosis: caspases or mitochondria? Trends Cell Biol 1998; 8: 267–271.
398. Green DR, Reed JC. Mitochondria and apoptosis. Science 1998; 281: 1309–1312.

399. Cassarino DS, Bennett JP Jr. An evaluation of the role of mitochondria in neurodegenerative diseases: mitochondrial mutations and oxidative pathology, protective nuclear responses, and cell death in neurodegeneration. Brain Res Brain Res Rev 1999; 29: 1–25.
400. Nieminen AL, Petrie TG, Lemasters JJ, et al. A delays mitochondrial depolarization induced by *N*-methyl-D-aspartate in cortical neurons: evidence of the mitochondrial permeability transition. Neuroscience 1996; 75: 993–997.
401. Khaspekov L, Friberg H, Halestrap A, et al. Cyclosporin A and its nonimmunosuppressive analogue N-Me-Val-4-cyclosporin A mitigate glucose/oxygen deprivation-induced damage to rat cultured hippocampal neurons. Eur J Neurosci 1999; 11: 3194–3198.
402. Uchino H, Elmer E, Uchino K, et al. Amelioration by cyclosporin A of brain damage in transient forebrain ischemia in the rat. Brain Res 1998; 812: 216–226.
403. Li PA, Uchino H, Elmer E, et al. Amelioration by cyclosporin A of brain damage following 5 or 10 min of ischemia in rats subjected to preischemic hyperglycemia. Brain Res 1997; 753: 133–140.
404. Uchino H, Elmer E, Uchino K, et al. Cyclosporin A dramatically ameliorates CA1 hippocampal damage following transient forebrain ischaemia in the rat. Acta Physiol Scand 1995; 155: 469–471.
405. Buki A, Okonkwo DO, Povlishock JT. Postinjury cyclosporin A administration limits axonal damage and disconnection in traumatic brain injury. J Neurotrauma 1999; 16: 511–521.
406. Nakatsuka H, Ohta S, Tanaka J, et al. Release of cytochrome c from mitochondria to cytosol in gerbil hippocampal CA1 neurons after transient forebrain ischemia. Brain Res 1999; 849: 216–219.
407. Kruman II, Mattson MP. Pivotal role of mitochondrial calcium uptake in neural cell apoptosis and necrosis. J Neurochem 1999; 72: 529–540.
408. Kroemer G, Dallaporta B, Resche-Rigon M. The mitochondrial death/life regulator in apoptosis and necrosis. Annu Rev Physiol 1998; 60: 619–642.
409. Daugas E, Susin SA, Zamzami N, et al. Mitochondrio-nuclear translocation of AIF in apoptosis and necrosis. FASEB J 2000; 14: 729–739.
410. Gunter TE, Pfeiffer DR. Mechanisms by which mitochondria transport calcium. Am J Physiol 1990; 258: C755–C786.
411. Crompton M. The mitochondrial permeability transition pore and its role in cell death. Biochem J 1999; 341: 233–249.
412. Brenner C, Cadiou H, Vieira HL, et al. Bcl-2 and Bax regulate the channel activity of the mitochondrial adenine nucleotide translocator. Oncogene 2000; 19: 329–336.
413. Anderson P. Kinase cascades regulating entry into apoptosis. Microbiol Mol Biol Rev 1997; 61: 33–46.
414. Sastry PS, Rao KS. Apoptosis and the nervous system. J Neurochem 2000; 74: 1–20.
415. Reed JC, Paternostro G. Postmitochondrial regulation of apoptosis during heart failure. Proc Natl Acad Sci USA 1999; 96: 7614–7616.

416. Gibson S, Widmann C, Johnson GL. Differential involvement of MEK kinase 1 (MEKK1) in the induction of apoptosis in response to microtubule-targeted drugs versus DNA damaging agents. J Biol Chem 1999; 274: 10916–10922.
417. Widmann C, Gibson S, Johnson GL. Caspase-dependent cleavage of signaling proteins during apoptosis. A turn-off mechanism for anti-apoptotic signals. J Biol Chem 1998; 273: 7141–7147.
418. Jarpe MB, Widmann C, Knall C, et al. Anti-apoptotic versus pro-apoptotic signal transduction: checkpoints and stop signs along the road to death. Oncogene 1998; 17: 1475–1482.
419. Wang HG, Reed JC. Bc1-2, Raf-1 and mitochondrial regulation of apoptosis. Biofactors 1998; 8: 13–16.
420. Majewski M, Nieborowska-Skorska M, Salomoni P, et al. Activation of mitochondrial Raf-1 is involved in the antiapoptotic effects of Akt. Cancer Res 1999; 59: 2815–2819.
421. Nicholson DW, Thornberry NA. Caspases: killer proteases. Trends Biochem Sci 1997; 22: 299–306.
422. Thornberry NA, Lazebnik Y. Caspases: enemies within. Science 1998; 281: 1312–1316.
423. Wang HG, Pathan N, Ethell IM, et al. Ca^{2+}-induced apoptosis through calcineurin dephosphorylation of BAD. Science 1999; 284: 339–343.
424. Wang CY, Mayo MW, Korneluk RG, et al. NF-kappaB antiapoptosis: induction of TRAF1 and TRAF2 and c-IAP1 and c-IAP2 to suppress caspase-8 activation. Science 1998; 281: 1680–1683.
425. Reuther JY, Baldwin AS Jr. Apoptosis promotes a caspase-induced amino-terminal truncation of IkappaBalpha that functions as a stable inhibitor of NF-kappaB. J Biol Chem 1999; 274: 20664–20670.
426. Levkau B, Scatena M, Giachelli CM, et al. Apoptosis overrides survival signals through a caspase-mediated dominant-negative NF-kappa B loop. Nat Cell Biol 1999; 1: 227–233.
427. Bennett MR. Mechanisms of p53-induced apoptosis. Biochem Pharmacol 1999; 58: 1089–1095.
428. Pochampally R, Fodera B, Chen L, et al. Activation of an MDM2-specific caspase by p53 in the absence of apoptosis. J Biol Chem 1999; 274: 15271–15277.
429. Kamata H, Hirata H. Redox regulation of cellular signalling. Cell Signal 1999; 11: 1–14.
430. Chan SL, Mattson MP. Caspase and calpain substrates: roles in synaptic plasticity and cell death. J Neurosci Res 1999; 58: 167–190.
431. Wang KK. Calpain and caspase: can you tell the difference? Trends Neurosci 2000; 23: 20–26.
432. Melloni E, Pontremoli S. The calpains. Trends Neurosci 1989; 12: 438–444.
433. Carafoli E, Molinari M. Calpain: a protease in search of a function? Biochem Biophys Res Commun 1998; 247: 193–203.
434. Perlmutter LS, Gall C, Baudry M, et al. Distribution of calcium-activated protease calpain in the rat brain. J Comp Neurol 1990; 296: 269–276.

435. Hamakubo T, Kannagi R, Murachi T, et al. Distribution of calpains I and II in rat brain. J Neurosci 1986; 6: 3103–3111.
436. Ray SK, Fidan M, Nowak MW, et al. Oxidative stress and Ca^{2+} influx up-regulate calpain and induce apoptosis in PC12 cells. Brain Res 2000 ; 852: 326–334.
437. Hewitt KE, Lesiuk HJ, Tauskela JS, et al. Selective coupling of mu-calpain activation with the NMDA receptor is independent of translocation and autolysis in primary cortical neurons. J Neurosci Res 1998; 54: 223–232.
438. Rami A, Ferger D, Krieglstein J. Blockade of calpain proteolytic activity rescues neurons from glutamate excitotoxicity. Neurosci Res 1997; 27: 93–97.
439. Bednarski E, Vanderklish P, Gall C, et al. Translational suppression of calpain I reduces NMDA-induced spectrin proteolysis and pathophysiology in cultured hippocampal slices. Brain Res 1995; 694: 147–157.
440. Brorson JR, Marcuccilli CJ, Miller RJ. Delayed antagonism of calpain reduces excitotoxicity in cultured neurons. Stroke 1995; 26: 1259–1266.
441. Bartus RT, Elliott PJ, Hayward NJ, et al. Calpain as a novel target for treating acute neurodegenerative disorders. Neurol Res 1995; 17: 249–258.
442. Hong SC, Goto Y, Lanzino G, et al. Neuroprotection with a calpain inhibitor in a model of focal cerebral ischemia. Stroke 1994; 25: 663–669.

6
But What About Ca^{2+} and Plants?

".......Ca^{2+} is, probably, a universal messenger in transducing external stimuli in plants."

— B.W. Poovaiah and A.S.N. Reddy[1]

"The Ca^{2+} hypothesis is alive and kicking in plant cells as in other eukaryotic systems."

— A. Trewavas et al.[2]

Plants don't have muscles to move them around or experience-sculpted brains to program and direct them. Nevertheless, they can sense and respond almost instantly to a wide variety of stresses. And Ca^{2+} signature-surges are the principal agents of their responsiveness.[3] The basic Ca^{2+}-signaling mechanisms appeared 2 Ga ago in the first Archeons, long before animals and plants went their separate ways. Therefore, we should expect that many, if not most, plant cell functions should be driven by Ca^{2+}, CaMs, calcineurin, the CAMKs, and other signaling mechanisms that are similar or identical to those that drive the many functions of our cells. As we shall now see from these few examples, this is indeed the case.

Ca^{2+} and blowing in the wind

Despite not having nervous systems, plants, like our bones, have a mechanosensory device(s) that can respond to the strains caused by touching, bending, and shaking by causing structural changes that minimize the strain. The morphological changes caused by touching has been called *thigmomorphogenesis,* and the response to shaking is called *seismomorphogenesis*.[2,3] When woody plants are being rocked by wind, they stop growing upward (which would exaggerate the bending strain produced by a given amount of stress) and redirect their carbohydrate resources to thickening and lignifying their stems to reduce the bending at the stem–root junction. The lack of a need to redirect resources from seed-making to strengthening stems is why seed yields are greater in greenhouse-grown plants that are sheltered from the wind strain.[2]

The ability of touching to trigger Ca^{2+} surges has been demonstrated by engineering *Nicotiana plumbaginafolia* to express apoaequorin (part of a jellyfish's anti-predator luminescent protein aequorin [coelenterazine·apoaequorin]) and then exposing seedlings to coelenterazine to make the complete aequorin Ca^{2+} sensor protein.[3,4] These aequorin-loaded seedlings light up almost immediately when touched or slightly rocked. The alternating compression and tension caused by touching or rocking trigger sharp gene-activating spikes of Ca^{2+}, the amplitudes of which depend on the amount of strain.[3–4] These strain-induced surges of Ca^{2+} from internal stores (possibly from the vacuole, the major Ca^{2+} storage site or maybe from the endoplasmic reticulum)[3,5] trigger a flurry of gene activity, the intensity of which depends on the amount of strain and is responsible for the strain-induced morphogenesis.[2] In *Arabidopsis*, touch-triggered Ca^{2+} spikes stimulate the expression of the so-called TCH genes, some of which encode at least three of the plant's six CaM genes.[3,3a,6–8] The TCH3 gene encodes a novel Ca^{2+}-binding protein, which complexes with six Ca^{2+}s, instead of CaM's four, and probably does not have the same targets as Ca^{2+}·CaM.[6] The mechanoinduced TCH3 protein accumulates at points of maximum strain where cell–wall reinforcement is needed.[7] This accumulated TCH3, responding to the strain-triggered spatio-temporal Ca^{2+} spikes, may stimulate, or collaborate with the product of, another TCH gene, TCH4, which encodes a xyloglucan endotransglycolase (XET).[7,9,10] When expressed, this enzyme would strengthen the cell walls at these strain hot spots by crosslinking the walls' cellulose microfibrils with xyloglucan.[7]

Calcium and light

Some plants have been found to respond to light red or blue light with changes in the cytoplasmic Ca^{2+} concentration.[3a] Ca^{2+} and CaM are involved in inducing the expression of one of the genes, *gsa*, involved in chlorophyll synthesis in *Chlamydomonas*.[3a] Moreover, external 3–5 mM Ca^{2+} can actually eliminate the need for light to trigger the formation of primary leaves by *Sorghum bicolor* grown in the dark.[3a]

Cold, drought, and Ca^{2+}

A major 'worry' for plants is their water supply. Freezing, drought, or excessive salination of their soil will restrict their ability to pick up and transport water through their vascular systems to their various parts, and a continuing loss of water through the small openings in the epidermis of their leaves, the stomata, becomes dangerous. And they use Ca^{2+} to trigger their responses to a water shortage.

A cold shock (e.g., 20°C→0°C) depolarizes the plasma membrane and causes the influx of Ca^{2+} into the cells of alfalfa, *Arabidopsis*, corn (maize), and tobacco.[3] The shock also activates PLC, which generates $InsP_3$ that causes the release of vacuolar Ca^{2+} to supplement the Ca^{2+} flowing in through the plasma membrane. The Ca^{2+} signature-surge stimulates the expression of various cold acclimation genes (*cas*[15], *cas*[18], *kin*, TCH2, 3, 4) by stimulating the genes for ABA (abscisic acid) synthesis. Of course, the Ca^{2+} surge generates Ca^{2+}·CaM, which stimulates among other things L-glutamate decarboxylase, which makes γ-aminobutyric acid (GABA).[3,3a] One of the indirect, possibly ethylene-mediated, effects of the Ca^{2+}- triggered ABA surge is the activation of hydrolytic enzymes that break the tethers holding the cells together in the leaf abscission zones, which causes the leaves to fall off (i.e., it causes *apoptosis* because this Greek word actually means the falling of leaves!).

When the roots of a plant in a drought-stricken or excessively saline soil sense the resulting water shortage, there is a burst of PLC activity and through this the production of $InsP_3$ that triggers the release of Ca^{2+} into the cytoplasm from the internal stores.[3] Obviously, the first thing to do is stop the 'hemorrhage' of the precious water through the leaves' stomata. This is done by the Ca^{2+}-stimulated expression of ABA that causes the guard cells regulating the size of the stomata to slam shut.[3] This phytohormone also triggers the cascade of mechanisms leading to the accumulation of osmocompatible solutes and osmoprotectors, promoting dormancy, and perhaps getting rid of leaves.[3]

Ca^{2+} as a driver of growth and proliferation

One of the earliest, if not the first, indications that Ca^{2+} controls the initiation of DNA replication in plant cells was provided by Setterfield,[11] using slices of quiescent Jerusalem artichoke (*Helianthus tuberosus*) tuber. The quiescent cells in these slices can either start expanding when activated by an auxin BTOA (benzthiazolyloxyacetic acid, a β-indoleacetic acid analog) and kinetin or start making DNA after a 20-h lag period and divide without expanding. It is Ca^{2+} that determines whether the cells will proliferate or just expand. In fact, the responses are exactly like the responses of hepatocytes to partial hepatectomy in normal and thyroparathyroidectomized rats that we talked about in Chap. 2. If the tuber cells are stimulated by addition of BTOA and kinetin in a Ca^{2+}-free medium, they will expand by 40% without initiating DNA replication. However, if the medium contains Ca^{2+} (30 meq/L) the cells can start making DNA between 20 and 24 h after adding the hormones, but they will not expand.

Ca^{2+} surges are involved in starting the growth and proliferation of quiescent cambial cells in the beech tree (*Fagus sylvatica*).[12] As the tree emerges from its winter quiescence, there are large transient Ca^{2+} surges in the cambium and phloem, which probably trigger events leading to cambial cell proliferation, with the production of secondary phloem and xylem and an increase in the stem's girth.[12]

Another example of Ca^{2+} and cell cycle starting is provided by the proliferogenic action of lipo-chito-oligosaccharide Nod (nodule) factors from bacterial *Rhizobium* species on root cells of leguminous plants. These factors trigger cytoplasmic and perinuclear Ca^{2+} surges by directly stimulating the entry of Ca^{2+} into the root hair cells through the cell wall and membrane and (or) by activating surface receptors which: (1) causes the root hair to curl and grab the bacterium by disrupting actin microfilaments; (2) switches on G_1 phase-specific, cyclin-dependent protein kinase(s) to stimulate the proliferation of cortical cells; and (3) through these changes causes the ultimate formation nitrogen-fixing nodules on the root hairs.[13–15]

Very little is known about the Ca^{2+} spikes and oscillations — the Ca^{2+} songs — and the Ca^{2+}-binding signal proteins that we suspect are involved in triggering the key transition points in plant cell cycles. However, we can see some light at the end of this tunnel of ignorance! One hint of Ca^{2+} being the same major player in driving plant cell proliferation as it is in animal cells is the high level of CaM mRNA in proliferatively active root and shoot meristems.[16] But higher plants do some things differently. Their predominant cell cycle-driving Ca^{2+}-binding/activated signaler(s) is not CaM. Instead, they use the ingeniously designed CaDPKs (Ca^{2+}-dependent protein kinases) that have a catalytic domain that is homologous to the mammalian Ca^{2+}·CaM-activated CaMKII plus a built-in Ca^{2+} sensor — a CaM-like domain.[3a] In other words, these enzymes can sense and respond to Ca^{2+} directly without having to wait for Ca^{2+}·CaM to appear. A proliferogen such as the auxin, 2,4-dichlorophenoxy acetic acid, but not the anti-proliferative ABA, stimulates the expression of an alfalfa CaDPK gene.[16] And CaDPK mRNA peaks in alfalfa cells first in the S phase and again in the mid-G_2.[16] Thus, it is probably safe to assume that the starting of a cell cycle and the initiation of DNA replication in plant cells are triggered by the same Ca^{2+} spikes and oscillations that we saw in the animal cells of Chap. 2. There is already much support for a surge of Ca^{2+} from spindle-associated stores triggering anaphase.[17–23] Ca^{2+} is also probably involved in cytokinesis but not as it is in animal cells. In higher plants, cytokinesis starts when the Ca^{2+} surge crashes at the end of anaphase and spindle interzonal microtubules, actin micofilaments, Golgi-derived vesicles loaded with cell wall matrix polysaccharides, and endoplasmic reticulum collect into a cell-partitioning, cell wall precursor, the phragmoplast, instead of the cytoplasm simply being pinched in two by the Ca^{2+}-driven tightening of a subcortical actomyosin belt as in the wall-less animal cells.[17–23] Ca^{2+} probably drives the delivery of components to the phragmoplast construction site, the formation of the tubulo-vesicular scaffold, and the correct positioning of the forming structure.[21]

It is also becoming increasingly certain that the plant cells drive their cycles with stage-specific Cdk·cyclin kinases, which do the same jobs as the homologous cycle engine kinases of animal cells.[24–26] Also, despite the once universally held belief that higher plants do not use cyclic AMP-dependent control mechanisms, it is now becoming ever clearer that they have adenylyl cyclase and use it and its cyclic AMP product to do key things.[27–30] Indeed, it seems that a small G_1 and a larger S-phase cyclic AMP surges work along with Ca^{2+} and the Cdk·cyclin kinases to initiate DNA replication and prepare for mitosis in plant cells such as BY-2 tobacco cells just as they do in animal cells.[27]

One of Ca^{2+}'s many jobs in both plants and animals, which dates back to the first pre-eukaryotes, is operating the motor the cell uses to crawl toward its targets. A striking example of this in plants is pollen tube growth. When a pollen grain lands on the stigma of a pistil of a flowering plant, it will start producing a tube, if it is recognized as being compatible and receives appropriate starting signals.[31,32] Then, the tube, with its precious cargo, two haploid male nuclei, pushes between the stigma cells and moves through the transmitting tract of the pistil's style at a rate that can be as high as 1 cm/h. As with a neuron's axonal growth cone, the direction-sensing engine (or 'brain') of the pollen tube is in its tip.[32] The navigation of the tip is directed by signals generated by the interaction of the tube with specific stylar guidepost glycoproteins. However, from our perspective the most interesting part of the motor's driving mechanism is the tube's apical Ca^{2+} gradient produced by Ca^{2+} surging through channels at the tip of the tube.[15,33] Also concentrated at the tip is another part of the driving mechanism, a CaDPK which is undoubtedly a mediator of the Ca^{2+} signal surges.[3] The driving ability of the Ca^{2+} pulses is dramatically demonstrated by a release of intracellularly 'caged' Ca^{2+} at one side of the tube apex, causing a redistribution of the CaDPK to the release site, the creation of a new tip, and the reorientation of the tube's movement.[15,33] It should be noted in passing that exactly the same apical Ca^{2+} gradients and pulses drive the extension of root tips.[15]

The tip of the pollen tube, with its Ca^{2+} pulses, eventually arrives at the ovule, crawls along the ovule's funiculus or stalk, and, guided by appropriate signalers, passes through the micropyle and into embryo sac containing the egg cell.[32] The tip of the tube ruptures and releases the two haploid male nuclei. One of these nuclei fuses with the egg. The other fuses with the primary diploid endosperm nucleus to form the triploid endosperm cell, which will start dividing to form the endosperm that will, in turn, accumulate 'yolk' for feeding the germinating seed. The fusion of a sperm cell with an animal egg cell triggers a Ca^{2+} 'tsunami' that spreads over the egg from the fusion site to start the proliferation of the zygote. A similarly spreading Ca^{2+} wave is triggered by the fusion of a *Zea mays* egg with a male nucleus.[34] It is probably safe to predict that a Ca^{2+} transient also triggers the division of the endosperm nucleus after the fusion of the other male nucleus with the primary endosperm nucleus.

Conclusion

We, and probably the reader, find this venture into the plant world rather cursory. We shall not venture further into this wonderful world mainly because of our all too obvious lack of credentials and most of all because the field has only just started to emerge. However, we have at least touched the buds that will very soon burst into a mass of exciting data, telling us how Ca^{2+} controls the various functions of higher plants. Given the vast antiquity of the ion's hold on eukaryotic cells, the themes will be generally familiar, yet at the same time fascinatingly different from those in the animal world.

References

1. Poovaiah BW, Reddy ASN. Calcium and signal transduction in plants. Crit Rev Plant Sci 1993; 12: 185–211.
2. Trewavas A, Read N, Campbell AK, et al. Transduction of Ca^{2+} signals in plant cells and compartmentalization of the Ca^{2+} signal. Biochem Soc Trans 1996; 24: 971–974.
3. Knight H. Calcium signaling during abiotic stress in plants. Int Rev Cytol 2000; 195: 269–324.

3*a*. Pandey, S, Tiwari SB, Upadhyaya KC, et al. Calcium signaling: linking environmental signals to cellular functions. Crit Revs Plant Sci 2000; 19: 291–318.

4. Trewavas A, Knight M. Mechanical signalling, calcium and plant form. Plant Mol Biol 1994; 26: 1329–1341.
5. Knight MR, Smith SM, Trewavas AJ. Wind-induced plant motion immediately increases cytosolic calcium. Proc Natl Acad Sci USA 1992; 89: 4967–4971.
6. Braam J. Regulation of expression of calmodulin and calmodulin-related genes by environmental stimuli in plants. Cell Calcium 1992; 13: 457–463.
7. Braam J, Sistrunk ML, Polisensky DH, et al. Plant responses to environmental stress: regulation and functions of the *Arabidopsis* TCH genes. Planta 1997; 203: S35–S41.
8. Snedden WA, Fromm H. Calmodulin, calmodulin-related proteins and plant responses to the environment. Trends Plant Sci 1998; 3: 299–304.
9. Antosiewicz DM, Purugganan MM, Polisenski DH, et al. Cellular localization of *Arabidopsis* xyloglucan endotransglycosylase-related proteins during development and after wind stimulation. Plant Physiol 1997; 115: 1319–1328.

10. Xu W, Campbell P, Vargheese AK, et al. The *Arabidopsis* XET-related gene family: environmental and hormonal regulation of expression. The Plant J. 1996; 9: 879–889.
11. Setterfield G. Growth regulation in excised slices of Jerusalem artichoke tuber tissue. Symp Soc Exper Biol 1963; 17: 98–126.
12. Follet-Gueye M-L, Verdus M-C, Demarty M, et al. Cambium pre-activation in beech correlates with a strong temporary increase of calcium in cambium and phloem but not in xylem cells. Cell Calcium 1998; 24: 205–211.
13. Cárdenas L, Holdaway-Clarke T, Sánchez F, et al. Ion changes in legume root hairs responding to Nod factors. Plant Physiol 2999; 123: 443–451.
14. Fehér A, Schultze M, Kondorosi E. Control of Nod-signal-induced cell division during root nodule formation. In: Francis D, Dudits D, Inzé D, editors. Plant Cell Division. London, Portland Press, 1998: 223–241.
15. Rudd JJ, Franklin-Tong VE. Calcium signaling in plants. Cell Mol Life Sci (CMLS) 1999; 55: 214–232.
16. Dudits D, Magyar Z, Deák M, et al. Cyclin-dependent and calcium-dependent protein kinase families: response of cell division cycle to hormone and stress signals. In: Francis D, Dudits D, Inzé D, editors. Plant Cell Division. London, Portland Press, 1998: 21–45.
17. Hepler PK. Calcium transients during mitosis: observations in flux. J Cell Biol; 109: 2567–2573.
18. Hepler PK. Calcium and mitosis. Int Rev Cytol 1992; 138: 239–268.
19. Hepler PK. The role of calcium in cell division. Cell Calcium 1994; 16: 322–330.
20. Hepler PK, Callaham DA. Free calcium increases during anaphase in stamen hair cells of *Tradescantia*. J Cell Biol 1987; 105: 2137–2143.
20*a*. Reddy ASN. Molecular motors and their functions in plants. Int Rev Cytol 2001; 204: 97–158.
21. Staehlin LA, Hepler PK. Cytokinesis in higher plants. Cell 1996; 84: 821–824.
22. Wolniak SM, Hepler PK, Jackson WT. Ionic changes in the mitotic apparatus at the metaphase/anaphase transition. J Cell Biol 1983; 96: 598–605.
23. Zhang DH, Callaham DA, Hepler PK. Regulation of anaphase chromosome motion in *Tadescantia* stamen hair cells by calcium and related signaling agents. J Cell Biol 1990; 111: 171–182.
24. Francis D, Dudits D, Inzé D. Plant Cell Division. London, Portland Press,1998.
25. Hemerly As, Ferriera PCG, Montagu MV, et al. Cell cycle control and plant morphogenesis: is there an essential link? BioEssays 1999; 21: 29–37.
26. Mironov V, De Veylder L, Van Montagu M, et al. Cyclin-dependent kinases and cell division in plants — the nexus. Plant Cell 1999; 11: 509–521.
27. Ehsan H, Reichheld J-P, Roef L, et al. Effect of indomethacin on cell cycle dependent cyclic AMP fluxes in tobacco BY-2 cells. FEBS Lett 1998; 422: 165–169.
28. Martelli P, Lusini P, Bovalini L, et al. Occurrence of cyclic AMP and related enzymes during germination of *Pinus Pinea* seeds. Ital J Biochem 1987; 36: 188–193.

29. Roef L, Winters E, Gadeyne J, et al. Analysis of 3′,5′-cAMP and adenylyl cyclase activity in higher plants using polyclonal chicken egg yolk antibodies. Anal Biochem 1996; 233: 188–196.
30. Sato S, Tabata S, Hotta Y. Changes in intracellular cAMP level and activities of adenylcyclase and phosphodiesterase during meiosis of lily microsporocytes. Cell Struct Funct 1992; 17: 335–339.
31. Wihelmi LK, Preuss D. The mating game: pollination and fertilization in flowering plants. Curr Opin Plant Biol 1999; 2: 18–22.
32. Palanivelu R, Preuss D. Pollen tube targeting and axon guidance: parallels in tip growth mechanisms. Trends Cell Biol 2000; 10: 517–524.
33. Moutinho A, Trewavas AJ, Malho R. Relocation of a Ca^{2+}-dependent protein kinase protein kinase activity during pollen tube reorientation. Plant Cell 1998; 10: 1499–1509.
34. Digonnet C, Aldon D, Leduc N, et al. First evidence of a calcium transient in flowering plants at fertilisation. Development 1997; 124: 2867–2874.

7
Summary

Our brief tour of the highlights of the parts of the current Ca^{2+} world concerned with cell cycles, differentiation, and death has ended. We have seen the Ca^{2+} oscillations that trigger protein synthesis and the delivery of transcription factors from the cell membrane and cytoplasm to the nucleus and that activate transcription factors already attached to their target genes' promoters in the nucleus. The products of these genes started the first wave of proliferogene activities that, in turn, started the parade of cyclin-dependent protein kinases beginning with Cdk4·cyclin D, which inactivated the gene transcription-suppressing Rb protein, initiated the buildup of the cation channels, receptors, and the autocrine/paracrine second signaler, such as IGF-I or IL-2, to use these devices. We saw the emergence of Cdk4·cyclin D's successor, Cdk2·cyclin E, that continued the inactivation of Rb family members to enable E2F·DP-1 transcription factor to stimulate the expression of key DNA replication enzymes. Then there was the downregulation of Cdk2·cyclin E and the cyclic AMP- and substrate adhesion (i.e., integrin signaling)-dependent emergence of Cdk2·cyclin A, which collaborates with another train of Ca^{2+} oscillations triggered by autocrine IGF-I and enhanced by c-Myb to activate the huge DNA replication factories attached to the chromosomal replication origins on the nuclear matrix. When all of the chromosomes were replicated and after the G_2 delay, there was the Ca^{2+} spike at the nuclear envelope that activated the first of the two emerging mitotic cylins, Cdk1·cyclin A, which triggered the chromosome condensation and nuclear envelope breakdown of mitotic prophase and then crashed to leave the rest of the job to Cdk1·cyclin B. This was followed by another Ca^{2+} spike when the chromosomes had lined up on the mitotic spindle, which, by activating the anaphase-promoting polyubiquitination-proteolysis mechanism that destroyed Cdk1·cyclin B, enabled the licensing of the sister chromatids for future replication, the removal of the residual DNA strand tangles holding the sister chromatids together, the destruction of the protein glues that also held the sisters together, and started the motors that moved the sister chromosomes to opposite spindle poles. We saw the cell divide when yet another, but this time more prolonged, Ca^{2+} surge drove the contraction of an equatorial belt of actomyosin filaments that pinched it into two daughters.

We then traveled to the skin where we saw Ca^{2+} drive the cycling of keratinocytes attached to the epidermal basement membrane and a Ca^{2+} surge trigger the spinous program when the basal keratinocytes lost their β1 integrins and with them their cycle-supporting, differentiation-suppressing signals and lifted off the basement membrane. We learned of the upwardly directed epidermal Ca^{2+} gradient, which near its top in the granular layer stimulates the divalent cation receptors on the cell's surface to signal the cell to generate an internal Ca^{2+} surge large enough to start the final cascade of differentiation/apoptosis-like events that transform the cell into a dead cornified shell packed with keratin macrofilaments. Here we learned one of the most important features (established mainly here in Ottawa) of epithelial cell carcinogenesis: loss of responsiveness to Ca^{2+} signals that would normally stop cycling and trigger differentiation and apoptosis (diffpoptosis).

When we moved from the skin to the colon, we again saw Ca^{2+} driving the cycling of the transit-amplifying progeny of stem cells in the depths of the colon crypts. We probed the possibility of there being an upwardly directed Ca^{2+} gradient like the gradient in epidermis which activates the divalent cation receptors on the colon cells' surfaces that send signals that drive, or at least substantially contribute to the driving of, the differentiation and ultimately apoptotic killing of the cells in the upper regions of the crypt and out on the flat mucosal surface. Here too an early loss of responsiveness to the Ca^{2+} signals commanding the cells to stop cycling, differentiate, and then die by apoptosis is a key part of the progression to malignancy. We then saw how Ca^{2+} might operate the brakes and accelerator of the engine that drives the axon growth cone as it advances toward its target. We also learned that the ion seems to similarly control the engine that drives the plant pollen tube toward its target egg cell. We saw Ca^{2+} and its client signal proteins driving the release of neurotransmitters and recording memories, but we also saw that given the right circumstances it can be an efficient neuron killer. Finally, we learned to look at plants in a very different way. We learned that although plants do not have brains and muscles for fight and flight, they nevertheless can rapidly respond with growth-altering Ca^{2+} signals to light, mechanical strain and various environmental challenges such as cold and drought. Indeed, the story of Ca^{2+} in plants is only now starting to be told.

Unfortunately for us, Ca^{2+} is a very busy ion with lots of different targets and mediators scattered throughout the animal and plant cell. But we can at least make out the dim, but tantalizing, outlines of the ion's diverse roles in cell cycles and diffpoptosis. However, the tools are now being invented to accurately map in cellular space and time the Ca^{2+} spikes, waves, and oscillations that start the several stages of the cell cycle and diffpoptosis. Clearly, these new more sophisticated tools will give us insights into the awesome networks used by Ca^{2+} in its roles of cell cycle driver, differentiator, and killer.

Index

A

B

C

F

G

H

I

O

P

R

S

T

U

V

W

Z